Green Chemistry and Sustainable Technology

Series Editors

Aims and Scope

The series Green Chemistry and Sustainable Technology aims to present cutting-edge research and important advances in green chemistry, green chemical engineering and sustainable industrial technology. The scope of coverage includes (but is not limited to):

- Environmentally benign chemical synthesis and processes (green catalysis, green solvents and reagents, atom-economy synthetic methods etc.)
- Green chemicals and energy produced from renewable resources (biomass, carbon dioxide etc.)
- Novel materials and technologies for energy production and storage (bio-fuels and bioenergies, hydrogen, fuel cells, solar cells, lithium-ion batteries etc.)
- Green chemical engineering processes (process integration, materials diversity, energy saving, waste minimization, efficient separation processes etc.)
- Green technologies for environmental sustainability (carbon dioxide capture, waste and harmful chemicals treatment, pollution prevention, environmental redemption etc.)

The series Green Chemistry and Sustainable Technology is intended to provide an accessible reference resource for postgraduate students, academic researchers and industrial professionals who are interested in green chemistry and technologies for sustainable development.

Mahmoud Nasr · Abdelazim Negm · Lai Peng
Editors

Artificial Intelligence Applications for a Sustainable Environment

Springer

Editors
Mahmoud Nasr
Department of Environmental Engineering
Egypt-Japan University of Science
and Technology
Alexandria, Egypt

Abdelazim Negm
Department of Water and Water Structures
Engineering
Faculty of Engineering
Zagazig University
Zagazig, Egypt

Lai Peng
Department of Resources
and Environmental Engineering
Wuhan University of Technology
Hubei, China

ISSN 2196-6982 ISSN 2196-6990 (electronic)
Green Chemistry and Sustainable Technology
ISBN 978-3-031-91198-9 ISBN 978-3-031-91199-6 (eBook)
https://doi.org/10.1007/978-3-031-91199-6

This Springer imprint is published by the registered company Springer Nature Switzerland AG
The registered company address is: Gewerbestrasse 11, 6330 Cham, Switzerland

Preface

This volume represents the applications of artificial intelligence (AI), including artificial neural network, deep neural network, support vector machine, genetic algorithm, fuzzy logic, adaptive-neuro fuzzy systems, sensors, software, and other technologies, to the field of environmental engineering. These AI tools have been applied in wastewater treatment processes for predicting the pollutants' removal efficiencies, optimizing the systems' operating conditions, and controlling the aeration, flowrates, and tank temperature. Moreover, AI-based digital techniques have been utilized to deploy real-time monitors and early warning systems to detect possible contamination in aquatic, terrestrial, and atmospheric environments. AI-supported soft-sensors are also used to collect, monitor, and analyze remote data related to renewable energy management and fossil fuel energy systems. Furthermore, AI is employed to define inputs/outputs nonlinear correlations in chemical and biological environmental-associated processes. Intelligent control systems are used for smart and wireless monitoring and collection of solid waste management in green cities. Regarding atmospheric environment sustainability, AI-based techniques are used to diagnose, manage, and forecast air pollution-related diseases. Different AI algorithms are utilized to create maps, showing the distribution of soil pollution in the area. These advantages are beneficial for fulfilling the sustainable development goals accompanied by pollution control and reduction, protection of human health, and climate change mitigation and adaptation.

Unique Aspects

- Ensures sustainable environment by Internet of things applications
- Controls environmental pollution by machine learning algorithms
- Reviews comprehensively artificial intelligence tools for cleaner environment

SEO Metadata for Volume I

This book represents the applications of Internet of things, machine learning tools, artificial neural network, deep neural network, support vector machine, genetic algorithms, sensors, software, and other artificial intelligence technologies to maintain renewable/cleaner energy and sustainable environment via pollution (air, water, soil, and noise) control.

Alexandria, Egypt Mahmoud Nasr
Zagazig, Egypt Abdelazim Negm
Hubei, China Lai Peng

Contents

Contributors

Aliu Adeniran Adebiyi Department of Statistics, University of Ibadan, Ibadan, Nigeria

Ayotunde Daniel Adegboyegun Environmental Quality Control Department, National Environmental Standards and Regulations Enforcement Agency (NESREA), Osun, Nigeria

Aina O. Adeogun Hydrobiology and Fisheries Unit, Department of Zoology, University of Ibadan, Ibadan, Nigeria

Saheed A. Afolabi Department of Industrial and Systems Engineering, College of Computing and Mathematics, King Fahd University of Petroleum and Minerals, Dhahran, Saudi Arabia

Farhan Ahmad School of the Environment and Safety Engineering, Biofuels Institute, Jiangsu University, Zhenjiang, China

Sharifah Rafidah Wan Alwi Process Systems Engineering Centre (UTM-PROSPECT), Research Institute of Sustainable Environment (RISE), Universiti Teknologi Malaysia, Bahru, Johor, Malaysia;
Faculty of Chemical and Energy Engineering, Universiti Teknologi Malaysia (UTM), Bahru, Johor, Malaysia

Andrés A. Amell Facultad de Ingeniería, Grupo de Ciencia y Tecnología de Gases y Uso Racional de la Energía, Universidad de Antioquia, Medellín, Colombia

Anjuman Ayub School of Bioengineering and Biosciences, Lovely Professional University, Jalandhar, Punjab, India

Kunjal Badrakiya Department of Civil Engineering, Sardar Vallabhbhai National Institute of Technology, Surat, India

Abdelfattah Belaid Centre de DÈveloppement Des Energies Renouvelables, CDER, UnitÈ de Recherche AppliquÈe en Energies Renouvelables, URAER, GhardaÔa, Algeria

Tayeb Boulmaiz Materials, Energy Systems Technology and Environment Laboratory, Ghardaia University, Ghardaia, Algeria

Hamouda Boutaghane Laboratory of Soil and Hydraulic, Badji Mokhtar Annaba University, Annaba, Algeria

Azubuike Victor Chukwuka Environmental Quality Control Department, National Environmental Standards and Regulations Enforcement Agency (NESREA), Osun, Nigeria

Azubuike Victor Chukwuka National Environmental Standards and Regulations Enforcement Agency (NESREA), Abuja, Nigeria

Oscar Corvacho-Ganahin Departamento de Ciencias Históricas y Geográficas, Universidad de Tarapacá, Arica, Chile

Mudasir A. Dar School of the Environment and Safety Engineering, Biofuels Institute, Jiangsu University, Zhenjiang, China

Kavita Dehalwar Department of Architecture and Planning, Maulana Azad National Institute of Technology, Bhopal, India

Khaled Ferkous Materials, Energy Systems Technology and Environment Laboratory, Ghardaia University, Ghardaia, Algeria

Abdulsamad Adeola Folorunso Department of Data and Information Science, University of Ibadan, Ibadan, Nigeria

Morufu Aderemi Folorunso Department of General Studies, Federal School of Statistics, Ibadan, Nigeria

Serifat A. Folorunso School of Computing, Engineering and Digital Technologies, Teesside University, Middlesbrough, UK

Shekoufeh Forouzan Department of Soil Science, College of Agriculture, Shiraz University, Shiraz, Iran

Camilo A. Franco Grupo de Investigación en Fenómenos de Superficie Michael Polanyi, Facultad de Minas, Universidad Nacional de Colombia Sede Medellín, Medellín, Colombia

Marcos Francos Department of Geography, Faculty of Geography and History, University of Salamanca, Salamanca, Spain

Lady Giraldo Grupo de Investigación en Fenómenos de Superficie Michael Polanyi, Facultad de Minas, Universidad Nacional de Colombia Sede Medellín, Medellín, Colombia

Babbiker Mohammed Taher Gorish School of the Environment and Safety Engineering, Biofuels Institute, Jiangsu University, Zhenjiang, China

Mawloud Guermoui Centre de DÈveloppement Des Energies Renouvelables, CDER, UnitÈ de Recherche AppliquÈe en Energies Renouvelables, URAER,

GhardaÔa, Algeria;
Telecommunications and Smart Systems Laboratory, University of ZianeAchour, Djelfa, Algeria

Mawloud Guermoui Centre de Développement Des Energies Renouvelables, CDER, Unité de Recherche Appliquée en Energies Renouvelables, URAER, Ghardaïa, Algeria;
Telecommunications and Smart Systems Laboratory, University of Ziane Achour, Djelfa, Algeria

Radia Hafsi Hydraulic and Civil Engineering Department, Ghardaia University, Ghardaia, Algeria

Ali Hashemizadeh Key Laboratory for Resilient Infrastructures of Coastal Cities, Ministry of Education, Shenzhen University, Shenzhen, China

Daniel H. de la Iglesia Department of Computer Science and Automation, Faculty of Science, University of Salamanca, Salamanca, Spain

Chinenye Adaobi Igwegbe Department of Chemical Engineering, Nnamdi Azikiwe University, Awka, Nigeria;
Department of Applied Bioeconomy, Wrocław University of Environmental and Life Sciences, Wrocław, Poland

Sarika Jain Indira Gandhi Delhi Technical University for Women, Delhi, India

Nur Fadzeelah Abu Kassim Chemical Engineering Studies, College of Engineering, Universiti Teknologi MARA, Cawangan Pulau Pinang, Permatang Pauh, Pulau Pinang, Malaysia

Richard O. Kehinde Department of Statistics, Federal College of Animal Health and Production Technology, Ibadan, Nigeria

Diana López Instituto de Química, Química de Recursos Energéticos y Medio Ambiente, Universidad de Antioquia UdeA, Medellín, Colombia

M. Mansoor Ahammed Department of Civil Engineering, Sardar Vallabhbhai National Institute of Technology, Surat, India

Oscar E. Medina Instituto de Química, Química de Recursos Energéticos y Medio Ambiente, Universidad de Antioquia UdeA, Medellín, Colombia;
Grupo de Ciencia y Tecnología de Gases y Uso Racional de La Energía, Facultad de Ingeniería, Universidad de Antioquia, Medellín, Colombia

Alejandro Moncayo-Riascos Grupo de Investigación en Fenómenos de Superficie Michael Polanyi, Facultad de Minas, Universidad Nacional de Colombia Sede Medellín, Medellín, Colombia

Iván Moncayo-Riascos Grupo de Investigación en Fenómenos de Superficie Michael Polanyi, Facultad de Minas, Universidad Nacional de Colombia Sede Medellín, Medellín, Colombia

Ali Akbar Moosavi Department of Soil Science, College of Agriculture, Shiraz University, Shiraz, Iran

Hasan Mozaffari Department of Soil Science, College of Agriculture, Shiraz University, Shiraz, Iran

Dinie Muhammad Safety and Health Research Group, Universiti Teknologi Malaysia, Bahru, Johor, Malaysia;
Faculty of Chemical and Energy Engineering, Universiti Teknologi Malaysia (UTM), Bahru, Johor, Malaysia

Pedro Narváez Facultad de Ingeniería, Grupo de Ciencia y Tecnología de Gases y Uso Racional de la Energía, Universidad de Antioquia, Medellín, Colombia

Mahmoud Nasr Environmental Engineering Department, Egypt-Japan University of Science and Technology (E-JUST), Alexandria, Egypt;
Sanitary Engineering Department, Faculty of Engineering, Alexandria University, Alexandria, Egypt

Mohammad Amin Nematollahi Department of Biosystems Engineering, College of Agriculture, Shiraz University, Shiraz, Iran

Niknik Nurhayati Research Centre for Genetic Engineering, National Research and Innovation Agency of Indonesia, Bogor, West Java, Indonesia

Chukwudi Nwabuisiaku Environmental Quality Control Department, National Environmental Standards and Regulations Enforcement Agency (NESREA), Osun, Nigeria;
Inspection and Enforcement Department, National Environmental Standards and Regulations Enforcement Agency (NESREA), Osun, Nigeria

Chukwudi Nwabuisiaku National Environmental Standards and Regulations Enforcement Agency (NESREA), Abuja, Nigeria

Mina Margaret Ogbanga Head of Department, Track2Training, New Delhi, India

Dami Emmanuel Omogbemi Department of Zoology, University of Ibadan, Ibadan, Nigeria

Olawale Dele Osanyintupin Department of Statistics, University of Ibadan, Ibadan, Nigeria

Timothy Tomiwa Oyelakin Department of Epidemiology and Medical Statistics, University of Ibadan, Ibadan, Nigeria

Morteza Pakjoo Department of Soil Science, College of Agriculture, Shiraz University, Shiraz, Iran

Ashutosh Kumar Pandey JECRC University, Jaipur, Rajasthan, India

Radityo Pangestu Research Centre for Genetic Engineering, National Research and Innovation Agency of Indonesia, Bogor, West Java, Indonesia

Saurab Patel Department of Civil Engineering, Sardar Vallabhbhai National Institute of Technology, Surat, India

Yugmi S. Patel Department of Civil Engineering, Sardar Vallabhbhai National Institute of Technology, Surat, India

Paulo Pereira Environmental Management Centre, Mykolas Romeris University, Vilnius, Lithuania

Abdelaziz Rabehi Telecommunications and Smart Systems Laboratory, University of ZianeAchour, Djelfa, Algeria

Kada Riche University of Sciences and Technology Houari Boumediene, Alger, Algeria

Fakhrony Sholahudin Rohman Process Systems Engineering Centre (UTM-PROSPECT), Research Institute of Sustainable Environment (RISE), Universiti Teknologi Malaysia, Bahru, Johor, Malaysia;
Faculty of Engineering, Universiti Malaysia Sabah, Jalan UMS, Kota Kinabalu, Sabah, Malaysia

Carlos Sánchez-García Department of Geography, Autonomous University of Madrid, Madrid, Spain

Sukurat Salam Department for Works and Pension, London, UK

Alexander Santamaría Instituto de Química, Química de Recursos Energéticos y Medio Ambiente, Universidad de Antioquia UdeA, Medellín, Colombia

Shashikant Nishant Sharma Department of Architecture and Planning, Maulana Azad National Institute of Technology, Bhopal, India

Iqra Sharmeen Department of Civil Engineering, Sardar Vallabhbhai National Institute of Technology, Surat, India

Pavitar Singh JECRC University, Jaipur, India

Reena Singh School of Bioengineering and Biosciences, Lovely Professional University, Jalandhar, Punjab, India

Jianzhong Sun School of the Environment and Safety Engineering, Biofuels Institute, Jiangsu University, Zhenjiang, China

Yurany Villada Villada Grupo de Investigación en Fenómenos de Superficie Michael Polanyi, Facultad de Minas, Universidad Nacional de Colombia Sede Medellín, Medellín, Colombia

Atif Khurshid Wani School of Bioengineering and Biosciences, Lovely Professional University, Jalandhar, Punjab, India

Rongrong Xie School of the Environment and Safety Engineering, Biofuels Institute, Jiangsu University, Zhenjiang, China

Bilel Zerouali Laboratory of Architecture, Cities and Environment, Department of Hydraulic, Faculty of Civil Engineering and Architecture, Hassiba Benbouali University of Chlef, Chlef, Algeria

Part I
AI in Pollution Management

Chapter 1
Internet of Things (IoT) and Environmental Monitoring: A Systematic Literature Review

Marcos Francos, Carlos Sánchez-García, Daniel H. de la Iglesia, Oscar Corvacho-Ganahin, and Paulo Pereira

Abstract The Internet of things (IoT) refers to the collection of physical devices that receive and transfer data over wireless networks. This paper presents the most important work using the IoT in environmental monitoring, specifically in the fields of land use, forest management, agriculture, urban environments, and climate. The text also examines how artificial intelligence (AI) enhances IoT systems by improving data analysis and decision-making processes, which are crucial for effective environmental monitoring and management. The studies have mainly focused on the application of this technology in urban and agricultural environments where humans live or where they derive the greatest economic benefit. The works generally value using the IoT for environmental monitoring because this technology may potentially be scaled up to permanently monitor many areas and improve environmental management. There are significant gaps in the practical application and existing studies regarding the IoT and land use cover change. Therefore, it is necessary to determine whether applying the IoT and improving its application for ongoing environmental monitoring in areas of special interest or high risk is a worthwhile goal.

M. Francos (✉)
Department of Geography, Faculty of Geography and History, University of Salamanca, Salamanca, Spain
e-mail: mfq@usal.es

C. Sánchez-García
Department of Geography, Autonomous University of Madrid, Madrid, Spain

D. H. de la Iglesia
Department of Computer Science and Automation, Faculty of Science, University of Salamanca, Salamanca, Spain

O. Corvacho-Ganahin
Departamento de Ciencias Históricas y Geográficas, Universidad de Tarapacá, Arica, Chile

P. Pereira
Environmental Management Centre, Mykolas Romeris University, Vilnius, Lithuania

M. Nasr et al. (eds.), *Artificial Intelligence Applications for a Sustainable Environment*, Green Chemistry and Sustainable Technology,
https://doi.org/10.1007/978-3-031-91199-6_1

Keywords Environmental control · Process optimisation · Environmental risk · Sensors · Weather conditions

1.1 Introduction

Sensor systems, in combination with intelligent information processing and management tools, are the basis on which the cities and urban environments of the future will be built [1]. Sensor systems represent a fundamental element in the collection and interpretation of information of their surroundings. This development in data acquisition has revolutionised the way we interact with our environment [2]. Thanks to automation and reduced technical costs, cost-effectiveness in monitoring and process automation in all types of environments can be achieved [3]. Based on this new paradigm, the concept of the Internet of things (IoT in advance) has emerged. The IoT can be defined as a dynamic network (from local to global scales) with self-configuring capabilities based on interoperable communication standards and protocols. Physical and virtual "things" in the network have identities and attributes and are operated using intelligent interfaces [4]. It can be said that, in general, IoT systems consist of three essential components. First, there are the devices or sensors, which are physical components designed to collect data from the environment or perform physical actions. These devices can range from temperature sensors to cameras or wearable devices. Second, there is connectivity, which enables these devices to communicate with each other and with other systems through various protocols such as Wi-Fi, Bluetooth, Zigbee, or LoRa. This connectivity is crucial for data transfer between devices and IoT servers or platforms. Lastly, there is the platform or backend, which consists of software responsible for collecting, processing, storing, and analysing data collected by IoT devices. This platform includes cloud services, databases, data analysis tools, APIs, and data visualisation applications, and it can manage aspects such as security and device management within the IoT infrastructure. A basic IoT ecosystem is illustrated in Fig. 1.1.

Because of the increase in the number of connected devices and the need for computing with simplified communications and reduced response times, the IoT has given rise to the edge computing paradigm. Edge computing allows processing to be distributed to the network devices themselves to reduce latency [5], reduce the increase in information traffic, and improve the energy efficiency and predictability of systems. Initiatives, such as open edge computing, which brings together the efforts of different companies in the search for new business opportunities, have been carried out on an international level [6]. Edge computing is related to different concepts associated with the IoT [7] since it is necessary to reduce the volume of information transmitted to the network and to make decisions with reduced uncertainty and transmission time. Thus, several reports have considered enhancing edge computing by incorporating traditional machine learning techniques [8] and newer deep learning techniques [9]. The Internet of things is related with everyday and common objects, interconnected with each other, that help to improve human lives.

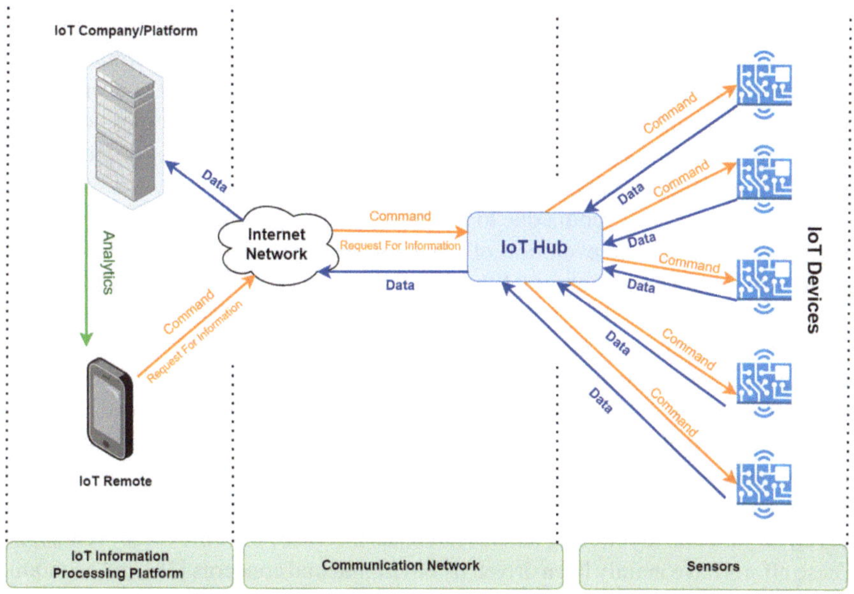

Fig. 1.1 General ecosystem of the Internet of things composed of three main elements: Sensorisation devices capable of capturing information directly from the environment; advanced data communication network for the exchange of data with the environment; platform for processing, management and analysis of the data obtained

In this sense, the use of these technological advances for monitoring, acquisition, management, analysis, and synthesis of ecological data allows us to know the ecological factors and environmental dynamics of a specific place, characterise it, and take measures accordingly. In terms of how IoT technology will improve the everyday lives of citizens, it is easy to introduce a few examples. IoT technology has been used in a wide variety of applications, from the smart home to health monitoring and precision agriculture as well as traffic accident beacons. In the smart home, IoT technology can automate household tasks such as energy management and lighting, control climate parameters, and enable greater control and convenience for users. In health care, IoT technology has enabled real-time monitoring of patients' health, allowing doctors to intervene in a timely manner and prevent complications. In addition, IoT technology has enabled the development of wearable health monitoring devices, such as smartwatches and activity tracking devices, which has improved people's quality of life. In agriculture, IoT technology has enabled real-time crop monitoring, allowing farmers to optimise irrigation and fertilisation to maximise crop yields. It has also enabled the implementation of precision irrigation systems, which reduce water wastage and lower production costs. There are some controversial aspects to this technology. As the number of IoT devices continues to grow, so does the need to pay attention to the security and ethics of the data being generated [10]. The amount of data generated by IoT devices is enormous and if not properly protected can be vulnerable to cyber-attacks, privacy breaches, and abuse by businesses and governments [11]. IoT devices can also collect personal information, such as behavioural patterns and preferences, which raises serious ethical concerns about privacy and data protection. A clear example of where such technologies pose a real risk to the security of citizens and conflict environments is the use of drones, which has increased in recent years, raising ethical and security issues [12]. Another implication to be taken into account due to the excess of data generated is the overloading of storage systems. The evolution of edge computing technology will bring relief in the management and storage of the data generated [13]. There are various types of IoT technologies used for environmental applications, including sensor networks, remote sensing devices, and mobile IoT platforms. These technologies are employed in different contexts, such as fixed sensor systems for continuous monitoring, mobile sensors for dynamic data collection, and satellite-based IoT systems for large-scale environmental analysis. It is also important to consider sustainability when using IoT technologies in the current global eco-crisis context [14].

The application of new technologies or informatic advances as IoT for ecological and environmental monitoring and research has increased over the recent years [15]. These efforts have mainly been driven by environmental concerns [16] and economic interests, such as process optimisation [17]. These scientific advances have allowed the IoT to be exploited in virtually every discipline for environmental monitoring [18]. The application of the IoT reached 3.8 billion devices in 2015 and are projected to reach 21.5 billion devices by 2025 [19]. Notably, massive data sets have been produced, and thus, it is necessary to develop models that can manage and interpret multiple interrelated variables. Technological progress can be very useful if it is accompanied by adequate knowledge. In this sense, it is fundamental how neatly a breakthrough is applied in a particular area and how absent it is in others. For this purpose, it is necessary to know the relevant studies on the subject and the possible gaps in the literature that offer niches for the application of this technology. This raises different questions that need to be answered. What are the most promising areas of application of this technology at present? What is the future potential of IoT for environmental monitoring?.

Therefore, the review is structured following the main topics related to the application of IoT in environmental monitoring. First, the first three sections are focused on land management, from land use and land cover, to forest and the application of some tools as a IoT in the agricultural technology. After that, we move on to anthropic environments such as urban context, where undoubtaly, IoT are used everyday. Finally, we analysed the implications and applications of IoT in climate studies, which are so important in the current global warming. At the end of this study and based on what we have analysed, we point out what we consider to be the potential applications of IoT in this area. This study aims to systemathic review the IoT application on environmental monitoring and identify literature gaps and future research needs. The specific objectives are to: (1) understand the structure and functioning of IoTs and their potential for environmental monitoring, (2) assess the proliferation of high quality studies and its results on the use of this technique in environmental monitoring, and (3) detect existing knowledge gaps related to environmental issues where the IoT may be particularly useful. To the best of our knowledge, no literature review encompasses the applicability of the IoT to all these environmental issues. We hypothesise that IoT have been used mainly in agriculture although its use can be extended to other disciplines where environmental monitoring is relevant. In these areas, the lack of studies and in particular case study articles shows that there is little follow-up on the implementation of this technology in land use cover and forestry issues.

1.2 Methodological Procedure and Literature Review

The systematic literature review was conducted following the framework developed by Preferred Reporting Items for Systematic Reviews and Meta-Analyses (PRISMA) [20] and is composed by six steps: (1) scoping, (2) planning, (3) identification and search, (4) screening articles, (5) eligibility assessment, and (6) presentation and interpretation. The same protocol have been applied in other bibliographic approaches [21–26] using Google Scholar, Scopus, or Web of Science (WOS) (or a combination of them) following the framework developed by Koutsos et al. [27]. In step 1 a review protocol was defined. In this review, the protocol followed was PRISMA statement protocol [28]. In step 2 (planning), the search for articles was performed online following Gusenbauer and Haddaway [29]. We performed a search using different combinations of relevant keywords and Boolean operators within the title, abstract, and keywords of each study: (1) "Internet of Things" AND "environmental monitoring"; (2) "Internet of Things" AND "environmental research"; (3) "Internet of Things" AND "land use"; (4) "Internet of Things" AND "land cover"; (5) "Internet of Things" AND "forestry"; (6) "Internet of Things" AND "forest management"; (7) "Internet of Things" AND "agriculture"; (8) "Internet of Things" AND "city"; (9) "Internet of Things" AND "urban"; (10) "Internet of Things" AND "global change"; (11) "Internet of Things" AND "climate"; (12) "IoT" AND "environmental monitoring"; (13) "IoT" AND "environmental research"; (14) "IoT" AND "land use"; (15) "IoT" AND "land cover"; (16) "IoT" AND "forestry"; (17) "IoT" AND "forest management"; (18) "IoT" AND "agriculture"; (19) "IoT" AND "city"; (20) "IoT" AND "urban"; (21) "IoT" AND "global change"; and (22) "IoT" AND "climate". Supplementary resources are used in addition to the principal resources for their specific qualities, which can help retrieve additional records and improve the evidence base further. Three databases were considered: Google Scholar, Scopus, and WOS. The search was conducted comprising the last 10 years (from 2013 to 2022). The time period is due to the fact that the first studies on these issues and the systematised application of IoT in environmental monitoring had significant advances by 2015 and showed the advantages of this tool, so studies began to be developed around those years, and it is expected that the growth in the application of IoT for environmental monitoring will grow exponentially and become massive by the middle of this decade. In step 3 (identification and search), 12,653 articles were identified and retrieved during second half of 2023. The first step of PRISMA protocol also includes excluding duplicate studies, which accounted for 6908 studies. All documents were filtered in step 4 (screening articles) based on two-stage criteria (Fig. 1.2). In the first stage (title and abstract reading) were identified and excluded the studies that not match with the criteria of being (1) double-blind peer review studies; (2) focused on related topics (title or abstract matching the research topics); (3) in English language; (4) case studies (avoiding as far as possible review and opinion papers); (5) full text manuscript availability; and (6) published between 2013 and 2022. During this first stage, 5745 studies were counted. After applying the criteria, 5296 texts were excluded, and a total of 449 articles and book chapters were included in the next stage. In the second stage (full text reading) and due to this paper analyses case studies published as scientific articles, extended congress proceedings, and book chapters were excluded the works that (1) were not complete studies but summaries or extended summaries that summarise other studies by the same authors without revision, (2) not directly related to the topic of study, (3) not match the sustainable development goals (SDG) main lines [30]. The application of this criteria produces the exclusion of 331 studies. In total, 118 articles were identified as eligible for review. Despite this, in some instances, earlier studies, reviews, and other types of studies (grey literature) were consulted and incorporated into the text for the explanation, contextualisation, and discussion of certain dynamics or concepts.

1.3 Study Limitation and Recommendation

There are some limitations in this type of studies because many of them, due to be scientific literature, are in English. However, there are many interesting studies written in other languages that are not covered in this paper because they are not in English. There are also many interesting papers and projects that end up as technical reports and are not published in journals or editorials, and they are not consulted by many users because their scope is very limited, so they are not included as eligible for this review.

A systematic review is a synthesis of the available evidence study, in which a review of primary scientific published works is carried out, with the aim of summarising the existing information on a particular topic. After collecting the articles of interest, researchers analyse them and compare the evidence they provide with that of similar articles. One of the recommendations and reasons for using this type of study is to provide evidence on the usefulness and practicality of a procedure, methodology or tool for the study of something specific, such as IoT for environmental monitoring.

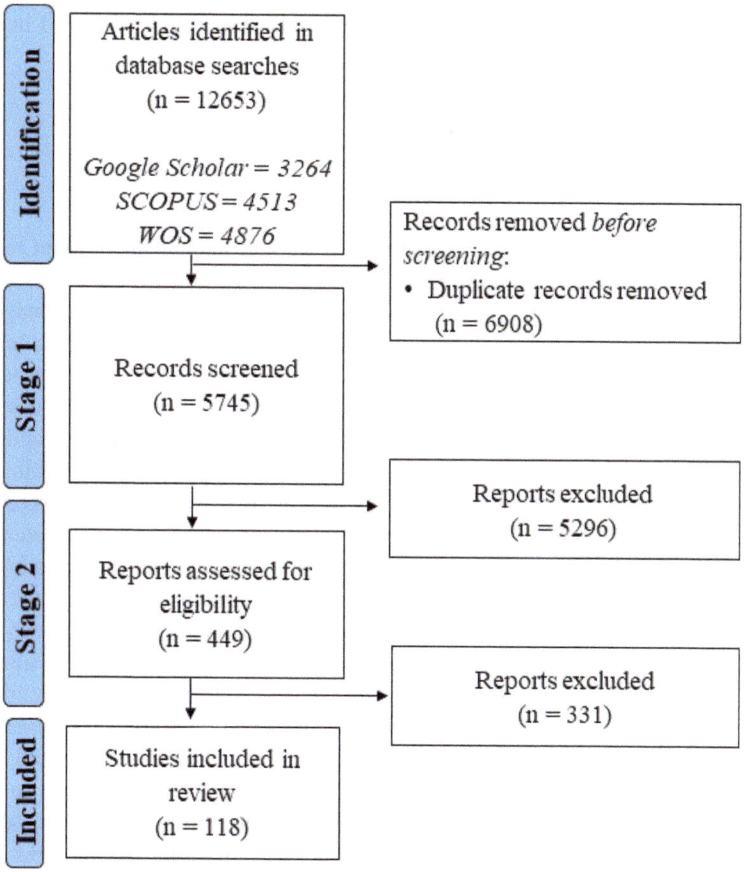

Fig. 1.2 Flow chart of the methodology followed in the systematic review of the literature

1.4 Land Use and Land Cover Change

Land use and land cover (LULC) is critical to understanding the changes observed in rapidly urbanising societies, which has happened recently in Europe, where most of the territory is still used for agricultural activities. The LULC factors appear to determine a wide range of goods and services, including food provision and biodiversity preservation, climate change mitigation, and landscape aesthetics [31]. The landscape, including LULC, reflects global trends, including the causes of landscape changes are urbanisation, agricultural intensification, land abandonment, forest expansion, world economic trends, and the development of renewable energy [32, 33]. The magnitude of these changes has given rise to concerns that landscape sustainability—the capacity of a landscape to consistently provide long-term, landscape-specific—is currently at risk [34]. The IoT can help deal with this risk and take measures to develop policies to mitigate it [35] and manage efficiently urban areas

[36]. However, there are few studies that point out the importance of the IoT in LULC [32], and most reports focus on the urban planning aspects.

1.5 Forest Management

The IoT has been used in forest management for soil management, plant regeneration monitoring of plant communities, and monitoring for extreme events such as forest fires. Singh et al. [1] pointed to the IoT as a useful tool in forest management for environmental monitoring, data acquisition, and analysis in the field. In terms of environmental variables, the IoT is useful for monitoring parameters such as temperature, humidity, and environmental quality [37]. The IoT is especially useful for transmitting data from sensors that capture forest information in real time and monitoring micro-climate change in the forest [38]. Thus, Bayne et al. [39] pointed out that small wireless sensor networks may change the way forests are managed in the future. Continuous monitoring of forest areas and environmental parameters is of great use for forest management, expanding the information collected, systematising this information, and analysing processes and phenomena that may affect the forest. Therefore, the IoT opens new opportunities for creating useful services in forest science, creating what is known as a "Smart Forest" [40]. This technology also optimises the marketing of timber products by carrying out chain-of-custody tracking and performing it and providing individual tree information to forest managers [41]. For this reason, the IoT and blockchain enable smarter and more economically beneficial forest management by acting with greater control, transparency, trust, and accuracy [42]. These technologies are not only applied in rural areas, but also for forest management in urban areas. The use of a tree structure to map devices that are connected to the IoT system simplifies the management and mapping of forest areas, thus complementing geographic information systems [43]. Sustainable forest management can even encourage the harvesting and use of these areas for recreational activities (e.g. sustainable tourism and recreation parks) [44].

Regarding vegetation in forested areas, the use of IoT is mainly focused on tree coverage. This application of new technologies in forestry helps with performance assessment, compliance, and maintenance supervision, and integrates safety systems in plants [41]. In addition, the systematisation of data on the vegetation that makes up the forest makes it possible to assess the current situation and the development of sustainable forest management [45]. The collected information must be uniform in terms of the distribution of sample units, and a constant periodicity of data collection must be maintained. Therefore, geomatic tools and the IoT can help ensure sustainable forest management. The implementation of IoT-related systems and innovations can benefit landowners by minimising downtime and decreasing tree damage losses [41]. Water and soil quality can be measured to establish thresholds at which intervention is needed to avoid environmental degradation [46]. Salam [47] pointed to the benefits that the IoT can have in forest management, among other things, to prevent soil degradation. In digital forestry, the IoT offers efficient operations, monitoring, and forecasting for soil erosion and avoiding or reducing the impact of extreme events such as forest fires.

The use of geomatics and the IoT is of great interest in detecting fires, fire-prone, and highly vulnerable areas and predicting fire dynamics [48]. In addition, real-time monitoring of environmental conditions in a forest area is possible [49, 50]. Zhang et al. [51] proposed an IoT-based acoustic detection system that could detect forest fires by canopy-to-surface fire variation. Moreover, the IoT allows the constant recording of data and progressive analysis over time of events such as illegal logging or deforestation [52]. Improvements in the monitoring of environmental variables, as well as improvements in data collection and communications, may enable reduce the uncertainties and faults detected in current systems (e.g. connectivity issues and detecting the fire at the edge). These developments may enhance the detection and prediction of forest fires, analysis of existing flora, and data acquisition of environmental parameters [53]. The development of the IoT implies a clear advance in the identification of potential hazards affecting forest areas. Older methods of preventing or detecting forest fires in their early stages of ignition are gradually becoming obsolete, and there is a growing need to manage forest fires more efficiently by detecting ignition as soon as it starts [54]. Previous systems are being replaced by intelligent early-warning systems that can not only detect and extinguish but control the fire so that it does not affect or endanger critical areas of special interest [55]. To protect forests further, from the growing population and infrastructure, more efficient data acquisition in forests and constant monitoring are needed [56]. The IoT can monitor and interpret the dynamics of a fire and its spread using sensors, such as DHT11 and flame sensors, that continuously detect fluctuations in the temperature and humidity, and using other devices, such as a NodeMCU microcontroller, which sends these values to the online database. In this way, thresholds are determined, and when they are exceeded, an alert message is sent to the corresponding forestry department or service [57]. In forested areas, all the information associated with the environment that the IoT can capture, store, analyse, and transmit helps in decision-making with regard to forest fires, pests, and tree diseases for forest management.

1.6 Agricultural Sector

The IoT has also contributed to improving agricultural aspects, such as smart irrigation, by making more efficient use of water and improving yields [41, 58]. The smart monitoring of soil variables and their suitability for agriculture has also been improved by the IoT [59]. This technological development also facilitates key tasks, such as the use of hyperspectral imaging, to determine crop production and fertiliser requirements, thereby limiting fertiliser waste and nutrient loss through runoff [60]. IoT solutions have also been used in some cases to determine the most suitable soil conditions for planting certain types of crops and increase crop productivity [61].

There are multiple approaches to the study of the IoT in agriculture and its utility. However, during the last decade, the developments in agriculture have advanced along two major dimensions. First, focus has been placed on issues related to monitoring, tracking, control, and automation of various processes aimed at improving the efficiency of agricultural activity, and second, studies have analysed the development of agricultural policies related to insights provided by the IoT [62, 63].

The development and application of the IoT for agriculture offer considerable advantages. Within the field of monitoring and tracking, for example, wireless sensor network systems are used to determine the behaviour of soil properties, electrical conductivity, ambient humidity, temperature, the intrusions of external agents in agricultural areas, and sunlight and leaf wetness have been mentioned [62, 64]. Furthermore, the IoT has enabled the development of various strategies and practices to address the challenges generated by the constant environmental changes that affect crop fields. A deeper understanding of environmental parameters inevitably provides insights for farmers because a diverse and temporal understanding of crop dynamics increases the efficiency of agricultural activity [65]. A concrete example of this has previously been seen in India, where the decline of agricultural sectors over time was strongly affecting production capacity. Real-time monitoring coupled with the availability of information throughout the agricultural cycle made it possible to overcome the limitations of traditional procedures and implement improvements in water efficiency and production costs [66]. In Europe, another notable application is that of winegrowers who managed to improve decision-making using support systems based on IoT-based monitoring and tracking, where information concerning the proper time for pesticide treatment became a determining factor in agricultural sustainability [67].

Alongside monitoring, tracking process control in agriculture is another contribution that has made a significant contribution to the application of IoT in agriculture. The informed development of interventions based on the inspection of parameters represents a fundamental area for the optimisation of processes and crop yields. The main focus of the research based in smart agriculture is the application of the IoT in the control of agricultural processes to resolve current issues [64, 67]. In particular, studies often highlight the changing environmental conditions, such as resource availability and atmospheric parameters, soil condition, water scarcity, electrical conductivity, and others, as measured by sensors and nodes, which can be used to establish limits, quantities, and schedules for the optimisation of agricultural activity. For example, in the case of hydroponic farming control using the IoT, the pH, which is critical to the nutrient uptake capacity of a plant, can be controlled within a range that is set by the user. Using a controller, it is possible to continuously detect when and by how much the pH exceeds the set limits and then take compensatory action [67]. Additionally, in cases where it has been required to establish irrigation quantities and availability in greenhouse conditions, the contribution of IoT controls has been highlighted. The possibility of establishing real-time control in greenhouses by incorporating an irrigation system connected to a network enables alternative solutions and protections for the problem of water scarcity [68].

In terms of process automation, several studies have addressed a diverse range of applications and challenges in the agricultural domain. Furthermore, the studies and applications of efficient water management and irrigation systems are the most predominant areas among the works related to smart agriculture [63, 69–71]. Another representative field that has had a notable presence in automated agriculture research in the last ten years is smart fertilisation systems. Overcoming the disadvantages of manual spraying, controlled product application adapted to specific conditions, and the proportion of fertiliser in exact amounts required over time for each specific crop area are the main subtopics that are being developed [69, 70, 72]. The third broad area is concerned with integrated systems for pest control, weed control, and disease detection. This area is regularly developed in relation to sub-themes of great interest, such as the study of pest diseases, fungicide applications, controlled delivery of herbicides, and early-warning systems for diseases or invasions. Mainly, the development of studies under this domain presents a common cross-cutting approach that aims at protecting and safeguarding the health and growth of the crop from sowing until the completion of its final production cycle and harvest [73–75]. The fourth area of research in automation also being considered frequently in various studies corresponds to the autonomy of vehicles and the operation of machinery in the service of agriculture. Recent specialised and general studies have focused on the main operations of ploughing, harvesting, transplanting, and autonomous spraying. Tractors with intelligent operation, unmanned aerial vehicles, and machinery for harvesting and transplanting receive the most attention [1, 70]. Essential opportunities, strengths, and challenges for advancements in agriculture are highlighted within this area of knowledge, with special emphasis on tractors as an essential and irreplaceable agricultural power unit. Across the board, it is emphasised that tractors can be monitored for power output, fuel consumption, draft, and wheel slip, which improves the traditional use of machinery in terms of sustainability and operational capacity. But challenges also arise with respect to advances in autonomous and assisted remote navigation, where the emphasis is placed on the current safety problems and perception concerns that have arisen [70]. In contrast, unmanned aerial vehicles appear as an emerging technological breakthrough in automation because of the versatility of operation they can provide in spraying and monitoring of variables (soil, vegetation, and moisture) on large and small scales [70].

There is also growing interest in the policies of countries and state administrators related to the IoT in agriculture. Although it seems to be a less developed topic in the scientific and research field, several recent studies and works have mentioned how countries adapt or react through regulations and policies to the changes and results in agricultural practices from the implementation of IoT solutions [76]. The main emphasis has been on nations taking advantage of the IoT to improve agricultural productivity. The development of regulatory frameworks and public policies as a result of the emergence of the IoT in smart agriculture has also been investigated. Subeesh and Mehta [70] discussed different nations that have made efforts to promote investment (Australia and the United States), support programmes for agricultural practices (Ireland), partnerships for innovation projects in agriculture (France, the United States, and the Philippines), integration of the IoT in agriculture to improve profitability (China), policies to improve crop growth and poverty alleviation through the development of an agricultural framework integrating IoT technologies into all phases of farming (Malaysia and India), technology centres for innovation in smart agriculture (Thailand), and multi-strategies for innovation in agriculture (Food and Agriculture Organization of the United Nations).

Regarding future challenges, there is a lack of technical knowledge among farmers (direct users) to use the new technology, making it necessary to develop user-friendly user interfaces. Furthermore, the cost and quality of devices and sensors can be a barrier, especially for the small farmers, which affects the democratisation of this technology. Another important point is the reliability of the technology. Data security in terms of guaranteed confidentiality and privacy is an issue that has not been clarified or investigated in depth. Ultimately, hardware maintenance, energy resources for the sustainability of the system, and the acceleration of data exchange are central concerns for the efficient technological development of agricultural activity [70, 77].

1.7 Urban Environment

The concept of smart cities has become increasingly important in the field of urban planning because of the growing need to improve the efficiency of resource management and the quality of life in the face of population growth and climate change. Technological advances in the field of IoT and artificial intelligence (AI) have enabled greater integration and monitoring of different urban systems such as transport, energy, and utilities [78]. In addition, greater community participation in decision-making and the implementation of innovative solutions are being promoted. For example, the use of sensors and Internet-connected devices has enabled real-time monitoring of water and energy consumption in buildings, which can reduce waste and improve efficiency. These advances have demonstrated the potential of the IoT in the creation of smarter and more sustainable cities [79].

One of the main IoT applications in urban environments is air quality monitoring (AQM). Air pollution has negative effects on people's health and the environment in general. Therefore, it is essential to develop efficient monitoring systems that can detect and address pollution problems. In smart cities, AQM is carried out using a wide range of existing environmental sensors that analyse factors such as the concentration of suspended particulate matter, carbon monoxide, nitrogen dioxide, and ozone, among others. These data are used to implement measures that improve air quality, such as controlling vehicle traffic, implementing renewable energy systems, or promoting non-motorised mobility. There are also different approaches to the deployment of IoT infrastructure for outdoor AQM. IoT systems can be fixed or mobile, and each of these approaches has advantages and disadvantages. Fixed sensor systems are installed at a specific point in the city and tend to be more costly, as well as more expensive to maintain. One of the advantages of this approach is that it allows constant and accurate monitoring of air quality in a specific area. In addition, these sensors are generally more accurate than mobile sensors, as they are not subject to movement or vibration. There are different techniques to locate these fixed infrastructures. Elkamel et al. [80] use a mathematical model based on the multi-cell approach to create monthly spatial distributions of the concentrations of pollutants emitted by different emission sources. The reports in the literature were subjected to a heuristic optimisation algorithm to identify the optimal configuration of a monitoring network. For Kang et al. [81], minimising the number of monitoring points to infer air quality across an urban area is essential for the solution to be economically viable. The authors have designed an efficient air quality inference model based on a proposed high-order graph convolution approach that can capture the spatiotemporal air quality distribution and extract features of influential external factors.

Conversely, mobile sensor systems are installed on devices, such as drones, vehicles, or people, that collect data as they move around the city. This allows for more detailed and dynamic monitoring of air quality because data can be collected at different points and at different times. In addition, these systems are often more accessible and cheaper than stationary systems. Elen et al. [82] stated that using bicycles is the most optimised method of mobile sampling in urban environments. Bicycles can be equipped with compact air quality measurement devices to monitor ultrafine particle counts, particle mass, and black carbon concentrations with high resolution. Other authors, such as Bales et al. [83] and Hu et al. [84], also use bicycles as mobile infrastructure for AQM in urban environments. Apte et al. [85], however, suggested that the best alternative is to use vehicles that make periodic routes through certain areas. In particular, the authors equipped Google Street View vehicles with a fast-response pollution measurement platform. They repeatedly sampled every street in a 30 km^2 area of Oakland (USA), developing the largest urban air quality dataset of its kind. Another alternative is to equip urban transport vehicles such as buses or trains [83]. Another alternative is to equip waste collection trucks with air quality sensing systems, which have been tested in the city of Cambridge, Massachusetts (USA) [86]. It is also possible to combine data from fixed infrastructure with data collected by low-cost IoT sensors that can be inserted into cars, bicycles, and even carried by joggers or pedestrians in the city [87]. In this regard, a machine learning model has been developed that combines sparse data from fixed stations with dense data from mobile sensors to estimate the surface area of air pollution at a given hour on any given day in the city of Sydney, Australia.

However, mobile sensor systems also have disadvantages. One is that they can be less accurate than fixed ones because vibrations and movements can affect the measurement data [83]. In addition, these systems can be more difficult to maintain and update than fixed systems [88]. As an alternative, satellite data can be useful for pollution level prediction in different urban areas and regions. For instance, Stebel et al. [89] use moderate resolution imaging spectroradiometer data together with meteorological factors to analyse their relationship with ground-based particulate matter stations. In addition, AQM can be conducted in a participatory way, through community collaboration. For example, mobile applications can allow people to report pollution problems in their immediate environment, allowing for more detailed monitoring (crowdsensing) [90]. Along these lines, Montori et al. [90] present an architecture, called SenSquare, that is capable of managing both heterogeneous data sources from the open IoT platform and crowdsensing campaigns for environmental monitoring.

Studies analysing indoor air quality make extensive use of IoT technologies. Different studies have shown that moderate levels of indoor carbon dioxide cause headaches, fatigue, and inattention. When this carbon dioxide concentration is higher, it causes vomiting, dizziness, and other symptoms [91], and therefore, it is necessary to monitor and control such concentrations in indoor environments. Several articles have been produced in recent years focusing on measuring the concentration levels of carbon dioxide [92, 93]. Additionally, considering the Covid19 situation, some authors have focused on the environmental factors that facilitate the proliferation of the virus [94].

Energy optimisation in infrastructures and buildings is another key issue in smart cities. In the literature, authors have proposed several IoT-based solutions that address the problem of energy optimisation. The main objective is to maintain a balance between user comfort and energy requirements. With this balance, the user can achieve the desired comfort level with minimum energy consumption. Current studies was based on combining thermal monitoring systems with control techniques [95]. While increasing the efficiency of thermal control systems in buildings, these systems still have disadvantages, such as inefficient algorithms, over-filtering, or incorrect configuration of the control strategy. A new evolution of these systems aims to overcome these problems. Fuzzy logic-based controllers have gained significant attention [96], which may improve comfort levels while further minimising energy consumption [97].

The integration of AI within IoT frameworks in smart cities is crucial for enhancing data processing, predictive analytics, and decision-making processes. AI algorithms can efficiently analyse the vast amounts of data generated by IoT devices, providing actionable insights that improve urban management and sustainability efforts. Furthermore, AI's ability to learn and adapt over time can lead to more proactive and adaptive urban systems, ultimately reducing costs and improving the overall quality of life in urban environments.

1.8 Climate Analysis and Implications

The application of the IoT in different sectors related to climate and global change has been widely discussed and reviewed in the literature [98]. The IoT can enhance the quality of life in different areas including medical services, smart cities, the construction industry, agriculture, and water management [99]. Some researchers have pointed out the need for investments in software, as well as in IoT devices and remote sensing techniques, to maintain and enhance carbon stocks and the provision of ecosystem services in threatened forests [100]. Regarding the proliferation of diseases, which may be spreading more frequently because of global changes, Franklinos et al. [101], Dash [102], and Ogiela et al. [103] suggested that the IoT is essential to generate an actionable connection between the health, science, and public sectors, providing scalable and flexible data that can be used together to manage this increasing global issues [104]. Buyya et al. [105] analysed all the implications that the IoT has on issues in sustainability, climate change, agriculture, policymakers, etc. They emphasised the importance of developing new technologies. Using new software, techniques, and methods, we can develop new management tools to decrease the effects of climate change, and in general, global change on the industry and society. This can allow us to measure, evaluate, understand, and take appropriate action in specific locations against global warming [106]. Sustainable agricultural production and smart farming and agriculture based on changing climatic conditions can be a key factor in determining its duration over time [107–110]. According to Silva et al. [111], novel software for the IoT is useful in the study of natural disasters, which appear to be increasing in future in strength and frequency [112]. Narmilan and Niroash [113] discussed the applications of IoT in agriculture to mitigate the problems and consequences of climate change, including floods, droughts, plant diseases, and greenhouse gas emissions. In this sense authors as El Maachi et al. [114] point to the need for sustainable technological and scientific development to help combat global climate change.

There are numerous possibilities that are not being used yet, and in the next few years, many of them will be implemented [115]. Some of the applications on climate monitoring and IoT are related to the prediction climate dynamic and remitting weather information to the Weather Forecast Centre and users [116–118]. The characteristics of the sensors used to collect information in the IoT not only allow them to measure air conditions, but also to determine water quality and values such as pH, temperature, and oxygen level, allowing to control and monitor the living conditions of aquatic plants and animals [119]. Climate monitoring represents one of the biggest challenges facing the world today [120, 121]. Consequently technological and scientific advances are required to climate monitoring [122]. In this sense, IoT can play a critical role to adapt us agains fatal challenges [123] avoiding climate vulnerability and monitoring climate risk [111, 124]. This application can help farmes to get the real-time data to carry out decisions [125–127] preventing overexploitation of resources and their degradation [128]. According to these climatic variables and other measurements of environmental characteristics and parameters (e.g. soil and water) a decision can be made to plant one species or another and to manage the plantation in a certain way [129, 130]. Monitoring the environmental variables that make up the climate (e.g. humidity, temperature, pressure, CO, air pollutants, etc.) can help raise awareness of climate change importance [131, 132] and its effect on cultural heritage monuments [133] and cities [134, 135]. This kind of studies can also predict indoor remote temperatures to understand environmental conditions, greenhouse micro-climate [136–138] and prevent environmental issues [139] or increment garden productivity [140, 141]. Through these sensors, anomalies that may affect plants, animal, or human health can be detected [142] in orchards and vertical garden walls [143–145]. The application of these sensors and IoT based on climate parameters control can conduct to carry out smart homes [146] and intelligent buildings that reduce electrical expenses [147, 148] and make easier and comfortable humans life [149]. In the studies mentioned to this point, several recurring and major essential components for system implementation have been identified (Fig. 1.3). These components are hardware (it includes the physical devices, sensors, actuators, and electronic components that form the basis of the IoT and is the tangible component that enables data collection and interaction with the physical environment), middleware (it is a software layer that facilitates communication and data management between IoT devices and applications and simplifies the integration of different devices and technologies, providing a common platform for interoperability and data management), visualisation and interpretation tools (software that enables graphical presentation and analysis of collected data and it facilitates the understanding and analysis of large volumes of data, enabling informed decision-making), data storage and analysis (it includes the databases and analytics systems that manage and process the data generated by IoT devices, and it is crucial for storing, organising, and extracting useful information from the massive data generated, enabling advanced analytics and artificial intelligence), addresing schemes (methods to assign unique addresses to each IoT device. Enable effective identification and communication between devices within the IoT network), and radio frequency identification and wireless sensor networks (multi-sensor technologies and networks that use radio waves for wireless communication between devices and data transmission). These components are fundamental because, together, they enable the collection, transmission, processing, and analysis of data, forming the basic backbone of IoT systems. Each component, therefore, plays a critical role in the functionality and efficiency of an IoT system. Therefore, these elements are outlined in Fig. 1.3 as the specificities of each of them will determine the power of IoT for environmental monitoring.

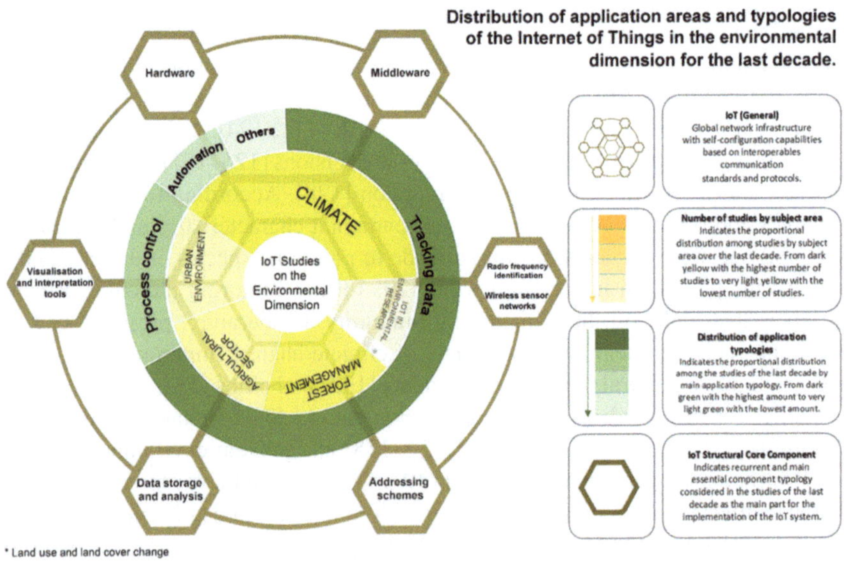

Fig. 1.3 Environmental IoT applications. Note: The interaction of hardware, middleware, visualisation and interpretation tools, data storage and analysis, addresing schemes, and radio frequency identification and wireless sensor networks together determine the successful implementation of an IoT system

1.9 Overview and Future Scenarios for the IoT

Applications in agriculture stand out for being widely studied and analysed, followed closely by the utilities that IoT can provide in urban environments, mainly oriented to smart cities and the improvement of elements of urban systems such as transport, energy, waste, citizen participation and integration, and public services. The area of forest management has a similar presence to that of urban environments, mainly soil management, vegetation regeneration, monitoring of environmental parameters of forests, and unusual events such as fires for example. In the case of global climate change and renewable energies, a wide range of works and thematic reviews are presented but mainly aimed at technological development in energy and improvement in the quality of processes to reduce the effects on environmental components that are essential for the maintenance of the planet and human life (climate, ecosystems, water, among others). Finally, there are changes in LULC, and although they are the focus of attention of the scientific community within other associated topics, the particular development of this dimension with the incorporation of IoT as the main focus is minor. Generally, there are few studies that specifically address the possibilities that IoT can provide in this area, especially if we consider that the study of this dimension provides insights into global trends and the integrated incorporation of intelligent monitoring systems could open a way to obtain a greater scope and understanding. For example, it can be applied to real-time monitoring of incident activities in the use and destination of land, changes in unusual periods (crises, disasters, and economic variations), or the possibility of having alternatives for sustainable projections that are obtained from the massive collection of data in this area. There are other environmental dimensions that have been mentioned over the last decade, such as IoT policies associated with the environment and the integration of industrial economic activity and its effects on the environment, although their particular development is minor compared to the other dimensions.

Although it appears that the main IoT application in the environment is monitoring, there are other uses of the IoT network infrastructure system in the environment with very specific functions. It has been observed that in all IoT application areas many of the studies were not validated by case studies with field measurements. This is even more evident in the case of climate where, due to the magnitude of the phenomena, studies are often not validated by case study fieldwork. Moreover, in this case, due to its novelty, many studies still focus on the architecture of the sensors and the transfer of environmental information and not so much on the implementation of these sensors in the field. It should be noted that, as shown below, many of these studies have been developed as a necessity for each country to adapt to its social, economic, and environmental context. In this sense, various countries, taking into account the existing population growth, have carried out numerous studies and scientific advances in terms of precision agriculture, optimisation of resources, and increases in agricultural productivity. Others, where priority has been given to cities and the well-being of citizens, have focused on this type of study. Finally, areas where the scarcity of natural resources is greater and the environmental impact of

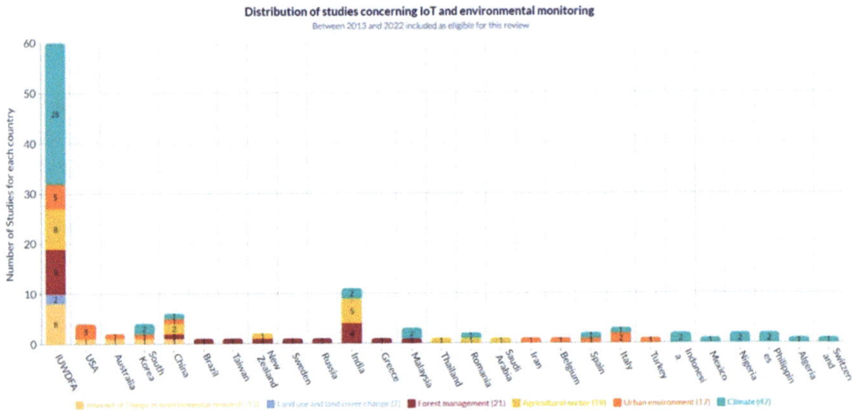

Fig. 1.4 Distribution of studies concerning IoT and environmental monitoring between 2013 and 2022 included as eligible for this review

climate global change and water stress can significantly affect their society have dedicated greater efforts to finding viable alternatives to traditional procedures and socio-environmental sustainability (Fig. 1.4).

IoT and smart automation technology will change the way in which the control and monitoring of environments is performed. One example of this future evolution is in forestry environments. The introduction of autonomous vehicles, such as drones and ground robots, will improve the efficiency and accuracy of data collection in forestry in the coming years. In the past, forest monitoring was an arduous and dangerous job for workers who had to venture into the forest to collect important data. With the introduction of drones and ground robots, forest monitoring will be done safely and accurately without the need to put workers at risk. Drones, for example, will be used on a massive scale for real-time forest monitoring and assessment. Remote sensing technology and the high-resolution camera on the drones will collect accurate information on tree density and growth, forest health, and the presence of pests or diseases. The automation of drones will enable the collection of massive data efficiently and over large areas, making it possible to identify critical areas that require special attention. On the other hand, autonomous terrestrial robots will be used for soil and vegetation sample collection, wildlife monitoring, and air quality assessment in forested areas. These robots can be programmed to collect information intelligently and send the data via IoT for further processing and analysis. The use of autonomous vehicles in forest monitoring will not only improve accuracy and efficiency, but also reduce the operational cost and environmental impact of traditional methods. In addition, the information collected by autonomous vehicles will be used for more effective and sustainable forest management.

In the agricultural field, the use of advanced techniques such as IoT monitoring is already generating significant benefits for farmers, enabling optimisations and significant savings in resources such as irrigation water. The way forward for the future is to further advance the application of IoT to achieve even more improvements in efficiency, sustainability, and productivity. One of the future applications of IoT in this area is the integration of artificial intelligence and machine learning technologies. This will allow farmers to collect and analyse large amounts of data, which would help them make more informed and accurate decisions on resource use and crop management. For example, AI systems will be able to help predict weather patterns and soil conditions, allowing farmers to adjust their cultivation practices accordingly. Another application where the application of IoT in agriculture will be key is the automation of harvesting and harvesting processes, through the use of robots and other autonomous devices. This will help farmers reduce labour costs and improve harvesting efficiency. In addition, autonomous vehicles will be used for the precise application of fertilisers and pesticides, thereby reducing the use of chemicals and increasing resource efficiency.

From a security and reliability point of view, IoT systems have evolved in recent years thanks to their global boom. The interest aroused by industry and governments has generated a line of work for the securitisation of IoT. Thus, there are currently different international standardisation and control committees that regulate the safe and responsible use of this technology. Within the realm of international standards, the ISO/IEC Joint Technical Committee 1 plays a pivotal role in establishing norms for information technologies, encompassing IoT systems. On the IEEE (Institute of Electrical and Electronics Engineers) side, the IEEE P2413 working group is dedicated to crafting an architectural framework specifically tailored for IoT systems. Furthermore, there are dedicated groups addressing security and interoperability concerns. For instance, the Internet Engineering Task Force (IETF) IoT Directorate is actively engaged in addressing IoT-related issues within the IETF, actively working on protocols and standards to ensure seamless interoperability and robust security in IoT systems. In Europe, the European Telecommunications Standards Institute (ETSI) TC CYBER focuses on establishing standards related to cybersecurity, extending its coverage to encompass IoT systems. Similarly, the U.S. National Institute of Standards and Technology (NIST) Cybersecurity Framework concentrates on addressing the cybersecurity dimensions applicable to IoT systems. Consequently, IoT emerges as a pertinent technology, well-grounded in standardisation, and fortified with protocols aimed at guaranteeing the requisite security and trust.

Apart from all the benefits of the application of IoT on environmental monitoring, there are some difficulties as well. The main difficulties are related to the economic cost that are intrinsically related to the application of high technologies such as IoT. The application of IoT can be difficult in some rural areas because of the cost, mostly in forest, land use and land cover, and agricultural context. In the urban areas and climate change context, these issues can be avoided because the actual politics are in-line with the application of new technologies to reduce the effects of climate change.

1.10 Conclusions

Based on the lack of articles that deal with the application of IoT in the study of land use changes, it may be helpful if this technology was used to collect information on land use changes, soil quality indexes, and ecosystem services. It has been shown that the IoT has mainly been applied and studied in urban areas with agriculture-related issues. This is mainly because these are the areas where most of the population lives and where most of the economic benefits are extracted. IoT is currently closely associated with the development of sensor and data transmission environments. Therefore, the use of IoT technologies is more widespread in environmental sensing environments. It is mainly focused on the measurement and monitoring of environmental variables such as air quality, temperature, humidity, atmospheric pressure, radiation, among others. The monitoring of these environmental variables is fundamental for the understanding of environmental behaviour and the identification of patterns and trends that can help to prevent and improve environments. For this reason, there are fields of study where advances in IoT are more integrated than in other fields of study where sensor systems are not as relevant when it comes to achieving the main objectives.

The IoT can be used not only to prevent catastrophes and risks (as in the case of reaching environmental variables conducive to a large forest fire) but also to monitor the recovery of an ecosystem after an environmental disturbance. Therefore, applying this technology to environmental aspects to prevent degradation and assess recovery can be fundamental to understanding the dynamics of an ecosystem on an ongoing basis and acting or managing land use, forest, or agricultural areas accordingly to mitigate environmental degradation.

1.11 Conflicts of Interest

The author declares no conflicts of interest.

Acknowledgements The authors would like to thank J. Lees and C. Leslie for their support for language revision.

Author Contributions All authors contributed to the study conception and design. Material preparation, data collection, and analysis were performed by Marcos Francos, Carlos Sáncez-García, Daniel H. de la Iglesia, and Oscar Corvacho-Ganahin. The first draft of the manuscript was written by Marcos Francos, Carlos Sáncez-García, Daniel H. de la Iglesia, Oscar Corvacho-Ganahin, and Paulo Pereira, and all authors commented on the previous versions of the manuscript. All authors read and approved the final manuscript.

Funding The authors have not received funding to carry out this study.

References

1. Singh PK, Naresh RK, Kumar L, Chandra MS, Kumar A (2021) Role of IoT technology in agriculture for reshaping the future of farming in India: a review. Int J Curr Microbiol App Sci 10(2):439–451. https://doi.org/10.20546/ijcmas.2021.1002.052
2. Vinod Kumar TM (2020) Smart environment for smart cities. Advances in 21st century human settlements, 1–53. https://doi.org/10.1007/978-981-13-6822-6_1
3. Khani NAKK, Fazal MJ, Arif U (2021) IoT based environment Analyzer. Int R Basic Appl Sci 9(3):62–70
4. Witkowski K (2017) Internet of things, big data, industry 4.0—innovative solutions in logistics and supply chains management. Procedia Eng 182:763–769. https://doi.org/10.1016/j.proeng.2017.03.197
5. Alamgir Hossain SK, Anisur Rahman M, Hossain MA (2018) Edge computing framework for enabling situation awareness in IoT based smart city. J Parallel Distr Com 122:226–237. https://doi.org/10.1016/j.jpdc.2018.08.009
6. Ai Y, Peng M, Zhang K (2018) Edge computing technologies for internet of things: a primer. Dig Comm Net 4(2):77–86. https://doi.org/10.1016/j.dcan.2017.07.001
7. Yang J, Lu Z, Wu J (2018) Smart-toy-edge-computing-oriented data exchange based on blockchain. J Syst Architect 87:36–48. https://doi.org/10.1016/j.sysarc.2018.05.001
8. Dao NN, Vu DN, Lee Y, Cho S, Cho C, Kim H (2018) Pattern-identified online task scheduling in multitier edge computing for industrial iot services. Mob Inf Syst, 2101206.https://doi.org/10.1155/2018/2101206
9. Park D, Kim S, An Y, Jung JY (2018) LiReD: a light-weight real-time fault detection system for edge computing using LSTM recurrent neural networks. Sensors 18(7):2110. https://doi.org/10.3390/s18072110
10. Ammar M, Russello G, Crispo B (2018) Internet of things: a survey on the security of IoT frameworks. J Inform Sec Appl 38:8–27. https://doi.org/10.1016/j.jisa.2017.11.002
11. Brous P, Janssen M, Herder P (2020) The dual effects of the Internet of Things (IoT): a systematic review of the benefits and risks of IoT adoption by organizations. Int J Infor Manage 51:101952. https://doi.org/10.1016/j.ijinfomgt.2019.05.008
12. Atlam HF, Wills GB (2020) IoT security, privacy, safety and ethics. Dig. twin technol. smart cities, 123–149.https://doi.org/10.1007/978-3-030-18732-3_8
13. Lucivero F (2020) Big data, big waste? A reflection on the environmental sustainability of big data initiatives. Sci Eng Ethics 26(2):1009–1030. https://doi.org/10.1007/s11948-019-00171-7
14. Zolotukhin O, Filatov V, Yerokhin A, Kudryavtseva M (2021) The methods for the prediction of climate control indicators in the internet of things systems. ICTERI 1, 391–400. https://icteri.org/icteri-2021/proceedings/volume-1/20210391.pdf
15. Trevathan J, Johnstone R (2018) Smart environmental monitoring and assessment technologies (SEMAT)-A new paradigm for low-cost, remote aquatic environmental monitoring. Sensors 18(7):2248.https://doi.org/10.3390/s18072248
16. Ahn EY, Lee J, Bae J, Kim JM (2020) Analysis of emerging geo-technologies and markets focusing on digital twin and environmental monitoring in response to digital and green new deal. Econ Environ Geol 53(5):609–617. https://doi.org/10.9719/EEG.2020.53.5.609
17. Millock K, Xabadia A, Zilberman D (2012) Policy for the adoption of new environmental monitoring technologies to manage stock externalities. J Environ Econ Manag 64(1):102–116. https://doi.org/10.1016/j.jeem.2012.02.004
18. Ma M (2021) Research on new methods and technologies of engineering and environmental geophysical exploration based on internet of things. J Phys Conf Ser 1744(2):022058. https://doi.org/10.1088/1742-6596/1744/2/022058
19. State of the IoT (2018) Number of IoT devices now at 7B—Market accelerating (n.d.). https://iot-analytics.com/state-of-the-iot-update-q1-q2-2018-number-of-iotdevices-now-7b/. Accessed March 22, 2023

20. Page MJ, McKenzie JE, Bossuyt PM, Boutron I, Hoffmann TC, Mulrow CD, Shamseer L, Tetzlaff JM, Akl EA, Brennan SE, Chou R, Glanville J, Grimshaw JM, Hróbjartsson A, Lalu MM, Li T, Loder EW, Mayo-Wilson E, McDonald S, McGuinness LA, Stewart LA, Thomas J, Tricco AC, Welch VA, Whiting P, Moher D (2021) The PRISMA 2020 statement: an updated guideline for reporting systematic reviews. PLoS Med 18:e1003583. https://doi.org/10.1016/j.ijsu.2021.105906
21. Perevochtchikova M, De, la Mora-De, la Mora G, Hernández Flores JÁ, Marín W, Langle Flores A, Ramos Bueno A, Rojo Negrete IA (2019) Systematic review of integrated studies on functional and thematic ecosystem services in Latin America, 1992–2017. Ecosyst Serv 36:100900. https://doi.org/10.1016/j.ecoser.2019.100900
22. Jones KW, Powlen K, Roberts R, Shinbrot X (2020) Participation in payments for ecosystem services programs in the global south: a systematic review. Ecosyst Serv 45:101159. https://doi.org/10.1016/j.ecoser.2020.101159
23. Inácio M, Barceló D, Zhao W, Pereira P (2022) Mapping lake ecosystem services: a systematic review. Sci Total Environ, 157561.https://doi.org/10.1016/j.scitotenv.2022.157561
24. Pinto LV, Inácio M, Ferreira CSS, Ferreira AD, Pereira P (2022) Ecosystem services and well-being dimensions related to urban green spaces–a systematic review. Sustain Cities Soc, 104072.https://doi.org/10.1016/j.scs.2022.104072
25. Viana CM, Freire D, Abrantes P, Rocha J, Pereira P (2022) Agricultural land systems importance for supporting food security and sustainable development goals: a systematic review. Sci Total Environ 806:150718. https://doi.org/10.1016/j.scitotenv.2021.150718
26. Kalinauskas M, Shuhani Y, Pinto LV, Inácio M, Pereira P (2024) Mapping ecosystem services in protected areas. A systematic review. Sci Total Environ 912:169248. https://doi.org/10.1016/j.scitotenv.2023.169248
27. Koutsos TM, Menexes GC, Dordas CA (2019) An efficient framework for conducting systematic literature reviews in agricultural sciences. Sci Total Environ 682:106–117. https://doi.org/10.1016/j.scitotenv.2019.04.354
28. Moher D, Liberati A, Tetzlaff J, Altman DG (2009) Preferred reporting items for systematic reviews and meta-analyses: the PRISMA statement. Ann Intern Med 7:370. https://doi.org/10.7326/0003-4819-151-4-200908180-00135
29. Gusenbauer M, Haddaway NR (2020) Which academic search systems are suitable for systematic reviews or meta-analyses? Evaluating retrieval qualities of Google Scholar, PubMed, and 26 other resources. Res Synth Methods 11(2):181–217. https://doi.org/10.1002/jrsm.1378
30. UN (2019) Special edition: progress towards the sustainable development goals; UN: New York, USA, 64 pp, https://sustainabledevelopment.un.org/content/documents/24978Report_of_the_SG_on_SDG_Progress_2019.pdf
31. Van Zanten BT, Verburg PH, Espinosa M, Gomez-y-Paloma S, Galimberti G, Kantelhardt J, Viaggi D (2014) European agricultural landscapes, common agricultural policy and ecosystem services: a review. Agron Sustain Dev 34(2):309–325. https://doi.org/10.1007/s13593-013-0183-4
32. Plieninger T, Draux H, Fagerholm N, Bieling C, Bürgi M, Kizos T, Kuemmerle T, Primdahl J, Verburg PH (2016) The driving forces of landscape change in Europe: a systematic review of the evidence. Land Use Policy 57:204–214. https://doi.org/10.1016/j.landusepol.2016.04.040
33. Sobana S, Prabha SK, Seeranguravar T, Sudha S (2022) Securing future autonomous applications using cyber-physical systems and the Internet of Things. Handbook of Research of Internet of Things and Cyber-Physical Systems. Apple Academic Press Inc., Canada, pp 81–148
34. Wu JG (2013) Landscape sustainability science: ecosystem services and human well-being in changing landscapes. Landsc Ecol 28:999–1023. https://doi.org/10.1007/s10980-013-9894-9
35. Jones M, Howard P, Olwig KR, Primdahl J, Herlin IS (2007) Multiple interfaces of the European landscape convention. Norwegian J Geogr 61:207–215. https://doi.org/10.1080/00291950701709176
36. Chen F, Wang Y (2021) Research and application of 3D visualization and Internet of Things technology in urban land use efficiency management. Displays 69:102050. https://doi.org/10.1016/j.displa.2021.102050

37. Assis TO, De Aguiar APD, Von Randow C, de Paula Gomes DM, Kury JN, Ometto JPH, Nobre CA (2020) CO2 emissions from forest degradation in Brazilian Amazon. Environ Res Lett 15(10):104035. https://doi.org/10.1088/1748-9326/ab9cfc
38. Chen ST, Hua CC, Chuang CC (2021) Forest management using internet of things in the Fushan botanical garden in Taiwan. J Adv Artif Life Robot 2(2):78–82. https://doi.org/10.57417/jaalr.2.2_78
39. Bayne K, Damesin S, Evans M (2017) The internet of things-Wireless sensor networks and their application to forestry. New Zeal J For 61(4):37–41
40. Tarjan-Tobolka A, Tarjan L, Galić Z (2018) "Smart forests": IoT in forest management in protected areas. Topola 201–202, 227–236. https://scindeks-clanci.ceon.rs/data/pdf/0563-9034/2018/0563-90341801227T.pdf
41. Hock B, Heaphy M, Evans M, Dunningham A, Graham B (2016) The internet of things for forestry: new concepts, new opportunities. NZ J For 60(4):25–28. http://www.nzjf.org/abstract.php?volume_issue=j60_4&first_page=25
42. Henriques J, Westerlund W (2020) Digitalization of forest management: Next generation unsupervised monitoring using Internet of Things and Blockchain. Master of Science Thesis TRITA-ITM-EX 2020:269, KTH Royal Institute of Technology, Stockholm, Sweden, 85 pp
43. Kolbina O, Istomin E, Yagotinceva N, Safonova T, Kalambet M (2021) Peculiarities of creating a database for the IoT system of urban forest management in the city of St. Petersburg. IOP C. Ser Earth Env 876, 012039. https://doi.org/10.1088/1755-1315/876/1/012039
44. Lin CC, Liu WY, Lu YW (2019) Three-dimensional internet-of-things deployment with optimal management service benefits for smart tourism services in forest recreation parks. IEEE Access 7:182366–182380. https://doi.org/10.1109/ACCESS.2019.2960212
45. Tewari VP, Kumar R, Gadow KV (2020) National forest inventory in India: developments toward a new design to meet emerging challenges. In: Chandra G, Nautiyal R, Chandra H (eds) Statistical methods and applications in forestry and environmental sciences (pp 13–33). Springer, Singapore. https://doi.org/10.1007/978-981-15-1476-0_2
46. Babu Loganathan G, Mohan E, Siva Kumar R (2019) IoT based water and soil quality monitoring system. Int J Mech Eng Tech 10(2):537–541. https://eprints.tiu.edu.iq/698/1/IJMET_10_02_055.pdf
47. Salam A (2020) Internet of things for sustainable community development. Springer Cham, ISBN: 978-3-030-35293-6, Switzerland, 334 pp. https://doi.org/10.1007/978-3-030-35291-2
48. Abedi Gheshlaghi H, Feizizadeh B, Blaschke T (2020) GIS-based forest fire risk mapping using the analytical network process and fuzzy logic. J Environ Plann Man 63(3):481–499. https://doi.org/10.1080/09640568.2019.1594726
49. Kaur H, Sood SK, Bhatia M (2020) Cloud-assisted green IoT-enabled comprehensive framework for wildfire monitoring. Cluster Comput 23(2):1149–1162. https://doi.org/10.1007/s10586-019-02981-7
50. Tsipis A, Papamichail A, Angelis I, Koufoudakis G, Tsoumanis G, Oikonomou K (2020) An alertness-adjustable cloud/fog IoT solution for timely environmental monitoring based on wildfire risk forecasting. Energies 13(14):3693. https://doi.org/10.3390/en13143693
51. Zhang S, Gao D, Lin H, Sun Q (2019) Wildfire detection using sound spectrum analysis based on the internet of things. Sensors 19(23):5093. https://doi.org/10.3390/s19235093
52. Saroja Devi H, Meghashree J, Shruthi D (2019) IoT based system for alerting forest fire and control of smuggling. Int J Adv Res Inn 7(2):91–93, https://ijari.org/assets/papers/7/2/IJARI-CS-19-06-102.pdf
53. Le Wong S (2022) IoT based fire detection system for peat soil. Prog Eng Appl Tech 3(1):402–410. https://publisher.uthm.edu.my/periodicals/index.php/peat/article/view/6217
54. Vikram R, Sinha D (2021) FogFire: fog assisted IoT enabled forest fire management. Evol Intell 16(1):329–350. https://doi.org/10.1007/s12065-021-00666-y
55. Cao S, Xiao J (2022) Forestry fire prevention detection and emergency command system based on internet of things and infrared light sensing technology. Int J Front Eng Tech 4(6):39–43. https://doi.org/10.25236/IJFET.2022.040606

56. Sharma M, Rastogi R, Arya N, Akram SV, Singh R, Gehlot A, Buddhi D, Joshi K (2021) LoED: LoRa and edge computing based system architecture for sustainable forest monitoring. Int J Eng Trends Tech 70(5):88–93. https://doi.org/10.14445/22315381/IJETT-V70I5P211
57. Deepthi S, Krishna SG, Sahana KB, Vandana HR, Latha M (2020) IOT enabled forest fire detection and management. Int J Eng Res Tech 8(11):196–199
58. Polepaka S, Swami Das M, Ram Kumar RP (2020) Internet of things and its applications: an overview. Adv Cybern Cogn Mach Learn Commun Techn, 67–75.https://doi.org/10.1007/978-981-15-3125-5_8
59. Sowmiya E, Sivaranjani S (2017) Smart system monitoring on soil using internet of things (IoT). Int Res J Eng Tech 4(2):1070
60. Grafton M, Yule I (2015) The role of technology transfer to improve fertiliser use efficiency. Food Secur 7:365–373. https://doi.org/10.1007/s12571-015-0434-0
61. Vani PD, Rao KR (2016) Measurement and monitoring of soil moisture using cloud IoT and android system. Indian J Sci Tech 9(31):1–8. https://doi.org/10.17485/ijst/2016/v9i31/95340
62. Suma N, Samson SR, Saranya S, Shanmugapriya G, Subhashri R (2017) IOT based smart agriculture monitoring system. Int J Rec Inn Trends Com Comm 5(2), 177–181. https://doi.org/10.17762/ijritcc.v5i2.193
63. Muangprathub J, Boonnam N, Kajornkasirat S, Lekbangpong N, Wanichsombat A, Nillaor P (2019) IoT and agriculture data analysis for smart farm. Comput Electron Agr 156:467–474. https://doi.org/10.1016/j.compag.2018.12.011
64. Sinha BB, Dhanalakshmi R (2022) Recent advancements and challenges of Internet of things in smart agriculture: a survey. Future Gen Comp Syst 126:169–184. https://doi.org/10.1016/j.future.2021.08.006
65. Mekala MS, Viswanathan P (2019) CLAY-MIST: IoT-cloud enabled CMM index for smart agriculture monitoring system. Measurement 134:236–244. https://doi.org/10.1016/j.measurement.2018.10.072
66. Mohanraj I, Ashokumar K, Naren J (2016) Field monitoring and automation using IOT in agriculture domain. Procedia Comput. Sci. 93:931–939. https://doi.org/10.1016/j.procs.2016.07.275
67. Charumathi S, Kaviya RM, Kumariyarasi J, Manisha R, Dhivya P (2017) Optimization and control of hydroponics agriculture using IOT. Asian J Appl Sci Technol 1(2):96–98. https://ssrn.com/abstract=2941105
68. Dobrescu R, Merezeanu D, Mocanu S (2019) Context-aware control and monitoring system with IoT and cloud support. Comput Electron Agr 160:91–99. https://doi.org/10.1016/j.compag.2019.03.005
69. Lin N, Wang X, Zhang Y, Hu X, Ruan J (2020) Fertigation management for sustainable precision agriculture based on Internet of Things. J Clean Prod 277:124119. https://doi.org/10.1016/j.jclepro.2020.124119
70. Subeesh A, Mehta CR (2021) Automation and digitization of agriculture using artificial intelligence and internet of things. Art Int Agr 5:278–291. https://doi.org/10.1016/j.aiia.2021.11.004
71. Ramachandran V, Ramalakshmi R, Kavin BP, Hussain I, Almaliki AH, Almaliki AA, Hussein EE (2022) Exploiting IoT and its enabled technologies for irrigation needs in agriculture. Water 14(5):719. https://doi.org/10.3390/w14050719
72. Lavanya G, Rani C, GaneshKumar P (2020) An automated low cost IoT based fertilizer intimation system for smart agriculture. Sustain Comput-Infor 28:100300. https://doi.org/10.1016/j.suscom.2019.01.002
73. Koubaa A, Aldawood A, Saeed B, Hadid A, Ahmed M, Saad A, Alkanhal M (2020) Smart Palm: an IoT framework for red palm weevil early detection. Agronomy 10(7):987. https://doi.org/10.3390/agronomy10070987
74. Ramalingam B, Mohan RE, Pookkuttath S, Gómez BF, Sairam Borusu CSC, Wee Teng T, Tamilselvam YK (2020) Remote insects trap monitoring system using deep learning framework and IoT. Sensors 20(18):5280. https://doi.org/10.3390/s20185280

75. Villa-Henriksen A, Edwards GT, Pesonen LA, Green O, Sørensen CAG (2020) Internet of Things in arable farming: Implementation, applications, challenges and potential. Biosyst Eng 191:60–84
76. Wu F, Ma J (2020) Evolution dynamics of agricultural internet of things technology promotion and adoption in China. Discrete Dyn Nat Soc, 1854193.https://doi.org/10.1155/2020/1854193
77. Tzounis A, Katsoulas N, Bartzanas T, Kittas C (2017) Internet of Things in agriculture, recent advances and future challenges. Biosyst Eng 164:31–48. https://doi.org/10.1016/j.biosystemseng.2017.09.007
78. Hossein Motlagh N, Mohammadrezaei M, Hunt J, Zakeri B (2020) Internet of Things (IoT) and the energy sector. Energies 13(2):494. https://doi.org/10.3390/en13020494
79. Bedi G, Venayagamoorthy GK, Singh R, Brooks RR, Wang KC (2018) Review of Internet of Things (IoT) in electric power and energy systems. IEEE Int Things J 5(2):847–870. https://doi.org/10.1109/JIOT.2018.2802704
80. Elkamel A, Fatehifar E, Taheri M, Al-Rashidi MS, Lohi A (2008) A heuristic optimization approach for Air Quality Monitoring Network design with the simultaneous consideration of multiple pollutants. J Environ Manag 88(3):507–516. https://doi.org/10.1016/j.jenvman.2007.03.029
81. Kang Y, Chen J, Cao Y, Xu Z (2021) A higher-order graph convolutional network for location recommendation of an air-quality-monitoring station. Remote Sens-Basel 13(8):1600. https://doi.org/10.3390/rs13081600
82. Elen B, Peters J, van Poppel M, Bleux N, Theunis J, Reggente M, Standaert A (2013) The Aeroflex: a bicycle for mobile air quality measurements. Sensors 13(1):221–240. https://doi.org/10.3390/s130100221
83. Bales E, Nikzad N, Quick N, Ziftci C, Patrick K, Griswold WG (2019) Personal pollution monitoring: mobile real-time air quality in daily life. Pers Ubiquit Comput 23:309–328. https://doi.org/10.1007/s00779-019-01206-3
84. Hu K, Sivaraman V, Luxan BG, Rahman A (2016) Design and evaluation of a metropolitan air pollution sensing system. IEEE Sens J 16(5):1448–1459. https://doi.org/10.1109/JSEN.2015.2499308
85. Apte JS, Messier KP, Gani S, Brauer M, Kirchstetter TW, Lunden MM, Marshall JD, Portier CJ, Vermeulen RCH, Hamburg SP (2017) High-resolution air pollution mapping with google street view cars: exploiting big data. Environ Sci Tech 51(12):6999–7008. https://doi.org/10.1021/acs.est.7b00891
86. deSouza P, Anjomshoaa A, Duarte F, Kahn R, Kumar P, Ratti C (2020) Air quality monitoring using mobile low-cost sensors mounted on trash-trucks: methods development and lessons learned. Sustain Cities Soc 60:102239. https://doi.org/10.1016/j.scs.2020.102239
87. Hu K, Rahman A, Bhrugubanda H, Sivaraman V (2017) HazeEst: machine learning based metropolitan air pollution estimation from fixed and mobile sensors. IEEE Sens J 17(11):3517–3525. https://doi.org/10.1109/JSEN.2017.2690975
88. Lanza J, Sánchez L, Muñoz L, Galache JA, Sotres P, Santana JR, Gutiérrez V (2015) Large-scale mobile sensing enabled internet-of-things testbed for smart city services. Int J Distrib Sens N 11(8):785061. https://doi.org/10.1155/2015/7850
89. Stebel K, Stachlewska IS, Nemuc A, Horálek J, Schneider P, Ajtai N, Diamandi A, Benešová N, Boldeanu M, Botezan C, Marková J, Dumitrache R, Iriza-Burcă A, Juras R, Nicolae D, Nicolae V, Novotný P, Ştefănie H, Vaněk L, Zehner C (2021) Samira-satellite based monitoring initiative for regional air quality. Remote Sens Basel 13(11):2219. https://doi.org/10.3390/rs13112219
90. Montori F, Bedogni L, Bononi L (2018) A collaborative internet of things architecture for smart cities and environmental monitoring. IEEE Internet Things 5(2):592–605. https://doi.org/10.1109/JIOT.2017.2720855
91. Asthana P, Mishra S (2018) IoT enabled real time bolt based indoor air quality monitoring system. Int Conf Comput Charact Tech Eng Sci 18:36–39. https://doi.org/10.1109/CCTES.2018.8674076

92. Jo J, Jo B, Kim J, Kim S, Han W (2020) Development of an IoT-Based indoor air quality monitoring platform. J Sensors 8749764. https://doi.org/10.1155/2020/8749764
93. L Tagliabue LC, Re Cecconi F, Rinaldi S, Ciribini ALC (2021) Data driven indoor air quality prediction in educational facilities based on IoT network. Energ Build 236:110782. https://doi.org/10.1016/j.enbuild.2021.110782
94. Mukhtar H, Rubaiee S, Krichen M, Alroobaea R (2021) An IoT framework for screening of covid-19 using real-time data from wearable sensors. Int J Env Res Pub He 18(8):4022. https://doi.org/10.3390/ijerph18084022
95. Moudgil V, Hewage K, Hussain SA, Sadiq R (2023) Integration of IoT in building energy infrastructure: a critical review on challenges and solutions. Renew. Sust Energ Rev 174:113121. https://doi.org/10.1016/j.rser.2022.113121
96. Chaouch H, Çeken C, Arı S (2021) Energy management of HVAC systems in smart buildings by using fuzzy logic and M2M communication. J Build Eng 44:102606. https://doi.org/10.1016/j.jobe.2021.102606
97. Talebi A, Hatami A (2020) Online fuzzy control of HVAC systems considering demand response and users' comfort. Energ Source Part B 15(7–9):403–422. https://doi.org/10.1080/15567249.2020.1825557
98. Ibarra-Esquer J, González-Navarro F, Flores-Rios B, Burtseva L, Astorga-Vargas M (2017) Tracking the evolution of the internet of things concept across different application domains. Sensors 17:1379. https://doi.org/10.3390/s17061379
99. Bandyopadhyay D, Sen J (2011) Internet of things: applications and challenges in technology and standardization. Wirel Pers Commun 58:49–69. https://doi.org/10.1007/s11277-011-0288-5
100. Castillo VM, Álvarez-Salgado XA, Camarero JJ, Campo ÁAD, García Palacios P, Gómez Peris A, Yufera M (2021) Challenge 5: impact of global change on managed ecosystems. In: García-González MB, Jordano P, Barbosa A (Eds), Global change impacts, CSIC Scientific Challenges: Towards 2030, Madrid, pp 130–151
101. Franklinos LHV, Jones KE, Redding DW, Abubakar I (2019) The effect of global change on mosquito-borne disease. Lancet Infect Dis 19(9):302–312. https://doi.org/10.1016/S1473-3099(19)30161-6
102. Dash SP (2020) The impact of IoT in healthcare: global technological change & the roadmap to a networked architecture in India. J Indian Inst Sci 100(4):773–785. https://doi.org/10.1007/s41745-020-00208-y
103. Ogiela L, Castiglione A, Gupta BB, Agrawal DP (2023) IoT-based health monitoring system to handle pandemic diseases using estimated computing. Neural Comput Appl 35(19):13709–13710. https://doi.org/10.1007/s00521-023-08625-7
104. M Fitrianawati M, Yuli V, Suwondo N, Sulisworo D (2019) An integrated internet of things (IoT) on the problem based learning strategy for climate issue: a preliminary design. ICERP 242–248
105. Buyya R, Garg L, Fortino G, Misra S (2022) New frontiers in cloud computing and internet of things. Springer Cham, ISSN: 2199–1073. 408 pp
106. Sharma G, Bade B (2020) IoT and machine learning for identifying correlation between factors causing climate change. J Electron Infor Syst 2(1):10–12. https://doi.org/10.30564/jeisr.v2i1.2020
107. Zecca F (2019) The use of internet of things for the sustainability of the agricultural sector: the case of climate smart agriculture. Int J Civil Eng Tech 10(3):494–501. https://iaeme.com/Home/article_id/IJCIET_10_03_050
108. Park J, Moon A, Lee E, Kim S (2021) Understanding IoT climate data based predictive model for outdoor smart farm. Int Conf Infor Comm Tech Converg 21:1892–1894. https://doi.org/10.1109/ICTC52510.2021.9620971
109. Ahmed RA, Hemdan EED, El-Shafai W, Ahmed ZA, El-Rabaie ESM, Abd El-Samie FE (2022) Climate-smart agriculture using intelligent techniques, blockchain and Internet of things: concepts, challenges, and opportunities. Trans Emerg Telecomm Techn 33(11):e4607. https://doi.org/10.1002/ett.4607

110. Rahman A, Wahyudi MP, Syamsu S (2022) IoT-based climate smart agriculture monitoring information system. Proceedings 1(2):19–29
111. Silva RF, Fava MC, Saraiva AM, Mediondo EM, Cugnasa CE, Delbem ACB (2021) A theoretical framework for multi-hazard risk mapping on agricultural areas considering artificial intelligence, IoT, and climate change scenarios. Eng Proc 9:39. https://doi.org/10.3390/engproc2021009039
112. Sánchez-García C, Francos F (2022) Human-environmental interaction with extreme hydrological events and climate change scenarios as background. Geogra. Sustain 3(3):232–236. https://doi.org/10.1016/j.geosus.2022.08.002
113. Narmilan A, Niroash G (2020) Reduction techniques for consequences of climate change by internet of things (IoT) with an emphasis on the agricultural production. Int J Sci Tech Eng Manag 2(3):06–13. http://ir.lib.seu.ac.lk/handle/123456789/5053
114. El Maachi S, Saadane R, Wahbi M, Chehri A, Badaoui A (2022) Vision of IoT, 5G and 6G data processing: applications in climate change mitigation. ISPACS 1-6.https://doi.org/10.1109/ISPACS57703.2022.10082846
115. Salam A, Salam A (2020) Internet of things for environmental sustainability and climate change. In: Salam A (Ed) Internet of things for sustainable community development: wireless communications, sensing, and systems. Springer Cham, Switzerland, pp 33–69. https://doi.org/10.1007/978-3-030-35291-2_2
116. Nallakaruppan MK, Kumaran US (2019) IoT based machine learning techniques for climate predictive analysis. Int J Recent Tech Eng 5:171–175
117. Ma J, Yu H, Xu Y, Deng K (2020) CDAM: conservative data analytical model for dynamic climate information evaluation using intelligent IoT environment-an application perspective. Com Commun 150:177–184. https://doi.org/10.1016/j.comcom.2019.11.014
118. Anindya DS, Yuliana M, Hadi MZS (2022) IoT based climate prediction system using long short-term memory (LSTM) algorithm as part of smart farming 4.0. IES 255–260. https://doi.org/10.1109/IES55876.2022.9888486
119. Dawa M, Lausa SM, Tibon MR (2022) Internet of things (IoT) based aquaponics management system adaptive to climate change. ICAMEROB 050004. https://doi.org/10.1063/5.0108681
120. Sunehra DD (2019) Raspberry pi based pollution and climate monitoring system using internet of things. Int J Adv Res Eng Tech 10(2):44–54. http://www.iaeme.com/IJARET/issues.asp?JType=IJARET&VType=10&IType=02
121. Saeed N, Al-Naffouri TY, Alouini MS (2020) Around the world of IoT/climate monitoring using internet of X-things. IEEE Int Things J 3(2):82–83. https://doi.org/10.1109/MIOT.2020.9125423
122. Muslim MA, Setyawan RA, Basuki A, Razak AA, Hario FP, Fernando E (2021) IOT based climate monitoring system. IOP Conf Series Earth Environ Sci 746(1):012044. https://doi.org/10.1088/1755-1315/746/1/012044
123. Abd El-Mawla N, Badawy M, Arafat H (2019) IoT for the failure of climate-change mitigation and adaptation and IIoT as a future solution. World J Environ Eng 6(1):7–16. https://doi.org/10.12691/wjee-6-1-2
124. Torres LE, Romero LC, Holguín EP, Ferro R, Aguirre EE (2022) Implementation of an IoT system for climate risk monitoring. Ing. Compet 24(2):1–11. https://doi.org/10.25100/iyc.v24i2.11356
125. Nazeri AHB, Thirunavakkarasu P (2021) Development of remote climate monitoring system using IOT. J Eng Tech 9(1):77–81. https://bmi.unikl.edu.my/wp-content/uploads/2021/11/77-81-DEVELOPMENT-OF-REMOTE-CLIMATE-MONITORING-SYSTEM-USING-IOT_20210616b.pdf
126. Nwankwo W, Olayinka DA, Umezuruike C (2019) Boosting self-sufficiency in maize crop production in Osisioma Ngwa Local Government with internet of things (IOT)-climate messaging: a model. Af J Agr Res 14(7):406–418. https://doi.org/10.2139/ssrn.3341619
127. Nwankwo W, Ukhurebor KE (2021) Big data analytics: a single window IoT-enabled climate variability system for all-year-round vegetable cultivation. IOP Conf Series Earth Environ Sci 655(1):012030. https://doi.org/10.1088/1755-1315/655/1/012030

128. Yu L, Ren Y, Tao S, Gao W, Song X, Zhang X, Yang S (2021) Eco-climate intelligent monitoring system of an agricultural science-and-technology park based on internet of things. IEEE ICAICA 708–715. https://doi.org/10.1109/ICAICA52286.2021.9498218
129. Chedea VS, Drăgulinescu AM, Tomoiagă LL, Bălăceanu C, Iliescu ML (2021) Climate change and internet of things technologies—sustainable premises of extending the culture of the Amurg cultivar in Transylvania—a use case for Târnave vineyard. Sustainability 13(15):8170. https://doi.org/10.3390/su13158170
130. López-Morales JA, Martínez JA, Caro M, Erena M, Skarmeta AF (2021) Climate-aware and IoT-enabled selection of the most suitable stone fruit tree variety. Sensors 21(11):3867. https://doi.org/10.3390/s21113867
131. Mitra S, Das P, Chakraborty N, Majumdar S, Barman T, Das P, Ghosh T (2022) IoT based saplings protection system against unsuitable natural climate. Int J Res Appl Sci Eng Tech 10(5):4921–4928. https://doi.org/10.22214/ijraset.2022.43526
132. Rusch A, Rösgen T (2022) An internet of things sensor array for spatially and temporally resolved indoor climate measurements. Sensors 22(12):4377. https://doi.org/10.3390/s22124377
133. Mitro N, Krommyda M, Amditis A (2022) Smart Tags: IoT sensors for monitoring the microclimate of cultural heritage monuments. App Sci 12(5):2315. https://doi.org/10.3390/app12052315
134. Golait SS, Kumar R, Pandey T, Yerpude S, Pandey K, Ghadole A (2020) Sensing harmful gases in industries using IOT and WSN. Int J Sci Res Comp Sci Engin Inf Tech 6(1):113–119. I. https://doi.org/10.32628/CSEIT206125
135. Khairi NA, Jambek AB (2021) Run-length encoding (RLE) data compression algorithm performance analysis on climate datasets for Internet of Things (IoT) application. Int J Nanoelectron Mater 14:191–197. https://ijneam.unimap.edu.my/images/PDF/InCAPE2021/Vol_14_SI_Dec_2021_191-197.pdf
136. Jayasuriya YP, Elvitigala CS, Wamakulasooriya K, Sudantha BH (2018) Low cost and IoT based greenhouse with climate monitoring and controlling system for tropical countries. ICSSE 1–6. https://doi.org/10.1109/ICSSE.2018.8519997
137. Sahour A, Boumehrez F, Benouaret M, Mokhneche A (2021) Greenhouse climate controller by using of internet of things technology and fuzzy logic. Instrum Mes Métrol 20(1):29–38. https://doi.org/10.18280/i2m.200105
138. Ullah I, Fayaz M, Aman M, Kim D (2022) An optimization scheme for IoT based smart greenhouse climate control with efficient energy consumption. Computing 104(2):433–457. https://doi.org/10.1007/s00607-021-00963-5
139. Ponce H, Gutiérrez S (2019) An indoor predicting climate conditions approach using Internet-of-Things and artificial hydrocarbon networks. Measurement 135:170–179. https://doi.org/10.1016/j.measurement.2018.11.043
140. Perwiratama R, Setiadi YK (2019) Smart hydroponic farming with IoT-based climate and nutrient manipulation system. Int Conf Art Intell Inf Tech 129–132. https://doi.org/10.1109/ICAIIT.2019.8834533
141. Mohd Ariffin MA, Ramli MI, Zainol Z, Mohd Amin MN, Ismail M, Adnan R, Jamil N (2021) Enhanced IoT-based climate control for oyster mushroom cultivation using fuzzy logic approach and NodeMCU microcontroller. Pertanika J Sci Tech 29(4):2863–2885. https://doi.org/10.47836/pjst.29.4.34
142. Rayan RA, Zafar I, Tsagkaris C, Romash I (2021) Internet of things for mitigating climate change impacts on health. In: Artificial intelligence and internet of things: applications in smart healthcare. CRC Press, USA, pp 317–330. https://doi.org/10.1201/9781003097204-14
143. Liu Y, Hassan KA, Karlsson M, Weister O, Gong S (2018) Active plant wall for green indoor climate based on cloud and Internet of Things. IEEE Access 6:33631–33644. https://doi.org/10.1109/ACCESS.2018.2847440
144. Kaburuan ER, Jayadi R (2019) A design of IoT-based monitoring system for intelligence indoor micro-climate horticulture farming in Indonesia. Pro Comp Sci 157:459–464. https://doi.org/10.1016/j.procs.2019.09.001

145. Liu Y, Pang Z, Karlsson M, Gong S (2020) Anomaly detection based on machine learning in IoT-based vertical plant wall for indoor climate control. Build Environ 183:107212. https://doi.org/10.1016/j.buildenv.2020.107212
146. Udrea I, Gheorghe VI, Cartal LA, Duminica D, Petrache S, Apostolescu TC, Kraus VF (2020) IoT solution for monitoring indoor climate parameters in open space offices. EDP Sci 180:02012. https://doi.org/10.1051/e3sconf/202018002012
147. Irshad K, Almalawi A, Khan AI, Alam MM, Zahir MH, Ali A (2020) An IoT-based thermoelectric air management framework for Smart building applications: a case study for tropical climate. Sustainability 12(4):1564. https://doi.org/10.3390/su12041564
148. Mendez-Monroy PE, Cruz May E, Jiménez Torres M, Gómez Hernández JL, Canto Romero M, Sanchez Dominguez I, Bassam A (2022) IoT system for the continuous electrical and environmental monitoring into Mexican social housing evaluated under tropical climate conditions. J Sens 5508713. https://doi.org/10.1155/2022/5508713
149. Thangamani A, Ganesh LS, Tanikella A, Meher Prasad A (2020) IoT based climate control systems diffusion in intelligent buildings-a system dynamics model. In: Sharma SK, Dwivedi YK, Metri B, Rana NP (Eds) Re-imagining diffusion and adoption of information technology and systems: a continuing conversation. Part I, Springer International Publishing, New York, USA, pp 673–684. https://doi.org/10.1007/978-3-030-64849-7_59

Chapter 2
Industrial Pollution Management in West Africa: NEEDS Assessment for AI-Enhanced Monitoring and Response Systems

Azubuike Victor Chukwuka and Chukwudi Nwabuisiaku

Abstract This study conducts a needs assessment to evaluate the critical gaps in industrial pollution monitoring and response systems in West Africa, with a focus on how artificial intelligence (AI) can address these deficiencies. Traditional methods such as manual sampling and laboratory analysis are constrained by infrastructure limitations, delayed response times, and low spatial coverage. In contrast, AI-driven technologies—including Internet of Things (IoT) sensors, machine learning algorithms, and autonomous drones—offer real-time monitoring, predictive analytics, and improved responsiveness. By synthesizing literature from 2015 to 2024 through a rapid review methodology, this study identifies key infrastructural, technical, and policy-related barriers hindering AI adoption in the region. It highlights the potential of integrating satellite-derived data, such as aerosol optical depth (AOD), to support continuous air quality tracking. The findings provide actionable strategies for deploying AI-enhanced systems to strengthen industrial pollution governance and support environmental sustainability. This review aims to provide valuable insights for policymakers and industry stakeholders, aligning with Sustainable Development Goals 12 (Responsible Consumption and Production) and 13 (Climate Action).

Keywords Artificial intelligence (AI) · Pollution monitoring · Internet of Things (IoT) · Machine learning · West Africa

A. V. Chukwuka (✉) · C. Nwabuisiaku
Environmental Quality Control Department, National Environmental Standards and Regulations Enforcement Agency (NESREA), Osun, Nigeria
e-mail: zubbydoo@gmail.com

C. Nwabuisiaku
Inspection and Enforcement Department, National Environmental Standards and Regulations Enforcement Agency (NESREA), Osun, Nigeria

M. Nasr et al. (eds.), *Artificial Intelligence Applications for a Sustainable Environment*, Green Chemistry and Sustainable Technology,
https://doi.org/10.1007/978-3-031-91199-6_2

2.1 Introduction

2.1.1 Background

Industrial pollution in West Africa has become a significant environmental and public health issue exacerbated by rapid industrialization, urban expansion, and insufficient regulatory frameworks. Countries such as Nigeria, Ivory Coast, and Burkina Faso are witnessing increasing pollution levels driven by industries such as manufacturing, mining, and energy production [1, 2]. The primary contributors to pollution in these countries include untreated industrial discharge into rivers, unregulated emissions from factories, and improper waste disposal. These activities have led to widespread environmental degradation, including contamination of water and soil, air quality deterioration, and significant biodiversity loss [3, 4]. Additionally, the lack of advanced pollution monitoring systems further hinders timely intervention and effective management strategies, worsening environmental and health outcomes in the region [5]. Addressing these challenges requires strengthened regulatory enforcement and the adoption of innovative technologies, such as AI- and IoT-based monitoring systems, to ensure sustainable industrial practices and environmental protection [6].

Air pollution levels that exceed the World Health Organization's recommended limits are contributing to a public health crisis in West Africa, with more than 70% of urban residents affected [7]. This critical situation has resulted in an increase in respiratory illnesses, waterborne diseases, and cancers due to toxic exposures, disproportionately affecting vulnerable groups such as children and the elderly [8]. The region's inadequate infrastructure for pollution detection and monitoring further complicates efforts to address these health and environmental challenges [9]. To mitigate these issues, satellite-derived data, such as aerosol optical depth (AOD), have become indispensable tools for understanding pollution dynamics in the region. The AOD offers information on aerosol concentrations in the atmosphere, which are often linked to industrial emissions, biomass burning, and dust storms—the primary contributors to air pollution in West Africa [10]. In the absence of comprehensive ground-based monitoring systems, time-averaged AOD data from satellite missions serve as a vital resource for real-time pollution monitoring and have emerged as a last resort for guiding interventions aimed at reducing the environmental and health impacts of industrial pollution in the region.

AOD data have proven to be an essential tool for assessing air pollution patterns in regions where ground-based monitoring infrastructure is limited. By analyzing AOD, researchers can identify pollution hotspots and track long-term trends, particularly in industrial areas. Higher AOD values are often indicative of increased aerosol presence, which is typically associated with industrial activities and other pollution sources [11]. However, while AOD data are invaluable for understanding pollution trends, the current study focuses on using AOD solely as a proxy to identify and assess the extent of industrial pollution in West Africa. The objectives of this study are to evaluate the state of industrial pollution monitoring in the region, identify major sources of pollution, and explore the potential for AI-enhanced monitoring systems to improve real-time detection and response to pollution events. The hypothesis is that AI-based monitoring systems, in conjunction with satellite data, can increase the accuracy of pollution management and improve policy enforcement, leading to better environmental and public health outcomes in West Africa.

2.1.2 Rationale for AI-Enhanced Systems

Current pollution monitoring systems in West Africa are limited by inadequate infrastructure, fragmented data collection, and delayed response mechanisms. Traditional methods, such as manual sampling and laboratory analysis, are time-consuming and lack real-time capabilities. Moreover, these systems are often underfunded and constrained by a lack of technical expertise [12]. Artificial intelligence (AI) offers transformative potential to address these limitations. AI-enabled systems leverage technologies such as Internet of Things (IoT) devices, machine learning, and autonomous drones to provide real-time pollution monitoring and predictive analytics. For example, IoT-enabled sensors can continuously track air and water quality parameters, transmitting data to centralized platforms for analysis [13]. Machine learning algorithms can analyze patterns in pollution data to predict future trends, enabling proactive measures to mitigate risks [14]. Additionally, autonomous drones equipped with remote sensing technologies can monitor hard-to-reach areas, increasing the scope and accuracy of data collection [15].

Automation through AI also reduces reliance on manual processes, increasing efficiency and reducing costs. By integrating AI into pollution management, policymakers and industries can make data-driven decisions, enhance regulatory compliance, and improve overall environmental governance [16].

2.1.3 Objectives of the Review

This review aims to achieve the following objectives:

1. **Assess AI technologies for industrial pollution monitoring**: Examine how IoT, machine learning, and drones are used to detect, monitor, and predict pollution levels in industrial settings.
2. **Identifying regional needs and challenges**: Evaluating the specific barriers to AI adoption in West Africa, including infrastructure deficits, funding limitations, and policy gaps.
3. **Recommend actionable strategies for AI integration**: Provide practical recommendations for policymakers, industry stakeholders, and researchers to facilitate the implementation of AI technologies in pollution management.

By addressing these objectives, this review contributes to the discourse on sustainable development and environmental resilience in West Africa, which aligns with global goals such as SDG 12 (Responsible Consumption and Production) and SDG 13 (Climate Action).

2.2 Methodology

Study Area

This study focuses on industrial pollution management in West Africa, a region with significant environmental challenges due to rapid industrialization, urbanization, and population growth (Fig. 2.1a). Key industrial hubs such as Lagos (Nigeria), Abidjan (Ivory Coast), and Ouagadougou (Burkina Faso) face severe pollution issues, including air, water, and soil contamination (Fig. 2.1b). In Nigeria, Lagos, the economic nerve center, experiences extensive air and water pollution due to oil and gas production, vehicular emissions, and industrial discharges, posing serious health and ecological risks [17–20]. Abidjan, as the commercial capital of the Ivory Coast, suffers from unchecked industrial waste disposal and marine pollution that threaten biodiversity and public health [21]. Ouagadougou faces unique challenges linked to mining byproducts, heavy metal contamination, and inadequate waste management systems, exacerbated by rapid urban growth and limited infrastructure [22].

These industrial centers reflect broader regional challenges, where weak regulatory frameworks, insufficient monitoring infrastructure, and socioeconomic constraints hinder effective pollution management [23]. West Africa's pollution crisis disproportionately impacts marginalized communities, with rising healthcare costs and declining agricultural productivity undermining livelihoods. This study highlights the urgent need for innovative solutions, such as artificial intelligence-enhanced monitoring and response systems, to bridge gaps in data collection, predictive analysis, and policy enforcement. By focusing on these representative sites, this research

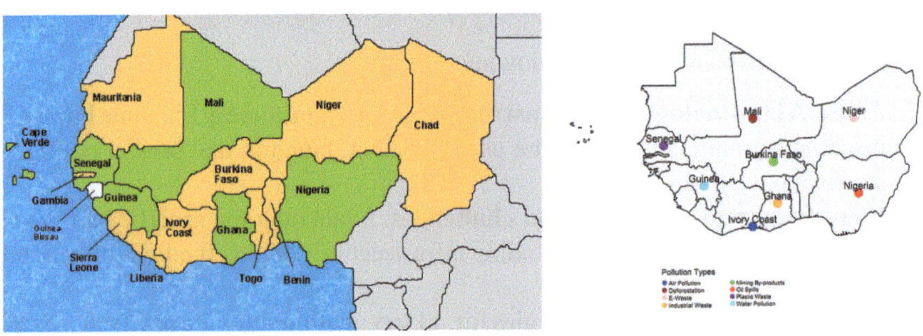

Fig. 2.1 **a** Map of West African countries, **b** Major pollution types across West African countries

aims to develop scalable strategies applicable across the region to address industrial pollution and foster sustainable development.

Table 2.1 highlights the major pollutants and their associated health and economic impacts across West Africa. Nigeria, with oil spills and gas flaring, faces $3.8 billion in annual costs from respiratory illnesses, cancer, and waterborne diseases. Ivory Coast's air pollution and industrial waste result in $2 billion annually in healthcare costs and lost productivity. Burkina Faso's mining byproducts lead to $1.5 billion in economic losses due to neurological issues and environmental cleanup. Ghana's mining pollution and deforestation contribute to $1.2 billion in costs from respiratory diseases and mental health issues. Senegal incurs $1.1 billion in annual costs from industrial and vehicle emissions, while Mali's desertification and air pollution result in $0.8 billion in losses. Togo, Liberia, and Sierra Leone face annual economic burdens of $0.5 billion, $0.4 billion, and $0.6 billion, respectively, due to waterborne diseases, malaria, and respiratory issues.

Prevalidation of Hypothesis

The hypothesis for this study was validated through the analysis of aerosol optical depth (AOD) data obtained from multiple satellite sources hosted on NASA Giovanni's platform. The dataset includes daily and monthly AOD measurements at various wavelengths (463.0 nm, 354.0 nm, 442.0 nm, 500.0 nm, etc.), offering detailed insights into aerosol characteristics across different regions of West Africa. These measurements provide a robust means of ascertaining and visualizing the spatial and temporal dynamics of aerosol pollution within a subregion, which is closely linked to industrial and natural pollution sources [33].

The AOD data, particularly at wavelengths such as 463 nm, 354 nm, and 388 nm, enable the identification of key aerosol types such as Saharan dust, biomass burning, and industrial emissions [34, 35]. Leveraging both daily (0.25° resolution) and monthly (0.5° × 0.625° resolution) datasets within the last 5 years (Table 2.2) not only confirms that there has been a gradient of industrial activity across the West African subregion but also supports the hypothesis that AI-enhanced monitoring systems can identify and map aerosol pollution sources in real time, providing valuable inputs for pollution control policies and health risk management [36]. This prevalidation process reinforces the study's focus on integrating AI systems with satellite-derived data to enhance industrial pollution management across the West African subcontinent (Fig. 2.2).

Prevalidation Inferences

Analysis of aerosol optical depth (AOD) variables across multiple wavelengths reveals that industrial pollution is a major contributor to aerosol dynamics in West Africa, particularly in urbanized and industrially concentrated coastal regions. Countries such as Nigeria, Côte d'Ivoire, and Sierra Leone consistently display elevated AOD values across fine and coarse aerosol bands (e.g., 354 nm, 388 nm, 500 nm), highlighting the cumulative effects of vehicular emissions, fossil fuel combustion, port-based industrial activity, and biomass burning. Notably, southern Nigeria—especially the South–South and South–West zones—exhibits persistently high AOD

Table 2.1 Major pollutants, health impacts, and economic costs of industrial pollution in West Africa

Country	Major pollutants	Health impacts	Economic costs	References
Nigeria	Oil spills, gas flaring	– Respiratory illnesses – Cancer – Waterborne diseases – Skin disorders from contaminated water	$3.8 billion annually (healthcare, lost productivity, environmental remediation)	[24, 25]
Ivory Coast	Air pollution, industrial waste	– Asthma – Respiratory infections – Cardiovascular diseases – Premature deaths from poor air quality	$2 billion annually (healthcare costs, reduced productivity)	[26, 27]
Burkina Faso	Mining byproducts, heavy metals	– Neurological issues – Water contamination – Developmental delays in children – Cancer risk	$1.5 billion annually (healthcare, environmental cleanup, agricultural loss)	[28, 29]
Ghana	Mining pollution, deforestation	– Respiratory diseases – Skin conditions – Mental health issues from environmental stress	$1.2 billion annually (environmental damage, healthcare)	[25]
Senegal	Industrial emissions, vehicle emissions	– Respiratory diseases – Cardiovascular diseases – Premature deaths from pollution	$1.1 billion annually (healthcare, loss of labor productivity)	[30]
Mali	Desertification, air pollution	– Respiratory diseases – Skin disorders – Impact on agricultural productivity	$0.8 billion annually (healthcare, agricultural loss)	[25]
Togo	Agricultural waste, deforestation	– Waterborne diseases – Malaria – Respiratory issues	$0.5 billion annually (healthcare, loss of ecosystem services)	[24, 31]
Liberia	Deforestation, agricultural runoff	– Malaria – Waterborne diseases – Respiratory diseases	$0.4 billion annually (healthcare costs, ecosystem disruption)	[32]
Sierra Leone	Mining, oil spills, industrial waste	– Respiratory illnesses – Cancer;—Waterborne diseases	$0.6 billion annually (healthcare, loss in agriculture)	[24, 25]

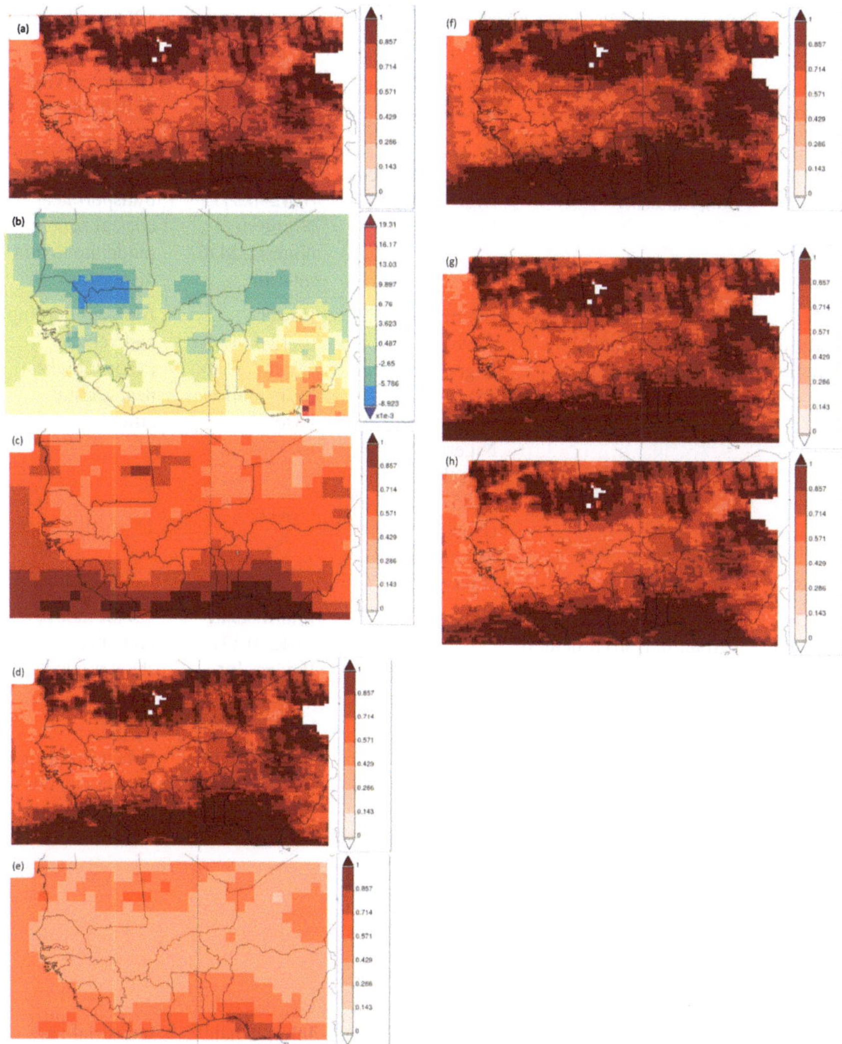

Fig. 2.2 Time-averaged maps of aerosol optical depth (AOD) over the West African subcontinent from 2019–01 to 2024–07-31 across various wavelengths and resolutions: **a** AOD at **463.0 nm**, daily, 0.25° (OMI OMAEROe v003); **b AOD** analysis increment, 0.5° × 0.625° (MERRA-2 Reanalysis M2IMNXGAS v5.12.4); **c AOD at 354.0 nm**, daily, 1° (OMI OMAERUVd v003); **d AOD at 442.0 nm**, daily, 0.25° (OMI OMAEROe v003); **e AOD at 500.0 nm**, daily, 1° (OMI OMAERUVd v003); **f AOD at 342.5 nm**, daily, 0.25° (OMI OMAEROe v003); **g AOD at 388.0 nm**, daily, 0.25° (OMI OMAEROe v003); **h AOD at 483.5 nm**, daily

Table 2.2 Preliminary insights into aerosol optical depth variables for assessing pollution dynamics in West Africa

AOD Variable	Pollutant Type / Focus	Coastal Regions (Examples)	Non-Coastal Regions (Examples)	Observed Values & Patterns
AOD at 463.0 nm	Coarse particles (e.g., Saharan dust)	Highest (>0.857): Nigeria–Sierra Leone coast	Intermediate (0.42–0.57): Northern Nigeria, Mali interior	High dust transport near coastal zones; moderate inland dispersion
AOD Analysis Increment	Long-term aerosol trend corrections (model + satellite data)	Intermediate (6.98–9.8): Nigeria–Liberia coast Highest (>13.03): Nigeria Middle Belt; Lowest (0.857): Lagos (Nigeria) to Monrovia (Liberia)	Lower (0.42–0.57): Central Togo, Mali, Burkina Faso	Biomass burning and vehicle exhaust highest near coastal cities
AOD at 442.0 nm	Mixed aerosols (urban-industrial-natural mix)	Highest (>0.857): Coastal Nigeria–Sierra Leone	Intermediate (0.42–0.57): Ibadan, Jos Plateau, Ouagadougou	Urban-industrial inland zones show moderate values
AOD at 500 nm	Moderate aerosol concentrations (natural + anthropogenic)	Highest (0.71–0.85): Lagos to Benin City corridor Intermediate (0.42–0.57): Ilorin, Minna; Lowest (0.857): Lagos–Freetown belt	Intermediate (0.42–0.57): Abuja, Kaduna, Eastern Ghana	Urban fossil fuel impact extends moderately inland
AOD at 388.0 nm	Fine anthropogenic particles (urban/ industrial)	Highest (>0.857): Nigeria–Sierra Leone coast	Intermediate (0.42–0.57): Kano, Tamale, Ouagadougou	Industrial fine particles penetrate non-coastal airspaces
AOD at 483.5 nm	Coarse aerosols (desert dust, marine aerosols)	Highest (>0.857): Coastal Nigeria–Sierra Leone	Intermediate (0.42–0.57): Niger interior, central Burkina Faso	Saharan dust influence strong both along coast and interior corridor

readings, driven by the synergistic impact of density of refineries, power plants, and transport corridors.

In contrast, non-coastal regions such as northern Nigeria, Burkina Faso, Mali, and inland Ghana show moderate to lower AOD levels, especially in UV-sensitive and anthropogenic bands, reflecting reduced industrial and vehicular activity. Here, aerosol loads are primarily influenced by natural sources, including Saharan dust intrusion (notably at 463.0 nm and 483.5 nm) and seasonal biomass burning, rather than intensive industrial emissions. However, emerging pollution hotspots in Nigeria's Middle Belt and urban centers like Kano and Ouagadougou show rising AOD trends, particularly in long-term aerosol analysis increments and black carbon wavelengths (e.g., 342.5 nm), suggesting an ongoing accumulation of pollutants (AOD increases) linked to unregulated industrial growth and urban expansion.

These spatial disparities underscore the need for region-specific pollution control strategies. AI-enhanced monitoring systems, integrating satellite-derived AOD data, IoT sensors, and predictive analytics, can offer real-time surveillance of both coastal industrial hubs and inland emerging hotspots. Such systems are essential for targeted policy interventions, enabling governments and industries to mitigate emissions, safeguard public health, and advance progress toward SDG 12 (Responsible Consumption and Production) and SDG 13 (Climate Action).

2.2.1 Review Design

This study adopts a rapid review design to provide timely and actionable insights into the potential of AI-enhanced systems for monitoring industrial pollution in West Africa. A rapid review is a streamlined version of a systematic review designed to synthesize evidence efficiently while maintaining methodological rigor [37]. The choice of a rapid review is justified by the urgent need for innovative solutions to address escalating environmental challenges in the region. It enables the identification of relevant technologies, barriers, and strategies to inform policymakers and stakeholders promptly, aligning with the pressing demands of sustainable development.

2.2.2 Search Strategy

A comprehensive search strategy was developed to identify relevant literature on the applications of AI for industrial pollution monitoring in West Africa. The search targeted peer-reviewed journal articles, conference papers, reports, and other scholarly sources published in English between 2015 and 2024. Searches were conducted across multiple academic databases, including Scopus, Web of Science, PubMed, IEEE Xplore, and Google Scholar. Key search terms and Boolean operators were employed, such as "AI in pollution monitoring" AND "West Africa," "Artificial

Table 2.3 Inclusion and exclusion criteria

Category	Inclusion criteria	Exclusion criteria
Focus	AI applications in pollution monitoring	Non-AI-related pollution studies
Region	West Africa	Studies outside West Africa
Publication date	2015–2024	Pre-2015 studies
Language	English	Non-English publications
Type of study	Peer-reviewed articles, gray literature	Blogs, non-scholarly articles

intelligence" AND "industrial pollution" AND "Machine learning" OR "IoT" AND "environmental monitoring," and "AI-enhanced systems" AND "pollution prediction." To ensure the inclusion of regionally relevant insights, gray literature, including government reports and publications from NGOs, was also reviewed.

2.2.3 Inclusion and Exclusion Criteria

A clear set of inclusion and exclusion criteria was applied to guarantee the relevance and quality of the selected studies. Studies were included if they focused on AI applications in pollution monitoring, such as detection, prediction, or response, and specifically addressed industrial pollution in West Africa, including countries such as Nigeria, Ivory Coast, and Burkina Faso. Only research published between 2015 and 2024 was considered, ensuring coverage of recent advancements in AI technologies. Additionally, the eligible materials included peer-reviewed articles, government reports, and gray literature published in English. Studies were excluded if they focused on general environmental pollution without an industrial-specific context, were conducted outside West Africa, were not region specific, did not relate to AI applications or pollution monitoring, or were published in languages other than English (Table 2.3).

2.2.4 Data Extraction and Synthesis

Data extraction focuses on identifying key aspects of AI applications in pollution monitoring, including detection, prediction, and response mechanisms. A standardized data extraction form was used to collect relevant information such as the following:

- **Study details**: Authors, publication year, and location.
- **AI technologies**: Specific tools and approaches used (e.g., IoT sensors, machine learning, predictive models).

- **Application domain**: Type of industrial pollution addressed (e.g., air, water, and soil).
- **Outcomes**: Effectiveness in pollution detection, prediction accuracy, and responsiveness.

The extracted data were synthesized via thematic analysis to identify recurring patterns and key insights. Studies have been categorized by their technological focus, industrial application, and regional context to provide a structured understanding of how AI is applied to address industrial pollution in West Africa. Descriptive statistics and narrative summaries were used to present findings effectively, highlighting both successes and challenges in implementing AI-based systems.

This methodology ensures a comprehensive and context-specific review, offering valuable insights into the potential for AI integration in industrial pollution monitoring in the region.

2.3 AI Technologies for Industrial Pollution Monitoring

Artificial intelligence (AI) has emerged as a transformative force in addressing the multifaceted challenges of industrial pollution monitoring, fundamentally redefining how environmental data are collected, analyzed, and acted upon. By integrating advanced tools such as IoT-enabled sensor networks, machine learning algorithms, and autonomous drones, AI enhances not only the precision and efficiency of pollution detection but also the scalability of monitoring systems in areas with limited resources or infrastructure [36]. IoT and sensor networks enable real-time tracking of critical pollutants, providing granular data on air, water, and soil quality, whereas machine learning algorithms analyze these vast datasets to identify trends, predict pollution hotspots, and propose targeted mitigation strategies [38, 39]. Autonomous drones complement these systems by accessing hard-to-reach areas, enabling comprehensive surveillance of industrial zones, and delivering actionable insights with minimal human intervention. Together, these AI-driven technologies not only bridge critical gaps in pollution management but also empower policymakers with tools to anticipate and respond to environmental risks, fostering a more sustainable balance between industrial development and environmental health.

2.3.1 IoT and Sensor Networks

The Internet of Things (IoT) has become a pivotal technology in industrial pollution monitoring, revolutionizing how environmental data are gathered, analyzed, and acted upon. IoT systems integrate advanced sensor networks that continuously monitor air, water, and soil quality, providing real-time, high-resolution data critical for understanding pollution dynamics. These sensors detect key pollutants such as particulate matter (PM2.5, PM10), heavy metals, and toxic gases, transmitting the data wirelessly to centralized platforms for analysis and visualization [40]. In industrial zones, IoT-driven solutions enable the identification of pollution hotspots, the detection of emerging contamination trends, and the triggering of alerts for immediate response. For example, smart air quality sensors deployed near manufacturing facilities can help enforce emission standards, whereas water quality sensors in river basins detect chemical discharges, preventing ecological damage. By combining real-time monitoring with remote access capabilities, the IoT offers a scalable and efficient framework to address the unique challenges of industrial pollution in resource-constrained regions such as West Africa, ultimately supporting informed decision-making and proactive environmental management.

Role of the IoT in Real-Time Monitoring

IoT sensors are equipped to measure a wide range of pollutants, including particulate matter (PM2.5, PM10), sulfur dioxide (SO_2), nitrogen oxides (NO_x), volatile organic compounds (VOCs), and heavy metals, in water systems. These sensors transmit data through robust communication networks such as Wi-Fi, ZigBee, or LoRaWAN to a centralized cloud platform, ensuring seamless integration, and rapid processing. This approach allows for real-time pollution tracking, early detection of threshold breaches, and the creation of predictive models to anticipate potential pollution spikes [41].

Figure 2.3 offers a detailed flowchart illustrating an IoT-based pollution monitoring system tailored for industrial pollution management in West Africa. The process begins with sensor nodes—air quality sensors for measuring gas and particulate emissions—and water quality monitors for parameters such as pH and heavy metal concentrations. Data from these sensors are transmitted through a communication gateway, which employs wireless technologies to ensure reliability even in regions with limited infrastructure. Once received, the data are processed on a **Cloud Platform**, where AI-powered models analyze patterns, detect anomalies, and generate actionable insights. Finally, insights are presented via an intuitive **Dashboard/Alerts System**, which delivers real-time notifications through mobile applications, SMS, or email alerts. These systems not only increase the precision of pollution monitoring but also allow policymakers and stakeholders to respond swiftly to environmental risks.

Examples of Successful Implementations

Real-world applications underscore the potential of IoT systems in industrial pollution management:

- **Air quality monitoring in China**: In urban and industrial zones, IoT-based air quality monitors have been deployed to track pollutants such as NO_2, PM2.5, and VOCs. These systems, integrated with cloud-based analytics platforms, enable real-time data collection and actionable insights, guiding regulatory measures and public health responses. The integration of the IoT and big data analytics has significantly contributed to air quality improvements in cities

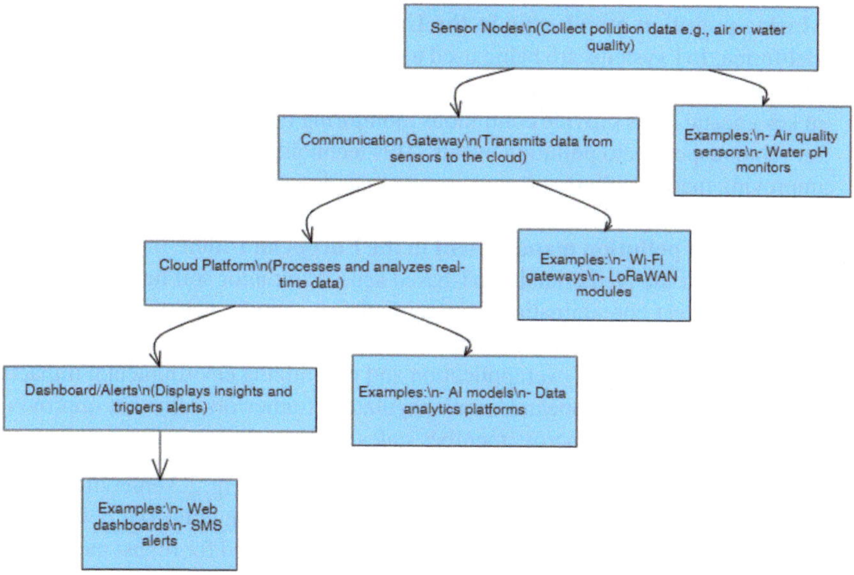

Fig. 2.3 IoT-based pollution monitoring system flowchart

such as Beijing, supporting effective pollution control strategies, such as stricter emission standards for industries and vehicles [42, 43]

- **Water quality monitoring in India**: Along the Ganges River, IoT systems with advanced electrochemical sensors continuously monitor critical water quality parameters such as pH, turbidity, and dissolved oxygen levels [44]. These real-time data help authorities manage the impact of industrial discharge and sewage, enabling timely interventions to protect public health and ecosystem integrity. The system has facilitated better regulation of industrial effluent discharge and promoted cleaner water bodies through coordinated government and community actions [44].
- **Smart waste management in Singapore**: IoT-powered waste management systems have been implemented to optimize waste collection processes in Singapore. Sensors embedded in waste bins monitor fill levels, enabling waste trucks to plan more efficient collection routes, thus reducing carbon emissions and operating costs [45]. The data collected also help track waste generation trends, supporting more targeted recycling and waste reduction initiatives and contributing to the country's zero-waste goals [45].
- **Forest monitoring in Brazil**: In the Amazon rainforest, IoT-based sensor networks monitor deforestation and forest health by measuring environmental factors such as temperature, humidity, and soil moisture [46]. These sensors transmit data to centralized platforms that allow for real-time tracking of illegal logging activities, offering authorities valuable insights to enforce conservation laws more effectively and protect biodiversity in the region [47].

- **Marine pollution monitoring in the United States**: In coastal regions such as California, IoT systems are being used to track marine pollution levels, including oil spills, plastic waste, and nutrient pollution in water bodies [48]. Sensors placed in key coastal areas provide continuous monitoring of water conditions, allowing for rapid responses to pollution events. This technology has been instrumental in improving marine pollution management strategies, especially in industrial zones near oceans [49].
- **Agricultural pollution management in the European Union**: In countries such as France and the Netherlands, IoT-based sensors monitor soil health and water quality to detect contamination from agricultural runoff, such as pesticides and fertilizers [50]. These real-time data help farmers optimize the use of fertilizers and pesticides, reducing overapplication and minimizing environmental impacts. It also supports the European Union's regulatory framework aimed at improving water quality standards across member states [51].

The IoT-based pollution monitoring system illustrated in Fig. 2.3 embodies a scalable and technologically sophisticated solution for managing industrial pollution in West Africa. This system addresses the region's pressing need for robust monitoring frameworks, particularly in areas lacking conventional infrastructure. By employing AI-powered analytics on the cloud platform, the system achieves unparalleled efficiency in processing large volumes of environmental data. Furthermore, the use of dashboards and automated alerts ensures that actionable insights are accessible to diverse stakeholders, from industrial regulators to community leaders. This integrated approach aligns with global best practices in pollution management, demonstrating how the IoT and AI can collectively bridge existing monitoring gaps while driving sustainable industrial development.

2.3.2 Machine Learning and Predictive Analytics

Machine learning (ML) has rapidly become a cornerstone of modern pollution management strategies, offering precise, data-driven insights that significantly improve the prediction and mitigation of industrial pollution. The graph in Fig. 2.4 illustrates the substantial improvements in predictive accuracy when ML models are applied to pollution management. It compares actual pollution levels to those predicted by ML algorithms, with shaded regions representing the margin of error between the two. The reduced prediction error shown in the graph highlights how ML enhances the reliability of real-time environmental monitoring, making it possible to predict pollution levels with much greater accuracy. This capability is particularly beneficial in industrial settings, where timely responses to pollution spikes are critical to minimizing health risks and ecological damage.

For example, pollutants such as particulate matter (PM2.5) and nitrogen dioxide (NO2), which are common byproducts of industrial activities such as manufacturing and energy production, pose significant threats to public health, particularly in rapidly

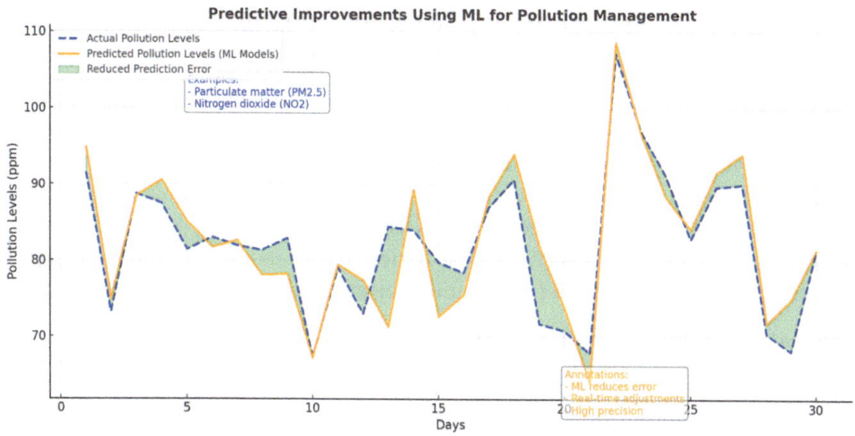

Fig. 2.4 Illustration of predictive improvements using machine learning (ML) for pollution management

industrializing regions such as West Africa [52, 53]. As shown in Fig. 2.4, the predictive models developed through ML algorithms allow for more accurate forecasting of pollutant concentrations. This is crucial for industries in regions such as Nigeria's Niger Delta, where high levels of particulate matter and toxic gases exacerbate respiratory illnesses, cardiovascular diseases, and environmental degradation [54]. By reducing the margin of error, ML not only supports regulatory compliance but also enables local authorities to intervene before pollution levels reach harmful thresholds.

In Fig. 2.4, the shaded regions illustrating reduced error emphasize the reliability and timeliness of real-time adjustments on the basis of predictive data, which enables swift interventions to mitigate industrial pollution. These enhancements not only help industries comply with regulatory standards but also support efforts to reduce their environmental impact and protect public health. The predictive capabilities of machine learning (ML) are especially valuable in developing regions, offering a sustainable path for industrial growth that balances environmental stewardship with public well-being. By leveraging advanced technologies, pollution management systems can evolve into smarter, more efficient solutions, contributing to cleaner, healthier environments.

This graph demonstrates the predictive power of ML models in industrial pollution management by comparing actual pollution levels, such as particulate matter (PM2.5) and nitrogen dioxide (NO2), with predictions made by ML algorithms over 30 days. The shaded areas highlight the reduced error, highlighting the accuracy of real-time monitoring and forecasting. These improvements allow for more effective emission management, particularly in regions such as West Africa, where adaptive strategies are essential for tackling environmental challenges. The data, derived from a simulated dataset, replicate pollution monitoring scenarios in industrial areas to illustrate the practical benefits of AI-driven systems.

2.3.3 *Autonomous Drones and Remote Sensing*

Autonomous drones and remote sensing technologies have emerged as cutting-edge tools for environmental monitoring, offering unique advantages in tracking air and water pollution over large, often inaccessible industrial zones [36]. Their ability to gather real-time, high-resolution data in areas that are difficult to monitor with traditional methods positions these technologies as game changers for pollution management, especially in regions such as West Africa, where industrial growth is rapidly accelerating but infrastructure for pollution detection remains underdeveloped.

Applications in Pollution Monitoring

Autonomous drones equipped with advanced sensors and imaging tools have become essential for monitoring air pollution in industrial zones. Drones can be outfitted with gas sensors and thermal cameras that capture the spatial distribution of pollutants such as carbon monoxide (CO), volatile organic compounds (VOCs), and particulate matter. For example, drones can fly over large industrial plants or mining sites to map pollutant concentrations across different altitudes and locations, providing valuable insights into the spread and intensity of emissions [15, 55]. These sensors allow for the detection of pollutants in real time, and their portability ensures that monitoring can be conducted continuously without the need to establish permanent, stationary monitoring stations. This mobility is particularly useful in monitoring transient sources of pollution, such as emissions from factories or spontaneous chemical spills, that traditional ground-based monitoring might miss.

In addition to air pollution, drones are also effective at monitoring water quality, particularly in industrial areas near rivers and reservoirs. By using specialized sensors or water samples, drones can assess the presence of hazardous substances such as heavy metals (e.g., lead and mercury) and monitor other critical indicators of water quality, such as biological oxygen demand (BOD). This capability is particularly important in regions where water contamination from industrial discharge poses significant health risks to local populations. Drones can fly over polluted rivers or lakes to gather data that would otherwise be difficult to access, enabling timely assessments and interventions [56]. For example, in regions affected by oil spills or mining runoff, drones equipped with water quality sensors can quickly identify contamination levels, providing actionable data for regulators and environmental agencies.

Another powerful tool for pollution monitoring is satellite imagery, which complements drone-based data by offering a broader, landscape-scale view of industrial impacts. Remote sensing via satellites provides high-resolution imagery that can detect deforestation, land degradation, and environmental destruction caused by industrial activities such as mining and agriculture. For example, satellites can capture images of large-scale oil spills, track the spread of toxic waste, or even observe changes in land cover resulting from unsustainable industrial practices [4, 11, 57]. Such imagery allows for large-area monitoring with reduced human intervention and can be used to detect trends over time, offering crucial data for long-term environmental planning and policy development [11]. Together, drones and satellite-based remote sensing technologies can provide a comprehensive, layered understanding of environmental pollution, enabling stakeholders to respond more effectively to industrial impacts.

Feasibility and Challenges in West Africa

Despite their advantages, the adoption of autonomous drones and remote sensing technologies in West Africa faces several significant challenges. One of the primary barriers is cost; the high initial investment required to procure drones and deploy sensor networks limits their scalability in regions with constrained budgets. Furthermore, the ongoing operational costs for maintaining drone fleets and the technical infrastructure needed for data analysis can be prohibitive, especially in low-income areas [58]. For many West African nations, such as Ghana and Nigeria, the upfront costs associated with drone technology may outweigh the perceived benefits unless supported by international funding or public–private partnerships.

Another challenge is the lack of reliable infrastructure in many parts of West Africa. Drones depend heavily on stable internet connectivity and a consistent power supply to transmit real-time data and perform autonomous operations. In rural or remote areas where power outages are frequent and internet access is limited, the effectiveness of drone-based monitoring can be severely compromised. For example, during extended power outages or poor internet coverage, the real-time transmission of critical pollution data may be delayed or interrupted, reducing the overall efficiency of the monitoring system [59]. Regulatory hurdles, such as unclear legal frameworks, privacy concerns, and liability issues, complicate the adoption of drones for environmental monitoring in West Africa, creating uncertainties around flight permissions and operational safety. To enable wider use of drone technology, governments must develop comprehensive guidelines that address these challenges while fostering innovation and safeguarding environmental and societal interests.

Figure 2.5 shows the transformative potential of drone technology in monitoring pollution across key industrial zones in West Africa, with a focus on Lagos State in Nigeria, Abidjan on the Ivory Coast, and Ouagadougou in Burkina Faso. These regions were selected because of their significant environmental challenges: Lagos, as Nigeria's economic hub, faces intense air and water pollution driven by industrial density and population pressures [19, 60–62]; Abidjan's industrial and coastal activities contribute to severe marine and atmospheric pollution [63–65]; and Ouagadougou contends with escalating air and waste pollution due to rapid urban growth [66]. The map illustrates how drones equipped with environmental sensors could provide comprehensive spatial coverage in these areas, offering real-time, high-resolution data on pollution levels. Drones offer a unique ability to address limitations posed by traditional monitoring systems, such as restricted access and inadequate infrastructure. By traversing diverse terrains and collecting granular environmental data, drones can identify pollution hotspots, model pollutant dispersal, and generate predictive insights. These capabilities could empower authorities to optimize resource allocation and implement targeted interventions with greater precision.

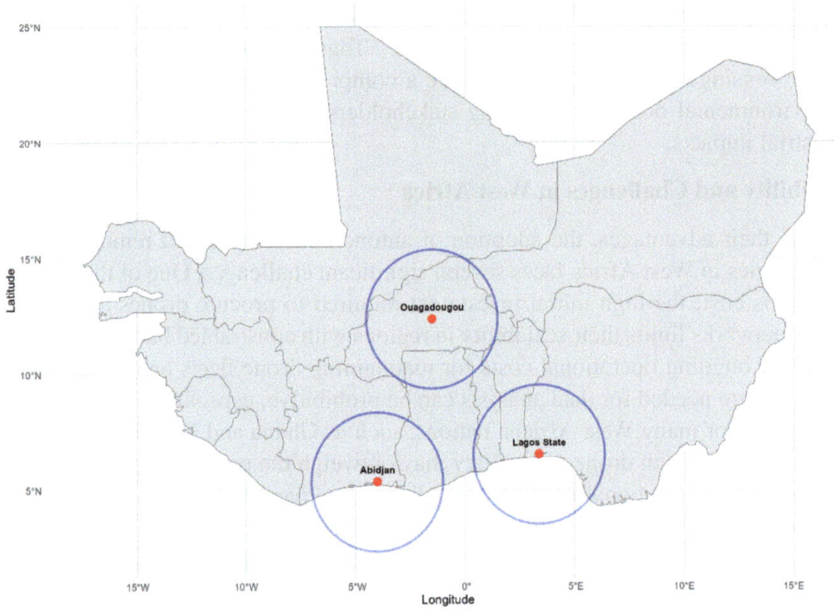

Fig. 2.5 Drone coverage map of West Africa: where red points = hotspot locations of industrial activity; blue circles = spatial possibilities of drone coverage

The deployment of drones in such regions not only bridges critical gaps in environmental management but also aligns with global sustainability goals by enabling scalable, data-driven strategies to mitigate pollution and enhance environmental governance.

The effectiveness of drones in pollution monitoring has already been demonstrated in Ghana, where they were deployed to monitor air pollution in mining zones [67]. These drones provide real-time spatial data on particulate matter, a key pollutant in such regions, revealing pollution hotspots that traditional methods have overlooked. These data could provide a basis for authorities to implement targeted pollution control measures, such as stricter industrial regulations and emission control adjustments [68]. As such, the potential application of drones in West Africa's industrial zones could revolutionize pollution management by offering rapid, accurate, and expansive data collection. When integrated with IoT networks, machine learning models, and satellite data, drones can provide a comprehensive picture of pollution levels, aiding policymakers in devising targeted strategies for environmental governance. By addressing limitations such as fragmented data and delayed responses inherent in traditional systems, drones and AI technologies have the potential to advance sustainability efforts, particularly in pollution-prone areas across the region [69].

2.4 Needs Assessment for AI-Enhanced Pollution Management

A comprehensive needs assessment for AI-enhanced pollution management is crucial to fully grasp the obstacles and opportunities within West Africa's environmental monitoring landscape. While AI holds immense potential to transform pollution management—offering real-time tracking, predictive analytics, and automated responses—the road to its successful integration is fraught with challenges. These include outdated infrastructure, fragmented data systems, and regulatory uncertainties, which undermine the effectiveness of existing monitoring efforts. Furthermore, significant barriers such as financial constraints, technical limitations, and socioeconomic disparities must be confronted to enable AI adoption at scale. This section delves into these gaps, critically examining the barriers to AI adoption while outlining the essential infrastructure, policy frameworks, and human resource development needed for sustainable implementation, aiming to pave the way for a future where AI-driven pollution management becomes a cornerstone of environmental resilience in the region.

2.4.1 Gaps in Current Practices

Despite significant progress in pollution monitoring, West Africa's industrial zones continue to face critical gaps in their environmental management systems. These gaps hinder the ability to track, analyze, and respond effectively to pollution, making it difficult to protect both the environment and public health. The primary challenges in current pollution monitoring systems include limited infrastructure, a lack of real-time tracking, and fragmented data collection processes, each of which contributes to inefficiencies in pollution management.

Limited Infrastructure: A key limitation in West Africa is the lack of robust infrastructure to support continuous and comprehensive pollution monitoring. Traditional monitoring systems, which often rely on manual sampling and periodic measurements, are inadequate for detecting sudden pollution events or fluctuations. These systems typically have fixed data collection points, making them ill equipped to capture the dynamic nature of industrial pollution, particularly in sprawling industrial zones. Without the infrastructure to support the deployment of real-time monitoring technologies such as IoT sensors and automated systems, it becomes increasingly difficult for authorities to obtain accurate and timely data, hindering their ability to respond effectively (Akinwande & Ojo, 2022). The lack of sufficient infrastructure results in slow, inefficient, and sometimes reactive pollution management rather than proactive solutions.

Lack of Real-Time Tracking: Another significant gap is the absence of real-time pollution tracking. Many pollution monitoring systems in West Africa still operate on outdated models that collect data at set intervals, making it challenging to track pollution dynamics, particularly during sudden industrial accidents or unexpected environmental changes. For example, during industrial accidents or rapid pollution spikes, having immediate access to accurate data can make a difference between timely intervention and long-term environmental degradation. The inability to track pollution levels in real time exacerbates this problem, leading to delayed responses from authorities and inadequate pollution control measures. Real-time tracking, enabled by AI technologies such as the IoT and machine learning, could drastically improve the effectiveness of pollution management by allowing authorities to make informed decisions on the basis of the latest available data [70].

Fragmented Data: Data fragmentation is another key issue that impedes effective pollution monitoring. Many existing monitoring systems operate in isolation, with separate stations collecting data without any integration or coordination. This leads to inconsistencies in data quality and gaps in coverage, which in turn make it difficult to generate a comprehensive understanding of pollution levels across a region. For example, different monitoring stations may collect data on different pollutants, and without proper integration, identifying pollution hotspots or trends becomes nearly impossible. A more integrated and holistic approach, where data from various sources are merged and analyzed via advanced AI techniques, could provide a clearer, more accurate picture of pollution patterns across industrial zones. AI-powered systems can bridge these gaps by enabling real-time data integration, improving the reliability and accessibility of environmental data [71].

AI Readiness Assessment: Addressing these gaps requires a thorough assessment of the region's readiness for AI adoption in pollution monitoring. This includes evaluating the technical, organizational, and policy factors that could influence the successful implementation of AI technologies. A readiness assessment will help identify the specific infrastructure, skills, and regulations needed to facilitate AI adoption and integration into pollution management practices.

2.4.2 Barriers to AI Adoption

While AI presents significant opportunities for improving pollution monitoring, its adoption in West Africa faces several barriers that must be addressed for successful integration into industrial pollution management. These barriers, which range from financial constraints to regulatory challenges, must be overcome to unlock the potential of AI technologies.

Financial Barriers One of the most significant obstacles to AI adoption in West Africa is the high cost associated with implementing AI technologies. AI systems, including sensors, software, and hardware, require substantial upfront investments that may be prohibitive for governments, industries, and local authorities in low-income regions. Beyond the initial purchase of hardware and software, there are also costs associated with setting up infrastructure, maintaining systems, and training staff. This financial burden can deter stakeholders from adopting AI solutions, particularly in resource-constrained environments. Furthermore, the lack of funding for pilot projects and research into the potential applications of AI in pollution management further limits its adoption. To address these barriers, governments and international partners must explore funding mechanisms and incentives to support AI adoption in environmental management [72, 73].

Technical Barriers The successful deployment of AI technologies also requires overcoming significant technical challenges. These include a lack of skilled personnel, unreliable internet connectivity, and inadequate computing infrastructure. In many regions, there is a shortage of technical expertise in AI, machine learning, and data science, which makes it difficult for local industries and governments to implement and maintain AI-driven pollution monitoring systems. Additionally, many areas in West Africa still struggle with inconsistent access to high-speed internet, which is essential for real-time data transmission and cloud-based AI applications. These technical limitations not only slow the adoption of AI technologies but also limit their ability to function effectively once deployed.

Regulatory Barriers The regulatory environment for AI adoption in West Africa is often underdeveloped, which further complicates the widespread use of AI in pollution monitoring. In many countries, there is a lack of clear policies on data privacy, security, and AI ethics, which are essential for guiding AI implementation in sensitive areas such as environmental management [74]. For example, concerns about surveillance and data privacy may prevent industries or governments from fully embracing AI-based monitoring solutions. To facilitate the successful adoption of AI, clear and supportive regulatory frameworks must be put in place, ensuring that ethical considerations and data privacy are adequately addressed [75].

Socioeconomic Barriers Socioeconomic challenges also play a significant role in hindering the adoption of AI in industrial pollution monitoring. Issues such as poverty, limited education, and restricted access to technology can impede the adoption of AI solutions, particularly in rural or underserved areas. There is also the risk of exacerbating the digital divide, where wealthier nations with advanced technological infrastructures can more easily implement AI solutions, while poorer regions are left behind. To ensure equitable access to AI technologies, it is crucial to provide training and education programs tailored to the local context and needs, as well as to develop affordable solutions that cater to low-income regions.

2.4.3 Requirements for Successful Implementation

To overcome the barriers and gaps discussed, several critical requirements must be met to ensure the successful implementation of AI technologies for industrial pollution monitoring in West Africa.

Essential Infrastructure

A strong and reliable infrastructure is the backbone of any AI-driven pollution monitoring system. This includes the deployment of IoT sensors across industrial zones, ensuring that data can be collected in real time. Additionally, reliable internet connectivity, cloud computing resources, and efficient data storage systems are essential for handling large volumes of environmental data. Without these infrastructural components, it would be impossible to support the advanced AI technologies required for real-time monitoring and predictive analytics [76]. In Table 2.4, the infrastructure requirements are clearly outlined, emphasizing the need for robust communication networks and data storage systems to ensure the scalability of AI-based pollution management solutions.

Technology Requirements: The adoption of AI technologies such as machine learning, IoT sensors, autonomous drones, and satellite imagery is crucial for effective pollution management. For example, machine learning models can be used to analyze real-time pollution data, detect anomalies, and predict future pollution trends. IoT sensors can continuously monitor pollutants such as particulate matter (PM2.5) and nitrogen oxides (NO_x), whereas drones can collect data from difficult-to-reach areas, offering a comprehensive view of pollution levels [76]. As outlined in Table 2.4, these technologies require substantial investment and expertise to ensure their successful deployment.

Human Resource Development: To support the effective use of AI technologies, local capacity-building efforts are necessary. This involves training engineers, data scientists, and environmental managers in the use of AI, machine learning, and IoT technologies. Tailored training programs ensure that local personnel can operate, maintain, and troubleshoot AI systems, making them more sustainable and adaptable to regional needs. These efforts will also help bridge the skills gap, ensuring that the local workforce is prepared for the future demands of AI-driven pollution monitoring [77].

Policy and Regulatory Framework: Finally, a clear and supportive policy and regulatory environment is essential for the successful implementation of AI technologies in pollution monitoring. This includes developing policies that protect data privacy, ensure the ethical use of AI, and incentivize industries to adopt AI-driven pollution monitoring solutions. Governments must also collaborate with international bodies to create frameworks that encourage innovation while ensuring the protection of public interests and the environment [78].

2.5 Comparative Analysis and Findings

AI-driven pollution monitoring technologies, such as Internet of Things (IoT) sensors, machine learning (ML) models, and autonomous drones, represent a paradigm shift in how industrial pollution is managed in West Africa [79]. These

Table 2.4 AI needs and resource requirements for industrial pollution monitoring

Gaps in current practices	AI needs	Examples	Resource requirements	Details
Limited infrastructure for monitoring	AI requires a robust infrastructure for sensor deployment and data transmission	Industrial zones in regions like West Africa often lack robust sensor networks, limiting pollution detection	Infrastructure	IoT sensors, communication networks, data storage, and cloud computing
Lack of real-time tracking	AI can provide real-time monitoring through IoT sensors and predictive analytics	IoT sensors integrated with AI can provide real-time air and water quality data in industries, such as mining or manufacturing	Technology	AI-driven algorithms (machine learning), drones, remote sensing devices
Fragmented data	AI can integrate data from diverse sources, offering comprehensive insights	Integration of data from drones, IoT sensors, and satellite imagery allows for a more holistic view of pollution trends	Human Resources	Training programs in AI, machine learning, IoT, and environmental management
Insufficient capacity for data analysis	AI enables the processing of large datasets for anomaly detection and trend forecasting	Machine learning algorithms can process historical air pollution data and predict future pollution levels in urban areas	Policy/ Regulatory Framework	Clear AI policies, data privacy regulations, incentives for AI adoption
Lack of actionable insights from data	AI can generate actionable insights for decision-makers through predictive analytics and visualizations	Predictive models for air pollution from AI can suggest adjustments in industrial practices to minimize environmental harm		

innovations promise to address long-standing challenges in pollution tracking, forecasting, and response by overcoming the limitations of traditional systems. In this section, a comparative analysis explores the effectiveness, scalability, and adaptability of these AI-based systems relative to conventional pollution monitoring approaches, focusing on factors such as accuracy, cost, responsiveness, and capacity to meet the unique needs of West Africa's industrial sectors.

2.5.1 Effectiveness of AI Solutions

AI-powered pollution monitoring systems provide several advantages over traditional methods, particularly in terms of real-time data collection, predictive capabilities, and responsiveness. These systems are designed to operate continuously, offering near-instantaneous data transmission and analysis, which is particularly vital in regions where industrial pollution can escalate rapidly.

1. **IoT and Sensor Networks**: IoT-enabled sensors represent a significant improvement over traditional monitoring by providing real-time data collection that can detect pollution events as they occur. These sensors can be deployed in remote, hard-to-reach industrial zones, where manual sampling or periodic measurements would be impractical. IoT systems, when calibrated accurately, offer high precision in detecting a broad range of pollutants, from particulate matter to volatile organic compounds. IoT-based systems offer rapid, real-time data transmission for quick pollution response, and despite high initial investment, their low operational costs and minimal maintenance make them cost-effective in the long term [80].
2. **Machine Learning and Predictive Analytics**: Machine learning models, particularly those focused on predictive analytics, provide another compelling advantage over traditional monitoring. These models can forecast pollution trends on the basis of historical and real-time data, allowing for the prediction of pollution events before they occur. This proactive approach marks a fundamental departure from traditional reactive methods, where responses are triggered only after pollution events are detected. Machine learning excels at detecting anomalies such as sudden pollution spikes, and while setup and training costs could also be high, it offers long-term cost savings through automation and improved forecasting accuracy [81, 82].
3. **Autonomous Drones and Remote Sensing**: Drones equipped with advanced sensors enhance pollution monitoring by providing high-resolution aerial data, particularly for air and water quality monitoring. Drones offer the unique ability to access remote or dangerous areas, delivering real-time data that can be transmitted to centralized systems for immediate analysis. Satellite imagery, when combined with drone data, allows for large-scale monitoring across expansive industrial zones, providing comprehensive insights into pollution sources. Drones are effective in hard-to-reach urban or industrial areas, offering rapid data collection with high initial costs but low long-term operational expenses, making them valuable for real-time pollution monitoring [15].

2.5.2 Scalability and Adaptation

Despite the proven effectiveness of AI-based pollution monitoring systems, scalability and adaptation to the diverse conditions of West Africa's industrial zones remain significant challenges. To ensure widespread adoption and long-term sustainability, these systems must be able to scale in response to the size and complexity of industrial zones and adapt to the region's unique environmental and socioeconomic conditions.

1. **Scalability**: AI-based systems are inherently scalable, with the ability to expand in response to growing needs. IoT and sensor networks can be scaled by adding additional sensors to the network, thereby increasing coverage across large industrial zones. The expansion of these systems is particularly feasible in regions with existing infrastructure, such as mobile networks or Wi-Fi connectivity. Machine learning systems, on the other hand, can be scaled by integrating more diverse data sources, although this often requires more advanced computational resources. The scalability of drones, however, is more constrained by cost. Deploying large fleets of drones to monitor vast industrial areas could prove prohibitive, particularly for governments or industries with limited budgets. Nonetheless, partnerships with local authorities, international organizations, and private sector actors could help reduce the financial burden associated with scaling drone operations [83].
2. **Adaptation to Local Needs**: The adaptation of AI systems to local needs is crucial for their successful implementation in West Africa. While AI technologies offer long-term cost efficiency by automating data collection and analysis, the initial investment required can be substantial [84]. To mitigate this, local partnerships between governments, industries, and international donors can help secure the necessary funding for infrastructure setup. Additionally, these collaborations can ensure that AI systems are tailored to the region's specific environmental and industrial contexts, taking into account local pollution sources and regulatory requirements. Capacity-building initiatives are essential for empowering local stakeholders, providing them with the necessary skills to maintain and operate AI systems effectively. Knowledge transfer through partnerships with local universities and technology providers will also be critical in ensuring that AI solutions are sustainable in the long run [85].

As such, AI technologies such as IoT sensors, machine learning models, and autonomous drones present compelling advantages over traditional pollution monitoring systems, particularly in the context of West Africa's industrial pollution challenges. However, the successful implementation of these systems will depend on overcoming financial, technical, and infrastructural barriers, as well as ensuring that AI solutions are scalable and adaptable to local needs. By addressing these challenges and fostering local partnerships, AI-driven pollution monitoring can play a transformative role in improving environmental management and public health outcomes across the region.

2.5.3 Comparison with Traditional Systems

The comparison between AI-enhanced pollution monitoring systems and traditional methods highlights key differences in accuracy, responsiveness, cost, and scalability (Table 2.5). Compared with traditional systems that rely on periodic, often delayed manual sampling, subject to human error, AI-enhanced pollution monitoring systems, which utilize IoT sensors, machine learning (ML), and drones, provide superior accuracy through real-time data collection and predictive analytics for pollutants such as air, water, and soil [86]. These systems also offer immediate responsiveness, using predictive models for early warnings, whereas traditional methods respond only after pollutant levels exceed predefined thresholds, missing sudden spikes [87]. Although AI systems involve high initial setup costs, they have lower operational costs in the long run, unlike traditional systems, which incur higher maintenance and labor costs [88]. AI systems provide seamless data integration across multiple platforms, whereas traditional systems often work with fragmented data [89].

Furthermore, AI systems are highly scalable, leveraging IoT networks and drone fleets for extensive coverage, whereas traditional systems face scalability challenges, requiring substantial manual effort [90]. AI technology is also more flexible, adapting to various pollution types and locations, including remote areas, whereas traditional systems are limited in adaptability [91]. Additionally, AI systems require modern infrastructure, such as reliable IoT networks and internet access for real-time data transfer, whereas traditional systems rely on outdated technology, hindering their ability to provide real-time analysis [92]

In essence, AI-enhanced systems for industrial pollution monitoring offer substantial advantages over traditional systems, particularly in terms of real-time tracking, accuracy, and predictive capabilities. While there are challenges in scaling and adapting these systems for West Africa, the potential benefits—especially with localized partnerships and capacity-building initiatives—are significant. A careful approach to integrate these AI technologies will ensure that they are both effective and sustainable in addressing the pollution challenges faced by industrial zones in the region.

2.6 Recommendations for Policymakers and Industry Stakeholders

To fully harness the potential of AI technologies for industrial pollution monitoring in West Africa, a coordinated effort between policymakers and industry stakeholders is essential. This section provides recommendations on frameworks for AI integration, strategies for industries, and how the adoption of AI aligns with global sustainability goals.

Table 2.5 Comparative analysis of AI-enhanced and traditional pollution monitoring systems, with examples of specific technologies used in each approach

Criteria	AI-enhanced systems (e.g., IoT, ML, Drones)	Traditional systems (e.g., manual sampling, fixed monitoring stations)
Accuracy	High; real-time data collection using IoT sensors (e.g., air quality sensors like PurpleAir, water quality probes) and predictive ML models for pollutants like PM2.5, NO2, and heavy metals	Moderate; periodic manual sampling of air and water pollutants, often delayed and prone to human error
Responsiveness	Immediate; predictive analytics (e.g., using ML algorithms) enable early warning systems for pollution events, with drone fleets detecting anomalies in real time	Slow; responses occur only after pollutant levels are confirmed to exceed thresholds, often missing sudden pollution spikes
Cost	High initial setup costs for IoT networks, drones (e.g., DJI Matrice for aerial surveillance), and cloud infrastructure, but lower operational costs over time	Low initial investment but high recurring costs for manual processes, including labor, equipment maintenance, and sampling
Data Integration	Seamless integration across IoT platforms, ML systems, and remote sensing technologies like satellite data (e.g., NASA MODIS, Sentinel-5P) for comprehensive pollution monitoring	Fragmented; data is isolated across different systems, often lacking real-time integration for actionable insights
Scalability	Highly scalable; IoT networks and drone fleets can cover expansive industrial zones, including remote and hard-to-reach areas	Limited scalability; expanding coverage requires significant manual effort, resources, and additional infrastructure
Flexibility	Highly adaptable to monitor various pollution types (air, water, soil) and geographical areas, including urban, rural, and industrial zones	Limited; often designed for specific pollutants or locations, requiring significant modifications for new applications
Infrastructure Requirements	Requires stable IoT networks, cloud computing, and reliable internet (e.g., 5G) for real-time data transfer and processing	Minimal; relies on traditional monitoring equipment and manual reporting, often using outdated technology

2.6.1 Policy Recommendations

1. **AI Integration Frameworks** To ensure the successful adoption and integration of AI in pollution monitoring, a robust policy framework must be established. This framework should focus on the following key areas:

Funding and Incentives

- Governments should provide financial incentives such as grants, subsidies, and tax rebates for industries adopting AI-based pollution monitoring technologies. Such incentives can ease the burden of high initial investments required for AI system setup, including IoT sensors, machine learning infrastructure, and drones. Additionally, governments can offer low-interest loans or public–private funding programs that support technological innovation [93].
- **Regulations and Standards**: Clear regulatory frameworks need to be set to guide the adoption of AI technologies. This includes defining standards for data quality, sensor calibration, and AI model validation. Policymakers should also ensure that AI systems comply with international environmental regulations and uphold data privacy and protection standards. Regulatory bodies could also work on setting AI-specific environmental standards to assess the effectiveness of pollution monitoring systems [94].
- **Capacity Building**: The adoption of AI technologies requires a skilled workforce. Policymakers should invest in educational programs and professional training to build local expertise in AI, data science, and environmental management [95]. Collaboration with international institutions can further enhance the knowledge base.

2. The role of public–private partnerships (PPPs) is critical for the successful deployment of AI technologies in pollution monitoring. These partnerships can help bridge the financial, technical, and infrastructural gaps faced by governments and industries [96, 97]. The key strategies for PPPs include the following:
   - **Collaborative Research and Development**: Governments and industries can collaborate on R&D to develop customized AI solutions tailored to a region's specific environmental and economic needs.
   - **Joint Infrastructure Projects**: The public and private sectors can coinvest in building the necessary infrastructure for AI-based monitoring systems, including sensor networks, the cloud computing infrastructure, and data centers.

2.6.2 Industry-Specific Strategies

For industries, the path to adopting AI in pollution monitoring requires careful planning and strategic implementation. The following steps are recommended for a smooth transition:

- **Pilot Programs**: Industries should begin with small-scale pilot programs to test AI technologies before full-scale implementation. These pilots allow industries to assess the feasibility, costs, and potential challenges of AI systems while refining the technology for their specific needs [98].
- **Phased Implementation**: Given the high initial costs and complex setup, industries can implement AI technologies in phases. In the first phase, industries can install IoT sensors in critical areas to monitor pollution in real time [70]. The second phase can involve the integration of machine learning models to predict

pollution trends and automate decision-making processes [99]. The final phase can expand to the use of autonomous drones for aerial monitoring of pollution in expansive industrial zones [100, 101].
- **Collaboration with Tech Providers**: Industries should collaborate with AI and technology providers to ensure that they have access to cutting-edge innovations in pollution monitoring. These collaborations can help industries stay at the forefront of AI advancements and maintain competitive advantages [102].

2.6.3 SDG Alignment

AI adoption in pollution monitoring aligns closely with the United Nations Sustainable Development Goals (SDGs), particularly:

- **SDG 12**: **Responsible Consumption and Production**: AI technologies contribute to responsible consumption and production by improving resource efficiency, reducing waste, and enabling real-time monitoring of industrial emissions. AI-enhanced systems allow industries to optimize their operations, detect inefficiencies, and minimize their environmental footprint. By providing accurate and timely pollution data, AI helps industries take proactive measures to prevent excessive waste generation and pollution [103].
- **SDG 13**: **Climate Action** AI technologies in pollution monitoring directly contribute to climate action by enabling industries to reduce their carbon footprint. By predicting pollution events and allowing for timely interventions, AI systems can reduce the release of harmful pollutants such as greenhouse gases [104, 105]. Furthermore, these systems enable better compliance with environmental regulations, which is critical for mitigating climate change.

Strategy Roadmap for AI Integration Aligned with SDGs:

1. **Short-Term (1–2 years)**:
    - Pilot projects should be conducted in select industries.
    - Develop policy frameworks to support AI integration.
    - Foster public–private partnerships for funding and technical support.
2. **Medium-Term (3–5 years)**:
    - Scale AI systems across major industrial sectors in West Africa.
    - Regulatory frameworks should be strengthened, and local capacity should be built.
    - AI solutions can be expanded to cover diverse pollution types (e.g., air, water, and soil).
3. **Long-Term (5–10 years)**:
    - The widespread adoption of AI-based pollution monitoring across West Africa has been achieved.

- Regularly update regulations to incorporate advancements in AI and environmental standards.
- Regional cooperation for pollution management and climate action should be promoted.

Study Limitations and Recommendations

The proposed AI-driven pollution management system offers significant potential but also presents several limitations that may hinder its widespread adoption, particularly in resource-constrained regions. A primary challenge is the high initial costs associated with deploying IoT sensors, drones, and cloud computing infrastructure, which may be prohibitive for developing areas with limited financial resources. Additionally, the system's effectiveness relies heavily on robust infrastructure, such as stable internet, electricity, and cloud-based data storage—resources that are often lacking in parts of West Africa. Data privacy concerns also arise, as the system involves real-time collection and transmission of sensitive environmental data, necessitating stringent safeguards against breaches. Furthermore, the system's reliance on skilled personnel to operate AI technologies and interpret complex data outputs may present a challenge in regions with limited technical expertise. While predictive analytics offer substantial benefits, they may not account for all local variables or unforeseen environmental factors, which could reduce the system's accuracy and adaptability in complex, real-world scenarios. Despite these challenges, AI's role in fostering responsible environmental management cannot be overlooked. By enabling real-time monitoring, providing predictive insights, and supporting proactive interventions, AI can encourage industries to adopt more sustainable practices, improve regulatory compliance, and reduce their environmental impact. Ultimately, AI contributes to creating a culture of accountability, transparency, and responsible environmental stewardship, offering a pathway to more sustainable and equitable industrial growth.

2.7 Conclusion

AI technologies, including IoT sensors, machine learning, and drones, offer transformative potential for revolutionizing pollution monitoring within West Africa's industrial sectors. These innovations enable the collection of more accurate, real-time data, providing predictive insights that support proactive pollution management and improved regulatory compliance. The integration of AI enhances environmental oversight, surpassing traditional monitoring methods and benefiting industries and governments alike. Moreover, AI supports sustainable industrial practices by helping industries minimize their environmental impact, optimize resource use, and contribute to global sustainability goals, fostering a culture of responsible environmental management. However, to fully realize the potential of AI, future research should focus on localized applications tailored to West Africa's unique environmental challenges, such as industrial diversity, pollution types, and regional geography. Additionally, future work should explore the socio-economic feasibility of adopting machine learning in pollution monitoring systems, including developing cost-effective financial models and assessing the long-term economic and environmental impacts. Addressing these research gaps will enhance the effectiveness, sustainability, and scalability of AI-based pollution monitoring systems, driving environmental, and economic progress in the region.

References

1. Seiyefa E, Oyosoro FI (2024) Green crime in West Africa: uncovering the threats to human security and ecosystem integrity in Cote d'Ivoire, Nigeria, and Senegal. GeoJournal 89(2):62
2. Kolawole AS, Iyiola AO (2023) Environmental pollution: threats, impact on biodiversity, and protection strategies. Sustainable utilization and conservation of Africa's biological resources and environment. Springer, pp 377–409
3. Okorondu J et al (2022) Anthropogenic activities as primary drivers of environmental pollution and loss of biodiversity: a review
4. Chukwuka AV et al (2023) Remotely sensed assessment of mining severity within buffer zones of Osun River: implications for drinking water quality, biodiversity protection, and status as UNESCO world heritage site. Water Air Soil Pollut 234(11):689
5. Amegah AK (2018) Proliferation of low-cost sensors. What prospects for air pollution epidemiologic research in Sub-Saharan Africa? Environ Pollut 241:1132–1137
6. Mukuyu P et al (2024) Innovations in water quality monitoring and management in Africa: towards developing an African Water Quality Program (AWaQ)
7. Vohra K et al (2022) Rapid rise in premature mortality due to anthropogenic air pollution in fast-growing tropical cities from 2005 to 2018. Sci Adv 8(14):eabm4435
8. Ayejoto DA et al (2023) Understanding the connections between climate change, air pollution, and human health in Africa: Insights from a literature review. J Environ Sci Health C 41(3–4):77–120
9. Mir Alvarez C et al (2020) A scoping review on air quality monitoring, policy and health in West African cities. Int J Environ Res Public Health 17(23):9151
10. Provençal S et al (2017) AOD distributions and trends of major aerosol species over a selection of the world's most populated cities based on the 1st version of NASA's MERRA Aerosol Reanalysis. Urban Climate 20:168–191
11. Chukwuka AV, Ogbeide O, Otomo PV (2023) Trend relationship between mountain normalized difference vegetation index (NDVI) and aerosol optical depth (AOD) across two decades: implication for water quality within the Lesotho Highlands, Drakensberg, South Africa. Environ Monit Assess 195(5):584
12. Firoozi AA, Firoozi AA (2023) Application of machine learning in geotechnical engineering for risk assessment
13. Mohammadi M et al (2024) An intelligent cloud-based IoT-enabled multimodal edge sensing device for automated, real-time, comprehensive, and standardized water quality monitoring and assessment process using multisensor data fusion technologies. J Comput Civ Eng 38(6):04024029
14. Essamlali I, Nhaila H, El Khaili M (2024) Supervised machine learning approaches for predicting key pollutants and for the sustainable enhancement of urban air quality: a systematic review. Sustainability 16(3):976

15. Asadzadeh S, de Oliveira WJ, de Souza Filho CR (2022) UAV-based remote sensing for the petroleum industry and environmental monitoring: State-of-the-art and perspectives. J Petrol Sci Eng 208:109633
16. Olawade DB et al (2024) Smart waste management: a paradigm shift enabled by artificial intelligence. Waste Manag Bull
17. Chukwuka AV et al (2023) Redox-active metals and oxidative stress–mediated myopathies in Callinectes amnicola, blue crab populations from impacted sites of the Lagos Lagoon: inferences for adverse ecological outcomes. Environ Sci Pollut Res 30(50):108565–108581
18. Jerome FC, Chukwuka AV (2016) Metal residues in flesh of edible blue crab, Callinectes amnicola, from a tropical coastal lagoon: health implications. Hum Ecol Risk Assess Int J 22(8):1708–1725
19. Jerome FC, Hassan A, Chukwuka AV (2020) Metalloestrogen uptake, antioxidant modulation and ovotestes development in Callinectes amnicola (blue crab): a first report of crustacea intersex in the Lagos lagoon (Nigeria). Sci Total Environ 704:135235
20. Chukwuka AV, Adeogun AO (2024) Urbanization effects on growth and otolith asymmetry in Chrysichthys nigrodigitatus and Oreochromis niloticus within tropical coastal lagoon watersheds. Chemosphere 359:142231
21. Ouattara ZA et al (2022) A review of sewerage and drainage systems management in sub-Saharan African Cities: Case of Abidjan, Côte d'Ivoire
22. Ofoezie EI et al (2022) Climate, urbanization and environmental pollution in West Africa. Sustainability 14(23):15602
23. Okereke C et al (2019) Governing green industrialisation in Africa: assessing key parameters for a sustainable socio-technical transition in the context of Ethiopia. World Dev 115:279–290
24. Unep UNEP (2022) Integrated assessment of air pollution and climate change for sustainable development in Africa: summary for decision makers. Available from: https://wedocs.unep.org/20.500.11822/41223
25. Bank W (2021) The world bank annual report 2021: towards a green, resilient, and inclusive recovery. Available from: https://www.worldbank.org/en/about/annual-report
26. Niohuru I (2023) Healthcare and disease burden in Africa: the impact of socioeconomic factors on public health. Springer Nature
27. Tiemele JA, Brou BRS, Djagoua EV (2023) Mapping of noise pollution and greenhouse gases from coastal protection works in grand-Lahou (Ivory Coast). In J Atmosph Oceanic Sci 7(2):50–54
28. Casari S et al (2022) Changing dietary habits: the impact of urbanization and rising socio-economic status in families from Burkina Faso in Sub-Saharan Africa. Nutrients 14(9):1782
29. Organization WH (2015) Chemicals of public health concern in the African Region and their management: regional assessment report
30. White R, Turpie J, Letley GL (2017) Greening Africa's cities: enhancing the relationship between urbanization, environmental assets, and ecosystem services. World Bank
31. Giovannetti G, Ticci E (2012) Sub-Saharan Africa in global trends of investment in renewable energy. Drivers and the challenge of the water-energy-land nexus
32. IM P (2015) Environmental sanitation
33. Dadashazar H, Ma L, Sorooshian A (2019) Sources of pollution and interrelationships between aerosol and precipitation chemistry at a central California site. Sci Total Environ 651:1776–1787
34. Mok J (2017) Multi-instrument approach for measuring spectral aerosol absorption properties in UV and VIS wavelengths. University of Maryland, College Park
35. Diner J et al, Light-absorbing black carbon and brown carbon components of smoke aerosol from 1 DSCOVR EPIC measurements over North America and Central Africa 2
36. Popescu SM et al (2024) Artificial intelligence and IoT driven technologies for environmental pollution monitoring and management. Front Environ Sci 12:1336088
37. Tricco AC et al (2022) Rapid reviews and the methodological rigor of evidence synthesis: a JBI position statement. JBI Evidence Synthesis 20(4):944–949

38. Nkinahamira F et al (2024) Machine learning approaches for monitoring environmental metal pollutants: recent advances in source apportionment, detection, quantification, and risk assessment. TrAC trends in analytical chemistry, p 117980
39. Hossain MS, Rahman MK, Dalim HM (2024) Leveraging AI for real-time monitoring and prediction of environmental health hazards: protecting public health in the USA. Revista de Inteligencia Artificial en Medicina 15(1):1117–1145
40. Seesaard T, Kamjornkittikoon K, Wongchoosuk C (2024) A comprehensive review on advancements in sensors for air pollution applications. Sci Total Environ, 175696
41. Ouaissa M et al (2024) Low-power wide area network for large scale internet of things: architectures, communication protocols and recent trends. CRC Press
42. Kaginalkar A et al (2023) Stakeholder analysis for designing an urban air quality data governance ecosystem in smart cities. Urban Climate 48:101403
43. Chen F et al (2024) Breathing in danger: Understanding the multifaceted impact of air pollution on health impacts. Ecotoxicol Environ Saf 280:116532
44. Yaroshenko I et al (2020) Real-time water quality monitoring with chemical sensors. Sensors 20(12):3432
45. Rahman MA et al. (2024) IoT-enabled intelligent garbage management system for smart City: a fairness perspective. IEEE Access
46. Zakari RY et al (2024) Internet of forestry things (IoFT) technologies and applications in Forest management. In: Advanced IoT technologies and applications in the industry 4.0 digital economy, CRC Press, pp 275–295
47. Singh R et al (2022) Forest 4.0: digitalization of forest using the Internet of Things (IoT). J King Saud Univ Comput Inform Sci 34(8):5587–5601
48. Hafeez S et al (2018) Detection and monitoring of marine pollution using remote sensing technologies. Monitoring of marine pollution
49. Ning J et al (2024) The Diversity of artificial intelligence applications in marine pollution: a systematic literature review. J Marine Sci Eng 12(7):1181
50. Ayaz M et al (2019) Internet-of-Things (IoT)-based smart agriculture: toward making the fields talk. IEEE Access 7:129551–129583
51. Singh G, Sharma S (2024) A comprehensive review on the internet of things in precision agriculture. Multimedia Tools and applications, pp 1–76
52. Sawyer WE et al (2024) Overview of air pollutants and their one health effects. Springer
53. Etim M-A et al (2021) Health risk and environmental assessment of cement production in Nigeria. Atmosphere 12(9):1111
54. Richard G, Izah SC, Ibrahim M (2023) Air pollution in the Niger Delta region of Nigeria: sources, health effects, and strategies for mitigation. J Environ Studies 29(1):1–15
55. Jońca J et al (2022) Drone-assisted monitoring of atmospheric pollution—a comprehensive review. Sustainability 14(18):11516
56. Fascista A (2022) Toward integrated large-scale environmental monitoring using WSN/UAV/Crowdsensing: a review of applications, signal processing, and future perspectives. Sensors 22(5):1824
57. Chukwuka AV et al (2024) Microplastic dynamics and risk projections in West African coastal areas: Developing a vulnerability index, adverse ecological pathways, and mitigation framework using remote-sensed oceanographic profiles. Science of the total environment, p 175963
58. Stierlin N, Risch M, Risch L (2024) Current advancements in drone technology for medical sample transportation. Logistics 8(4):104
59. Mandloi D, Arya R, Verma AK (2024) Internet of Drones. Recent Trends in Artificial Intelligence Towards a Smart World: Applications in Industries and Sectors. Springer, pp 353–373
60. Merem E et al (2018) Analyzing emerging environmental issues in major areas: the case of Lagos in South West Nigeria. Architect Res 8(1):19–38
61. Akangbe OA et al (2024) Gonad pathology, sex hormone modulation and vitellogenin expression in Chrysichthys nigrodigitatus from Lagos and Epe lagoons within the southern-lagoon system. Nigeria. Front Toxicol 6:1336916

62. Chukwuka AV et al (2023) Redox-active metals and oxidative stress–mediated myopathies in Callinectes amnicola, blue crab populations from impacted sites of the Lagos Lagoon: inferences for adverse ecological outcomes. Environ Sci Pollut Res, 1–17
63. Chukwuka AV, Adeogun AO (2024) Predicting the dynamics of per-and polyfluoroalkyl substances in coastal regions of Africa: vulnerability index and adverse ecological pathways from remote-sensed variables. Environ Monit Assess 196(6):573
64. Chukwuka AV, Omogbemi ED, Adeogun AO (2024) Habitat sensitivity in the West African coastal area: inferences and implications for regional adaptations to climate change and ocean acidification. Environ Monit Assess 196(1):79
65. Barnes-Dabban H, Karlsson-Vinkhuyzen S (2018) The influence of the regional coordinating unit of the Abidjan convention: implementing multilateral environmental agreements to prevent shipping pollution in West and Central Africa. Int Environ Agreements Politics Law Econ 18(4):469–489
66. Yiougo LS et al (2013) Urban cities and waste generation in developing countries: a gis evaluation of two cities in burkina faso. J Urban Environ Eng 7(2):280–285
67. Nyamekye C et al (2021) Examining the performances of true color RGB bands from Landsat-8, Sentinel-2 and UAV as stand-alone data for mapping artisanal and Small-Scale Mining (ASM). Remote Sensing Appl Soc Environ 24:100655
68. Asmal E (2021) Constructing intelligent drone systems to monitor environmental conditions
69. Alsamhi SH et al (2019) Survey on collaborative smart drones and internet of things for improving smartness of smart cities. IEEE Access 7:128125–128152
70. Ramadan MN et al (2024) Real-time IoT-powered AI system for monitoring and forecasting of air pollution in industrial environment. Ecotoxicol Environ Saf 283:116856
71. Mahule A et al, Enhancing environmental resilience: precision in air quality monitoring through AI-driven real-time systems. Artificial intelligence for air quality monitoring and prediction, pp 48–74
72. Dhakal KP, Chevalier LR (2017) Managing urban stormwater for urban sustainability: barriers and policy solutions for green infrastructure application. J Environ Manage 203:171–181
73. Dwivedi YK et al (2021) Artificial Intelligence (AI): Multidisciplinary perspectives on emerging challenges, opportunities, and agenda for research, practice and policy. Int J Inf Manage 57:101994
74. Wakunuma K, Jiya T, Aliyu S (2020) Socio-ethical implications of using AI in accelerating SDG3 in least developed countries. J Respon Technol 4:100006
75. de Almeida PGR, dos Santos CD, Farias JS (2021) Artificial intelligence regulation: a framework for governance. Ethics Inf Technol 23(3):505–525
76. Adewale AA et al (2024) Achieving net zero emissions in sustainable environmental remediation through the integration of IoT and Big Data. World J Adv Res Rev 23(3):1991–2013
77. Bibri SE, Huang J, Krogstie J (2024) Artificial intelligence of things for synergizing smarter eco-city brain, metabolism, and platform: Pioneering data-driven environmental governance. Sustain Cities Soc 108:105516
78. Voegtlin C, Scherer AG (2017) Responsible innovation and the innovation of responsibility: governing sustainable development in a globalized world. J Bus Ethics 143:227–243
79. Farghali M, Osman AI (2024) Revolutionizing waste management: unleashing the power of artificial intelligence and machine learning. Advances in Energy from Waste. Elsevier, pp 225–279
80. Maksimovic M (2018) Greening the future: Green Internet of Things (G-IoT) as a key technological enabler of sustainable development. Internet of things and big data analytics toward next-generation intelligence, pp 283–313
81. Hussain A et al (2020) Waste management and prediction of air pollutants using IoT and machine learning approach. Energies 13(15):3930
82. Akinosho TD et al (2022) A scalable deep learning system for monitoring and forecasting pollutant concentration levels on UK highways. Eco Inform 69:101609
83. Stokenberga A, Ochoa MC (2021) Unlocking the lower skies: the costs and benefits of deploying drones across use cases in East Africa. 2021: World Bank Publications

84. Kim J et al (2020) Scaling up disruptive agricultural technologies in Africa. World Bank Publications
85. Kolog EA et al (2022) Rethinking the implementation of artificial intelligence for a sustainable education in Africa: challenges and solutions. In: Management and information technology in the digital era: Challenges and perspectives. Emerald Publishing Limited, pp 27–46
86. Bayomi N, Fernandez JE (2023) Eyes in the sky: drones applications in the built environment under climate change challenges. Drones 7(10):637
87. Mohanty A, Mohanty SK, Mohapatra AG (2024) Real-time monitoring and fault. The AI cleanse: transforming wastewater treatment through artificial intelligence, p 165
88. Mithas S et al (2022) How will artificial intelligence and Industry 4.0 emerging technologies transform operations management? Prod Opera Manag 31(12):4475–4487
89. Rozony FZ et al (2024) A systematic review of big data integration challenges and solutions for heterogeneous data sources. Acad J Bus Administr Innov Sustain 4(04):1–18
90. Heidari A et al (2023) Machine learning applications in internet-of-drones: systematic review, recent deployments, and open issues. ACM Comput Surv 55(12):1–45
91. Ye Z et al (2020) Tackling environmental challenges in pollution controls using artificial intelligence: a review. Sci Total Environ 699:134279
92. Nižetić S et al (2020) Internet of Things (IoT): Opportunities, issues and challenges towards a smart and sustainable future. J Clean Prod 274:122877
93. Prasad KR et al (2024) AI in public-private partnership for IT infrastructure development. J High Technol Managem Res 35(1):100496
94. Christine KE, Aidonojie PA, Owoche AG, Harnessing artificial intelligence for environmental waste management in Uganda: legal and policy implications
95. Mikhaylov SJ, Esteve M, Campion A (2018) Artificial intelligence for the public sector: opportunities and challenges of cross-sector collaboration. Philos Trans Royal Soc Mathem Phys Eng Sci 376(2128):20170357
96. Ugwu MC, Adewusi AO, Nwokolo NE (2024) The role of public-private partnerships in building clean energy infrastructure in The United States And Nigeria. Int J Manag Entrepreneurship Res 6(4):1049–1068
97. Lam PT, Yang W (2020) Factors influencing the consideration of Public-Private Partnerships (PPP) for smart city projects: evidence from Hong Kong. Cities 99:102606
98. Hruby V (2024) Exploring AI adoption dynamics and entrepreneurial orientation in czech chemical SMEs: a pilot study perspective. Scientific papers of the university of Pardubice, series D: faculty of economics and administration 32(1)
99. Rane N, Kaya O, Rane J (2024) Artificial intelligence, machine learning, and deep learning applications in smart and sustainable industry transformation. Artif Intell Mach Learn Deep Learn Sustain Indus 5:2–29
100. Bakirci M (2024) Efficient air pollution mapping in extensive regions with fully autonomous unmanned aerial vehicles: a numerical perspective. Sci Total Environ 909:168606
101. Fumian F et al (2020) Application of miniaturized sensors to unmanned aerial systems, a new pathway for the survey of polluted areas: preliminary results. Atmosphere 11(5):471
102. Aldoseri A, Al-Khalifa KN, Hamouda AM (2024) AI-powered innovation in digital transformation: key pillars and industry impact. Sustainability 16(5):1790
103. Rane N, Kaya O, Rane J (2024) Advancing the sustainable development goals (SDGs) through artificial intelligence, machine learning, and deep learning. Artif Intell Mach Learn Deep Learn Sustain Indus 5:2–74
104. Kumari N, Pandey S (2023) Application of artificial intelligence in environmental sustainability and climate change. Visualization techniques for climate change with machine learning and artificial intelligence. Elsevier, pp 293–316
105. Bachmann N et al (2022) The contribution of data-driven technologies in achieving the sustainable development goals. Sustainability 14(5):2497

Chapter 3
Enhancing Coastal Resilience: Current Innovations in AI-Driven Pollution Detection Systems and Feasibility for West African Marine Environments

Azubuike Victor Chukwuka, Dami Emmanuel Omogbemi, Ayotunde Daniel Adegboyegun, and Aina O. Adeogun

Abstract This chapter explores the transformative potential of artificial intelligence (AI) and machine learning (ML) in enhancing marine pollution detection and management within West Africa's coastal regions. These regions, which are vital for their rich biodiversity and economies reliant on fisheries and tourism, face escalating environmental challenges such as marine pollution, coastal degradation, and climate change. However, effective pollution management is constrained by inadequate infrastructure, limited monitoring systems, and insufficient financial resources. By conducting a rapid review of peer-reviewed studies published between 2015 and 2024, this chapter examines how AI-driven technologies—particularly ML algorithms and remote sensing—can address these challenges. The findings reveal that AI applications improve pollution detection by monitoring sources such as industrial discharges, plastics, and oil spills, enabling real-time data analysis and timely responses to mitigate environmental impacts. Case studies from key areas, including Lagos (Nigeria) and other West African coastal stretches, highlight the practical applications of AI in strengthening resilience, enhancing biodiversity protection, and supporting sustainable resource management. Despite challenges such as limited infrastructure and funding, AI-driven systems show significant promise in advancing pollution control, improving environmental governance, and offering actionable insights for policymakers and stakeholders. This work contributes to the growing body of knowledge on how innovative technologies can transform pollution management and promote the sustainability of coastal ecosystems in West Africa.

A. V. Chukwuka (✉) · A. D. Adegboyegun
Environmental Quality Control Department, National Environmental Standards and Regulations Enforcement Agency (NESREA), Osun, Nigeria
e-mail: zubbydoo@gmail.com

D. E. Omogbemi · A. O. Adeogun
Department of Zoology, University of Ibadan, Ibadan, Nigeria

M. Nasr et al. (eds.), *Artificial Intelligence Applications for a Sustainable Environment*, Green Chemistry and Sustainable Technology,
https://doi.org/10.1007/978-3-031-91199-6_3

Keywords Coastal resilience · Remote sensing · Artificial intelligence · Pollution management · West Africa

3.1 Introduction

The West African coastline, which is rich in marine biodiversity and vital to the livelihoods of millions of people, faces increasing pollution pressures from rapid urbanization, industrialization, and inadequate waste management [1, 2]. From the Niger Delta's oil spills to Ghana's industrial zones and Senegal's urban centers, coastal ecosystems are threatened by a range of pollutants, including plastics, oil, agricultural runoff, and chemical discharges [3]. This environmental degradation not only harms marine life but also jeopardizes the socioeconomic stability of communities reliant on fishing, tourism, and coastal resources [4]. Traditional pollution monitoring methods often fall short in addressing the growing environmental challenges faced by African countries. However, the gradual adoption of artificial intelligence (AI) evidence in more recent publications offers a promising solution [5]. AI's ability to process real-time data and generate predictive models facilitates the identification of pollution hotspots, forecasts environmental impacts, and enhances coastal communities' resilience, enabling developing African nations to improve pollution detection and management through timely, data-driven insights for more effective environmental protection [6, 7].

The integration of artificial intelligence (AI) with remote sensing is revolutionizing environmental monitoring, particularly for marine pollution and coastal resilience. AI-driven image recognition technologies enable the rapid detection of environmental issues, such as oil spills and algal blooms, from satellite imagery, significantly improving monitoring accuracy and frequency [8]. Machine learning algorithms further enhance this by identifying subtle patterns in environmental data, allowing for predictive modeling that can anticipate threats such as harmful algal blooms or oil spill trajectories (Zhao et al. 2022). Additionally, AI integrates diverse data streams, including satellite imagery, oceanographic sensors, and meteorological data, to provide comprehensive risk assessments and early warning systems, which are crucial for areas with limited infrastructure, such as West Africa [9, 10]. This integration supports proactive pollution management, strengthens coastal resilience, and enables sustainable resource management by providing real-time data and predictive insights, offering significant benefits for both local communities and global environmental health [11].

Coastal resilience refers to the ability of coastal ecosystems and communities to anticipate, prepare for, respond to, and recover from environmental stresses such as pollution, climate change, and natural disasters [12]. This process involves safeguarding the integrity of vital ecosystems such as mangroves, seagrasses, and coral reefs, which provide crucial services, including biodiversity support, carbon sequestration, and coastal protection [13]. Resilience also entails the capacity of communities to adapt to socioeconomic impacts, such as the loss of livelihoods and infrastructure [12, 14]. The integration of innovative solutions such as artificial intelligence (AI) into pollution monitoring and management is key to strengthening coastal resilience, offering real-time data and predictive insights that improve decision-making and support adaptive strategies [15]. This review examines current AI-driven pollution detection systems and their potential to increase coastal resilience in West Africa, focusing on their effectiveness, scalability, and adaptability for managing pollutants in Nigeria, Ghana, and Senegal, while exploring how AI can be applied to improve local environmental protection and sustainability.

This chapter aims to investigate the role of artificial intelligence (AI) and machine learning in enhancing pollution detection and management in the coastal regions of West Africa. Specifically, it explores the effectiveness, scalability, and adaptability of AI-driven systems for monitoring and mitigating pollutants in Nigeria, Ghana, and Senegal. The chapter also seeks to evaluate how these innovative technologies can support coastal resilience by providing real-time data, predictive insights, and decision-making tools. Ultimately, this work aims to demonstrate how AI can contribute to local environmental protection and sustainability while addressing critical challenges posed by pollution and climate change in these regions.

3.2 Methodology

Study Area

This study focuses on the coastal regions of West Africa, specifically Nigeria, Ghana, and Senegal, which face significant environmental challenges such as marine pollution, coastal degradation, and climate change impacts (Fig. 3.1). With extensive coastlines along the Gulf of Guinea, these countries are critical to the region's biodiversity and economy, especially in terms of fisheries and tourism [4]. However, they face constraints such as limited infrastructure, inadequate monitoring systems, and insufficient financial resources, hindering their ability to effectively manage and mitigate pollution [16]. Rapid urbanization and industrial activities further exacerbate these challenges, making West Africa an ideal region for exploring the potential of AI-driven pollution monitoring systems for enhancing coastal resilience and promoting sustainable resource management [17]. The integration of AI technologies can support pollution mitigation efforts and strengthen environmental governance, ultimately benefiting coastal communities and ecosystems [18].

The coastal regions of these countries experience diverse pollution sources that significantly impact key ecosystems and areas of interest (Table 3.1). In Lagos, Nigeria, industrial waste, untreated sewage, oil spills, and plastics threaten mangroves, coastal wetlands, and coral reefs, with major concerns around Lagos Lagoon and Lagos Ports [4]. Ghana's Tema faces pollution from port activities, ship

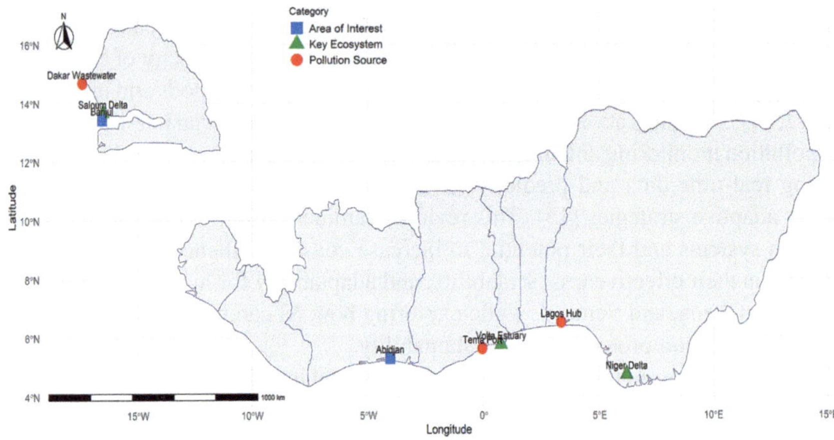

Fig. 3.1 Pollution sources, key ecosystems, and areas of interest in West African coastal areas

waste, plastic pollution, and oil discharge, affecting coastal wetlands, estuaries, and beaches, particularly around Tema Port [19, 20]. Dakar, Senegal, is impacted by oil and gas discharge, industrial runoff, and plastics, with Dakar Port being a significant concern for tourism and fishing [3, 21]. The Niger Delta, Nigeria, faces severe ecological damage from oil spills and agricultural runoff, which threatens biodiversity and mangroves [22, 23]. Similarly, Abidjan, Ivory Coast, and regions in Gambia and Sierra Leone suffer from plastic waste, oil discharges, and urban runoff, which impact wetlands and marine habitats [24, 25]. These cases underscore the urgent need for sustainable pollution control and ecosystem management to support biodiversity and regional development.

The coastal areas of West Africa are home to diverse ecosystems that support both environmental and economic activities (Table 3.1). Lagos, Nigeria, is characterized by vital ecosystems such as mangroves, coastal wetlands, and coral reefs, with areas of interest including the Lagos Lagoon, Lagos Port, and the bustling urban hub. Tema, Ghana, features coastal wetlands, estuaries, and sandy beaches, with key areas like Tema Port and a growing urban fishing industry. Dakar, Senegal, is known for its mangrove forests and marine fish habitats, with Dakar Port and the local fishing community being significant areas of focus. The Niger Delta, Nigeria, is rich in mangroves, swamp forests, and marine biodiversity, with oil exploration sites and vulnerable fishing regions being of particular interest. Abidjan, Ivory Coast, has estuaries, lagoon ecosystems, and marine life, with the Abidjan Port being a key commercial and economic center. Banjul, Gambia, is home to mangroves, saltwater marshes, and wetlands, with the River Gambia mouth and tourism and fishing industries as focal areas. Sierra Leone's coast features coastal forests, mangroves, and coral reefs, with Freetown Port and nearby fishing and tourism sites being of particular importance. The Volta Estuary in Ghana contains coastal wetlands and estuarine ecosystems, with biodiversity hotspots and fishing grounds. Senegal's Saloum Delta has wetlands, mangrove forests, and bird habitats, with the delta region and

Table 3.1 Pollution sources, key ecosystems, and areas of interest in West African coastal regions

Coastal area	Pollution sources	Key ecosystems	Areas of interest	References
Lagos (Nigeria)	Industrial waste, untreated sewage, oil spills, plastics	Mangroves, coastal wetlands, coral reefs	Lagos Lagoon, Lagos Port, busy urban hub	Yahaya, Muhammad [26], Kadiri [27], Chukwuka and Adeogun [28], Chukwuka, Jerome [29], Jerome, Hassan [30]
Tema (Ghana)	Port activities, ship waste, plastic pollution, oil discharge	Coastal wetlands, estuaries, sandy beaches	Tema Port, fishing industry, growing urbanization	Takyi, El Mahrad [20], Agbemabiese [19], Dzakpasu [31]
Dakar (Senegal)	Oil and gas-related discharge, industrial runoff, plastics	Mangrove forests, marine fish habitats	Dakar Port, fishing community, tourism hub	Sian Davies-Vollum, Koomson [32], Kojo [21], Alves, Bapentire [33]
Niger Delta (Nigeria)	Oil spills, gas flaring, agricultural runoff, industrial discharge	Mangroves, swamp forests, marine biodiversity	Oil exploration sites, fishing communities, vulnerable regions	Numbere and Maduike [23], Akani, Amuzie [22], Chukwuka and Adeogun [34], Chukwuka [35]
Abidjan (Ivory Coast)	Plastic waste, oil discharge, urban runoff	Estuaries, lagoon ecosystems, marine life	Abidjan Port, commercial and economic center	N'Goran, Yao [24], Sian Davies-Vollum, Koomson [32], Ibo Guei, Kouassi Aka [36]
Banjul (Gambia)	Solid waste, untreated sewage, agricultural runoff	Mangroves, saltwater marshes, wetlands	River Gambia mouth, tourism and fishing industry	Corcoran, Ravilious [37], Njie [25], Chuku, Okyere [38], Sian Davies-Vollum, Koomson [32]
Sierra Leone Coast	Mining waste (especially from bauxite), sewage, plastic pollution	Coastal forests, mangroves, coral reefs	Freetown Port, fishing communities, tourism sites	Pilkey, Longo [39], Kamara, Azizi [40], Sesay [41], Norman, Bebeley [42]
Volta Estuary (Ghana)	Agricultural runoff, plastic pollution, industrial discharge	Coastal wetlands, estuarine ecosystems	Estuary, fishing grounds, biodiversity hotspot	Onsrud [43], Boanu [44], Larbi, Nukpezah [45], Nunoo and Agyekumhene [46]

(continued)

Table 3.1 (continued)

Coastal area	Pollution sources	Key ecosystems	Areas of interest	References
Saloum Delta (Senegal)	Agricultural runoff, wastewater discharge, plastics	Wetlands, mangrove forests, bird habitats	Delta region, UNESCO Biosphere Reserve	Niang-Diop, Dansokho [47], Bodin, N'Gom-Kâ [48], Mathieu, Taïbou [49], Amara, Diop [50]
Cotonou (Benin)	Port-related pollution, agricultural runoff, plastics	Coastal wetlands, mangroves, marine ecosystems	Port of Cotonou, urban and industrial growth	Deguenon, Hounmenou [51], Deguenon, Baguere [52], Goussard and Ducrocq [53]

UNESCO Biosphere Reserve being key points of interest. Cotonou, Benin, is known for its coastal wetlands, mangroves, and marine ecosystems, with the Port of Cotonou being central to urban and industrial growth.

Review Approach

A rapid literature review was conducted to synthesize key findings on AI applications in marine and coastal pollution detection [54]. This approach provides a concise overview of high-quality studies, highlighting the effectiveness of AI technologies, such as machine learning and remote sensing, in enhancing environmental monitoring. The review emphasized AI's potential to improve pollution detection, offering actionable insights for future research, policy, and management. By streamlining the analysis compared with traditional systematic reviews, this method ensures credible, unbiased findings while delivering timely and relevant conclusions [55].

Search Strategy

To identify relevant literature, reports, and case studies, several academic databases, including Scopus, Web of Science, Google Scholar, IEEE Xplore, PubMed, and SpringerLink, were searched (Table 3.2). A combination of core and expanded keywords was used to guide the search, with core terms such as "artificial intelligence," "pollution detection," "marine ecosystems," and "coastal resilience" serving as the foundation. Expanded terms such as "remote sensing," "machine learning," "AI applications," "West African coasts," "environmental monitoring," and "pollution monitoring" were also included to broaden the scope. Boolean operators, such as AND, OR, and NOT, were employed to refine the search results for greater precision. Filters were applied to focus on publications from 2015 to 2024 in English, with a geographic focus on studies related to West African coasts or similar coastal regions.

Inclusion and Exclusion Criteria

To ensure the relevance and quality of the selected studies, the following criteria were applied:

Table 3.2 Inclusion and exclusion criteria for selecting studies on AI in marine pollution monitoring

Inclusion criteria	Exclusion criteria
Studies focusing on AI applications in marine pollution	Studies unrelated to AI or pollution monitoring
Research conducted in coastal/marine ecosystems	Studies focused on terrestrial pollution exclusively
Studies evaluating AI technologies like ML, remote sensing, or analytics	Publications without clear methodologies or results
Peer-reviewed journals, conference proceedings, reports	Non-peer-reviewed articles, opinion pieces, or editorials
Studies published between 2015 and 2024	Studies published before 2015

Data Extraction Process

Data were extracted via a standardized template to ensure consistency and focus on the study objectives. The following variables were recorded (Table 3.3):

- **AI Type**: Machine learning, remote sensing, or other technologies.
- **Application Scope**: Focus area, e.g., detecting oil spills, plastic waste, or industrial discharges.
- **Environmental Impact**: Measured outcomes such as pollution reduction, ecosystem recovery, or biodiversity protection.
- **Feasibility**: Cost-effectiveness, scalability, and ease of implementation.

Each study was assessed independently by two reviewers to minimize bias, with disagreements resolved through discussion.

Table 3.3 Methodological details

Category	Details
Search databases	Scopus, Web of Science, Google Scholar, IEEE Xplore, PubMed, SpringerLink
Keywords used	"Artificial Intelligence," "pollution detection," "marine ecosystems," "coastal resilience," etc
Filters applied	Publication dates (2015–2024), English language, focus on West African coastal regions
Inclusion criteria	Peer-reviewed studies, focus on AI in marine/coastal pollution detection, 2015–2024 publications
Exclusion criteria	Non-peer-reviewed, terrestrial focus, pre-2015 publications
Data variables	AI type, application scope, environmental impact, feasibility
Data reviewers	Two independent reviewers; disagreements resolved through discussion

3.3 Results: Current AI Innovations in Coastal Pollution Detection

The growing threat of coastal pollution in West Africa, driven by factors such as industrial effluents, oil spills, plastic waste, and agricultural runoff, calls for advanced technological solutions to increase environmental resilience. In this context, artificial intelligence (AI) innovations, particularly those leveraging remote sensing, machine learning (ML), and real-time monitoring systems, offer promising tools for pollution detection and management. By integrating satellite-based data, predictive modeling, and sensor networks, AI can enable the continuous monitoring of pollution trends, early identification of hotspots, and formulation of targeted interventions (Table 3.4). This section explores current AI advancements in coastal pollution detection and their applicability to West African coastal regions, highlighting the role of remote sensing technologies, ML algorithms, real-time monitoring systems, and predictive analytics in combating pollution and safeguarding marine ecosystems.

3.3.1 Remote Sensing and Satellite Monitoring

Satellite-based remote sensing has emerged as an indispensable tool for large-scale environmental monitoring, offering unparalleled capabilities in detecting and tracking pollution across vast and often inaccessible coastal regions. By leveraging advanced optical and multispectral sensors, remote sensing technologies capture high-resolution, real-time data on critical environmental parameters such as turbidity, chlorophyll concentrations, oil slicks, and suspended particulate matter. These datasets are instrumental in identifying pollution hotspots, assessing ecosystem health, and informing targeted mitigation strategies [34, 56].

In West Africa, where the vast coastline faces significant pollution from industrial effluents, oil exploration, plastic waste, and agricultural runoff, satellite remote sensing provides a cost-effective alternative to labor-intensive ground surveys, which are often constrained by logistical and financial barriers [34]. Additionally, remote sensing allows for continuous monitoring over time, enabling stakeholders to detect seasonal trends, evaluate policy effectiveness, and design adaptive management strategies tailored to the region's socioecological context.

Application Examples

Moderate Resolution Imaging Spectroradiometer (MODIS) is highly efficient at monitoring marine pollution, detecting oil spills, eutrophication, and sediment transport with precision, as demonstrated in the Gulf of Mexico and its potential for adaptation in West Africa, particularly in the Niger Delta, where oil spills threaten marine biodiversity [4, 57]. In West Africa, MODIS has been utilized to assess habitat sensitivity, climate change adaptation, and ocean acidification impacts [4]. It has also proven valuable in studying urbanization effects on fish growth and otolith asymmetry

Table 3.4 Tools, techniques, and applications in monitoring and managing aquatic pollution

Major section	Technique/Tool	Application/ description	Case studies/ examples	References
Remote Sensing and Satellite Monitoring	MODIS (Moderate Resolution Imaging Spectroradiometer)	High-resolution monitoring of marine pollution, including oil spills, sediment transport, and eutrophication. Adaptable for assessing habitat sensitivity and climate change impacts	Used in the Niger Delta to detect oil spills threatening biodiversity and assess fish kills in Bonny-Andoni, where strong winds and high water temperatures were implicated	Chukwuka [35, 57]
	Sentinel-2	High-resolution multispectral imaging to map plastic debris, suspended solids, and groundwater contamination	Applied in Nigeria's Lagos Lagoon to track plastic dispersal pathways and in Onitsha for groundwater contamination detection. Also used in Ghana's Volta Delta for monitoring plastic accumulation	Pamungkas, Manessa [58]; Chinye-Ikejiunor, Iloegbunam [59]
	Landsat Program	Long-term ecosystem monitoring (sediment transport, water quality, habitat degradation). Floating Plastic Index (FPI) developed to detect macroplastics	Landsat imagery assessed vegetation health in the Lesotho Highlands and mining impacts in Osun River. FPI revealed growing microplastic pollution in West Africa, linked to poor waste management	Chukwuka, Ogbeide [61], Chukwuka, Adegboyegun [1], Chukwuka, Nwabuisiaku [62]

(continued)

Table 3.4 (continued)

Major section	Technique/Tool	Application/ description	Case studies/ examples	References
Machine Learning Algorithms for Pollution Pattern Analysis		ML algorithms process large datasets (e.g., satellite imagery, sensors) to detect pollution patterns, forecast trends, and identify hotspots	Predictive models in the Ganges Basin, India, reduced pollutant loads. In Chesapeake Bay, USA, Random Forest algorithms monitored nutrient pollution, aiding better land-use practices. Deep learning models analyzed coral bleaching linked to pollution in the Great Barrier Reef	Somisetty, Pachore [84]; Zhang, Bostic [85]; Mao, Turner [86]; Patricio-Valerio, Schroeder [87]
Real-Time Monitoring with Sensor Networks	Chemical Sensors	Detect hydrocarbons during oil spills and monitor water turbidity	Combined with satellite imagery to track the Deepwater Horizon oil spill in the Gulf of Mexico, enabling targeted cleanup operations	Younos and Heyer [70]; Temitope Yekeen and Balogun [75]
	Underwater Acoustic Sensors	Detect chemical leaks and sediment movement, aiding in pollution tracking	Used during pollution events to complement optical and chemical sensors	Rizzo [71];Ferrari, Carrington [72]
	Integrated Sensor Systems	Combine data from satellites, weather stations, and in situ sensors for comprehensive monitoring	Enhanced predictive accuracy in pollution events like algal blooms and oil spills	Glaviano, Esposito [68]); Mahrad, Newton [74]

(continued)

Table 3.4 (continued)

Major section	Technique/Tool	Application/ description	Case studies/ examples	References
Predictive Analytics for Early Warning Systems	Statistical Models	Predict pollution incidents by integrating data from weather forecasts, sensor networks, and historical records	Early warning systems in the North Sea forecast oil spills using real-time oceanographic data. In New Zealand, models predict harmful algal blooms, reducing health risks and fisheries impacts	Gruiz [77]; Fetissov, Aps [80]; Pinkerton, Bell [81]
	ML-Driven Predictive Analytics	Forecast pollutant dispersal (e.g., oil spills, nutrient loadings) using ocean current and wind simulations	The Baltic Sea uses predictive modeling for nutrient load forecasting to mitigate eutrophication impacts from agricultural runoff	Yekeen and Balogun [79]; Thodsen, Farkas [82]

in coastal lagoons [28] and identifying causes of fish kills in Bonny-Andoni, linking them to high water temperatures and strong winds, thereby demonstrating its critical role in mitigating environmental impacts [35].

Sentinel-2: Sentinel-2's high-resolution multispectral imaging capabilities have been effectively applied in the Mediterranean Sea to map plastic debris and suspended solids, offering a valuable model for addressing similar environmental challenges in West Africa [58]. In regions such as Ghana's Volta Delta and Nigeria's Lagos Lagoon, Sentinel-2 could play a pivotal role in monitoring plastic accumulation zones and tracking waste dispersal pathways, thereby guiding interventions such as waste segregation and anti-litter campaigns [57]. Chinye-Ikejiunor, Iloegbunam [59] demonstrated the utility of Sentinel-2 imagery in assessing groundwater contamination in urban areas such as Onitsha. These studies highlighted Sentinel-2's ability to detect subtle spectral variations in water bodies, enabling the identification of emerging pollution threats. Additionally, Chukwuka and Adeogun [60] employed Sentinel-3 data to examine the bathymetric profile of tropical coastal lagoons and implications for adaptations for local fish populations, further demonstrating the adaptability of Sentinel imagery for comprehensive environmental monitoring.

Landsat Program: The Landsat program, with nearly five decades of data, has been essential for monitoring coastal ecosystems, assisting in trend analyses of sediment transport, water quality, and habitat degradation. For example, Landsat imagery correlates the NDVI with AOD to reveal vegetation health as an indicator of water quality changes in the Lesotho Highlands [61]. It also assesses the impact of mining on ecosystems in the Osun River buffer zones and supports the creation of vulnerability indices for monitoring PFAS in coastal regions [62]. These applications highlight Landsat's role in monitoring critical ecosystems such as mangroves and seagrass beds. Additionally, Chukwuka, Adegboyegun [1] advanced the use of satellite data to discern coastal microplastic pollution with the floating plastic index (FPI), which uses optical satellite data to detect and quantify floating macroplastics. Their research addresses the gap left by traditional methods, revealing the growing plastic pollution in West Africa, driven by rapid population growth and poor waste management, and underscores the need for improved monitoring and intervention strategies.

3.3.2 Machine Learning Algorithms for Pollution Pattern Analysis

Machine learning (ML) algorithms revolutionize pollution pattern analysis by enabling the rapid processing of vast, heterogeneous environmental datasets. These datasets often include satellite imagery, sensor data, and weather models, which are otherwise too complex for traditional analysis [63]. By leveraging techniques such as neural networks, clustering, and predictive modeling, ML systems can detect hidden patterns, identify pollution hotspots, and forecast pollution trends [64, 65]. This capacity to generate actionable insights not only enhances environmental resilience but also empowers stakeholders with tools for targeted interventions and policy formulation.

For example, ML-driven insights can be applied to prioritize areas for cleanup, optimize waste management strategies, and forecast pollutant dispersal, aiding in the preservation of coastal ecosystems and their socioeconomic benefits [66, 67]. The deployment of ML in coastal resilience planning directly aligns with the manuscript's focus on AI innovations tailored to the challenges faced by West African marine environments.

3.3.3 Real-Time Monitoring with Sensor Networks

Real-time monitoring via sensor networks is crucial for detecting and managing marine pollution, particularly for pollutants such as oil spills, heavy metals, and chemicals. These networks use technologies such as remote sensing systems, underwater acoustic sensors, and electrochemical sensors to collect environmental data continuously [68, 69]. For example, chemical sensors are effective at detecting hydrocarbons in water during oil spills, whereas optical sensors monitor changes in water turbidity caused by pollutants such as suspended solids or algal blooms [69, 70]. Underwater acoustic sensors can detect disturbances caused by chemical leaks or sediment movement, aiding in the identification of pollutants that are difficult to track visually [71]. In cases of heavy metal contamination, electrochemical sensors are widely used because of their ability to measure trace metal concentrations with high accuracy [72]. This continuous data collection provides authorities with actionable insights, enabling quick responses to pollution incidents and reducing environmental and economic damage.

In addition to individual sensors, real-time monitoring systems often integrate data from multiple sources to increase their effectiveness. For example, satellite imagery can complement in situ sensor networks by offering large-scale views of oil slicks or algal blooms, which can then be tracked in greater detail via localized sensors [73, 74]. Weather station data, such as wind speed and ocean current data, can further refine predictions of pollutant dispersion, improving response strategies. During the Deepwater Horizon oil spill, a combination of satellite data, chemical sensors, and acoustic monitoring was used to track the spread of the oil across the Gulf of Mexico, enabling targeted cleanup operations [75, 76]. Such integrated systems provide a comprehensive picture of pollution events, enhancing the accuracy and timeliness of environmental monitoring and decision-making.

3.3.4 Predictive Analytics for Early Warning Systems

Predictive analytics has become an invaluable tool for environmental management, particularly in providing early warnings for pollution events. By analyzing historical and real-time data from multiple sources—such as weather forecasts, satellite imagery, and pollution sensor networks—predictive models can forecast potential pollution incidents, including oil spills, algal blooms, and chemical discharges [77, 78] These models rely on advanced algorithms such as machine learning (ML) and statistical modeling to predict pollution patterns, enabling early intervention to mitigate environmental damage. For example, predictive analytics can anticipate the spread of oil spills by simulating how ocean currents and winds may carry pollutants. Similarly, predictive models can forecast harmful algal blooms, a common issue in coastal zones affected by nutrient runoff [79]. By integrating multiple data inputs into a unified system, predictive analytics provides actionable insights that allow decision-makers to take timely action, such as issuing alerts, deploying response teams, or closing fisheries in advance of pollution impacts.

Several coastal regions worldwide have successfully implemented early warning systems via predictive analytics. For example, the **North Sea** has adopted predictive models for oil spill response, combining real-time sensor data with oceanographic data to forecast spill movement and recommend response actions [80]. In **New Zealand**, predictive models based on ocean circulation and water quality data have been used to issue early warnings for harmful algal blooms, reducing the health risks to local populations and the impacts on fisheries [81]. Similarly, the **Baltic**

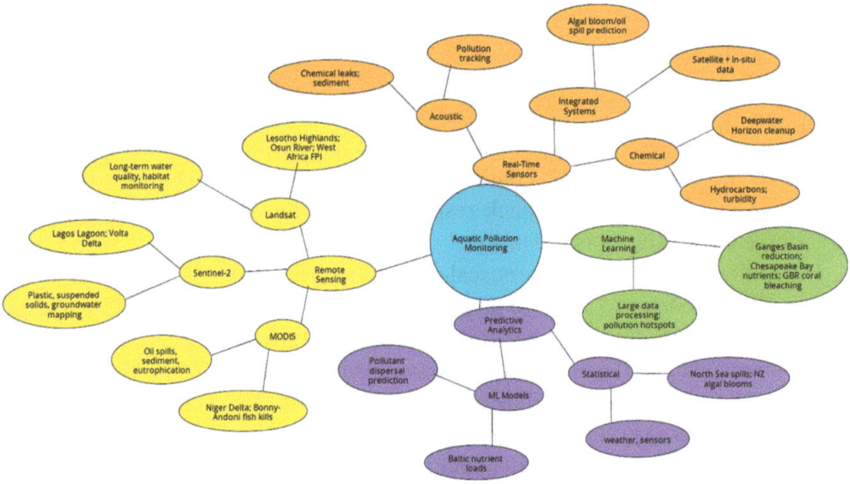

Fig. 3.2 AI-related tools and techniques for monitoring and managing aquatic pollution

Sea has adopted predictive modeling techniques to forecast nutrient loadings and potential eutrophication events, helping mitigate the effects of agricultural runoff on coastal ecosystems [82, 83]. These systems have proven successful in not only providing early warnings but also guiding resource allocation and response efforts in a more efficient manner (Fig. 3.2).

3.4 Discussion and Synthesis

AI Implementation in Coastal Pollution Management

Remote sensing technologies, including MODIS, Sentinel-2, and Landsat, are increasingly recognized as transformative tools for enhancing coastal resilience in West Africa. By generating dynamic pollution maps and supporting early warning systems, these technologies play a pivotal role in guiding targeted interventions to mitigate the impacts of pollution on coastal ecosystems. These satellite-based systems not only enable the monitoring of pollution over vast areas but also facilitate data sharing among regional stakeholders, thereby fostering collaboration between local governments and international agencies to address transboundary pollution issues effectively [1, 4]. In tandem, machine learning (ML) applications offer scalable, cost-effective solutions to complex environmental challenges such as industrial pollution, agricultural runoff, and plastic waste, all of which threaten a region's coastal ecosystems and economies [88].

The integration of predictive ML models into pollution management strategies holds great promise for West Africa's coastal zones. For example, in Lagos Lagoon, ML algorithms could utilize hydrodynamic data and land-use patterns to forecast pollutant dispersal, allowing authorities to pinpoint pollution hotspots and prioritize remediation efforts [66, 67]. Similarly, in the Niger Delta, ML models such as random forests could be employed to predict the accumulation of nutrients from agricultural runoff and oil spills, providing critical insights for developing sustainable farming practices and improving spill response strategies [57]. In Senegal, deep learning techniques applied to satellite imagery could monitor coastal ecosystem health, identify early signs of degradation, and support community-driven conservation efforts [4, 59].

However, the effectiveness of these AI-driven solutions is contingent upon overcoming significant socioeconomic and environmental challenges. Localized predictive models and sensor networks are essential for addressing the unique conditions of West Africa's coastal regions, and their success requires close collaboration among governments, research institutions, and local communities [1]. By ensuring the relevance and actionability of these AI technologies and investing in the necessary infrastructure and capacity-building efforts, these tools can significantly enhance pollution management strategies and improve the resilience of coastal ecosystems in West Africa.

Challenges and Opportunities for West Africa

AI and machine learning (ML) present significant opportunities for addressing coastal pollution in West Africa, particularly for improving monitoring, prediction, and intervention strategies (Table 3.5). These technologies, by integrating satellite imagery, in situ environmental data, and socioeconomic factors, enable a comprehensive understanding of pollution dynamics and support targeted interventions. However, scaling these solutions across the region faces several challenges. Data sparsity, inconsistent monitoring frameworks, and limited computational infrastructure are major barriers to effective implementation. Overcoming these problems requires strategic investments in robust data collection systems, affordable sensor technologies, and computational resources [89]. Moreover, local stakeholder engagement, including that of policymakers, researchers, and community members, is essential for ensuring that AI outputs are contextually relevant and actionable. Capacity-building efforts, such as workshops and training programs, can bridge the gap between AI-generated insights and practical policy applications [17].

Predictive analytics, another key AI tool, has vast potential in forecasting pollution events such as industrial discharges, plastic debris, and nutrient loading in West African coastal waters. By tracking pollutant pathways and anticipating seasonal variations, predictive models can improve early warning systems and inform targeted actions to address specific risks, such as those caused by agricultural runoff or plastic waste accumulation [90]. However, the effectiveness of these models depends on the integration of localized data, which requires collaboration among governments, research institutions, and local communities. Furthermore, real-time monitoring through sensor networks offers another promising solution for pollution detection, but challenges related to cost, maintenance, and data connectivity persist. Affordable, durable sensor technologies and partnerships for international funding could

Table 3.5 Challenges and opportunities for AI implementation in coastal regions

Category	Challenges	Opportunities/Solutions
Scaling ML for regional relevance	– Data sparsity and inconsistent monitoring frameworks – Limited computational infrastructure and resources	– Invest in robust data collection systems and affordable sensor technologies – Build capacity for localized ML use
Feasibility for sensor networks	– High costs of acquisition, maintenance, and analysis – Sparse infrastructure and limited connectivity in remote areas	– Secure funding from international organizations and public–private partnerships – Use cost-effective, durable sensors
Predictive analytics	– Need for localized data and socioeconomic integration – Lack of collaboration among governments, institutions, and communities	– Adapt models to regional contexts using historical and local data – Encourage partnerships for tailored solutions
Infrastructure needs	– Frequent power outages – Limited broadband coverage in rural areas – Dependency on external data servers	– Invest in solar-powered systems – Expand broadband and leverage satellite internet – Develop regional data centers
Economic constraints	– High hardware and software costs – Limited local manufacturing and reliance on imports	– Seek international funding (e.g., World Bank, African Development Bank) – Provide tax incentives for AI systems
Regulatory gaps	– Fragmented and poorly enforced environmental laws – Unclear policies on data privacy and AI ethics	– Harmonize policies under regional organizations (e.g., ECOWAS) – Develop clear regulatory frameworks for AI
Skills and technical expertise	– Insufficiently trained workforce for AI system management	– Establish training programs for environmental data professionals

help address these challenges, with an emphasis on building local capacity for sensor management and data analysis.

The feasibility of AI-driven pollution monitoring in West Africa also hinges on overcoming infrastructure and capacity limitations. Frequent power outages, inadequate renewable energy adoption, and poor internet connectivity, particularly in rural and coastal areas, impede the continuous operation of AI systems [91]. Solutions such as solar-powered sensors with battery backups and expanded broadband coverage via satellite internet services could address these infrastructural deficits. Additionally, developing local data centers and training technical personnel will ensure the sustainability and accessibility of AI technologies. Financial barriers, including the high costs of AI hardware and software, further exacerbate these challenges. Governments often prioritize immediate needs over environmental needs, but international funding and public–private partnerships could help distribute costs and encourage local ownership of AI initiatives [92]. Regulatory challenges also affect the adoption of AI in environmental monitoring. Fragmented and poorly enforced environmental policies, coupled with the lack of clear guidelines on data privacy, AI ethics, and environmental monitoring, create uncertainties in implementation [93]. To address these issues, regional organizations such as ECOWAS can play a key role in harmonizing policies and establishing comprehensive frameworks for AI governance. In addition to financial incentives such as tax breaks and subsidies, strengthened enforcement mechanisms could further incentivize the adoption of AI technologies for pollution management in West Africa.

IoT Technologies for Climate Change-Related Environmental Applications in West Africa

The Internet of Things (IoT) is increasingly being deployed in West Africa to address climate change-related environmental challenges. IoT technologies offer innovative solutions by integrating sensors, data transmission systems, and analytical tools to monitor and manage environmental conditions in real-time. One notable application is the use of sensor networks for air and water quality monitoring. For example, low-cost IoT sensors are used to measure pollutant concentrations in urban and coastal areas, providing critical data to mitigate pollution and assess its impact on public health and ecosystems (Akyildiz et al. 2020). Another widely adopted technology is smart weather stations, which track meteorological parameters such as temperature, humidity, rainfall, and wind speed. These systems help predict extreme weather events and support agricultural planning, thereby reducing vulnerabilities to droughts and floods (Faye et al. 2021). Additionally, IoT-enabled drones are utilized for environmental mapping, including deforestation tracking, coastal erosion assessment, and mangrove monitoring, offering high-resolution data that inform conservation efforts (Mensah et al. 2020). IoT-based irrigation systems are also prominent in addressing water scarcity. These systems use soil moisture sensors and remote monitoring platforms to optimize water use, reduce wastage, and improve crop resilience against climate stressors (Adeoti 2022). Furthermore, smart waste management systems leverage IoT-enabled bins with sensors to monitor waste levels, optimize collection routes, and minimize landfill emissions, contributing to sustainable waste management practices (Ogunleye and Akinrinade 2021).

Despite these advances, challenges such as limited infrastructure, high costs, and unreliable internet connectivity remain barriers to the broader adoption of IoT technologies in the region. Addressing these barriers through regional partnerships and investments could significantly enhance the impact of IoT in mitigating climate change and supporting environmental sustainability in West Africa.

Relevance of Global AI Applications to Coastal Pollution Management in West Africa

AI technologies have demonstrated their potential in managing coastal pollution across the globe, providing valuable lessons for their application in West Africa. Convolutional neural networks (CNNs) paired with satellite imagery have been used for real-time oil spill detection, allowing for quick responses and minimizing environmental damage [94, 95]. Similarly, random forest algorithms have been employed to monitor agricultural runoff, enabling predictive insights that inform targeted interventions to address nutrient pollution [96]. These approaches can be directly applied to regions such as the Niger Delta, where frequent oil spills require more efficient detection systems, and West African coastal areas, which are impacted by agricultural runoff.

The Mediterranean Sea's use of AI-powered drones and autonomous vehicles to track plastic debris also provides a relevant model. Real-time AI sensor networks help improve waste cleanup and pinpoint pollution hotspots [15]. This approach could be adopted for coastal areas such as Lagos Lagoon, where plastic pollution is increasingly problematic. Similarly, support vector machines (SVMs) have been used to predict industrial discharge patterns, improve waste management and reduce pollution [97]. This technology could be adapted for industrialized coastal zones in Ghana and Ivory Coast, helping predict and manage industrial pollution more effectively.

AI-driven systems have been proven to offer several advantages over traditional methods, such as providing faster, more accurate, and scalable solutions for pollution monitoring. Neural networks, for example, can quickly detect anomalies in satellite imagery, which traditional methods often fail to do because of slower response times [98]. In contrast, traditional methods are often constrained by limited resources and geographic coverage, which AI systems can overcome through large-scale, data-driven approaches. However, AI's dependence on infrastructure, including a stable internet and power supply, poses challenges for low-resource environments in West Africa [99].

To integrate AI successfully into West Africa's coastal pollution management, the region must address several technological gaps. The lack of high-quality, localized datasets limits the accuracy of AI models [100]. Additionally, affordable and durable sensors are necessary for deployment under challenging environmental conditions. Future research should focus on building region-specific datasets, creating low-cost sensor technologies, and establishing frameworks for integrating AI insights into policy. Investments in infrastructure, particularly renewable energy and improved internet connectivity, are essential to overcome these barriers and ensure that AI can maximize its potential in enhancing environmental resilience in West Africa.

Roadmap for Policy and Implementation

A phased roadmap is necessary for the effective integration of AI-driven monitoring systems (Table 3.6). The first phase involves assessing infrastructure gaps and developing AI-focused policy frameworks. This is followed by capacity-building efforts, including training programs and the establishment of regional AI hubs. The deployment phase initiates pilot projects such as oil spill detection and plastic pollution tracking while robust data-sharing frameworks are constructed. Successful pilots can then be scaled up through public–private partnerships. Finally, a comprehensive evaluation phase helps refine policies and improve outcomes on the basis of the lessons learned, ensuring long-term sustainability. Below is a proposed roadmap for integrating AI-driven pollution monitoring in West African coastal regions.

3.5 Recommendations for Policymakers and Environmental Managers

Integrating AI-driven pollution monitoring systems into West African coastal regions requires a strategic focus on regulatory reforms, economic incentives, and collaborative frameworks. Policymakers must update environmental laws to provide clear guidelines for AI use, encompassing data collection standards, ethical considerations, and compliance mechanisms. For example, enforcing AI-assisted monitoring for oil spills and industrial discharges can significantly improve accountability and environmental protection [101]. Economic measures such as financial incentives and tax breaks for industries adopting AI-powered sensors and software could further accelerate the uptake of these technologies. Collaborative public–private partnerships (PPPs), such as those demonstrated in India's Ganges River Cleanup initiative, where governments, private entities, and research institutions effectively pool resources, offer a proven model for fostering impactful solutions [102]. Establishing regional DGFs is equally essential, ensuring interoperable and secure datasets that enable AI systems to operate efficiently while safeguarding privacy.

Capacity building is fundamental to the long-term sustainability of AI integration in pollution management. Training programs and certification workshops for environmental professionals can equip them with the skills to utilize AI tools such as predictive models and real-time monitoring systems effectively [103]. Partnerships with global AI research institutions and international organizations can facilitate the transfer of expertise and drive innovation tailored to West Africa's unique environmental challenges [104]. Furthermore, engaging local communities in roles such as data collection and sensor maintenance promotes awareness while creating job opportunities, ensuring broader societal benefits. Establishing regional AI knowledge hubs would also encourage collaboration among researchers, practitioners, and policymakers, advancing innovations and disseminating best practices to build local expertise.

These efforts align closely with the United Nations Sustainable Development Goals (SDGs), underscoring their broader significance. AI-driven monitoring systems directly contribute to **SDG 14 (Life Below Water)** by reducing marine pollution (Target 14.1) and protecting marine ecosystems to enhance biodiversity (Target 14.2). They also support **SDG 6 (Clean Water and Sanitation)** by improving water

Table 3.6 Phased action plan for implementing AI-driven environmental solutions in West Africa's coastal regions

Phase	Key actions	Stakeholders involved
Phase 1: Planning	• Conduct comprehensive assessments of existing infrastructure gaps, including internet connectivity, power supply, and environmental monitoring systems • Develop AI-focused policy frameworks addressing data collection, ethical use, and regulatory compliance • Identify funding sources and form steering committees to oversee implementation	Governments, research institutions, international funding organizations, environmental policymakers, technology consultants, community leaders
Phase 2: Capacity building	• Develop targeted training programs for environmental managers and stakeholders, emphasizing predictive analytics, real-time monitoring, and AI ethics • Establish regional AI knowledge hubs with training facilities, data repositories, and simulation tools • Facilitate international exchange programs to strengthen technical expertise	Local environmental managers, AI experts, universities, vocational training centers, international organizations (e.g., UNEP), NGOs, and technology companies specializing in AI solutions
Phase 3: Deployment	• Initiate pilot projects addressing key environmental issues, such as oil spill detection, plastic waste tracking, and agricultural runoff management • Implement AI-driven sensor networks and establish interoperable data-sharing frameworks to facilitate cross-border collaboration • Test scalability and robustness of deployed systems in real-world coastal environments	Governments, private sector stakeholders (e.g., technology firms, environmental consultants), NGOs, local environmental enforcement agencies, and community representatives

(continued)

Table 3.6 (continued)

Phase	Key actions	Stakeholders involved
Phase 4: Scaling up	• Analyze pilot projects results and identify best practices for expansion • Develop a roadmap for scaling up successful AI systems across additional coastal regions • Strengthen public–private partnerships (PPPs) to share expertise, funding, and technology • Advocate for legislative amendments to support large-scale AI deployment and adoption	Policymakers, industries (e.g., fisheries, oil and gas, tourism), financial institutions, local communities, regional environmental organizations, and private investors
Phase 5: Evaluation	• Establish robust monitoring and evaluation mechanisms to assess the environmental, social, and economic impacts of AI systems • Use insights to refine policy frameworks, improve data collection protocols, and address unforeseen challenges • Publish findings to promote transparency and knowledge sharing • Integrate lessons learned into regional environmental strategies and future initiatives	Environmental agencies, research institutions, governments, independent evaluators, think tanks, international donors, and local stakeholders impacted by pollution monitoring outcomes

quality through better monitoring and reducing pollution levels (Target 6.3). Furthermore, fostering public–private partnerships and engaging communities directly align with **SDG 17 (Partnerships for the Goals)** by promoting multi-stakeholder collaboration to mobilize resources and implement shared strategies for sustainable environmental management (Target 17.16). By leveraging AI technologies in alignment with these specific targets, West Africa can enhance environmental resilience, advance socioeconomic development, and sustainably tackle critical coastal pollution challenges.

Study Limitation

This study acknowledges several limitations that may affect the generalizability and scalability of its findings. First, the reliance on a rapid literature review means that the scope of analysis was confined to peer-reviewed studies published between 2015 and 2024, potentially excluding relevant gray literature or unpublished data. Second, while the study highlights the application of artificial intelligence (AI) and machine learning (ML) technologies in marine pollution detection, practical implementation challenges such as high costs, limited infrastructure, and a lack of technical expertise in many West African regions were not fully addressed. Additionally, the environmental, socio-economic, and regulatory diversity across countries like Nigeria, Ghana, and Senegal may limit the uniform applicability of AI solutions. The study also does not include primary data or field validation, which could further support the effectiveness of the proposed AI-driven systems. Future research should incorporate on-ground testing, assess the economic feasibility of these technologies, and explore collaborative frameworks to address infrastructural and financial constraints.

3.6 Conclusion

This review highlights the transformative potential of AI-driven pollution monitoring systems in addressing the environmental challenges faced by West African coastal regions. AI technologies, including predictive analytics, sensor networks, and remote sensing, provide significant advantages in detecting and managing pollution by delivering real-time data, improving monitoring accuracy, and enabling faster responses to mitigate environmental damage. These tools enhance coastal management while supporting sustainable development by facilitating early warning of pollution events, reducing environmental harm, and prioritizing resources for pollution hotspots and vulnerable ecosystems. This, in turn, aids effective policymaking and improves the resilience of communities reliant on marine resources, such as those in the fishing and tourism sectors. Future work should explore the economic feasibility of adopting advanced AI technologies, such as convolutional neural networks (CNNs), for waste management in resource-constrained settings. Research should focus on assessing the cost–benefit ratio of implementing CNNs and other AI models, considering factors such as affordability, durability, and scalability in West Africa's unique climatic and socio-economic conditions. Additionally, the development of localized, high-quality datasets for AI model training and the establishment of regional data-sharing platforms are vital to enhancing system performance. Investigating the social and economic implications of AI adoption, including job creation, capacity building, and equitable access, remains an important area for further exploration. By integrating AI systems into existing environmental policies and fostering collaboration among governments, private companies, and local communities, West Africa can effectively leverage these technologies to increase coastal resilience, conserve marine ecosystems, and promote sustainable socio-economic growth.

References

1. Chukwuka AV et al (2024) Microplastic dynamics and risk projections in West African coastal areas: Developing a vulnerability index, adverse ecological pathways, and mitigation framework using remote-sensed oceanographic profiles. Sci Total Environ, 175963
2. Dada O et al (2021) Towards West African coastal social-ecosystems sustainability: interdisciplinary approaches. Ocean Coast Manag 211:105746
3. Davies-Vollum KS, Koomson D, Raha D (2024) Coastal lagoons of West Africa: a scoping study of environmental status and management challenges. Anthropocene Coasts 7(1):7
4. Chukwuka AV, Omogbemi ED, Adeogun AO (2024) Habitat sensitivity in the West African coastal area: inferences and implications for regional adaptations to climate change and ocean acidification. Environ Monit Assess 196(1):79
5. Jaldi A (2023) Artificial intelligence revolution in Africa: economic opportunities and legal challenges. Policy Cent. New South
6. Samaei SR, Ghahfarrokhi MA (2023) AI-enhanced GIS solutions for sustainable coastal management: navigating erosion prediction and infrastructure resilience. In: 2th International conference on creative achievements of architecture, urban planning, civil engineering and environment in the sustainable development of the middle east
7. Rane N, Choudhary S, Rane J (2024) Artificial intelligence for enhancing resilience. J Appl Artif Intell 5(2):1–33
8. Esiri AE, Babayeju OA, Ekemezie IO (2024) Advancements in remote sensing technologies for oil spill detection: policy and implementation. Eng Sci Technol J 5(6):2016–2026
9. Chen G et al (2022) Deep blue AI: a new bridge from data to knowledge for the ocean science. Deep Sea Res Part I 190:103886
10. Dong C et al (2022) Recent developments in artificial intelligence in oceanography. Ocean-Land-Atmosph Res
11. Wang L (2024) Advances in monitoring and managing aquatic ecosystem health: integrating technology and policy. Int J Aquac, 14
12. Cinner JE et al (2018) Building adaptive capacity to climate change in tropical coastal communities. Nat Clim Chang 8(2):117–123
13. Ayyam V et al (2019) Coastal ecosystems and services. Coastal ecosystems of the Tropics-Adaptive management, pp 21–47
14. Uddin MS et al (2021) "Disasters threaten livelihoods, and people cope, adapt and make transformational changes": community resilience and livelihoods reconstruction in coastal communities of Bangladesh. Int J Disaster Risk Reduction 63:102444
15. Sultana M et al (2024) Artificial intelligence for sustainable ocean health. Artificial intelligence and edge computing for sustainable ocean health. Springer, pp 3–38
16. Hauser L et al (2023) A scoping study on coastal vulnerability to relative sea-level rise in the Gulf of Guinea. AFD Res Papers 283:1–42
17. Gesami BK, Nunoo J (2024) Artificial intelligence in marine ecosystem management: addressing climate threats to Kenya's blue economy. Front Mar Sci 11:1404104
18. Bibri SE et al (2024) Smarter eco-cities and their leading-edge artificial intelligence of things solutions for environmental sustainability: a comprehensive systematic review. Environ Sci Ecotechnol 19:100330
19. Agbemabiese N (2020) Characterization and quantification of litter on selected beaches in the central region of Ghana: towards the management and proper disposal of solid waste. University of Cape Coast
20. Takyi R et al (2023) Assessment of coastal and marine ecosystems in West Africa: the case of Ghana. Mar Pollut Bull 197:115735
21. Kojo AA (2020) Information-analytical system and research on the impact of coastal pollution and degradation on West Africa
22. Akani GC et al (2022) Factors militating against biodiversity conservation in the Niger Delta, Nigeria: the way out. Biodiversity in Africa: Potentials, Threats and Conservation. Springer, pp 573–600

23. Numbere AO, Maduike EM (2023) Ecosystem management of polluted forest and its implication on biodiversity conservation in the Niger delta. Ecorestoration for sustainability, pp 295–315
24. N'Goran KM, Yao KM, Trokourey A (2023) Distribution of polycyclic aromatic hydrocarbons (PAHs) in M'Badon Bay (Ebrie Lagoon, Abidjan-Cote d'Ivoire)
25. Njie M (2017) Climate change mitigation technologies in the Gambian energy, transport and waste sectors. Technol. Needs Assess. https://tech-action.unepdtu.org/wp-content/uploads/sites/2/2019/03/mitigation-tna-report-finalreport-myversion11.pdf
26. Yahaya T et al (2022) Health risks of ecosystem services in Ologe Lagoon, Lagos. Southwest Nigeria. Pollution 8(2):681–692
27. Kadiri TE (2015) Regulating land-based sources and activities causing pollution of the coastal and marine environment in South Africa, Kenya and Nigeria within the context of integrated coastal zone management
28. Chukwuka AV, Adeogun AO (2024) Urbanization effects on growth and otolith asymmetry in Chrysichthys nigrodigitatus and Oreochromis niloticus within tropical coastal lagoon watersheds. Chemosphere 359:142231
29. Chukwuka AV et al (2023) Redox-active metals and oxidative stress–mediated myopathies in Callinectes amnicola, blue crab populations from impacted sites of the Lagos Lagoon: inferences for adverse ecological outcomes. Environ Sci Pollut Res, 1–17
30. Jerome FC, Hassan A, Chukwuka AV (2020) Metalloestrogen uptake, antioxidant modulation and ovotestes development in Callinectes amnicola (blue crab): a first report of crustacea intersex in the Lagos lagoon (Nigeria). Sci Total Environ 704:135235
31. Dzakpasu MFA (2019) Ecological assessment of some coastal lagoons and estuaries in Ghana: abiotic and biotic approaches. University of Cap e Coast, Ghana
32. Sian Davies-Vollum K, Koomson D, Raha D (2024), Coastal lagoons of West Africa: a scoping study of environmental status and management challenges
33. Alves RB et al (2022) Compendium: Coastal management practices in West Africa: existing and potential solutions to control coastal erosion, prevent flooding and mitigate damage to society
34. Chukwuka AV, Adeogun AO (2024) Predicting the dynamics of per-and polyfluoroalkyl substances in coastal regions of Africa: vulnerability index and adverse ecological pathways from remote-sensed variables. Environ Monit Assess 196(6):573
35. Chukwuka AV (2023) First Report of Fishkill Incidence in Bonny-Andoni Coastal Area. Nigeria. Croatian Journal of Fisheries: Ribarstvo 81(2):83–91
36. Ibo Guei J et al (2002) Cote D'ivoire Coastal Zone: Phase 1: Integrated environmental problem analysis
37. Corcoran E, Ravilious C, Skuja M (2007) Mangroves of western and central Africa. UNEP/Earthprint
38. Chuku E et al (2022) Site-based assessment of oyster shellfisheries and associated bio-physical conditions in Ghana and the Gambia
39. Pilkey OH, Longo NJ, Neal WJ (2023) Vanishing sands: losing beaches to mining. Duke University Press
40. Kamara M et al (2024) Assessing the impact and management of plastic pollution in freetown: a comprehensive situational analysis. Int J Environ Climate Change 14(9):718–736
41. Sesay DK (2020) The risk of oil spill in the exclusive economic zone of sierra Leone
42. Norman PE et al (2018) Biodiversity in Sierra Leone, in Global Biodiversity. Apple Academic Press, pp 213–246
43. Onsrud A (2023) Microplastic Pollution in the Volta River Delta's Blue Carbon Mangrove Ecosystem in Ghana. NTNU
44. Boanu NY (2020) Assessment of water pollution and biodiversity change of the Muni-Pomadze lagoon and its catchment in Winneba, Ghana. University of Education, Winneba
45. Larbi L et al (2018) An integrated assessment of the ecological health status of coastal aquatic ecosystems of Ada in Ghana. West African J Appl Ecol 26(1):89–107

46. Nunoo FK, Agyekumhene A (2022) Mangrove degradation and management practices along the coast of Ghana. Agric Sci 13(10):1057–1079
47. Niang-Diop I et al (2002) Senegal national report phase 1: integrated problem analysis. GEF MSP Sub-Saharan Africa Project (GF/6010–0016):" Development and protection of the coastal and marine environment in Sub-Saharan Africa
48. Bodin N et al (2013) Assessment of trace metal contamination in mangrove ecosystems from Senegal. West Africa. Chemosphere 90(2):150–157
49. Mathieu G, Taïbou B, Aminata DS (2018) Biodiversity in Senegal: terrestrial ecosystems flora and Great Fauna. In: Global Biodiversity. Apple Academic Press, pp 173–212
50. Amara R et al (2019) The Senegalese coastal and marine environment. World seas: an environmental evaluation. Elsevier, pp 855–873
51. Deguenon SDDM et al (2023) Simulation of the impacts of sea-level rise on coastal ecosystems in Benin using a combined approach of machine learning and the sea level affecting marshes model. Sustainability 15(22):16001
52. Deguenon SDDM et al (2022) Using of intensity analysis approach in Benin coastal zone (West Africa) to assess land use/land cover change for further decision making. Heliyon 8(12)
53. Goussard J-J, Ducrocq M (2014) West African coastal area: challenges and outlook. The land/ ocean interactions in the coastal zone of West and Central Africa, pp 9–21
54. Tricco AC et al (2015) A scoping review of rapid review methods. BMC Med 13:1–15
55. Patnode CD et al (2018) The use of rapid review methods for the US preventive services task force. Am J Prev Med 54(1):S19–S25
56. Mohanty S et al (2025) Challenges and future implications in monitoring and assessment of aquatic ecosystems. Aquatic Ecosystems Monitoring. CRC Press, pp 297–318
57. Almar R et al (2023) Coastal zone changes in West Africa: challenges and opportunities for satellite earth observations. Surv Geophys 44(1):249–275
58. Pamungkas FD et al (2024) Multispectral insights into turbidity variations over time in the Derawan Island. In: IOP Conference series: earth and environmental science. IOP Publishing
59. Chinye-Ikejiunor N et al (2021) Groundwater contamination and health risk assessment across an urban gradient: case study of Onitcha metropolis, south-eastern Nigeria. Groundw Sustain Dev 14:100642
60. Chukwuka AV, Adeogun AO (2024) Habitat-driven otolith shape morphology in Oreochromis niloticus (pelagic) and Chrysichthys nigrodigitatus (benthic) from distinct zones of a tropical coastal lagoon system. Regional Studies Marine Sci 78:103760
61. Chukwuka AV, Ogbeide O, Otomo PV (2023) Trend relationship between mountain normalized difference vegetation index (NDVI) and aerosol optical depth (AOD) across two decades: implication for water quality within the Lesotho Highlands, Drakensberg, South Africa. Environ Monit Assess 195(5):584
62. Chukwuka AV et al (2023) Remotely sensed assessment of mining severity within buffer zones of Osun River: implications for drinking water quality, biodiversity protection, and status as UNESCO world heritage site. Water Air Soil Pollut 234(11):689
63. Mukonza SS, Chiang J-L (2023) Meta-analysis of satellite observations for united nations sustainable development goals: exploring the potential of machine learning for water quality monitoring. Environments 10(10):170
64. Nkinahamira F et al (2024) Machine learning approaches for monitoring environmental metal pollutants: recent advances in source apportionment, detection, quantification, and risk assessment. TrAC Trends Analyt Chem, 117980
65. Bi Z et al (2024) Machine learning-driven source identification and ecological risk prediction of heavy metal pollution in cultivated soils. J Hazard Mater 476:135109
66. Huang Y et al (2024) Dynamic patterns and potential drivers of river water quality in a coastal city: insights from a machine-learning-based framework and water management. J Environ Manage 370:122911
67. Deng T, Chau K-W, Duan H-F (2021) Machine learning based marine water quality prediction for coastal hydro-environment management. J Environ Manage 284:112051

68. Glaviano F et al (2022) Management and sustainable exploitation of marine environments through smart monitoring and automation. J Marine Sci Eng 10(2):297
69. Ramakrishnan S et al (2024) Water pollutants, sensor types, and their advantages and challenges. In: Sensors for environmental monitoring, identification, and assessment. IGI Global, pp 78–101
70. Younos T, Heyer CJ (2015) Advances in water sensor technologies and real-time water monitoring. Adv Watershed Sci Assess, 171–203
71. Rizzo P (2022) Sensing solutions for assessing and monitoring underwater systems. Sensor technologies for civil infrastructures, pp 355–376
72. Ferrari AG-M et al (2020) Recent advances in portable heavy metal electrochemical sensing platforms. Environ Sci Water Res Technol 6(10):2676–2690
73. Yang Z et al (2022) UAV remote sensing applications in marine monitoring: knowledge visualization and review. Sci Total Environ 838:155939
74. Mahrad BE et al (2020) Contribution of remote sensing technologies to a holistic coastal and marine environmental management framework: a review. Remote Sensing 12(14):2313
75. Temitope Yekeen S, Balogun A-L (2020) Advances in remote sensing technology, machine learning and deep learning for marine oil spill detection, prediction and vulnerability assessment. Remote Sensing 12(20):3416
76. Hu C et al (2018) Remote sensing estimation of surface oil volume during the 2010 Deepwater Horizon oil blowout in the Gulf of Mexico: scaling up AVIRIS observations with MODIS measurements. J Appl Remote Sens 12(2):026008
77. Gruiz K (2016) Monitoring and early warning in environmental management. Site Assessment and Monitoring Tools. CRC Press, Boca Raton, FL, USA, pp 255–259
78. Popescu SM et al (2024) Artificial intelligence and IoT driven technologies for environmental pollution monitoring and management. Front Environ Sci 12:1336088
79. Yekeen ST, Balogun A-L (2020) Advances in remote sensing technology, machine learning and deep learning for marine oil spill detection, prediction and vulnerability assessment. Remote Sens 12(20):1–31
80. Fetissov M et al (2021) Next-generation smart response web (Ng-srw): an operational spatial decision support system for maritime oil spill emergency response in the gulf of Finland (baltic sea). Sustainability 13(12):6585
81. Pinkerton MH et al (2015) Reporting on the state of the New Zealand marine environment: recommendations for ocean indicators as part of the Atmospheric and Ocean Climate Change Tier 1 Statistic. 2015: Ministry for Primary Industries
82. Thodsen H et al (2017) Modelling nutrient load changes from fertilizer application scenarios in six catchments around the Baltic Sea. Agriculture 7(5):41
83. Kalinowska D et al (2020) Model of nutrient and pesticide outflow with surface water to puck bay (Southern Baltic Sea). Water 12(3):809
84. Somisetty A et al (2022) Multi-model assessment of streamflow simulations under climate and anthropogenic changes exemplified in two Indian river basins. Water 14(2):194
85. Zhang Q, Bostic JT, Sabo RD (2023) Effects of point and nonpoint source controls on total phosphorus load trends across the Chesapeake Bay watershed, USA. Environ Res Lett 19(1):014012
86. Mao Y et al (2024) Predicting ground cover with deep learning models—an application of spatio-temporal prediction methods to satellite-derived ground cover maps in the great barrier reef catchments. Remote Sensing 16(17):3193
87. Patricio-Valerio L et al (2023) Meteorological satellite observations reveal diurnal exceedance of water quality guideline thresholds in the coastal great barrier reef. Remote Sensing 15(9):2335
88. Kumari N, Pandey S (2023) Application of artificial intelligence in environmental sustainability and climate change. Visualization techniques for climate change with machine learning and artificial intelligence. Elsevier, pp 293–316
89. Samarinas N et al (2023) Synergistic use of earth observation driven techniques to support the implementation of water framework directive in europe: a review. Remote Sensing 15(8):1983

90. Sudriani Y, Sebestyén V, Abonyi J (2023) Surface water monitoring systems—the importance of integrating information sources for sustainable watershed management. IEEE Access 11:36421–36451
91. Lazaroiu C, Roscia M (2018) Smart resilient city and IoT towards sustainability of Africa. In: 2018 7th international conference on renewable energy research and applications (ICRERA). IEEE
92. Nwokolo SC et al (2023) Africa's path to sustainability: harnessing technology, policy, and collaboration. Trends Renew Energy 10(1):98–131
93. Barzelay A, Ng J, Romanoff M (2023) Governing artificial intelligence responsibility in low to middle income countries: enabling pathways to sustainable development. Cal W Int'l LJ 54:415
94. Sathya S et al (2024) Deep attention transformer nets for accurate analysis of oil spilled images to minimize pollution in the marine environment. J Intell Fuzzy Syst (Preprint), 1–13
95. Ghorbani Z, Behzadan AH (2021) Monitoring offshore oil pollution using multi-class convolutional neural networks. Environ Pollut 289:117884
96. Pandit A (2024) Deep learning and aquatic sensing reveal nitrate dynamics in the Mississippi River Basin. University of Kansas
97. Gani A et al (2023) Groundwater quality index development using the ANN model of Delhi Metropolitan City, India. Environ Sci Pollu Res, 1–16
98. Peterson KT, Sagan V, Sloan JJ (2020) Deep learning-based water quality estimation and anomaly detection using Landsat-8/Sentinel-2 virtual constellation and cloud computing. GIScience & Remote Sensing 57(4):510–525
99. Sey A, Mudongo O (2021) Case studies on AI skills capacity building and AI in workforce development in Africa. Research ICT Africa
100. Alam M et al (2024) Artificial intelligence as key enabler for safeguarding the marine resources. Artificial Intelligence and Edge Computing for Sustainable Ocean Health. Springer, pp 409–451
101. Hunter TS, Taylor M, Selvadurai N (2023) Emerging technologies in oil and gas development: Regulatory and policy perspectives. Research handbook on oil and gas law, pp 345–372
102. Rajeev PN (2023) Being an impact Chapion: enacting corporate social consciousness. Penguin Random House India Private Limited
103. Abulibdeh A, Zaidan E, Abulibdeh R (2024) Navigating the confluence of artificial intelligence and education for sustainable development in the era of industry 4.0: Challenges, opportunities, and ethical dimensions. J Cleaner Prod, 140527
104. Amankwah-Amoah J, Lu Y (2024) Harnessing AI for business development: a review of drivers and challenges in Africa. Production Planning & Control 35(13):1551–1560

Chapter 4
An Assessment of the Applications and Prospects of AI Tools in Solid Waste Management

Shashikant Nishant Sharma, Kavita Dehalwar, Sarika Jain, and Ashutosh Kumar Pandey

Abstract The chapter explores the transformative potential of artificial intelligence (AI) in enhancing the efficiency and sustainability of solid waste management (SWM) systems. As urbanization and population growth escalate waste generation, traditional waste management approaches are increasingly inadequate. This chapter examines the integration of AI technologies, such as machine learning, computer vision, and IoT, to optimize various SWM processes, including waste collection, sorting, recycling, and disposal. The objective of the chapter is to assess the current applications and prospects of AI tools in solid waste management in the global context. This study assesses current technologies, their performance across waste collection, sorting, recycling, and disposal, and future opportunities for advancement. The chapter objectives are interlinked to provide a structured exploration, beginning with a conceptual foundation of SWM challenges and AI's role in addressing them. The chapter begins by discussing the challenges faced in solid waste management, such as inefficient waste segregation, high operational costs, and environmental impacts. It then delves into how AI tools can address these issues by automating tasks, predicting waste generation patterns, and improving resource recovery rates. Case studies highlight successful implementations of AI-driven solutions in smart waste bins, robotic sorters, and route optimization for waste collection vehicles, demonstrating significant cost savings and reduced carbon footprints. Furthermore, the chapter addresses the barriers to widespread adoption of AI in SWM, including data privacy concerns, lack of infrastructure, and the need for skilled personnel. It concludes with a discussion on future prospects, emphasizing the role of AI in

S. N. Sharma (✉) · K. Dehalwar
Department of Architecture and Planning, Maulana Azad National Institute of Technology, Bhopal, India
e-mail: urp2025@gmail.com

S. Jain
Indira Gandhi Delhi Technical University for Women, Delhi, India

A. K. Pandey
JECRC University, Jaipur, Rajasthan, India

M. Nasr et al. (eds.), *Artificial Intelligence Applications for a Sustainable Environment*, Green Chemistry and Sustainable Technology,
https://doi.org/10.1007/978-3-031-91199-6_4

enabling circular economy practices and promoting sustainable urban development. This comprehensive assessment underscores the critical role of AI in revolutionizing solid waste management, paving the way for greener and more resilient cities.

Keywords Artificial intelligence · Solid waste management · Waste minimizations · Sustainable development

4.1 Introduction

Solid waste management (SWM) has emerged as one of the most significant challenges facing urban areas worldwide [1]. The escalating volume of waste generated by burgeoning populations, combined with increasing environmental awareness, necessitates the development of efficient, sustainable, and technologically advanced solutions. Urbanization and industrialization are primary drivers of the mounting waste crisis, placing immense pressure on existing waste management systems and highlighting the urgent need for innovation [2].

The adverse environmental impacts of improper waste management, such as pollution, greenhouse gas emissions, and habitat destruction, further amplify the need for transformative approaches [3]. Traditional methods, reliant on manual sorting and static waste collection systems, often prove inadequate in addressing the complexities of modern waste streams. This inadequacy has catalyzed the exploration of advanced technologies to improve operational efficiency and environmental sustainability.

Artificial intelligence (AI) has emerged as a game changer in this context, leveraging computational capabilities to address multifaceted challenges in SWM [4–6]. By automating processes, enhancing precision, and providing predictive insights, AI tools offer a plethora of applications that extend across various aspects of waste management [7–10]. These technologies have the potential to minimize human intervention, reduce costs, and create a more circular economy where resources are reused and waste generation is minimized.

One of the most promising aspects of AI in SWM lies in its ability to integrate with other cutting-edge technologies, such as the Internet of Things (IoT), robotics, and big data analytics [11–14]. This integration enables the creation of smart waste management systems that are capable of real-time monitoring, decision-making, and adaptability to changing conditions. Such advancements pave the way for a more efficient and sustainable approach to managing the waste lifecycle.

Moreover, the application of AI is not limited to technological enhancements alone; it also has significant socio-economic implications. For instance, AI-driven systems can help municipalities allocate resources more effectively, reduce operational inefficiencies, and promote community engagement through data-driven insights [15–17]. In developing regions, where waste management infrastructure often lags behind, AI can bridge critical gaps and support sustainable development goals.

This chapter delves into the current applications of AI tools in solid waste management, evaluates their effectiveness, and explores their future prospects. By examining case studies and emerging trends, the discussion aims to shed light on the transformative potential of AI in addressing one of the most pressing environmental challenges of our time.

The objectives of this chapter are structured to provide a comprehensive assessment of how artificial intelligence can enhance the efficiency, effectiveness, and sustainability of solid waste management systems. It first aims to explore the current challenges in waste management, focusing on inefficiencies in collection, sorting, recycling, and disposal processes. The subsequent part of the chapter focuses on AI tools and techniques that improve these operations by lowering operational costs, allocating resources better, and minimizing the imprint on the environment through machine learning algorithms, robotics, and predictive analytics. Following this, it covers case studies and applications of AI on real-world waste management systems. The last part of the chapter deals with future potential and hurdles of wider adoption of AI on the basis of issues such as economic feasibility, regulatory framework, and technological advance. In the context of integrating these objectives, this chapter delivers a holistic overview of AI's transformative potential to create smarter and more sustainable waste management practices while identifying key areas for future research and innovation.

4.2 Methodology

The literature review for this research on the applications and prospects of artificial intelligence (AI) tools in solid waste management (SWM) follows a systematic approach to ensure comprehensive coverage of the existing body of knowledge. The review process includes the steps shown in the figure of inclusion and exclusion criteria used as shown in Fig. 4.1.

The literature review method for gathering relevant sources on the topic of "AI in Solid Waste Management" included a systematic approach wherein the writer systematically carried out the search for relevant literature sources. The search term used was "AI in Solid Waste Management," which resulted in 336 sources. Further filtering based on the publication year (2014–2024) resulted in 277 sources. Focus narrowed it down to articles, books, book chapters, and conference papers, thereby limiting the scope to 223. Language limits then restricted it further to the use of the English language alone, thus further narrowing it to 217 in total. In order to obtain 64 sources, a limit of open-access material was introduced. These were then narrowed based on abstract relevance, yielding 48 sources, which were subjected to a detailed study to assess their alignment with the research focus, ultimately finalizing 43 highly relevant documents.

The review further synthesizes findings into a thematic summary to present the current state of AI applications in solid waste management. Some key themes such as how AI helps in improving the efficiency of the operations, making the whole thing

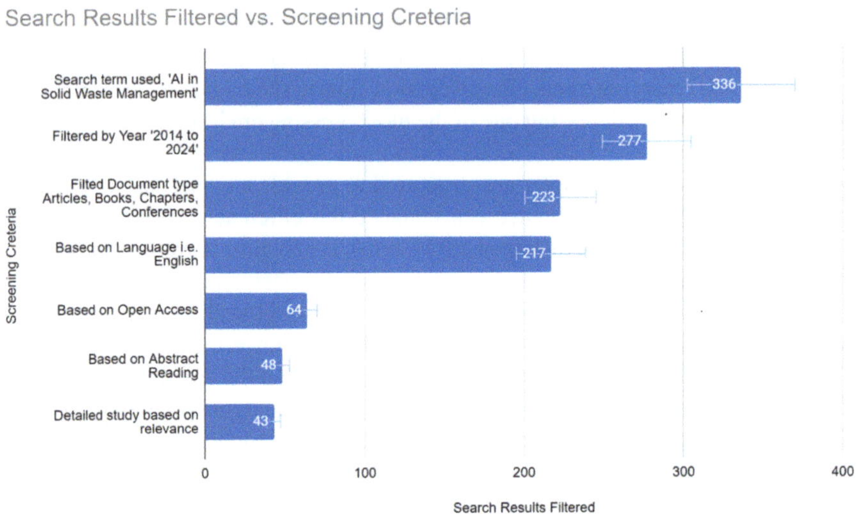

Fig. 4.1 Literature study methods adopted for research

more sustainable, improving public engagement, and getting rid of some challenges related to data and infrastructure, are covered. Finally, it ends by providing some suggestions for future research opportunities especially on accuracy in AI model development, cost feasibility, and the challenges concerning policies and regulations.

4.3 Current Applications of AI in Solid Waste Management

1. Waste Collection and Routing

AI-driven systems optimize waste collection by analyzing historical data, traffic patterns, and waste generation trends to design efficient collection routes. This approach addresses longstanding inefficiencies in waste collection logistics, leading to both cost savings and environmental benefits.

Dynamic Routing

Dynamic routing powered by AI algorithms enables municipal bodies and private companies to minimize fuel consumption and operational costs. These algorithms consider variables such as real-time traffic conditions, vehicle capacity, and waste accumulation rates. For instance, waste collection vehicles equipped with GPS and IoT sensors can dynamically adjust their routes to avoid congested areas, ensuring timely pickups while reducing fuel usage. This not only enhances service efficiency but also contributes to a significant reduction in carbon emissions.

Predictive Maintenance

AI-driven predictive maintenance systems enhance the reliability of waste collection fleets by identifying potential mechanical failures before they occur. Sensors embedded in vehicles monitor key performance indicators (KPIs), such as engine temperature, tire pressure, and oil levels. AI models analyze this data to predict when maintenance is required, thereby preventing costly breakdowns and reducing downtime. For example, predictive algorithms can flag early signs of wear and tear in garbage compactor systems, ensuring uninterrupted operations and extending vehicle lifespans.

Resource Allocation and Cost Efficiency

AI tools also aid in the strategic allocation of resources for waste collection. By analyzing patterns of waste generation across neighborhoods, these systems can determine the optimal frequency and timing of pickups [10]. This reduces instances of underfilled or overflowing bins, ensuring that resources are utilized effectively. Moreover, the automation of route planning and fleet management minimizes administrative overhead, allowing waste management teams to focus on other critical tasks.

Environmental and Social Benefits

Optimized collection systems supported by AI contribute to broader environmental and social goals. Reducing fuel consumption and emissions aligns with climate action plans, while timely and efficient waste collection improves urban cleanliness and reduces public health risks. Additionally, these systems foster transparency and accountability, as data-driven operations allow for the tracking of service quality and performance metrics.

4.4 Case Studies and Real-World Implementations

Several cities and private companies have successfully implemented AI-driven waste collection systems. For example, in Barcelona, smart waste management solutions integrate AI with IoT-enabled bins to monitor fill levels and optimize collection schedules. Similarly, companies like Rubicon leverage AI platforms to provide municipalities with actionable insights, streamlining waste collection and improving overall efficiency. These examples demonstrate the scalability and adaptability of AI technologies in diverse urban and economic contexts.

Barcelona, Spain

In Barcelona, smart waste management solutions integrate AI with IoT-enabled bins to monitor fill levels and optimize collection schedules [18]. The city employs dynamic routing systems that have significantly reduced collection costs and improved service efficiency. These innovations also support Barcelona's broader sustainability goals by minimizing vehicle emissions and enhancing waste sorting practices.

San Francisco, USA

San Francisco has implemented AI-powered systems as part of its ambitious zero-waste initiative [19]. San Francisco's zero-waste initiative has taken a giant leap forward with the introduction of AI-powered systems that leverage data analytics and machine learning to optimize waste management processes. It has analyzed real-time data from waste collection points to develop intelligent algorithms that optimize collection routes, thereby reducing fuel consumption, lowering carbon emissions, and enhancing operational efficiency. Besides this, AI is used to track and predict waste patterns, hence making better resource allocation and higher recycling rates. AI has also helped in integrating with public outreach efforts, playing a pivotal role in increasing citizen engagement and educating residents on best practices for waste sorting and encouraging participation in composting and recycling programs. It streamlined the management of waste operations while creating a community that is not only more sustainable and environmentally conscious, but is in line with the vision to ensure San Francisco attains zero waste by 2030.

Singapore

Singapore's National Environment Agency (NEA) employs AI-driven tools to monitor waste generation and optimize collection schedules [20]. Singapore's National Environment Agency has leveraged AI-powered tools to revamp its waste management system to meet the "smart nation" vision of the country. Equipped with sensors, smart waste bins collect real-time data on fill levels, thus enabling the NEA to monitor waste generation more effectively and optimize waste collection schedules. This data-driven approach allows for the dynamic allocation of resources, meaning that waste collection is carried out only when necessary, reducing unnecessary trips and operational costs. In addition, AI algorithms analyze waste trends to predict future patterns, which further enhances the efficiency and sustainability of waste management. By integrating these latest technologies, Singapore not only answers the operational inefficiency but also addresses a major portion of its footprints on the earth, thus achieving its commitment to make it cleaner, smarter, and more.

Dubai, UAE

Dubai leverages AI and IoT technologies to streamline waste collection and recycling processes [21]. The city has introduced AI-powered robotic systems for sorting recyclables, improving the accuracy and efficiency of its recycling efforts. Additionally, predictive maintenance tools are used to maintain the reliability of its waste collection fleet, reducing costs and environmental impact. Dubai has embraced AI and IoT technologies to enable tremendous improvement in its waste management systems, where the processes in both collection and recycling are optimized. Smart waste bins equipped with IoT sensors monitor fill levels in real-time, allowing the city to create data-driven collection schedules to reduce unnecessary pickups and ensure efficient resource use. It will analyze patterns in waste by forecasting future levels of waste to help optimize truck routes, using fuel and minimally emitting carbon dioxide. Recycling plants are incorporating AI to help sort items efficiently and precisely as well as making the separating of recyclables into different recyclables more perfect. Dubai achieves its broader objectives of sustainability goals through this technology integration, for a cleaner urban environment and moving along the path that will make the city a worldwide hub for smart cities.

By revolutionizing waste collection and routing, AI is not only addressing operational inefficiencies but also laying the groundwork for sustainable urban development. The integration of AI into this aspect of SWM underscores its potential to transform traditionally resource-intensive processes into streamlined, environmentally friendly systems.

1. Waste Sorting and Recycling

Advanced AI tools have revolutionized waste management, particularly in sorting and recycling processes. By integrating cutting-edge technology such as **computer vision systems** and **robotic sorting mechanisms**, cities and industries are achieving higher efficiency and sustainability. Here's an expanded discussion with real-world examples:

1. Computer Vision Systems

AI-powered **image recognition** systems analyze and classify waste with high precision. These systems use deep learning algorithms to detect different materials based on their shape, size, and texture. Cameras and sensors capture images of waste items moving along a conveyor belt. AI models trained on diverse datasets identify and categorize these items into distinct types (e.g., PET plastic bottles, aluminum cans, and cardboard).

Oslo, Norway: Oslo employs AI-driven optical sorting machines in their recycling centers. These machines achieve a near 95% sorting accuracy, significantly reducing contamination in recyclable streams [22]. Oslo, Norway, has installed AI-driven optical sorting machines in its recycling centers. These machines can sort waste materials with a level of precision unmatched by human ability, using AI algorithms and optical sensors to detect and separate recyclable materials. The technology attains nearly 95% accuracy in sorting and minimizes contamination in recyclable streams, making the recyclables of better quality and the processing more efficient. Oslo's recycling centers can handle a wide variety of the different waste types as Oslo's recycling centers make use of AI. This initiative supports the city's sustainability goals and contributes to a circular economy, thereby increasing the amount of material that can be reused or recycled.

San Francisco, USA: Recology, a waste management company, uses AI vision systems to streamline waste sorting, contributing to the city's zero-waste goal [23]. In San Francisco, the waste management company Recology has adopted AI vision systems to streamline the sorting of waste, playing a pivotal role in the city's ambitious zero-waste goal. These AI-powered systems use advanced cameras and machine learning algorithms to automatically identify and sort different types of waste, such as recyclables, compostables, and trash. By improving sorting accuracy and speed, the technology helps ensure that more materials are diverted from landfills and processed appropriately for recycling or composting. This innovation not only enhances operational efficiency but also contributes to reducing contamination in recycling streams, advancing San Francisco's efforts to achieve zero waste and promoting a more sustainable waste management system.

2. Robotic Sorting

Robots equipped with AI and machine learning capabilities enhance the speed and precision of waste sorting operations. These systems can process waste 24/7, reducing dependency on manual labor and human error. Robots with grippers or suction arms use AI to identify and pick recyclable materials from mixed waste streams. Some are even trained to recognize new materials over time, adapting to changes in waste compositions.

Tokyo, Japan: Tokyo's recycling centers utilize robotic sorting systems to handle the city's massive waste output efficiently [24]. These robots identify and separate high-value recyclables like electronics and metals. Tokyo, Japan, has the robotic sorting of recyclables as part of recycling centers to facilitate the management of large waste volumes in the city. The sorters are equipped with advanced AI and machine learning for the identification of valuable recyclable materials, for example, electronics, metals, and other valuable materials that are likely to have a high rate of reuse. The robotic systems can efficiently sort through vast volumes of waste within a very short time while maximizing the recovery of valuable materials that can be reused or recycled. This automation results in the process being much faster and accurate and saves on labor costs and decreases contamination levels within recyclable streams. The innovation of such systems enhances Tokyo's recycling processes to help in making the city achieve its sustainability vision and the transition toward the circular economy.

Amsterdam, Netherlands: A facility operated by the company Waste Robotics uses robotic arms to extract recyclables from construction and demolition waste, significantly reducing landfill contributions [25]. In Amsterdam, Netherlands, a facility operated by Waste Robotics has introduced robotic arms to handle the sorting of construction and demolition waste. The robotic arms are equipped with advanced sensors and AI algorithms that allow them to efficiently identify and extract valuable recyclables such as metals, wood, and concrete from the waste stream. By automating the process, the facility reduces its human labor input and increases accuracy in sorting so that more material is recovered for reuse and recycling. This innovation contributes to reducing waste in landfills and supports the city in sustainability efforts toward construction, which is helping close the loop of the circular economy by diverting waste from landfills and increasing the recycling rate of construction materials.

2. Waste Monitoring and Reduction

AI technologies are playing a pivotal role in transforming waste management by enabling smarter monitoring and facilitating reduction strategies. By leveraging tools such as **smart bins** and **data analytics**, cities and organizations are gaining deeper insights into waste generation patterns and are effectively addressing challenges associated with waste accumulation and disposal.

Smart Bins: Intelligent Waste Management

Smart bins are AI-enabled devices equipped with sensors and algorithms that enhance waste collection and disposal processes. These bins are designed to monitor waste levels, identify the types of waste being disposed of, and even encourage proper recycling habits.

Sensors embedded in the bins measure the volume of waste, and AI algorithms analyze the data to optimize collection schedules. Some bins are equipped with features such as automatic sorting or QR codes for user interaction.

Singapore: The city-state employs smart bins across its urban areas as part of its Smart Nation initiative [26]. These bins use sensors to notify waste management teams when they need to be emptied, reducing unnecessary trips and optimizing logistics. The Smart Nation initiative by the government of Singapore has also used this opportunity to introduce smart bins in its urban city areas. Being used on smart bins fitted with sensors that can monitor the fill level, which sends a notification when the bins are near full capacity to the waste management teams, it can help minimize the number of unwanted journeys undertaken by waste collection trucks since it optimizes collection schedules and minimizes fuel. The use of smart bins streamlined logistics in managing waste and enhanced efforts toward sustainability as it diminishes carbon footprint emissions and even operation costs. Thus, embedding this IoT solution promotes Singapore's dream to become one of the pioneering global smart cities as it encourages and brings about greater efficiency and responsible city waste management practices.

Dubai, UAE: Smart bins in Dubai not only monitor fill levels but are also integrated with solar panels to compact waste, allowing for more efficient use of space and energy [27]. In Dubai, smart bins play an important role in enhancing the waste management system in the city. They are fitted with sensors that will monitor fill levels in real-time, giving the waste management teams the signal when they should be emptied. It is different because it integrates with solar panels to power a compaction mechanism within the bins. This enables the bins to compress the waste and maximize the usage of available space, thus reducing collection trips. Therefore, the city can optimize its collection schedules, reduce fuel consumption, and increase the efficiency of its waste management operations. The innovation is consistent with Dubai's pursuit of sustainability and its vision to become a global leader in smart city technologies.

San Francisco, USA: The city has piloted smart bins equipped with AI cameras that analyze the content of waste to educate residents on proper disposal practices [28]. San Francisco made a giant stride in improving the waste management system by introducing pilot smart bins containing AI-powered cameras. These cameras scan the material in the bins in real time and identify contents that are put in the bins inappropriately or are to be recycled. The AI responds to the camera's findings, giving feedback to the residents in terms of learning proper waste disposal and encouraging good disposal habits in the community. This initiative, on one hand, helps decrease contamination in the recycling stream, while also building a culture of environmental responsibility among residents of the city. With the use of AI technology in waste management, San Francisco has been improving recycling rates and reducing landfill waste toward zero-waste goals that further make for a more sustainable urban environment.

3. Data Analytics: Insights for Waste Reduction

Waste reduction through traditional measures like uses of waste material in construction of roads [29] is now being coupled with AI-powered data analytics tools help analyze large volumes of waste-related data, uncovering trends and actionable insights. These insights guide policymakers and businesses in designing strategies to minimize waste generation at the source. AI systems collect data from sources like smart bins, IoT devices, and municipal records. This data is processed to identify patterns, such as peak waste generation times, common waste types, and recycling rates.

London, UK: The city uses AI analytics tools to assess waste generation across neighborhoods. Insights from this data have led to the implementation of waste reduction campaigns tailored to specific demographics [30]. London has incorporated artificial intelligence analytics technologies into its solid waste management programs to monitor changes in waste patterns across different neighborhoods. Its analysis of these data on produced waste helps create a better and more targeted sense of the volumes and types that are being dumped in different locales. These insights have informed the design of waste reduction campaigns tailored to specific demographics, ensuring that outreach efforts resonate with local communities. For instance, areas with high levels of single-use plastic waste may receive campaigns promoting reusable alternatives, while others may focus on improving recycling practices. This data-driven approach reduces waste and supports the broader environmental sustainability goals of London, encouraging responsible consumption and waste disposal practices throughout the city.

Seoul, South Korea: Seoul's AI-driven waste monitoring system helps track food waste patterns, enabling the city to implement a successful "pay-as-you-throw" program that has significantly reduced food waste [31]. Seoul, South Korea, has installed a novel AI-based monitoring system for waste tracking, which captures food waste patterns across the city. Analyzing the amount and types of food waste generated at source yields insights that have been instrumental for the design of its successful "pay-as-you-throw" program. The charge on residents was based on volumes of food wastes disposed of from the house to encourage residents in waste reduction practices and more environmentally friendly consumption practices. The system monitored compliance besides providing valuable feed-forward information back to the household, which makes it encourage reductions in food waste. As a consequence, Seoul achieved substantial food waste reduction with a positive outcome for the broader citywide sustainability and waste reduction plans. This program is an example of how AI can be used to address the complexities of urban waste while promoting greener living.

Copenhagen, Denmark: The city integrates AI analytics with its waste management systems to predict waste generation trends, helping optimize collection routes and reduce fuel consumption [32]. San Francisco has also integrated AI systems into its ambitious zero-waste initiative, utilizing data analytics and machine learning algorithms to improve waste management practices. The AI systems optimize the routes for waste collection by analyzing real-time data on waste levels, ensuring more efficient pickups and reducing fuel consumption. Machine learning algorithms also help identify trends in waste generation and improve recycling rates by sorting materials more effectively. Public outreach programs along with AI will have been vital for citizen engagement in San Francisco; this is going to help train residents on recycling in the correct manner and build sustainability. Therefore, these modernizations are believed to reduce the landfill waste while recycling and working toward a city of zero wastes by 2030.

4. Landfill Management

Artificial intelligence (AI) is transforming landfill management by providing tools to monitor environmental parameters, optimize operations, and ensure compliance with regulatory standards. By integrating AI-driven technologies, cities and organizations can minimize the environmental impact of landfills and enhance sustainability through predictive analytics and real-time monitoring.

Leachate Management: Preventing Environmental Contamination

Leachate, the liquid that drains from landfills, poses significant environmental risks if not managed properly. AI tools predict leachate generation patterns, allowing landfill operators to design and implement effective mitigation systems. AI models analyze variables such as rainfall, waste composition, and landfill design to forecast the volume and composition of leachate. These insights guide the construction and operation of leachate collection and treatment systems.

Berlin, Germany: Berlin employs AI-driven leachate monitoring systems in its landfills [33]. These systems predict leachate accumulation based on weather and waste input data, ensuring timely collection and treatment. Berlin, Germany, has introduced AI-based leachate monitoring in its landfills to improve waste management and protect the environment. The AI algorithm predicts when the leachate will accumulate through weather conditions and the type of waste being disposed of. After processing data regarding rainfall, temperature, and composition of waste, the AI-based systems make correct predictions about when leachate would reach critical levels, thus helping in its timely collection and treatment. This proactive method avoids contamination of soil and groundwaters around it. The landfills are hence designed to follow all environmental safety standards. Berlin using AI in managing the landfills showcases the will and commitment toward adopting technology for reducing the environmental damage due to improper waste disposal in its city.

Bangalore, India: AI is used to monitor leachate levels and quality in the city's landfill sites [34]. The insights have helped improve the efficiency of treatment plants, reducing the risk of groundwater contamination. In Bangalore, India, AI is used to monitor the quantity and quality of leachate in the city's landfills, providing critical insights into waste management and environmental protection. The city can track the amounts of leachate that are being produced and predict potential risks if there are changes in the weather and composition of waste through the use of cutting-edge AI algorithms. This data-based strategy has resulted in making treatment plants highly efficient and thus ensures leachate is collected and treated before it is able to become harmful to the surrounding environment. As a consequence, there is lesser chances of contaminating the ground water and helps Bangalore handle its landfills more efficiently with reduced public health hazards as well as to the environment. It depicts how AI can be applied for improvement of waste management strategies in cities.

5. Methane Emission Monitoring: Supporting Energy Recovery and GHG Mitigation

Landfills are a significant source of methane, a potent greenhouse gas. AI technologies enable the continuous monitoring and analysis of methane emissions, supporting both environmental compliance and energy recovery efforts. AI models use data from sensors and satellite imagery to detect methane leaks, quantify emissions, and predict future trends. These tools help operators capture methane for use in energy production while reducing the overall greenhouse gas footprint.

Toronto, Canada: The Green Lane Landfill uses AI systems to monitor methane emissions and optimize gas capture for renewable energy generation. This initiative powers nearby communities and reduces greenhouse gas emissions [35]. In Toronto, Canada, at the Green Lane Landfill, AI systems monitor methane emissions and optimize gas capture to support the city's effort to convert waste into renewable energy. The AI technology tracks methane levels and adjusts the gas collection system to ensure maximum efficiency in capturing and utilizing the gas for energy generation. This process not only helps power local communities but also greatly reduces the release of greenhouse gases, toward achieving Toronto's sustainability goals. With AI added to landfill management, the city does not just reduce environmental impacts but also creates energy from what once was viewed as waste—while providing another excellent example of an application of AI in waste-to-energy projects.

Copenhagen, Denmark: Landfills in Copenhagen leverage AI-powered drones equipped with methane-detecting sensors to identify leaks and enhance the efficiency of gas recovery systems [36]. In Copenhagen, Denmark, landfills are using AI-powered drones with methane-detecting sensors to enhance environmental monitoring and gas recovery efforts. The drones are deployed to fly over landfill sites, detecting methane leaks that might otherwise go unnoticed. The AI system processes the sensor data in real-time, pinpointing areas where gas emissions are higher than expected, allowing for prompt intervention. The city can optimize its gas recovery systems, improving the efficiency of methane capture and reducing the environmental impact of landfill operations by identifying leaks quickly. This innovative approach not only helps reduce greenhouse gas emissions but also supports Copenhagen's broader sustainability goals, contributing to cleaner air and a more sustainable waste management system.

San Jose, USA: The Newby Island landfill uses AI-based analytics to monitor and mitigate methane emissions while generating biogas for energy use [37]. At the Newby Island landfill in San Jose, USA, AI-based analytics is very vital in the tracking and curbing of emissions of methane while concurrently generating energy from biogas. The AI system continuously analyses data from sensors positioned at various points within the landfill to monitor methane levels in order to trace the actual source of emissions. This helps optimize the capturing of biogas more effectively by predicting the accumulation of methane. It later utilizes captured methane as renewable energy sources, helping run local communities with minimal damage caused to the environment from landfills. Thus, through such creative application of AI, San Jose is spearheading its journey into sustainability by saving on the overall emission of greenhouses, creating a by-product that serves to supply energy from this waste material.

BOD (Biochemical Oxygen Demand), COD (Chemical Oxygen Demand), and Color Removal

In the context of solid waste management, the assessment of BOD (Biochemical Oxygen Demand), COD (Chemical Oxygen Demand), and color removal is crucial for evaluating the effectiveness of waste treatment processes. These parameters are indicative of the organic matter, pollutants, and aesthetic impact of waste water, which are vital concerns in the management of waste. BOD and COD are both measures of the oxygen consumption required for the biological and chemical degradation of organic pollutants, respectively, with lower values signifying better treatment outcomes. Color removal is also an essential aspect, as it affects the visual quality of treated effluents. The integration of artificial intelligence (AI) tools in solid waste management has shown promising prospects in enhancing the efficiency of waste treatment processes. AI-powered systems can optimize the treatment conditions, predict contamination levels, and enhance the real-time monitoring of BOD, COD, and color removal, leading to more effective and sustainable waste management practices. Through AI-driven models, waste treatment facilities can improve their ability to meet regulatory standards, reduce operational costs, and enhance environmental sustainability, thus providing a holistic solution to modern waste management challenges.

Visualizations done to help readers understand the efficiency of the BOD (Biochemical Oxygen Demand), COD (Chemical Oxygen Demand), and color removal as shown in Fig. 4.2. The BOD removal efficiency has a mean value of

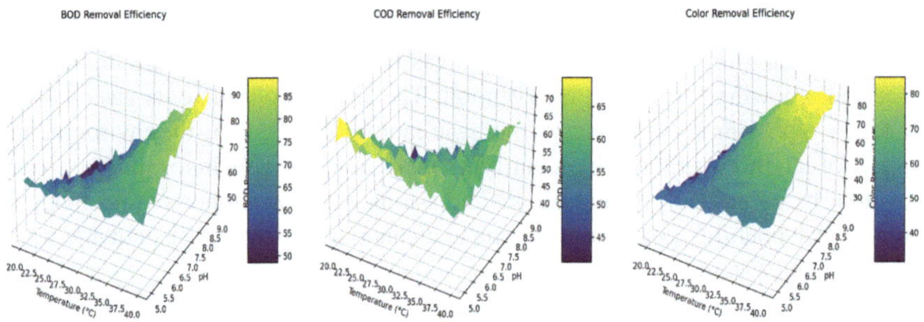

Fig. 4.2 BOD (Biochemical Oxygen Demand), COD (Chemical Oxygen Demand), and color removal (Visualization based on secondary data from different sources)

68.19%, with a standard deviation of 9.71%, a minimum of 45.82%, and a maximum of 91.21% [38]. For COD removal efficiency, the mean is 57.24%, the standard deviation is 6.32%, with a minimum value of 38.80% and a maximum value of 72.03% [39]. The color removal efficiency shows a mean of 59.62%, a standard deviation of 13.80%, with the minimum value recorded at 24.52% and the maximum at 88.11% [39]. These statistics provide a clear overview of the variation in removal efficiencies for BOD, COD, and color.

4.5 Prospects of AI in Solid Waste Management

1. Enhanced Circular Economy Initiatives

Artificial intelligence (AI) holds immense potential to drive circular economy goals by enhancing recycling rates and promoting efficient resource recovery. By analyzing life-cycle data, AI can aid in designing products optimized for easier recycling, ensuring that materials can be reused or repurposed effectively. This innovative approach not only reduces waste but also extends the life span of valuable resources, aligning with the principles of a circular economy. The Internet of Things (IoT) has a crucial role in assessing and optimizing waste management technologies that are contributing to the circular economy through improvements in efficiency, data collection, and real-time monitoring across the various stages. There are many types of IoT used in different stages of implementing waste management. Smart Sensors are integrated in waste collection and segregation services, which indicate real-time statistics on waste collection levels, temperatures, and moisture levels, all of which can further optimize the collection schedule and result in proper and accurate segregation. IoT-enabled bins with a sensor can report the volume of the waste, generate alerts when full, and significantly reduce unnecessary wasteful trips or movements, allowing for more productive operations. Tracking and Monitoring Systems, utilizing GPS and RFID technology, allow for the real-time tracking of waste transport, helping in efficient routing and ensuring proper disposal or recycling. Additionally, IoT-based recycling systems monitor the status and quality of recyclables, assisting in sorting and processing materials effectively. The recycling phase is powered by IoT technologies that use sensors in Smart Sorting Machines to identify materials for separation, thus enhancing the quality of the recycling stream. Such IoT-enabled systems contribute to a more sustainable, efficient, and transparent waste management process that promotes the principles of a circular economy through reducing waste, reusing materials, and recycling efficiently.

Future AI tools could also play a critical role in stabilizing supply chains by predicting market demand for recycled materials. This capability allows industries to anticipate and meet requirements for secondary raw materials, reducing reliance on virgin resources. By integrating AI-driven insights into recycling processes and material recovery strategies, organizations can build sustainable systems that support environmental and economic resilience.

2. Integration with IoT and Smart Cities

The integration of artificial intelligence (AI) with Internet of Things (IoT) devices and smart city infrastructure offers transformative possibilities for solid waste management (SWM). By leveraging real-time data from IoT-enabled sensors, AI can facilitate efficient waste tracking and management. This allows municipalities and waste management companies to monitor waste levels, optimize collection schedules, and reduce unnecessary trips, resulting in cost savings and a lower environmental footprint.

Additionally, autonomous vehicles powered by AI can revolutionize waste collection and sorting processes. These vehicles can navigate urban environments to collect waste efficiently, while advanced AI algorithms can sort recyclable materials with precision, minimizing contamination, and maximizing resource recovery. Together, these technologies pave the way for smarter, greener, and more sustainable waste management systems in urban areas.

3. Public Awareness and Behavioral Change

AI-powered platforms have the potential to actively engage communities in adopting sustainable practices, making them key drivers of environmental awareness and action. For instance, gamified educational apps powered by AI can motivate users to practice proper waste segregation by turning it into an interactive and rewarding experience. These apps can provide real-time feedback, quizzes, and challenges, fostering a sense of responsibility while making sustainability engaging and fun.

Additionally, AI can offer personalized recommendations tailored to individual waste habits. By analyzing data such as consumption patterns and disposal behaviors, AI can suggest actionable steps, such as reducing single-use plastics, composting organic waste, or adopting reusable alternatives. These targeted interventions empower individuals to make meaningful contributions toward sustainability while fostering a culture of environmental stewardship within communities.

4. Policy Formulation and Compliance

AI tools can significantly enhance regulatory compliance and policymaking in waste management by providing data-driven insights and streamlining processes. For example, AI can simulate the environmental impacts of proposed waste management policies, enabling policymakers to evaluate potential outcomes before implementation. This predictive capability supports the development of strategies that maximize environmental benefits while minimizing economic and social costs.

Additionally, AI can automate compliance reporting, reducing administrative burdens for organizations and ensuring adherence to regulations. By analyzing operational data, AI tools can generate accurate and timely reports, identify areas of non-compliance, and recommend corrective actions. This not only increases efficiency but also fosters transparency and accountability in waste management systems, paving the way for more robust and evidence-based policymaking (Table 4.1).

Artificial intelligence has been applied across different stages of solid waste management, and this has been very promising in efficiency and sustainability. For instance, the generation stage in waste can use predictive models like random forest and support vector machine to predict the volumes of wastes with a 95% accuracy, which enables better resource planning through smart bin systems with IoT-based

Table 4.1 Optimum conditions and model validation for waste management

Parameter	Optimal condition	AI Model/ Technique	Performance metric	Application example	References
Route optimization	Minimized travel distance and time	Genetic algorithm, A* Algorithm	Reduced fuel consumption (20%)	Dynamic routing for urban waste collection	[40]
Waste volume prediction	High accuracy (95% prediction accuracy)	Random forest, support vector machine	RMSE: 0.02	Smart bin systems with IoT-based monitoring	[41]
Collection frequency	Adaptive scheduling based on fill-level sensors	Reinforcement learning	Improved collection efficiency	Automated scheduling for dynamic waste pickup	[42]
Fleet utilization	Maximized vehicle usage with reduced downtime	Linear programming	Cost savings of 15%	Allocation of trucks for large metropolitan areas	[42, 43]
Environmental impact	CO2 emissions reduction	Neural networks	Emissions reduced by 18%	Eco-friendly routing for waste transportation	[44, 45]
Recycling	Materials separation	Convolutional neural networks	98%	Materials separate and recycling	[43]

monitoring [41]. For waste collection, the use of Genetic and A* algorithms in route optimization reduces travel distances and fuel consumption by up to 20% [46], while adaptive scheduling based on fill-level sensors, powered by reinforcement learning, enhances collection efficiency [42]. In transportation, linear programming models maximize fleet utilization, achieving cost savings of 15% by optimizing vehicle usage in large urban areas [43]. AI-based waste treatment uses convolutional neural networks for material separation with 98% classification results in recycling centers [47]. Finally, environmental concerns due to the contribution of neural networks to green routing where CO_2 emissions are reduced by 18% [45]. All these AI applications will form a whole system that can help optimize SWM operations, reducing costs, and inviting healthy waste handling practices.

4.6 Challenges and Considerations

While the benefits of AI in SWM are manifold, several challenges remain:

- **High Initial Costs**: Deploying AI systems requires substantial investment in hardware, software, and training. One significant challenge in adopting AI systems is the high initial cost, which includes substantial investments in hardware, specialized software, and workforce training. Organizations, particularly in developing regions, may find it difficult to allocate the necessary resources for these expenses. This financial barrier can delay or limit the implementation of AI-driven solutions, hindering their potential to address critical issues like waste management and sustainability.
- **Data Privacy Concerns**: The collection and processing of waste data may raise privacy issues. Data privacy concerns pose a significant challenge in the adoption of AI-driven waste management solutions. The collection and processing of waste-related data, particularly in residential areas, may inadvertently capture sensitive information about individuals or households. Ensuring robust data protection measures and compliance with privacy regulations is essential to address these concerns and build public trust in AI systems.
- **Technological Integration**: Achieving seamless integration with existing infrastructure is complex. Technological integration is another critical challenge in implementing AI systems for waste management. Many existing infrastructures are not designed to accommodate advanced technologies, making seamless integration complex and costly. Upgrading or retrofitting legacy systems to work with AI-powered tools often requires significant time, expertise, and resources, which can slow down adoption and limit the effectiveness of these innovations.
- **Skill Gaps**: Workforce training is necessary to operate and maintain AI systems effectively. Skill gaps represent a significant barrier to the successful implementation of AI systems. Many organizations lack a workforce equipped with the technical expertise required to operate, manage, and maintain advanced AI technologies. Addressing this challenge necessitates comprehensive training

programs and capacity-building initiatives, which can be time-consuming and resource-intensive, further complicating the adoption process.

4.7 Study Limitations and Recommendation

This study is limited to the literature review of the published research around the world on the topic of the AI applications in solid waste management. Despite the promise of artificial intelligence in solid waste management, various limitations have to be addressed to apply AI at an expansive scale effectively. The absence of standard high-quality data used for training AI models hampers the development of precise and dependable predictive systems. Data heterogeneity among different areas, inconstant categorization procedures of wastes, and poor deployment of sensors degrades the AI-driven solution's performance. Additionally, the economic feasibility of AI technologies remains a challenge, especially in low- and middle-income countries where financial resources for advanced infrastructure and technology investments are limited. High initial costs for implementing AI-based systems, including smart bins, IoT devices, and machine learning platforms, often deter adoption despite long-term cost-saving potential.

Future research directions to address the abovementioned challenges are through improvement in data collection frameworks that would involve integrating IoT-enabled sensors with centralized data management systems for sharing standardized data. Cooperation among governments, technology providers, and research institutions would play a pivotal role in providing cost-effective and scalable AI-based solutions suited to the specific local waste management requirements. Investment in training programs and capacity-building initiatives for waste management professionals will also promote AI literacy and allow more effective technology deployment. Policymakers should encourage public–private partnerships that can help reduce the financial burden on municipalities and promote innovation in sustainable waste management.

From an economic standpoint, AI tools significantly contribute to reducing operational costs, optimizing resource allocation, and increasing efficiency in collection, sorting, and recycling. Route optimization algorithms reduce fuel consumption and travel time, while predictive maintenance reduces the cost of fleet management. Automated sorting systems powered by AI reduce labor costs and increase recycling rates. In developed countries, these technologies have shown savings of up to 20% in urban waste collection. In developing countries, AI can be used to enhance service delivery within limited budgets by providing affordable, context-specific solutions that focus on high-impact areas. By integrating AI in SWM practices, national governments will promote circular economies, minimize environmental impacts, and achieve long-term economic sustainability. Overcoming these limitations and playing the economic potential of AI will accelerate innovation and efficiency in the global waste management system.

4.8 Conclusion

AI tools represent the future of solid waste management in a par today of paradigm shifts in providing unique solutions for complex problems. From route optimization in collections to enhancing recycling efficiency and environmental sustainability, the list is endless. To achieve these ends, however, means working through key implementation challenges, collaboration with the stakeholders, and investments in research and development. As cities grow, and waste management challenges escalate, the need for AI to generate sustainable, efficient, and scalable solutions will only intensify.

Future work should consider exploring the socio-economic feasibility of using machine learning models in raising public awareness about sustainable waste management practices. Specifically, using AI-based learning platforms can bring to the fore the environmental and economic advantages of using ANN-based algorithms to prevent the movement of harmful landfill leachate into rivers and oceans. Such efforts can promote proactive measures, enhancing both policy compliance and public engagement. Furthermore, cost–benefit analyses of AI deployment in low- and middle-income countries should be conducted to identify scalable solutions that balance technological sophistication with affordability. In addressing these emerging areas, AI can continue to drive transformative advances in waste management, fostering resilience and sustainability on a global scale.

Funding Statement There is no funding for this research.

Declarations

Consent to Publish The authors give their consent for the publication of this manuscript in book and confirm that it is original and does not infringe any existing copyrights.

Other Declarations Conflicts of Interest: The authors declare no conflict of interest.

References

1. Sharma SN, Dehalwar K, Singh J (2024) Emerging techniques of solid waste management for sustainable and safe living environment. In: Nasr M, Negm A (eds) Solid waste management. Sustainable development goals series. Cham: Springer Nature Switzerland, pp 29–51. https://doi.org/10.1007/978-3-031-60684-7_3
2. Kumar G, Vyas S, Sharma SN, Dehalwar K (2024) Challenges of environmental health in waste management for Peri-urban Areas. In: Nasr M, Negm A (eds) Solid waste management. Sustainable development goals series. Cham: Springer Nature Switzerland, pp 149–168. https://doi.org/10.1007/978-3-031-60684-7_9
3. Shen M, Huang W, Chen M, Song B, Zeng G, Zhang Y (2020) (Micro)plastic crisis: Un-ignorable contribution to global greenhouse gas emissions and climate change. J Clean Prod 254:120138. https://doi.org/10.1016/j.jclepro.2020.120138
4. Alshaikh R, Abdelfatah A (2024) Optimization techniques in municipal solid waste management: a systematic review. Sustain Switz 16(15). https://doi.org/10.3390/su16156585

5. Bui T-D, Tseng J-W, Tseng M-L, Wu K-J, Lim MK (2023) Municipal solid waste management technological barriers: a hierarchical structure approach in Taiwan. Resour Conserv Recycl 190. https://doi.org/10.1016/j.resconrec.2022.106842
6. Lin L, Xu F, Ge X, Li Y (2018) Improving the sustainability of organic waste management practices in the food-energy-water nexus: a comparative review of anaerobic digestion and composting. Renew Sustain Energy Rev 89:151–167. https://doi.org/10.1016/j.rser.2018.03.025
7. Faruqi MHZ, Siddiqui FZ (2020) A mini review of construction and demolition waste management in India. Waste Manag Res J Sustain Circ Econ 38(7):708–716. https://doi.org/10.1177/0734242X20916828
8. Koul B, Yakoob M, Shah MP (2022) Agricultural waste management strategies for environmental sustainability. Environ Res 206:112285. https://doi.org/10.1016/j.envres.2021.112285
9. Pinhal Luqueci Thomaz I, Fernando Mahler C, Pereira Calôba L (2023) Artificial intelligence (AI) applied to waste management: a contingency measure to fill out the lack of information resulting from restrictions on field sampling. Waste Manag Bull 1(3):11–17. https://doi.org/10.1016/j.wmb.2023.06.002
10. Rawat S, Goel V, Chowdhary A (2022) Efficient municipal solid waste management: a case study of Dehradun city. In: Singh VP, Li Z, Siddiqui NA, Patel H (eds) Sustainable infrastructure development, vol 199. Lecture notes in civil engineering, vol 199. Singapore: Springer Singapore, pp 221–230. https://doi.org/10.1007/978-981-16-6647-6_21
11. Bibri SE (2020) The eco-city and its core environmental dimension of sustainability: green energy technologies and their integration with data-driven smart solutions. Energy Inform 3(1):4. https://doi.org/10.1186/s42162-020-00107-7
12. Hui CX, Dan G, Alamri S, Toghraie D (2023) Greening smart cities: an investigation of the integration of urban natural resources and smart city technologies for promoting environmental sustainability. Sustain Cities Soc 99:104985. https://doi.org/10.1016/j.scs.2023.104985
13. Manea A, Dolci G, Grosso M (2024) Life cycle assessment and cost analysis of an innovative automatic system for sorting municipal solid waste: a case study at Milan Malpensa airport. Waste Manag 183:63–73. https://doi.org/10.1016/j.wasman.2024.04.049
14. Namoun A, Tufail A, Khan MY, Alrehaili A, Syed TA, BenRhouma O (2022) Solid waste generation and disposal using machine learning approaches: a survey of solutions and challenges. Sustain Switz 14(20). https://doi.org/10.3390/su142013578
15. Berawi MA, Suwartha N, Salim AV, Saroji G, Sari M (2021) Developing mobile application for land value capture scheme to finance urban rail transit projects. Int J Technol 12(7):1448–1457. https://doi.org/10.14716/IJTECH.V12I7.5332
16. Thomaz IPL, Mahler CF (2024) Artificial neural network models to support urban waste management: a technological resource that drives the achievement of Sustainable Development Goals. Eng Sanit E Ambient 29. https://doi.org/10.1590/S1413-415220230085
17. Xu X, Wang S, Li J, Qiao T (2024) Environmental regulatory intensity, green finance and corporate green sustainable development performance. Heliyon 10(9):e30114. https://doi.org/10.1016/j.heliyon.2024.e30114
18. Gulzar B, Sofi SA, Sholla S (2024) Exploring personalized internet of things (PIoT), social connectivity, and artificial social intelligence (ASI): a survey. High-Confid. Comput. 4(3):100242. https://doi.org/10.1016/j.hcc.2024.100242
19. Okonta DE, Vukovic V (2024) Smart cities software applications for sustainability and resilience. Heliyon 10(12):e32654. https://doi.org/10.1016/j.heliyon.2024.e32654
20. Agamuthu P, Babel S (2023) Waste management developments in the last five decades: Asian perspective. Waste Manag Res J Sustain Circ Econ 41(12):1699–1716. https://doi.org/10.1177/0734242X231199938
21. Noori N, Hoppe T, De Jong M (2020) Classifying pathways for smart city development: comparing design, governance and implementation in Amsterdam, Barcelona, Dubai, and Abu Dhabi. Sustainability 12(10):4030. https://doi.org/10.3390/su12104030
22. Malinauskaite J et al (2017) Municipal solid waste management and waste-to-energy in the context of a circular economy and energy recycling in Europe. Energy 141:2013–2044. https://doi.org/10.1016/j.energy.2017.11.128

23. Ejimofor MI, Aniagor CO, Oba SN, Menkiti MC, Ugonabo VI (2022) Artificial intelligence in the reduction and management of land pollution. In: Current trends and advances in computer-aided intelligent environmental data engineering, Elsevier, pp 319–333. https://doi.org/10.1016/B978-0-323-85597-6.00009-4
24. Cheng T, Kojima D, Hu H, Onoda H, Pandyaswargo AH (2024) Optimizing waste sorting for sustainability: an AI-powered robotic solution for beverage container recycling. Sustainability 16(23):10155. https://doi.org/10.3390/su162310155
25. Setaki F, Van Timmeren A (2022) Disruptive technologies for a circular building industry. Build Environ 223:109394. https://doi.org/10.1016/j.buildenv.2022.109394
26. Huseien GF, Shah KW (2022) A review on 5G technology for smart energy management and smart buildings in Singapore. Energy AI 7:100116. https://doi.org/10.1016/j.egyai.2021.100116
27. Al-Shihabi S, Aydin R, Al Nahlawi M (2023) Forecasting the effects of municipal solid plastic waste generation and streaming using system dynamics: a case study in Dubai. Sci Total Environ 897:165204. https://doi.org/10.1016/j.scitotenv.2023.165204
28. Demirel D (2023) The impact of managing diversity on building the smart city a comparison of smart city strategies: cases from Europe, America, and Asia. SAGE Open 13(3):21582440231184972. https://doi.org/10.1177/21582440231184971
29. Sharma SN, Lodhi AS, Dehalwar K, Jaiswal A (2024) Life cycle assessment (LCA) of recycled & secondary materials in the construction of roads. IOP Conf Ser Earth Environ Sci 1326(1):012102. https://doi.org/10.1088/1755-1315/1326/1/012102
30. Minhas N, Gerassimidou S, Iacovidou E (2023) Residual waste management in London, England: a reality check. Environ Monit Assess 195(11):1283. https://doi.org/10.1007/s10661-023-11760-2
31. Zamri GB et al (2020) Delivery, impact and approach of household food waste reduction campaigns. J Clean Prod 246:118969. https://doi.org/10.1016/j.jclepro.2019.118969
32. Elassy M, Al-Hattab M, Takruri M, Badawi S (2024) Intelligent transportation systems for sustainable smart cities. Transp Eng 16:100252. https://doi.org/10.1016/j.treng.2024.100252
33. Mepaiyeda S, Madi K, Gwavava O, Baiyegunhi C (2020) Geological and geophysical assessment of groundwater contamination at the Roundhill landfill site, Berlin, Eastern Cape, South Africa. Heliyon 6(7):e04249. https://doi.org/10.1016/j.heliyon.2020.e04249
34. Wdowczyk A, Szymańska-Pulikowska A, Gupta A (2024) Application of selected indicators to assess contamination of municipal landfill leachate and its impact on groundwater. Water Resour Ind 32:100265. https://doi.org/10.1016/j.wri.2024.100265
35. Zhang S, Ma J, Zhang X, Guo C (2023) Atmospheric remote sensing for anthropogenic methane emissions: Applications and research opportunities. Sci Total Environ 893:164701. https://doi.org/10.1016/j.scitotenv.2023.164701
36. Ashik AS et al (2024) Integrated-path multi-gas sensor using near-infrared diode lasers: An alternative to vehicle-driven point gas analyzer. Sens Actuators B Chem 412:135855. https://doi.org/10.1016/j.snb.2024.135855
37. Ayasse AK et al (2019) Methane mapping with future satellite imaging spectrometers. Remote Sens 11(24):3054. https://doi.org/10.3390/rs11243054
38. Hami ML, Al-Hashimi MA, Al-Doori MM (2007) Effect of activated carbon on BOD and COD removal in a dissolved air flotation unit treating refinery wastewater. Desalination 216(1–3):116–122. https://doi.org/10.1016/j.desal.2007.01.003
39. Pourgholi M, Masoomi Jahandizi R, Miranzadeh M, Hassan Beigi O, Dehghan S (2018) Removal of dye and COD from textile wastewater using AOP (UV/O3, UV/H2O2, O3/H2O2 and UV/H2O2/O3). J Environ Health Sustain Dev 621–629. https://doi.org/10.18502/jehsd.v3i4.223
40. Kona VVS, Subramoniam M (2024) A smart Iot-based waste management system using vehicle shortest path routing and trashcan visiting decision making based on deep convolutional neural network. Peer-Peer Netw Appl 17(3):1051–1074. https://doi.org/10.1007/s12083-024-01623-z
41. Fang B et al (2023) Artificial intelligence for waste management in smart cities: a review. Environ Chem Lett 21(4):1959–1989. https://doi.org/10.1007/s10311-023-01604-3

42. Xia W, Jiang Y, Chen X, Zhao R (2022) Application of machine learning algorithms in municipal solid waste management: a mini review. Waste Manag Res J Sustain Circ Econ 40(6):609–624. https://doi.org/10.1177/0734242X211033716
43. Alberdi E, Urrutia L, Goti A, Oyarbide-Zubillaga A (2020) Modeling the municipal waste collection using genetic algorithms. Processes 8(5):513. https://doi.org/10.3390/pr8050513
44. Cha G-W, Choi S-H, Hong W-H, Park C-W (2022) Development of machine learning model for prediction of demolition waste generation rate of buildings in redevelopment areas. Int J Environ Res Public Health 20(1):107. https://doi.org/10.3390/ijerph20010107
45. Ma D, Zhang S, Zhao J (2022) The high-speed railway opening and audit fees: evidence from China. Sustainability 14(20):13353. https://doi.org/10.3390/su142013353
46. Karadimas NV, Papatzelou K, Loumos VG (2007) Genetic algorithms for municipal solid waste collection and routing optimization. In: Boukis C, Pnevmatikakis A, Polymenakos L (eds) Artificial Intelligence and Innovations 2007: from Theory to Applications, vol 247. IFIP The International Federation for Information Processing, vol 247. Boston, MA: Springer US, pp 223–231. https://doi.org/10.1007/978-0-387-74161-1_24
47. Yang L et al (2020) On the utilization of waste fried oil as flotation collector to remove carbon from coal fly ash. Waste Manag 113:62–69. https://doi.org/10.1016/j.wasman.2020.05.045

Chapter 5
Green Intelligence for Sustainable Plastic Waste Management in Africa: A Comprehensive Framework for Policy Innovation, ICT Solutions, and Public–Private Partnerships

Azubuike Victor Chukwuka, Chukwudi Nwabuisiaku, Ayotunde Daniel Adegboyegun, and Aina O. Adeogun

Abstract The growing plastic waste crisis in Africa demands innovative and sustainable management strategies. This review explores the concept of green intelligence, which integrates information and communication technologies (ICTs) with environmental management practices, as a viable solution to Africa's plastic waste challenges. Green intelligence facilitates data-driven decision-making, promotes technological innovations, and enhances policy development, making it an essential tool for addressing a continent's waste management issues. Using a rapid review approach, this study analyses the literature on plastic waste management in Africa, focusing on policy frameworks, community engagement, and technological advancements. The review also introduces a strategic framework that incorporates green intelligence into Africa's waste management systems. This framework emphasizes the importance of collaboration among governments, businesses, and communities, integrating technology, policy reforms, and behavioral change to improve plastic waste management outcomes. Key findings suggest that green intelligence holds significant potential for transforming waste management in Africa by enabling efficient recycling technologies, enhancing waste-to-energy conversion, and promoting environmental awareness. Countries with higher green intelligence scores tend to have more effective waste management practices, underscoring the value of technology and data-driven policies in fostering sustainability. The review concludes that actionable policies, such as plastic bans, extended producer responsibility programs, and community engagement initiatives, are critical for successful waste management. It

A. V. Chukwuka (✉) · C. Nwabuisiaku · A. D. Adegboyegun
National Environmental Standards and Regulations Enforcement Agency (NESREA), Abuja, Nigeria
e-mail: zubbydoo@gmail.com

A. O. Adeogun
Hydrobiology and Fisheries Unit, Department of Zoology, University of Ibadan, Ibadan, Nigeria

M. Nasr et al. (eds.), *Artificial Intelligence Applications for a Sustainable Environment*, Green Chemistry and Sustainable Technology,
https://doi.org/10.1007/978-3-031-91199-6_5

calls for a coordinated effort among governments, the private sector, civil society, and international organizations to implement green intelligence-driven strategies. This collaboration is essential to mitigate the growing plastic waste problem and ensure sustainable environmental practices across the African continent.

Keywords Green intelligence · Plastic waste management · Recycling · Africa · Sustainable development

5.1 Introduction

Plastic waste has become one of the most pressing global environmental challenges, with the world producing up to 430 million tonnes annually [1]. This staggering volume leads to substantial environmental degradation, as large amounts of plastic waste enter landfills and marine environments [2, 3]. According to recent projections, plastic pollution in oceans could outweigh that of fish by 2050 [4, 5]. Single-use plastics, such as packaging and bottles, account for the bulk of this waste, posing serious ecological and public health risks because of their long-lasting presence in ecosystems [6].

Africa faces significant challenges in managing plastic waste, driven by rapid urbanization, growing populations, and the increasing consumption of single-use plastics [7]. Despite contributing only approximately 5% to global plastic production, Africa disproportionately bears the brunt of plastic pollution due to underdeveloped waste collection and recycling systems [8]. Inadequate infrastructure exacerbates these issues, with some African cities collecting less than 50% of their waste and less than 12% being recycled, leading to environmental degradation and public health concerns [9]. For example, it is estimated that only 44% of waste is collected across many African cities, with less than 10% of this waste being recycled [10]. The informal waste sector, crucial to waste management in Africa, often operates under hazardous conditions with limited government support, which impacts the efficiency and effectiveness of waste management [11]. Furthermore, improper disposal practices lead to significant environmental problems, such as plastic accumulation in rivers and coastal areas, which degrades ecosystems and threatens marine life [2, 12].

Africa's policy landscape adds complexity to its plastic waste management crisis. Although countries such as Kenya and Rwanda have taken bold steps with single-use plastic bans, enforcement remains inconsistent, and broader, integrated waste management strategies are lacking [13, 14]. In Kenya, despite the 2017 ban being one of the strictest globally, challenges such as insufficient monitoring and persistent illegal plastic usage limit its effectiveness [15]. Moreover, the rapid rise in plastic consumption across Africa, which increased by 95% between 2010 and 2020, has outpaced infrastructure development, contributing to widespread plastic waste mismanagement [16]. Additionally, the rapid growth in plastic consumption, which increased by 92% from 2010 to 2020, outpaced the development of waste management infrastructure [17], contributing to a cycle of ineffective waste management, environmental degradation, and persistent pollution challenges. The combination of inadequate infrastructure, weak policy enforcement, and low public awareness creates a formidable barrier to addressing the plastic waste crisis effectively in Africa.

Significance of Green Intelligence

As Africa confronts its growing plastic waste crisis, green intelligence (GI) has emerged as a pivotal, data-centric solution. GI leverages advanced information and communication technologies (ICTs) alongside cutting-edge analytics to enhance waste management systems [18]. This approach, through tools such as geographic information systems (GISs) and sensor technology, enables real-time monitoring of waste generation, recycling efficiency, and pollution hotspots, offering vital insights for policymakers [19, 20]. By optimizing waste collection and promoting recycling initiatives, GI not only improves operational efficiency but also supports the transition toward circular economy practices; as such, by utilizing tools such as geographic information systems, sensors, and smart waste management technologies, GI provides real-time data on waste generation, recycling rates, and pollution hotspots [21, 22]. This data-driven framework enables policymakers to develop targeted interventions, optimize waste collection, promote recycling, and explore waste-to-energy solutions. For example, geographical information systems techniques can track waste movement and pinpoint pollution hotspots, particularly in urban areas and coastal regions, whereas sensors can monitor waste levels and recycling efficiency [21].

The significance of green intelligence lies in its ability to transform plastic waste management by integrating advanced digital tools into decision-making processes. By offering insights into the lifecycle of plastic waste, GI helps develop targeted strategies to address specific problem areas; this includes reducing illegal dumping, enhancing recycling programs, and implementing waste-to-energy technologies that contribute to a circular economy [23]. Furthermore, GI fosters cross-sector collaboration by connecting governments, businesses, and communities through a unified data platform. Such collaboration drives the development of policies that encourage reductions in plastic production and support sustainable practices. Ultimately, green intelligence holds the potential to revolutionize Africa's waste management landscape, improve operational efficiency, promote sustainability, and empower stakeholders to effectively combat plastic pollution.

Study Objectives

The main goal of this review is to develop a comprehensive strategic framework that incorporates green intelligence to address the growing plastic waste crisis in Africa. The framework aims to integrate technological innovations, policy interventions, and community engagement strategies to create a holistic approach to waste management.

The specific objectives include the following:

1. To identify key challenges in Africa's plastic waste management, including infrastructural deficiencies, policy gaps, and public awareness issues.
2. To evaluate the role of ICT-based tools in optimizing plastic waste management and promoting sustainable practices.

3. To propose a strategic framework that combines data-driven decision-making, technological innovations, and policy reforms tailored to Africa's socioeconomic and environmental context.
4. To highlight the importance of cross-sector collaboration among governments, the private sector, and local communities in reducing plastic waste.
5. To offer actionable policy recommendations aimed at improving plastic waste management systems, focusing on sustainable development goals and long-term environmental health in Africa.

Through these objectives, this study seeks to provide a clear roadmap for leveraging green intelligence to create sustainable and effective waste management practices across Africa, addressing both immediate and long-term challenges associated with plastic waste.

5.2 Literature Review: Plastic Waste Management

5.2.1 Global Plastic Waste Management Strategies

Global plastic waste management strategies reflect a diverse range of approaches tailored to different regional contexts. In developed countries, advanced recycling technologies and strict regulatory frameworks are key to managing plastic waste effectively. For example, Germany boasts one of the highest recycling rates globally, with 67% of its waste being recycled, primarily due to its "Green Dot" system, which incentivizes manufacturers to reduce packaging waste [24]. Germany's extended producer responsibility (EPR) program compels producers to take responsibility for the disposal of their products, leading to eco-friendly packaging design [25]. Similarly, Sweden utilizes waste-to-energy technology, converting nearly 50% of household waste into energy, which significantly reduces landfill usage [26]. The country's stringent regulations on single-use plastics and promotion of a circular economy further bolster its waste management efforts.

In contrast, developing countries face considerable challenges due to inadequate infrastructure and enforcement. In India, approximately 60% of the 26,000 tonnes of plastic waste generated daily are collected, while the remainder contribute to environmental pollution through waterways and landfills [27]. Despite initiatives such as single-use plastic bans and community-based recycling programs, enforcement is limited, particularly in rural areas. India's efforts include partnerships with NGOs to advance waste-to-energy projects, although scaling these initiatives remains problematic.

Emerging technologies such as green intelligence (GI) offer innovative solutions for plastic waste management. GI incorporates artificial intelligence (AI) and machine learning to enhance waste management practices. For example, predictive analytics using AI can forecast waste generation trends, allowing for better resource allocation [28, 29]. AI-driven smart recycling systems improve sorting accuracy and recycling efficiency by identifying and separating different types of plastics [30, 31]. Real-time monitoring through AI can streamline waste management operations by identifying inefficiencies and providing actionable data [32]. These technological advancements are not only applicable to waste management but also effective in other environmental sectors, such as water and air quality monitoring, where AI aids in predicting contamination and pollution events [33]. Triangulation across various studies and reports validates these findings, demonstrating the transformative potential of GI in addressing plastic waste challenges globally.

5.2.2 *Plastic Waste in Africa*

Africa's struggle with plastic waste management is intensifying due to rapid urbanization, population growth, inadequate infrastructure, and insufficient policy enforcement. Projections indicate that by 2025, annual urban waste in sub-Saharan Africa will more than double from 174 million tonnes to approximately 360 million tonnes, exacerbating the pressure on already overstretched waste management systems [34]. The capacity of waste management infrastructure, particularly in rapidly growing cities such as Accra, Nairobi, and Dar es Salaam, has failed to keep pace with this expansion. In these cities, infrastructure development has struggled to address the escalating demand, leading to an overflow of uncollected plastic waste [35, 36].

Conversely, Rwanda has demonstrated significant success with its plastic bag ban enacted in 2008, resulting in a notable reduction in plastic pollution [37, 38]. This success underscores the importance of stringent policy enforcement and public engagement in waste management. Notably, Egypt, Nigeria, South Africa, and Kenya (Fig. 5.1a) are among the highest contributors to plastic waste generation on the continent, collectively producing more than 6 million tonnes of plastic waste annually (Fig. 5.1b). Egypt alone generates approximately 3 million tonnes, followed by Algeria, Nigeria, and South Africa, each contributing significantly to the continent's plastic pollution problem. Weak policy enforcement and gaps in recycling infrastructure further hinder effective waste management in these regions, underscoring the need for robust policy reforms and technological innovations to address this growing environmental crisis.

One critical issue is the lack of adequate waste management infrastructure. In many African cities, only 44% of waste is collected, and less than 10% of this collected waste is recycled [10]. For example, in Lagos, Nigeria, only 40% of the city's 13,000 tonnes of daily waste are collected, leaving a substantial portion of plastic waste unaccounted for [41]. This uncollected waste frequently ends up in waterways, contributing to flooding and water contamination. The informal waste sector, which handles up to 90% of plastic waste collection in some cities, often operates without formal support or protection. Workers in this sector face poor working conditions and are exposed to hazardous materials, undermining the effectiveness and scalability of waste management efforts [42].

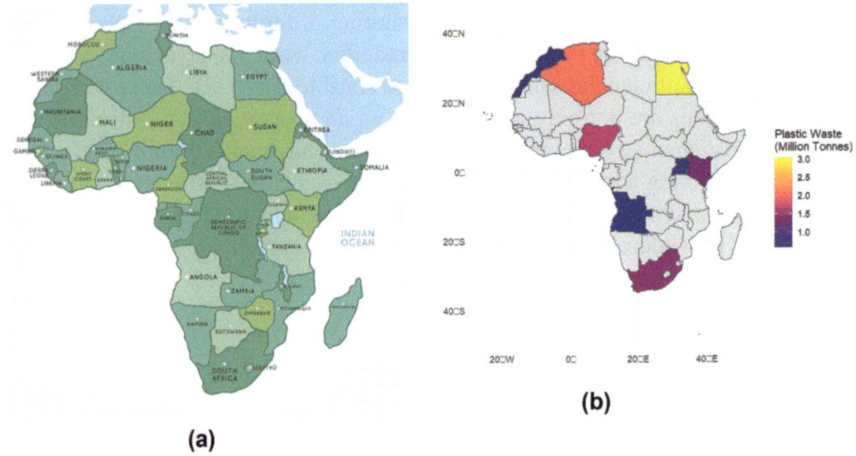

Fig. 5.1 **a** Map of Africa showing countries. **b** Plastic waste generation across Africa. *Data sources* Babayemi, Nnorom [39]; Jambeck, Hardesty [40]

The trends and patterns illustrated in the combined graph (Fig. 5.2) reveal a significant increase in both plastic waste and urban waste in Sub-Saharan Africa over time, reflecting the region's increasing consumption and rapid urbanization. Historically, plastic waste has shown a sharp upward trajectory (forecast) from 1990 to 2060, driven by industrial growth and increased plastic use, which is projected to continue rising, reaching critical levels by 2060. Similarly, urban waste generation is projected to increase from 2000 to 2050, mirroring urban population growth and economic development. Practically, these trends are interrelated, as both types of waste are exacerbated by the same factors of population increase and economic expansion, leading to overburdened waste management systems. In principle, the rising figures for both plastic and urban waste underscore the urgent need for integrated waste management solutions and sustainable practices to mitigate environmental impacts and enhance waste management infrastructure in the region.

Policy enforcement remains a major hurdle despite the introduction of regulations in several African countries. For example, Kenya's nationwide ban on plastic bags, implemented in 2017 and considered one of the world's strictest, has faced challenges with weak enforcement and ongoing black-market activity due to insufficient monitoring [15, 43]. In contrast, Rwanda's successful ban on plastic bags, supported by public awareness campaigns and strict enforcement, highlights the importance of effective policy implementation and community engagement in managing plastic waste [44, 45]. Overall, the interplay between rapid urbanization, inadequate infrastructure, weak policy enforcement, and low public awareness continues to challenge Africa's ability to manage plastic waste effectively.

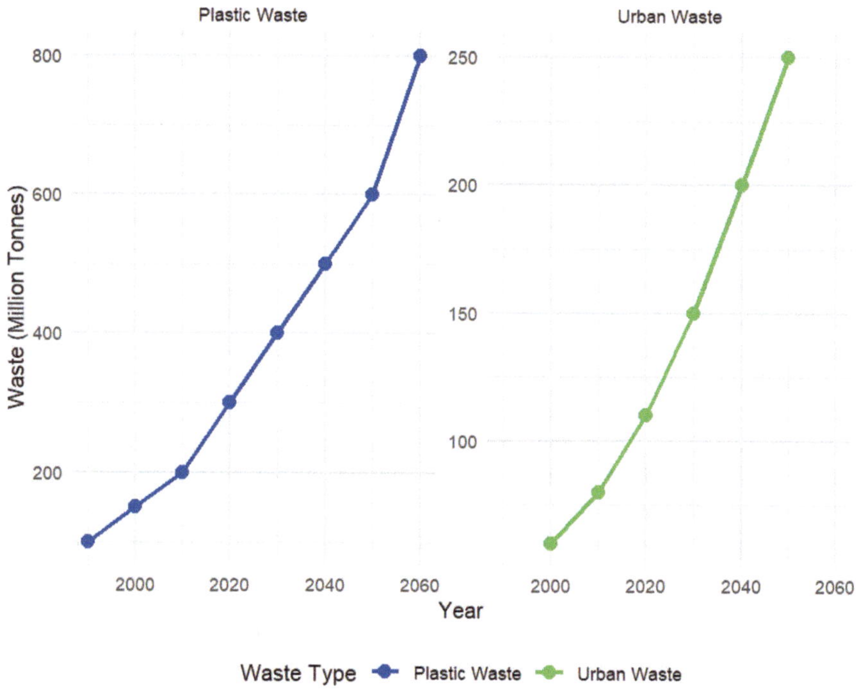

Fig. 5.2 Historic and future projections of plastic and urban waste outputs in Sub-Saharan Africa. *Source* OECD, 2024

5.2.3 *Green Intelligence in Environmental Management*

Green intelligence (GI) is revolutionizing environmental management by leveraging information and communication technologies (ICTs) to enhance sustainable practices. This innovative approach integrates data analytics, real-time monitoring, and digital platforms to support data-driven decision-making aimed at improving waste management systems and reducing plastic waste. GI employs advanced tools to track and manage waste generation, collection, disposal, and recycling. For example, smart waste bins equipped with sensors can monitor fill levels and alert collection services when they need to be emptied, minimizing overflow and illegal dumping [46]. Additionally, geographic information system (GIS) technology helps identify plastic waste hotspots, allowing for more effective resource allocation and policy interventions [3, 21].

The integration of AI technologies into solid waste management across Africa is a critical component of this green intelligence movement (Fig. 5.3). Countries such as South Africa and Kenya are leading the way with innovative AI-driven solutions, such as smart waste sorting systems that optimize recycling processes and predictive analytics that increase waste collection efficiency. These advancements not only improve operational performance but also contribute to sustainable practices by reducing landfill overflow and increasing recycling rates. In contrast, nations such as Nigeria and Egypt exhibit moderate levels of AI adoption, indicating ongoing efforts to integrate digital technologies into their waste management systems; however, these countries face challenges such as limited infrastructure and regulatory hurdles. This disparity highlights the need for a concerted effort to invest in AI and digital infrastructure across the continent, ensuring that the benefits of these technologies can be realized in transforming waste management practices and addressing the pressing challenges of plastic waste.

In other environmental sectors, GI has demonstrated its effectiveness in enhancing resource management. For example, in water management, GIS technology monitors water usage and detects leaks, optimizing resource distribution [47]. Similarly, in agriculture, smart farming technologies use data from IoT sensors and satellite imagery to track soil conditions and crop growth and optimize the use of fertilizers and water [48]. These applications highlight how data-driven technologies can provide real-time insights and improve decision-making, which is crucial for effective environmental management.

Applying GI to plastic waste management could significantly address Africa's waste challenges. AI-driven smart recycling systems can increase recycling efficiency by accurately sorting different types of plastics, improving recycling rates and reducing environmental impact [49]. Additionally, waste-to-energy systems, which convert plastic waste into usable energy, offer a dual solution for waste management

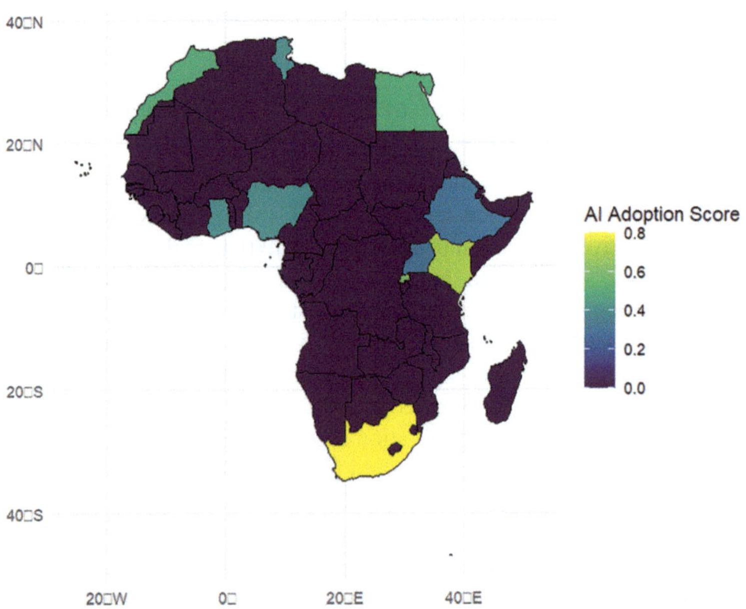

Fig. 5.3 Digital adoption and AI-technology initiatives in waste management across Africa *Sources* Africa Sustainability Matters, 2023; Impacting Africa, 2023; Waste Management World, 2023; Ajala, 2023 [39–51]

and energy production, as seen in countries such as Sweden and Japan [50]. Despite these advancements, gaps remain in the African context, including limited public awareness, inadequate infrastructure, and a lack of reliable data. Addressing these challenges requires a comprehensive approach that incorporates advanced technologies, improved infrastructure, and robust policies to effectively manage plastic waste and foster sustainable practices.

5.2.4 Gaps in Current Approaches

Despite increased awareness of the plastic waste crisis in Africa, several critical gaps persist in current waste management strategies. One major issue is the lack of public awareness, particularly in rural and low-income urban areas where access to education and resources is limited. While countries such as Rwanda and Kenya have implemented national bans on single-use plastics, many regions still have insufficient understanding of the environmental and health risks associated with plastic waste [52, 53]. Additionally, waste management strategies often focus disproportionately on urban centers, neglecting the needs of rural areas where informal waste disposal methods, such as burning or dumping plastic waste, are common [54]. For example, Ghana's waste management policies are concentrated in cities such as Accra and Kumasi, leaving approximately 43% of the population living in rural areas with inadequate waste management infrastructure [55, 56]. This oversight exacerbates environmental pollution and negatively impacts agricultural productivity by contaminating soil and water sources.

Moreover, the lack of reliable data and inadequate recycling infrastructure further complicate this issue. Many African nations struggle with insufficient technological infrastructure and financial resources, resulting in poor data collection on waste generation and management. This lack of comprehensive data hampers the development of targeted interventions and the measurement of policy effectiveness. Furthermore, while some progress has been made in recycling, the majority of African countries recycle less than 10% of their plastic waste because of underfunded and unsupported recycling initiatives [7, 57]. The dominance of the informal sector, which lacks the necessary technology and equipment, also limits recycling efforts. Compounding these issues is weak policy enforcement, where even well-intentioned regulations are undermined by inconsistent application and corruption, as seen in Kenya's ongoing challenges with its plastic bag ban [58]. Addressing these gaps requires a multifaceted approach that includes improving public awareness, enhancing rural waste management, investing in data collection and recycling infrastructure, and strengthening policy enforcement, potentially through the integration of innovative solutions such as green intelligence, to improve overall efficiency and accountability in waste management.

5.3 Methodology

5.3.1 *Scope and Approach*

This review aims to develop a strategic framework for managing plastic waste in Africa by leveraging green intelligence (GI), an ICT-based approach. The decision to focus on Africa's specific context arises from the continent's unique socioeconomic challenges, such as rapid urbanization, limited waste management infrastructure, and weak policy enforcement, which exacerbate plastic pollution. Africa's plastic waste crisis is not only an environmental concern but also a socioeconomic concern, given the significant impacts on public health, agriculture, and marine ecosystems [40].

By narrowing the scope to Africa, the review specifically addresses the distinct needs and barriers faced by African nations. It seeks to evaluate existing waste management strategies and propose data-driven, technology-based solutions through GI that are contextually appropriate for African countries. Additionally, the review aims to highlight successful case studies from African nations, such as Rwanda and Kenya, that have made strides in addressing plastic pollution through policy reforms and public engagement [37].

Inclusion Criteria

This review draws on a wide range of sources, focusing on studies from the past ten years (2014–2023) to capture the latest advances in waste management technologies and policies. Priority was given to peer-reviewed journals, government reports, and conference papers addressing plastic waste management in Africa or scalable global innovations. The key themes included green intelligence, waste management technologies, African policy frameworks, and sustainability practices. Studies on waste management in other regions, such as Southeast Asia and Latin America, were excluded, as were non-English articles, to ensure the relevance of the focus on African innovations and green technologies.

Data Collection and Synthesis

Data collection for this review involved a comprehensive search of academic databases, including Google Scholar, Scopus, and Web of Science, to gather relevant studies, reports, and empirical data on plastic waste management in Africa. Keywords such as "plastic waste management in Africa," "green intelligence," "waste-to-energy technologies," and "recycling in Africa" were used to identify studies that offered insights into existing waste management practices and the potential application of green intelligence in the region. A total of 101 sources, including 47 academic journal articles, 9 NGO reports, 5 government publications, and 14 online articles, were analyzed.

Once relevant studies were identified, data synthesis was carried out to integrate findings on plastic waste generation, collection, and recycling efforts across different African nations. Key metrics such as the green intelligence score and waste management score were introduced to evaluate the effectiveness of plastic waste management strategies. The synthesis process involved coding data from the sources to categorize them according to themes such as policy enforcement, infrastructure gaps, technological innovations, and community engagement efforts.

5.3.2 *Sustainability Metric*

This study adopts a mixed-methods approach, combining quantitative and qualitative data to evaluate plastic waste management strategies across selected African countries. The green intelligence score, a composite metric developed for this study, forms the quantitative backbone, offering a standardized comparison of each country's waste management effectiveness. The selection of countries and calculation of the GI scores are based on multiple factors, including plastic waste output, recycling rates, and policy or technological adoption. This is complemented by qualitative insights drawn from case studies of policy interventions in these countries.

5.3.2.1 Country Selection

The countries included in the analysis—Nigeria, South Africa, Kenya, Rwanda, Ghana, and Egypt—were selected on the basis of several key criteria relevant to understanding plastic waste management practices across Africa:

- **Geographical Representation**: Countries were selected from different regions of Africa, including West, East, South, and North Africa, to ensure diverse geographic perspectives.
- **Plastic Waste Output**: Nations with varying levels of plastic waste generation were chosen to facilitate comparisons of waste management approaches on the basis of the scale of the problem.
- **Recycling Rate**: Countries with different recycling capabilities were included to explore the correlation between waste output and recycling performance.
- **Policy and Technological Adoption**: Countries with varying degrees of policy enforcement and adoption of ICT and green technologies were selected to assess their impact on waste management practices.

5.3.2.2 Green Intelligence (GI) Score Calculation and Data Standardization

The green intelligence (GI) score was developed to quantitatively assess each country's performance in terms of plastic waste management. This composite score integrates three key metrics: plastic waste output (PWO), the recycling rate (RR), and policy and technological adoption (PTA).

The formula for computing the GI score is as follows:

$$\text{GI score} = (w_1 \times m_1) + (w_2 \times m_2) + \ldots + (w_n \times m_n)$$

where $w_1, w_2, w_3, \ldots, w_n$ represent the weights assigned to each metric, and $m_1, m_2, m_3, \ldots, m_n$ are the standardized scores for those metrics. The assigned weights reflect the relative importance of each factor in determining the effectiveness of plastic waste management.

Metric Weight Assignments

The percentage weights assigned to each metric were derived through a systematic process:

- **Plastic Waste Output (PWO)**: 30%—This weight acknowledges the fundamental role of reducing waste generation as a critical first step in managing plastic pollution.
- **Recycling Rate (RR)**: 40%—Given its central importance in mitigating waste impact, this metric received the highest weight, emphasizing effective recycling strategies as crucial for overall waste management.
- **Policy and Technological Adoption (PTA)**: 30%—This weight recognizes the significance of robust policies and innovative technologies in driving improvements in both PWO and RR.

Data Sources

- **Plastic Waste Output**: Measured in million tonnes per year, with data sourced from publicly available databases, government reports, and peer-reviewed studies. Countries with lower plastic waste outputs receive higher standardized scores.
- **Recycling Rate**: This metric represents the proportion of plastic waste recycled, where higher recycling rates lead to higher scores. Data were also sourced from government reports and academic studies.
- **Policy and Technological Adoption (PTA)**: Evaluated qualitatively, this metric considers national policies, the adoption of ICT, and green technology initiatives. Countries such as Rwanda and Kenya, known for their strong policies and innovative approaches, scored higher in this category.

Standardization Process

To ensure comparability across countries, the raw data for both the plastic waste output (PWO) and recycling rate (RR) were standardized on a scale from 0 to 1. A clearer breakdown of the process is as follows:

Individual Standardization

Plastic Waste Output (PWO): Each country's plastic waste output was measured in million tonnes per year. The standardization process assigns a score of 1 to the country with the lowest plastic waste output and a score of 0 to the country with the highest output. Intermediate values were scaled proportionally via the following formula:

The formula for calculating the standardized PWO score is as follows:

$$\text{Standardized PWO score} = 1 - \left(\text{Country's PWO} - \text{Min PWO}\right)/(\text{Max PWO} - \text{Min PWO})$$

This formula assigns a score of 1 to the country with the lowest plastic waste output and a score of 0 to the country with the highest output, while intermediate values are scaled proportionally.

Recycling Rate (RR): Similarly, the recycling rate was standardized. The country with the highest recycling rate received a score of 1, whereas the country with the lowest recycling rate received a score of 0. The formula for this standardization is as follows:

$$\text{Standardized RR score} = 1 - \left(\text{Country's RR} - \text{Min RR}\right)/(\text{Max RR} - \text{Min RR})$$

Combined Influence

In cases where a country has both the highest PWO and the highest RR, the standardized scores reflect their relative positions within their respective metrics. For instance, the country with the highest PWO would receive a score of 0, whereas the country with the highest RR would receive a score of 1. This indicates that although a country may excel in recycling, its high waste output negatively impacts its overall green intelligence score.

Qualitative Assessment of Policy and Technological Adoption (PTA)

The PTA was evaluated through qualitative rankings on the basis of an analysis of each country's policy initiatives and technological advancements. This qualitative assessment complements the quantitative scores from PWO and RR, ensuring a comprehensive evaluation of plastic waste management performance. By standardizing PWO and RR separately, we can accurately reflect each metric's contribution to effective plastic waste management. The final GI score is a weighted combination of these standardized scores, facilitating a thorough comparison across countries.

Final GI Score Calculation

The final GI score for each country was calculated via the following formula:

GI score = 0.3 * (plastic waste output score) + 0.4 * (recycling rate score) + 0.3 * (policy and technological adoption score).

This composite score facilitates a comprehensive comparison of waste management effectiveness across the selected countries, providing valuable insights into their performance and areas for improvement (Table 5.1).

Table 5.1 Distribution of literature sources used in the review

Source type	Number of sources
Academic journals	47
Conference proceedings	12
Books	8
Government reports	5
NGO reports	9
Industry publications	6
Online articles	14
Total	101

5.4 Results

5.4.1 Overview of Green Intelligence Scores

The green intelligence score provides a standardized comparison of the effectiveness of plastic waste management in the selected African countries. Table 5.2 presents the breakdown of plastic waste output, recycling rates, and the resulting GI score for each country. The GI score highlights the overall performance of each country in managing plastic waste through the integration of three primary components: waste output, recycling rates, and policy or technological adoption.

Table 5.2 Contributors to the green intelligence (GI) score

Country	Plastic waste output (Million Tonnes/ Year)	Recycling rate (%)	PTA score (0–1)	Standardized PWO score (0–1)	Standardized RR score (0–1)	Green intelligence (GI) score
Rwanda	0.15	35	0.9	1.0	1.00	0.97
Kenya	0.20	30	0.8	0.9	0.85	0.85
South Africa	1.60	25	0.6	0.5	0.70	0.61
Nigeria	2.50	10	0.3	0.0	0.20	0.17
Ghana	1.00	12	0.4	0.7	0.25	0.43
Egypt	3.02	20	0.5	0.6	0.60	0.57

5.4.2 Plastic Waste Output and Recycling Rates

Plastic waste output varies significantly across countries. Nigeria, with the second highest waste generation of 2.5 million tonnes per year, received the lowest standardized score for this metric (0), whereas Rwanda, with the smallest output of 0.15 million tonnes per year, scored the highest (1). In terms of recycling rates, Rwanda also led with 35%, followed by Kenya with 30%, whereas Nigeria lagged behind with just 10%.

The variation in recycling performance illustrates the range of waste management strategies employed. Countries such as Rwanda and Kenya have adopted progressive waste management policies and invested in recycling infrastructure, leading to higher recycling rates. In contrast, countries such as Nigeria and Ghana face challenges related to insufficient waste management infrastructure and lower policy enforcement, reflected in their lower recycling rates.

5.4.3 Policy and Technological Adoption

Rwanda and Kenya stand out for their strong policy frameworks and technological innovation in waste management. Rwanda's ban on plastic bags and the integration of ICT in waste tracking systems have contributed to a greater GI score. Kenya, with its enforcement of plastic bag bans and growing adoption of green technologies, also performed well in this category. On the other hand, Nigeria and Ghana, while having policies in place, suffer from weaker enforcement and limited technological adoption, resulting in lower scores.

5.4.4 Green Intelligence (GI) Score Ranking

Table 5.2 presents the final GI rankings, highlighting the performance of various countries in terms of plastic waste management. Rwanda achieved a GI score of 0.97, which indicates a strong commitment to reducing plastic waste through effective policies and recycling efforts. However, Rwanda's low plastic waste output of only 0.15 million tonnes per year significantly influences this score. While Rwanda's proactive measures are commendable, the very low level of plastic waste generated may inflate its GI score compared with countries with higher outputs. Kenya follows with a GI score of 0.85, reflecting its relatively high recycling rate and progressive policy measures. Like Rwanda, Kenya's GI score may also be influenced by its low plastic waste output of 0.20 million tonnes per year, which suggests that while recycling efforts are noteworthy, the overall impact may not be as significant when considering the total waste produced.

In contrast, Nigeria has the lowest GI score of 0.17, highlighting significant challenges in managing plastic waste, as indicated by its high output of 2.50 million tonnes per year and a low recycling rate of only 10%. This score underscores the need for substantial improvements in waste management strategies. The GI scores for South Africa, Ghana, and Egypt further illustrate varying degrees of effectiveness in addressing plastic waste, with South Africa scoring 0.61, Ghana scoring 0.43, and Egypt scoring 0.57, each reflecting unique challenges and strategies in waste management.

5.4.5 Comparative Analysis

Countries with high green intelligence scores, such as Rwanda and Kenya, present effective plastic waste management strategies driven by robust policies and technological adoption. Rwanda, with a GI of 0.97, led primarily because of its comprehensive plastic bag ban implemented in 2008 and a recycling rate of 35%. However, its low plastic waste output of just 0.15 million tonnes per year may inflate this score, indicating that while Rwanda's initiatives are commendable, their overall impact should be viewed in the context of its minimal waste generation. Kenya follows closely with a GI of 0.85, bolstered by a stringent plastic bag ban enacted in 2017 and the integration of technological innovations such as GIS mapping technologies and smart bins. These measures have improved waste management efficiency and increased the recycling rate to 30%. Like in Rwanda, Kenya's relatively low plastic waste output of 0.20 million tonnes per year may also influence its GI score, suggesting that while its efforts are significant, the scale of waste management challenges may differ from those of larger waste-producing nations. Both countries exemplify how effective policies and technology can address plastic waste challenges, but their high GI scores must be interpreted alongside their low plastic waste outputs.

On the other hand, countries with lower GI scores, such as Nigeria and Ghana, face significant challenges in terms of plastic waste management because of their high waste outputs and low recycling rates. Nigeria, with a GI of 0.17, struggles with a plastic waste output of 2.50 million tonnes and a recycling rate of just 10%, hindered by inadequate infrastructure and weak policy enforcement. Ghana, with a GI of 0.43, has a slightly better recycling rate of 12%, but it also suffers from ineffective waste collection systems and limited public awareness. Despite some progress in policy and technological adoption, such as Ghana's PTA score of 0.4, both countries need to strengthen their waste management strategies to achieve sustainable outcomes. South Africa (GI of 0.61) and Egypt (GI of 0.57) also face challenges, although they have moderate recycling rates of 25% and 20%, respectively, and need further innovation and policy enforcement to improve their scores.

5.4.6 Key Challenges Identified

Several key challenges were identified in countries with lower green intelligence scores, particularly Nigeria, Ghana, and Egypt.

1. **Inadequate Waste Collection Infrastructure**: Many African countries, especially those with low GI scores, suffer from inefficient waste collection systems. In Nigeria, for example, only 40% of waste is collected in major cities, whereas rural areas often have no formal waste collection services (**World Bank, 2023**). This results in large quantities of plastic waste being dumped in landfills, rivers, and oceans, contributing to environmental degradation (**United Nations Environment Programme, 2022**).
2. **Low Recycling Rates**: Countries with low GI scores also have poor recycling infrastructures. For example, in Nigeria and Ghana, less than 12% of plastic waste is recycled (**World Bank, 2023**). The lack of investment in recycling facilities and the informal nature of waste collection further hinder efforts to improve recycling rates (**United Nations Environment Programme, 2022**). In contrast, countries such as Rwanda and Kenya, which have higher GI scores, have made significant investments in recycling infrastructure (**Waste Management World, 2021**).
3. **Weak Policy Enforcement**: The inconsistent enforcement of waste management policies, particularly those related to plastic waste, remains a significant obstacle for many African nations. Despite the implementation of various regulations, including plastic bans, effective enforcement is hindered by a combination of factors. Corruption, inadequate funding, and a lack of politics often undermine the implementation of these policies. For example, in Nigeria, while regulations for plastic waste management exist, their enforcement is confined primarily to urban areas, leaving rural regions largely unregulated. This is in line with findings from the World Bank's report on "Plastic Waste Management in Africa: Challenges and Opportunities" (2023), which highlights the pervasive issue of weak enforcement in the region.
4. **Lack of Public Awareness and Engagement**: In countries with lower green intelligence scores, there is often a lack of public awareness of the environmental and health impacts of plastic waste. In Nigeria and Ghana, for example, many communities are not adequately informed about proper waste disposal practices, recycling, or the long-term effects of plastic pollution [59]. Without widespread education and awareness campaigns, efforts to reduce plastic waste and improve recycling rates are hindered [7]. In contrast, countries such as Rwanda have invested heavily in public engagement initiatives, ensuring that communities are involved in waste reduction efforts [60], which may explain their higher GI scores.
5. **Limited Adoption of Technology**: A significant barrier for low-performing countries is the limited use of technology in waste management. In countries such as Kenya and Rwanda, where GIS mapping, smart waste collection systems, and digital platforms are being utilized, waste management efficiency has improved

significantly [51, 61]. However, in countries with lower GI scores, such as Ghana and Nigeria, the adoption of such technologies remains minimal due to limited funding, technical expertise, and infrastructure.

The results of this review reveal significant disparities in plastic waste management across African nations. Countries with higher green intelligence scores, such as Rwanda and Kenya, demonstrate that strong policy enforcement, public awareness, and the use of technology are critical to effective waste management. In contrast, countries with lower GI scores, such as Nigeria and Ghana, face challenges related to weak infrastructure, low recycling rates, and limited public engagement. Addressing these challenges will require targeted investments in waste management infrastructure, the adoption of innovative technologies, and the development of public awareness campaigns to promote sustainable waste practices.

5.5 Discussion

5.5.1 Summary of Findings

The **green intelligence (GI)** score offers a comprehensive measure of plastic waste management performance across African countries, focusing on three critical aspects: **plastic waste output**, **recycling rates**, and the adoption of **policies and technologies** designed to mitigate plastic pollution. Countries with higher GI scores, such as **Rwanda** and **Kenya**, exhibit lower plastic waste outputs, higher recycling rates, and strong enforcement of progressive waste management policies. On the other hand, countries with lower GI scores, such as **Nigeria** and **Ghana**, face challenges with higher plastic waste outputs, minimal recycling efforts, and weaker policy frameworks.

In terms of **plastic waste output**, nations with high GI scores have introduced measures to reduce plastic waste generation, such as **bans on single-use plastics** and the promotion of alternative materials. Lower-scoring countries still grapple with uncontrolled plastic consumption and insufficient waste disposal strategies. **Recycling rates** are another key factor, with higher GI scores linked to efficient recycling systems. For example, **Kenya's** investment in recycling facilities and public–private partnerships has significantly improved its recycling rates. In contrast, nations with lower scores often lack the necessary infrastructure for efficient recycling, resulting in increased environmental pollution. Finally, in the area of **policy and technology adoption**, high-GI-scoring countries have embraced advanced waste management technologies and progressive policies, such as **GIS mapping** for waste tracking and **smart waste bins**. Conversely, lower-scoring countries often struggle with policy enforcement and lag in technological integration.

5.5.2 *Key Observations*

A clear pattern emerges when countries with high and low GI scores are compared. High-scoring countries have successfully reduced **plastic waste generation** through the implementation of **strict regulations** and the promotion of alternative materials. They also boast higher **recycling rates**, driven by government incentives, efficient recycling infrastructure, and **public awareness campaigns**. Additionally, these countries are leaders in the **adoption of advanced technologies**, such as smart waste management systems and GIS-based monitoring tools, allowing for more efficient waste control and reduction.

In contrast, countries with lower GI scores produce more plastic waste due to limited regulatory measures for single-use plastics and a lack of **comprehensive recycling programs**, which results in minimal recovery of plastic materials. These nations often struggle with **policy enforcement** and face significant barriers to implementing innovative waste management technologies. These differences underscore the critical need for coordinated policy efforts, community involvement, and technological investments to promote sustainable waste management practices across the continent.

5.5.3 *Interpretation of Results*

The adoption of technology and robust policies plays a significant role in enhancing a country's green intelligence score, as seen in examples such as **Rwanda** and **Kenya**. These nations have leveraged innovations such as **GIS mapping** for tracking waste management processes and the implementation of **smart waste bins** that optimize waste collection, thereby reducing inefficiencies. The integration of these technologies not only streamlines waste disposal but also promotes accountability in waste management systems. For example, Rwanda's strict enforcement of **plastic bans** and Kenya's focus on **public–private partnerships** to promote recycling have led to substantial improvements in waste management outcomes, resulting in higher GI scores. In both countries, the alignment of policy frameworks with technological advancements has proven crucial for reducing plastic waste output and increasing recycling rates.

In contrast, low-scoring countries such as **Nigeria** and **Ghana** face significant challenges due to **inadequate infrastructure** and weak policy enforcement. In these nations, the lack of investment in recycling facilities and the absence of effective waste collection systems hinder the management of plastic waste [62]. For example, Nigeria continues to struggle with enforcing its environmental policies, which results in low recycling rates and the proliferation of plastic pollution in urban and rural areas. Similarly, Ghana faces difficulties in implementing policies aimed at reducing plastic waste due to **limited governmental oversight** and low public awareness of waste management practices [63]. The **absence of technological integration**, such as smart waste systems, exacerbates these problems, contributing to overall lower GI scores.

These disparities between high- and low-GI-scoring nations underscore the critical role of **technology adoption** and **policy enforcement** in advancing sustainable waste management. High-scoring countries demonstrate the benefits of a combined approach, whereas lower-scoring nations highlight the need for more **investment in infrastructure** and **stronger regulatory frameworks** to improve plastic waste management.

5.5.4 Green Intelligence Applications

Integrating green intelligence (GI) into waste management systems has the potential to transform how African countries handle plastic waste, driving improvements in **efficiency**, **sustainability**, and **community participation**. By fostering **community engagement**, GI can empower local populations to take ownership of waste management efforts, encouraging better recycling practices and waste segregation at the source. In regions where **public–private partnerships** are utilized, such as Kenya, the integration of **GIS-based waste tracking** has significantly enhanced waste collection efficiency, reduced illegal dumping, and increased recycling rates. Technology adoption plays a central role in this transformation. The implementation of **smart waste management systems**, including **sensor-equipped bins** and **data analytics**, can improve waste collection scheduling, optimize routes, and reduce operational costs.

Rwanda exemplified the potential success of GI, where robust **public engagement campaigns** have raised awareness about the importance of sustainable waste management. This has been complemented by Rwanda's strict **plastic bans** and effective policy enforcement, leading to cleaner urban environments and higher recycling rates [64]. Kenya has also embraced **GIS-based solutions** to map waste collection patterns and track recycling progress, further demonstrating the efficiency gains and environmental benefits of integrating green intelligence into waste management systems.

5.5.5 Policy and Technological Recommendations

To build on the success of high-scoring countries, expanding regulations such as **plastic bans** and introducing **extended producer responsibility (EPR)** programs can drive significant improvements in waste management across Africa. EPR programs place responsibility on producers to manage the end-of-life of their products, encouraging businesses to design more sustainable packaging and invest in recycling solutions. However, it is crucial that governments enforce these regulations rigorously. Many countries, such as Nigeria, struggle with weak policy implementation, which limits the effectiveness of existing waste management frameworks [65].

In addition to stricter enforcement, providing **incentives for recycling** and supporting **technology adoption** will be key to accelerating progress. Governments could offer **tax breaks** or **grants** to companies that invest in **smart waste management systems** or develop **AI-powered sorting technologies**, both of which can improve recycling rates and reduce contamination. For example, **AI-powered sorting** systems are already being used in countries such as South Africa to identify and separate waste streams more efficiently, ensuring that a greater percentage of materials are recycled [66].

Furthermore, the implementation of **waste-to-energy systems** can provide an innovative solution to address the challenge of landfill overuse while generating renewable energy. Waste-to-energy technology could help African countries reduce their dependence on fossil fuels and address the issue of unmanaged waste, providing both environmental and economic benefits. By investing in **waste-to-energy plants** and further advancing **AI-based waste management technologies**, African nations could develop a circular economy model that both reduces waste and drives energy production [67].

In the context of green intelligence for sustainable plastic waste management in Africa, various IoT technologies play a critical role in enabling waste-to-energy (WTE) solutions that align with circular economy principles. **Smart waste management systems**, such as IoT-enabled bins with fill-level sensors, facilitate efficient waste collection by optimizing routes and reducing fuel consumption, contributing to lower greenhouse gas emissions [68]. **IoT-integrated sorting technologies**, including AI-driven optical sorters and RFID-enabled waste tracking, ensure accurate separation of plastics suitable for energy recovery. During the energy conversion phase, **IoT-based monitoring systems** track critical parameters like combustion efficiency, emissions, and energy output in real-time, ensuring compliance with environmental standards and optimal performance [69]. Additionally, **blockchain-enabled IoT platforms** promote transparency by tracing plastic waste from collection to energy recovery, fostering trust among stakeholders and enhancing public–private partnerships. These IoT applications, embedded within green intelligence frameworks, enhance the efficiency and scalability of WTE technologies, driving sustainable plastic waste management practices across Africa.

Community and Collaborative Efforts

Community engagement plays a pivotal role in enhancing waste management practices across Africa. Public education campaigns on proper waste disposal and recycling are essential for fostering a culture of sustainability within communities. Rwanda offers a successful model in which **nationwide public awareness campaigns** have significantly improved waste management outcomes, leading to higher recycling rates and cleaner urban environments. By engaging the public through schools, local media, and community leaders, Rwanda has cultivated a sense of shared responsibility for environmental protection [70]. African countries can adopt similar strategies, incorporating digital tools such as mobile apps and social media platforms to reach a wider audience.

Supporting **informal waste pickers** is also crucial, as they play a significant role in the recycling ecosystem in many African countries. These workers can be integrated into formal waste management systems through digital tools that track their contributions, providing incentives such as direct payments, health benefits, or safety training. By organizing waste pickers into cooperatives and using technology to ensure transparency and accountability, countries can increase recycling efforts while improving the livelihoods of marginalized populations [71].

Cross-border cooperation among African countries is equally vital for the sharing of best practices and resources in waste management. Regional collaborations, such as the **African Union's environmental programs**, can help harmonize waste management policies and coordinate efforts to combat plastic waste across borders. Partnerships with international organizations and the private sector can facilitate the exchange of technological innovations and policy frameworks that have proven successful in countries such as Rwanda and Kenya. **Collaboration with global private-sector players** can also help African nations gain access to advanced waste management technologies and funding for large-scale infrastructure projects [72].

International bodies such as the **United Nations Environment Programme (UNEP)** and various **NGOs** play a significant role in providing **technical and financial support**. They can assist African governments in building **capacity** for better waste management practices by offering training programs, technological assistance, and research funding. Additionally, these organizations can help establish **public–private partnerships** that are vital for implementing large-scale waste management solutions, ensuring long-term sustainability. The UNEP's financial support for Rwanda's zero-waste initiatives is a prime example of how international cooperation can drive national progress [73, 74].

Moving forward, there is a need for further research into the economic and environmental impacts of new waste management technologies, such as AI-powered waste sorting and waste-to-energy systems. Additionally, longitudinal studies tracking the effectiveness of public education campaigns and community engagement strategies would provide valuable insights into improving waste management at the grassroots level. More research should also explore funding mechanisms and partnerships that can accelerate the implementation of smart waste management systems, particularly in countries with low GI scores. Expanding international cooperation, as well as fostering innovative public–private partnerships, is critical for addressing Africa's plastic waste crisis on a sustainable, long-term basis.

5.5.6 *Proposed Strategic Framework for Plastic Waste Management*

Framework Structure

The proposed strategic framework for plastic waste management in Africa integrates green intelligence (GI) as a core component, along with policy reforms, technological innovations, community engagement, and cross-sector collaboration. This framework aims to create a holistic and sustainable approach to managing plastic waste by addressing the key challenges identified across African nations. The framework is designed to adapt to the unique socioeconomic and environmental conditions of both urban and rural settings, allowing for flexibility in its implementation.

Keywords of the Framework

Technological Integration Through Green Intelligence: The use of information and communication technology (ICT) tools is crucial for enhancing plastic waste management. Geographic information systems (GISs) can map waste generation hotspots, enabling targeted interventions [21]. Smart waste collection systems equipped with sensors optimize collection routes and reduce operational costs [75]. Additionally, AI-powered sorting technologies improve the efficiency and accuracy of recycling processes by automating the separation of plastic types, increasing the effectiveness of recycling [30]. Leveraging these technologies allows for data-driven decision-making and more efficient waste management [76].

Policy Reforms and Enforcement: Strong regulatory frameworks are essential for reducing plastic waste. Policies such as plastic bans and extended producer responsibility (EPR) programs can significantly decrease plastic pollution [77]. For example, Rwanda's comprehensive ban on plastic bags, implemented in 2008, has substantially reduced plastic waste and set a global example for effective enforcement (Kigali Declaration, 2022). Similarly, providing incentives for companies to adopt sustainable practices, such as tax breaks for the use of eco-friendly materials, encourages broader compliance and innovation in waste management [78].

Community Engagement and Education: Engaging communities through education and awareness is key to fostering sustainable waste management practices. Initiatives such as waste collection drives and school-based recycling competitions can drive behavioral changes and increase recycling rates [79]. In rural areas, organizing informal waste collectors into structured waste management systems can improve recycling efficiency and provide economic benefits [11]. Community participation not only enhances waste management but also builds local capacity for ongoing sustainability [80].

Cross-sector Collaboration: Effective waste management requires collaboration across multiple sectors. Public–private partnerships (PPPs) can mobilize resources and expertise to enhance both technological adoption and policy implementation [81]. Governments, private companies, NGOs, and international organizations can work together to share best practices and provide the necessary support for successful waste management initiatives [11]. For example, international support from organizations such as UNEP and the World Bank can assist in scaling up local efforts and implementing advanced technologies.

To visualize the framework, the Sankey diagram (Fig. 5.4) places green intelligence at its core, serving as the central hub from which four key pillars emerge: technological integration, policy reforms, community engagement, and cross-sector collaboration. Each pillar is intricately linked to targeted strategies that drive sustainable plastic waste management. For example, technological integration includes innovations such as ICT tools and smart waste collection systems, whereas policy reforms encompass measures such as plastic bans and extended producer responsibility (EPR) programs. Community engagement highlights education and local involvement, and cross-sector collaboration emphasizes the synergy between public–private partnerships and international cooperation. Together, these interconnected elements create a holistic and adaptable framework aimed at addressing plastic waste across diverse contexts.

Practical Application

Urban Areas: In cities such as Lagos and Nairobi, the challenges of waste management are characterized by high population density and significant waste generation. Upgrading waste management infrastructure is crucial for improving efficiency and

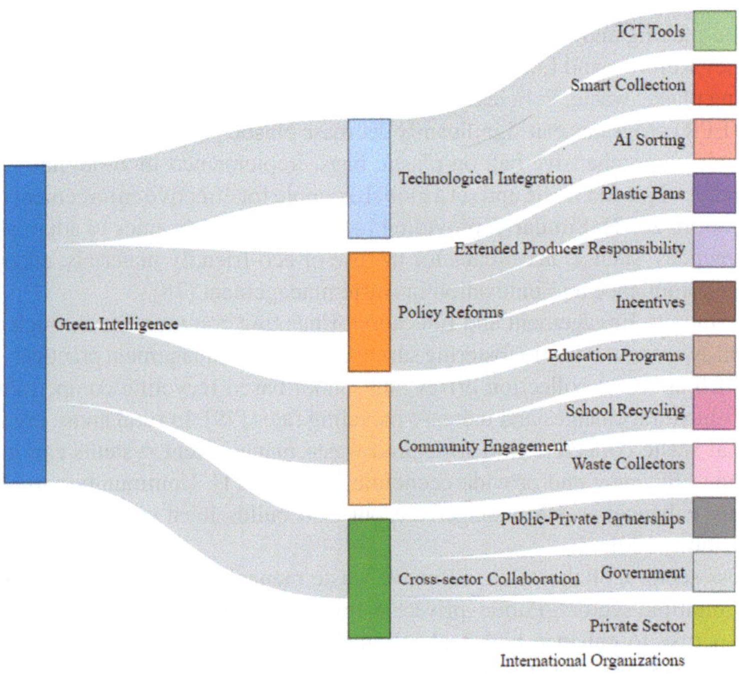

Fig. 5.4 Proposed green intelligence framework for Africa

effectiveness. The implementation of smart technologies, such as sensor-equipped waste bins and real-time monitoring systems, can significantly increase waste collection processes and reduce operational costs [82]. Geographic information system (GIS) mapping can identify waste hotspots and optimize collection routes, leading to more efficient waste management [83]. Additionally, public–private partnerships (PPPs) play a vital role in advancing recycling technologies and developing waste-to-energy solutions. These collaborations can facilitate the deployment of state-of-the-art recycling facilities and waste-to-energy plants, transforming plastic waste into valuable resources and reducing reliance on landfills [84]. Such partnerships provide both financial and technical support while driving innovation in urban waste management practices.

Rural Areas: In contrast, rural regions face distinct challenges, including limited formal waste management systems and lower waste generation rates. The development of community-based programs is essential for addressing these challenges effectively. Integrating informal waste collectors into formal systems can increase waste collection efficiency and create economic opportunities for local workers [85]. Mobile applications can play a significant role in improving waste tracking and management by allowing rural communities to report waste generation and monitor disposal practices [86]. Furthermore, employing low-cost technologies, such as manual sorting stations and small-scale waste-to-energy units, offers practical solutions for waste management in areas with minimal infrastructure [87]. These approaches, combined with local training programs and community engagement initiatives, can significantly improve recycling rates and promote sustainable waste management practices in rural settings [88].

In essence, the contrasting challenges and solutions between urban and rural areas underscore the necessity for tailored approaches in waste management. Urban strategies should focus on advanced technologies and infrastructure improvements, whereas rural strategies should emphasize community involvement and low-cost solutions. Recognizing these differences is vital for developing effective waste management systems that cater to the unique needs of each area.

Benefits of the Framework

The integration of effective policies and advanced technologies plays a crucial role in mitigating plastic pollution. The implementation of stringent regulations, such as plastic bans and extended producer responsibility (EPR) programs, combined with innovative technologies such as smart waste management systems, has proven effective in significantly reducing plastic waste. Research indicates that these measures decrease the volume of plastic entering landfills and the natural environment, thereby improving overall environmental conditions [89].

Examples of Successful Implementation

- **Rwanda's Plastic Bag Ban**: Since its comprehensive plastic bag ban was implemented in 2008, Rwanda has successfully curtailed plastic waste, serving as a model for other regions. The ban is complemented by strong enforcement measures and public education campaigns, leading to a notable decrease in plastic waste littering the environment [44].

- **Kenya's EPR Program**: Kenya has adopted an EPR program, which encourages producers to manage the lifecycle of their plastic products. The collaboration between the government and manufacturers has resulted in increased recycling rates and reduced plastic waste in urban areas such as Nairobi [15].
- **Smart Waste Management in South Africa**: Cities such as Cape Town utilize smart waste management technologies, including sensor-equipped bins and real-time monitoring systems. These innovations have improved waste collection efficiency and reduced operational costs, leading to more sustainable waste management practices [90].

Public Health Benefits

Proper waste management directly contributes to improved public health by reducing pollution-related risks. Effective disposal and recycling practices mitigate health hazards caused by improper plastic waste, such as flooding and waterborne diseases. For example, in cities such as Accra, where waste management systems have been enhanced, a significant decline in health issues related to plastic pollution has been observed, underscoring the importance of comprehensive strategies in safeguarding public health [91].

Economic Opportunities and the Circular Economy

Adopting recycling technologies and waste-to-energy systems reduces environmental impacts while creating economic opportunities. For example, the waste-to-energy project in Dar es Salaam, Tanzania, has successfully converted organic waste into energy, providing new business ventures and job opportunities for local communities [92]. Moreover, integrating informal waste pickers into the formal economy not only improves working conditions but also drives sustainable growth in these communities.

Collaborative Approaches

Strengthening collaboration among governments, the private sector, and NGOs is essential for scaling up waste management solutions. Public–private partnerships (PPPs) in countries such as Kenya demonstrate how collaborative efforts can lead to innovative and efficient waste management strategies [93]. These partnerships enable resource sharing and knowledge exchange, which are critical to addressing the complex challenges of plastic waste management.

Alignment with Sustainable Development Goals (SDGs)

These approaches align with sustainable development goals (SDGs), particularly SDG 11 (Sustainable Cities and Communities) and SDG 12 (Responsible Consumption and Production). By promoting effective waste management practices, African nations can align their policies with international sustainability benchmarks, contributing to global efforts to combat plastic pollution and support sustainable development [94].

The findings of this study highlight the urgent need for innovative solutions to address plastic waste management in Africa. Countries with higher green intelligence (GI) scores, such as Rwanda and Kenya, demonstrate the effectiveness of integrating technology, robust policy enforcement, and community engagement in improving waste management outcomes. Conversely, nations such as Nigeria and Ghana face significant challenges due to inadequate infrastructure, weak policy enforcement, and low public awareness, leading to poor waste collection and recycling rates.

The implementation of a strategic framework that combines green intelligence with policy reforms, public–private partnerships, and community-driven initiatives presents a viable solution to Africa's growing plastic waste crisis. It is crucial for African governments, businesses, and civil society to collaborate and adopt green intelligence-based strategies for managing plastic waste. Policymakers should prioritize the development and enforcement of regulations, such as plastic bans and EPR programs, while investing in ICT-based solutions to increase waste management efficiency.

The private sector must be incentivized to invest in recycling technologies and sustainable materials, whereas communities should be empowered through education and participatory waste management initiatives. By fostering collaboration across sectors and leveraging technological innovations, African nations can significantly reduce plastic pollution and improve public health and economic outcomes.

Future Research Directions

Future research should focus on several key areas. First, more detailed studies on rural waste management are needed to understand the unique challenges faced by rural communities and to develop scalable solutions. Second, the validation of green intelligence metrics in real-world African contexts will help refine the measurement tools and improve their applicability across diverse regions. Finally, research into the impact of policy interventions on plastic waste reduction—such as the effectiveness of plastic bans and recycling incentives—will provide valuable insights into the long-term sustainability of these strategies. These future research directions will contribute to more effective, inclusive, and sustainable waste management solutions for Africa.

The integration of effective policies and advanced technologies plays a crucial role in mitigating plastic pollution. Through the implementation of stringent regulations such as **plastic bans** and **extended producer responsibility (EPR)** programs, combined with technologies such as **smart waste management systems**, plastic waste can be significantly reduced. Research shows that these measures help decrease the volume of plastic entering landfills and the natural environment, thus improving overall environmental conditions [77, 95]. A prime example is **Rwanda's plastic bag ban**, which has been highly effective in curbing plastic waste, providing a model for other regions [44].

Proper waste management also contributes directly to **improved public health** by reducing pollution-related risks. Effective disposal and recycling practices help mitigate the health hazards caused by improper plastic waste, including flooding and waterborne diseases [96]. In cities with well-implemented waste management systems, there has been a significant decline in health issues associated with plastic pollution, highlighting the importance of comprehensive strategies in safeguarding public health [97].

Adopting **recycling technologies** and **waste-to-energy systems** not only reduces the environmental impact but also creates **economic opportunities**. These technologies promote a **circular economy** by providing new business ventures and employment opportunities. Additionally, waste-to-energy systems help reduce reliance on nonrenewable energy sources while improving working conditions for informal waste pickers, who are integrated into the formal economy, driving sustainable growth [98].

Strengthening collaboration among governments, the private sector, and NGOs is essential for scaling up waste management solutions. **Public–private partnerships (PPPs)** in countries such as **Kenya** demonstrate how collaborative efforts can lead to innovative and efficient waste management strategies [93]. These partnerships enable resource sharing and knowledge exchange, which are critical to tackling the complex challenges of plastic waste management.

Finally, these approaches align with **sustainable development goals (SDGs)**, especially SDG 11 (Sustainable Cities and Communities) and SDG 12 (Responsible Consumption and Production). By promoting effective waste management practices, African nations can align their policies with international sustainability benchmarks, contributing to global efforts to combat plastic pollution and support sustainable development [99, 100].

Study Limitation

The study acknowledges several limitations in implementing green intelligence for sustainable plastic waste management in Africa, particularly regarding the development and application of the green intelligence (GI) score. Limited access to reliable and consistent data on plastic waste generation, recycling rates, and waste-to-energy technologies challenges the accuracy and effectiveness of the GI score in guiding decision-making. Variability in technological infrastructure, digital capacity, and policy enforcement across African nations further complicates the use of the GI score as a standardized tool. Additionally, the socio-economic diversity and fragmented governance structures across the continent hinder the uniform implementation of green intelligence strategies. While the GI score highlights the relationship between technological advancements and effective waste management, gaps in funding, stakeholder collaboration, and community-level engagement limit its practical application. Addressing these challenges will require investments in data collection and technological capacity, improved coordination among stakeholders, and localized strategies to ensure the GI score's relevance and impact in fostering sustainable plastic waste management.

5.6 Conclusion

This study reveals a critical need for innovative approaches to plastic waste management in Africa. Countries such as Rwanda and Kenya, which boast high green intelligence (GI) scores, effectively integrate technology, robust policy enforcement, and community engagement, leading to improved waste management outcomes. In contrast, nations such as Nigeria and Ghana face significant challenges stemming from inadequate infrastructure, weak policy enforcement, and low public awareness, resulting in poor waste collection and recycling rates. Successful models, such as Rwanda's plastic bag ban and Kenya's extended producer responsibility (EPR) program, serve as examples that other countries can adopt. The implementation of a strategic framework that combines green intelligence, public–private partnerships, and community-driven initiatives is essential for addressing Africa's escalating plastic waste crisis. Policymakers should prioritize the development of stringent regulations and invest in ICT solutions that increase waste management efficiency. Moreover, fostering private sector investment in recycling technologies and empowering communities through education and participatory initiatives will be crucial in promoting sustainable waste management practices.

Future research should focus on understanding rural waste management challenges and developing scalable solutions tailored to these communities. Validating green intelligence metrics in real-world African contexts will also help refine measurement tools, enhancing their applicability across diverse regions. Furthermore, the socio-economic feasibility of using advanced machine learning techniques, such as ANN-based algorithms, to raise awareness and minimize plastic waste dumping in landfills, rivers, and oceans presents an important area for exploration. Investigating the impact of policy interventions on waste reduction, particularly regarding the effectiveness of plastic bans and recycling incentives, will provide valuable insights into the long-term sustainability of these strategies. These actions are vital for advancing effective and inclusive waste management practices across Africa, ultimately reducing plastic pollution and improving public health.

Appendix

Green intelligence (GI) score based on the provided data from Table 5.2. The GI score is calculated via the weights for each component:

- Plastic waste output (PWO): 30%
- Recycling rate (RR): 40%
- Policy and Technological Adoption (PTA): 30%

The formula for the GI score is as follows:
GI score = (0.3 × PWO Score) + (0.4 × RR Score) + (0.3 × PTA Score).

Step 1: Collect and standardize the data

Country	Plastic waste output (Million Tonnes/ Year)	Recycling rate (%)	PTA score (0–1)	Standardized PWO score (0–1)	Standardized RR score (0–1)	GI score
Rwanda	0.15	35	0.9	1.0	1.00	0.97
Kenya	0.20	30	0.8	0.9	0.85	0.85
South Africa	1.60	25	0.6	0.5	0.70	0.61
Nigeria	2.50	10	0.3	0.0	0.20	0.17
Ghana	1.00	12	0.4	0.7	0.25	0.43
Egypt	3.02	20	0.5	0.6	0.60	0.57

Step 2: Calculate the GI score for each country

1. Rwanda: GI score $= (0.3 \times 1.0) + (0.4 \times 1.00) + (0.3 \times 0.9) = 0.30 + 0.40 + 0.27 = 0.97$
2. Kenya: GI score $= (0.3 \times 0.9) + (0.4 \times 0.85) + (0.3 \times 0.8) = 0.27 + 0.34 + 0.24 = 0.85$
3. South Africa: GI score $= (0.3 \times 0.5) + (0.4 \times 0.70) + (0.3 \times 0.6) = 0.15 + 0.28 + 0.18 = 0.61$
4. Nigeria: GI score $= (0.3 \times 0.0) + (0.4 \times 0.20) + (0.3 \times 0.3) = 0.00 + 0.08 + 0.09 = 0.17$
5. Ghana: GI score $= (0.3 \times 0.7) + (0.4 \times 0.25) + (0.3 \times 0.4) = 0.21 + 0.10 + 0.12 = 0.43$
6. Egypt: GI score $= (0.3 \times 0.6) + (0.4 \times 0.60) + (0.3 \times 0.5) = 0.18 + 0.24 + 0.15 = 0.57$

Step 3: Final Green Intelligence Score Table

Country	GI score
Rwanda	0.97
Kenya	0.85
South Africa	0.61
Nigeria	0.17
Ghana	0.43
Egypt	0.57

Analysis

- **Rwanda** has the highest GI score at 0.97, largely because of its strong performance across all three metrics—low plastic waste output, high recycling rate, and advanced policy and technological adoption.

- **Kenya** follows with a GI of 0.85, demonstrating solid waste management practices.
- **Nigeria** scores the lowest (0.17) due to its high plastic waste output and poor recycling performance, despite some progress in policy and technological adoption.

References

1. Azoulay D, et al (2019) Plastic & health: the hidden costs of a plastic planet. CIEL Geneva
2. Adeogun AO et al (2020) Detection and occurrence of microplastics in the stomach of commercial fish species from a municipal water supply lake in southwestern Nigeria. Environ Sci Pollut Res 27:31035–31045
3. Chukwuka AV, et al (2024) Microplastic dynamics and risk projections in West African coastal areas: developing a vulnerability index, adverse ecological pathways, and mitigation framework using remote-sensed oceanographic profiles. Science of The Total Environment, p 175963
4. Jambeck JR et al (2015) Plastic waste inputs from land into the ocean. Science 347(6223):768–771
5. Jambeck J et al (2023) Leveraging multi-target strategies to address plastic pollution in the context of an already stressed ocean. The Blue Compendium: From Knowledge to Action for a Sustainable Ocean Economy. Springer, pp 141–184
6. Maurya A, et al (2025) Plastic waste: a long-lasting threat to the environment. Glob Waste Manag, 335–358
7. Ayeleru OO et al (2020) Challenges of plastic waste generation and management in sub-Saharan Africa: a review. Waste Manage 110:24–42
8. Raza A et al (2024) Impact of plastic packaging on waste accumulation and global tax regimes to mitigate the plastic pollution: multifaceted crisis and sustainable solutions. Plastic and the COVID-19 Pandemic: Innovative Solutions to Mitigate Plastic Pollution. Springer, pp 83–97
9. Gonçalves LR et al (2024) Against the clock to address plastic pollution: critical challenges to elaborate a comprehensive and ambitious plastic treaty. Camb Prism: Plast 2:e26
10. Godfrey L, et al (2019) Solid waste management in Africa: governance failure or development opportunity. RegNal Dev Afr **235**:(10.5772)
11. Aparcana S (2017) Approaches to formalization of the informal waste sector into municipal solid waste management systems in low-and middle-income countries: review of barriers and success factors. Waste Manage 61:593–607
12. Kibria MG et al (2023) Plastic waste: challenges and opportunities to mitigate pollution and effective management. Int J Environ Res 17(1):20
13. Kolade O, Oyinlola M, Rawn B (2023) A multi-stakeholder, multi-sectoral approach to a circular plastic economy in Eastern Africa. Digital Innovations for a Circular Plastic Economy in Africa. Routledge, pp 143–155
14. Enge CNG (2018) The Kenyan ban on plastic bags: a study of attitudes and adaptation in Nairobi. Norwegian University of Life Sciences, Ås
15. Katunge L (2019) Prohibition of plastics in Kenya: moving towards a constitutionally compliant governance regime for single-use plastics. University of Nairobi
16. Irambona AD, Mkhize I, Gbadeyan OJ (2024) Plastic pollution in Africa: a global concern. Biomass-based Bioplastic and Films: Preparation, Characterization, and Application. Springer, pp 1–26
17. Agnelli A, Tortora P (2022) The role of development co-operation in tackling plastic pollution: Key trends, instruments, and opportunities to scale up action

18. Chaudhary A, Sharan P (2022) Innovative computing for green technologies: potential paths for improved livelihood. Notion Press
19. Bibri SE, et al (2024) The synergistic interplay of artificial intelligence and digital twin in environmentally planning sustainable smart cities: a comprehensive systematic review. Environ Sci Ecotechnology 100433
20. Fragulis F, et al (2024) A survey of geographic information systems (GIS) and internet of things (IoT) technologies in brownfield transformation. In: AIP Conference Proceedings. AIP Publishing
21. Graham CC (2024) The role of geographic information systems in mitigating plastics pollution in the global South—a spatial analysis of recycling facilities in Costa Rica. Sci Total Environ, 173396
22. Dano UL et al (2020) Transformative urban governance: Confronting urbanization challenges with geospatial technologies in Lagos Nigeria. GeoJournal 85:1039–1056
23. Fang B et al (2023) Artificial intelligence for waste management in smart cities: a review. Environ Chem Lett 21(4):1959–1989
24. Szaky T (2019) The future of packaging: from linear to circular. Berrett-Koehler Publishers
25. Auer M, Detter K (2023) Eco-modulation of extended producer responsibility in the EU: impacts on the packaging industry and for the transition to a circular economy
26. Malinauskaite J et al (2017) Municipal solid waste management and waste-to-energy in the context of a circular economy and energy recycling in Europe. Energy 141:2013–2044
27. Hossain R et al (2022) Plastic waste management in India: Challenges, opportunities, and roadmap for circular economy. Sustainability 14(8):4425
28. Hernandez D et al (2024) The role of artificial intelligence in sustainable agriculture and waste management: towards a green future. Int Trans Artif Intell 2(2):150–157
29. Hoang T-D et al (2022) Artificial intelligence in pollution control and management: status and future prospects. Artif Intell Environ Sustain: ChallS Solut Era Ind 4:23–43
30. Reza M, Hassan M (2023) AI-Driven solutions for enhanced waste management and recycling in urban areas. Int J Sustain Infrastruct Cities Soc 8(2):1–13
31. Lubongo C, Alexandridis P (2022) Assessment of performance and challenges in use of commercial automated sorting technology for plastic waste. Recycling 7(2):11
32. Olawade DB, et al (2024) Smart waste management: a paradigm shift enabled by artificial intelligence. Waste Manag Bull
33. Popescu SM et al (2024) Artificial intelligence and IoT driven technologies for environmental pollution monitoring and management. Front Environ Sci 12:1336088
34. Thorleifsson T (2020) Barriers for private sector companies operating in waste management and recycling in Sub-Saharan Africa: a qualitative exploration
35. Worrall L, et al (2017) Better urban growth in Tanzania: preliminary exploration of the opportunities and challenges
36. Cartwright A (2015) Better growth, better cities: Rethinking and redirecting urbanisation in Africa. The New Climate Economy Working Paper
37. Uche OL (2023) Plastic waste regime in Rwanda, Kenya and South Africa: a comparative case study. Am J Law 5(2):54–85
38. Twagirayezu G, et al (2024) A critical review and analysis of plastic waste management practices in Rwanda. Environ Sci Pollut Res, 1–21
39. Babayemi JO et al (2019) Ensuring sustainability in plastics use in Africa: consumption, waste generation, and projections. Environ Sci Eur 31(1):1–20
40. Jambeck J et al (2018) Challenges and emerging solutions to the land-based plastic waste issue in Africa. Mar Policy 96:256–263
41. Mamman A, Gambo M (2024) The role of indigenous knowledge on plastic waste management in Nigeria. Technical University of Kenya
42. Harfadli MMa, et al (2024) Challenges and characteristics of the informal waste sector in developing countries: an overview. J Mater Cycles Waste Manag 26(3):1294–1309
43. Paul WM, Mironga J (2020) Effectiveness of the implementation of plastic bags ban: empirical evidence from Kenya. IOSR J Environ Sci 14:53–61

44. Behuria P (2021) Ban the (plastic) bag? Explaining variation in the implementation of plastic bag bans in Rwanda, Kenya and Uganda. Environ Plan C: Polit Space 39(8):1791–1808
45. Ogutu MO et al (2023) Implementing circular economy and sustainability policies in Rwanda: experiences of Rwandan manufacturers with the plastic ban policy. Front Sustain 4:1092107
46. Abba S, Light CI (2020) IoT-based framework for smart waste monitoring and control system: a case study for smart cities. Eng Proc 2(1):90
47. Ayad A et al (2021) An integrated approach for non-revenue water reduction in water distribution networks based on field activities, optimisation, and GIS applications. Ain Shams Eng J 12(4):3509–3520
48. Dhanaraju M et al (2022) Smart farming: Internet of Things (IoT)-based sustainable agriculture. Agriculture 12(10):1745
49. Lubongo C, Bin Daej MA, Alexandridis P (2024) Recent developments in technology for sorting plastic for recycling: the emergence of artificial intelligence and the rise of the robots. Recycling 9(4):59
50. Rezania S et al (2023) Review on waste-to-energy approaches toward a circular economy in developed and developing countries. Processes 11(9):2566
51. Ajala O (2023) Assessing plastic circular economy policies and the use of digital technology in Africa. Digital Innovations for a Circular Plastic Economy in Africa p 222
52. Kibii CJ (2021) Effectiveness of single-use plastic bag bans in Africa: barriers and potential solutions
53. Hira A et al (2022) Plastic waste mitigation strategies: a review of lessons from developing countries. J Dev Soc 38(3):336–359
54. Maalouf A, Agamuthu P (2023) Waste management evolution in the last five decades in developing countries–a review. Waste Manage Res 41(9):1420–1434
55. Cobbinah PB, Addaney M, Agyeman KO (2017) Locating the role of urbanites in solid waste management in Ghana. Environ Dev 24:9–21
56. Amoah ST, Kosoe EA (2014) Solid waste management in urban areas of Ghana: issues and experiences from Wa. J Environ Pollut Hum Health 2(5):110–117
57. Ebhaleme OA (2022) Regulating single-use plastic marine pollution: a case study of Nigeria
58. Niño N (2021) Kenya's plastic bag ban
59. Anokye K et al (2024) Understanding the perception and awareness of senior high school teachers on the environmental impacts of plastic waste: implications for sustainable waste education and management. Soc Sci & Humlties Open 10:100999
60. Kabera T, Wilson DC, Nishimwe H (2019) Benchmarking performance of solid waste management and recycling systems in East Africa: comparing Kigali Rwanda with other major cities. Waste Manag & Res 37(1_suppl):58–72
61. Mboup G, Oyelaran-Oyeyinka B (2019) Relevance of smart economy in smart cities in Africa. Smart economy in smart African cities: Sustainable, inclusive, resilient and prosperous, p 1–49
62. Tejaswini M et al (2022) A comprehensive review on integrative approach for sustainable management of plastic waste and its associated externalities. Sci Total Environ 825:153973
63. Lissah SY et al (2021) Managing urban solid waste in Ghana: Perspectives and experiences of municipal waste company managers and supervisors in an urban municipality. PLoS ONE 16(3):e0248392
64. Danielsson M (2017) The plastic bag ban in Rwanda: local procedures and successful outcomes
65. Ezeudu OB, Ezeudu TS (2019) Implementation of circular economy principles in industrial solid waste management: Case studies from a developing economy (Nigeria). Recycling 4(4):42
66. Onyeaka H et al (2023) Using artificial intelligence to tackle food waste and enhance the circular economy: maximising resource efficiency and Minimising environmental impact: a review. Sustainability 15(13):10482
67. Farghali M, Osman AI (2024) Revolutionizing waste management: unleashing the power of artificial intelligence and machine learning. Advances in Energy from Waste. Elsevier, pp 225–279

68. Lakhouit A (2025) Revolutionizing Urban Solid waste management with AI and IoT: a review of smart solutions for waste collection, sorting, and recycling. Results Eng, 104018
69. Nižetić S et al (2019) Smart technologies for promotion of energy efficiency, utilization of sustainable resources and waste management. J Clean Prod 231:565–591
70. Andre ODdS (2022) The issue of plastic pollution in the city of Luanda
71. Gutberlet J (2021) Grassroots waste picker organizations addressing the UN sustainable development goals. World Dev 138:105195
72. Taron A, et al (2023) Public-private partnerships for the circular bio-economy in the Global South: lessons learned
73. Marcussen IØ (2020) Urban solid waste management, environmental governance, and private sector initiatives: a case from Nairobi, Kenya. Norwegian University of Life Sciences, Ås
74. Hakuzimana J (2021) Break free from plastics: environmental perspectives and evidence from rwanda. Environ Ecosyst Sci 5(1):27–36
75. Ramos TRP, de Morais CS, Barbosa-Povoa AP (2018) The smart waste collection routing problem: alternative operational management approaches. Expert Syst Appl 103:146–158
76. Farjana M et al (2023) An iot-and cloud-based e-waste management system for resource reclamation with a data-driven decision-making process. IoT 4(3):202–220
77. Diggle A, Walker TR (2020) Implementation of harmonized extended producer responsibility strategies to incentivize recovery of single-use plastic packaging waste in Canada. Waste Manage 110:20–23
78. Purwandani JA, Michaud G (2021) What are the drivers and barriers for green business practice adoption for SMEs? Environ Syst Decis 41(4):577–593
79. Thomas RM (2018) Environmental engagement through behaviour change interventions: a case study of litter reduction in New Zealand Schools. Open Access Te Herenga Waka-Victoria University of Wellington
80. Fraser ED et al (2006) Bottom up and top down: Analysis of participatory processes for sustainability indicator identification as a pathway to community empowerment and sustainable environmental management. J Environ Manage 78(2):114–127
81. Kang S et al (2019) Public-private partnerships in developing countries: factors for successful adoption and implementation. Int J Public Sect Manag 32(4):334–351
82. Mishra A, Kumar Ray A (2020) IoT cloud-based cyber-physical system for efficient solid waste management in smart cities: a novel cost function based route optimisation technique for waste collection vehicles using dustbin sensors and real-time road traffic informatics. IET Cyber-Physical Syst: Theory & Appl 5(4):330–341
83. Slavík J, Dolejš M, Rybová K (2021) Mixed-method approach incorporating Geographic information system (GIS) tools for optimizing collection costs and convenience of the biowaste separate collection. Waste Manage 134:177–186
84. Kerdlap P, Low JSC, Ramakrishna S (2019) Zero waste manufacturing: a framework and review of technology, research, and implementation barriers for enabling a circular economy transition in Singapore. Resour Conserv Recycl 151:104438
85. Wilson DC, Velis C, Cheeseman C (2006) Role of informal sector recycling in waste management in developing countries. Habitat Int 30(4):797–808
86. Adeniran AA, Shakantu W, Ayesu-Koranteng E (2022) A proposed digital control system using a mobile application for municipal solid waste management in South Africa. Waste Technol 10(1):30–42
87. Hoang AT et al (2022) Perspective review on municipal solid waste-to-energy route: characteristics, management strategy, and role in circular economy. J Clean Prod 359:131897
88. Das S et al (2019) Solid waste management: scope and the challenge of sustainability. J Clean Prod 228:658–678
89. Mihai F-C, et al (2024) Plastic waste management for zero waste to landfills: potential, challenges, and opportunities. Zero Waste Management Technologies, p 97–136
90. Samuriwo TN, et al (2024) Comparative review of remote sensing methods for ocean plastic litter detection. IEEE Access

91. Mudu P, et al (2021) Solid waste management and health in Accra, Ghana. World Health Organization
92. Doukali I et al (2024) Advancing circular economy and waste management in Zanzibar through youth entrepreneurship and innovation. Generation of Energy from Municipal Solid Waste: Circular Economy and Sustainability. Springer, pp 345–373
93. Ogutu FA, Kimata DM, Kweyu RM (2021) Partnerships for sustainable cities as options for improving solid waste management in Nairobi city. Waste Manage Res 39(1):25–31
94. Awewomom J et al (2024) Addressing global environmental pollution using environmental control techniques: a focus on environmental policy and preventive environmental management. Discov Environ 2(1):8
95. Lorang S et al (2022) Achievements and policy trends of extended producer responsibility for plastic packaging waste in Europe. Waste Dispos & Sustain Energy 4(2):91–103
96. Chireshe A, Shabani T, Shabani T (2023) Safety and health risks associated with illegal municipal solid waste disposal in urban Zimbabwe."A case of Masvingo City". Saf Extrem Environ 5(3):243–252
97. Silva ALP et al (2021) An urgent call to think globally and act locally on landfill disposable plastics under and after covid-19 pandemic: pollution prevention and technological (Bio) remediation solutions. Chem Eng J 426:131201
98. Williams PA, et al (2023) Review of barriers to effective implementation of waste and energy management policies in Ghana: implications for the promotion of waste-to-energy technologies. Waste. MDPI
99. Muheirwe F, Kombe W, Kihila JM (2022) The paradox of solid waste management: a regulatory discourse from Sub-Saharan Africa. Habitat Int 119:102491
100. Rodić L, Wilson DC (2017) Resolving governance issues to achieve priority sustainable development goals related to solid waste management in developing countries. Sustainability 9(3):404

Part II
AI in Water and Wastewater Treatment

Chapter 6
Application Bootstrap Machine Learnings for Modeling of Sonocatalytic Degradation of Caffeine in Wastewater Treatment

Fakhrony Sholahudin Rohman, Dinie Muhammad, Sharifah Rafidah Wan Alwi, and Nur Fadzeelah Abu Kassim

Abstract Sonocatalysis, an efficient technique for the degradation of organic contaminants, synergistically combines ultrasonic irradiation with catalysts such as CeO_2 nanorods (NRs). This study investigates the utilization of CeO_2 NRs in this process to degrade caffeine in water. The CeO_2 NRs was synthesized via the hydrothermal method and characterized using XRD, high-resolution transmission electron microscope (HRTEM), Fourier-transform infrared spectroscopy (FTIR), and Raman spectroscopies. Then, the effect of the process parameter on the sonocatalytic degradation of caffeine was evaluated by varying the initial pH of the solution, initial caffeine concentrations, and catalyst dosages. Machine learning (ML) models, including support vector machines (SVM), artificial neural networks (ANN), and Gaussian process regression (GPR) with bootstrap resampling, were employed to analyze and optimize the process. The bootstrap technique improved prediction accuracy by generating additional data, with the GPR-bootstrap model achieving an R^2 of 0.999 and RMSE of 0.001. The highest caffeine degradation (91.5%) was observed at an initial pH of the solution of 7.5, 1 g L^{-1} of CeO_2 NRs dosage, 5 mg L^{-1} of

F. S. Rohman · S. R. W. Alwi (✉)
Process Systems Engineering Centre (UTM-PROSPECT), Research Institute of Sustainable Environment (RISE), Universiti Teknologi Malaysia, Bahru, Johor, Malaysia
e-mail: syarifah@utm.my

F. S. Rohman
Faculty of Engineering, Universiti Malaysia Sabah, Jalan UMS, Kota Kinabalu, Sabah, Malaysia

D. Muhammad
Safety and Health Research Group, Universiti Teknologi Malaysia, Bahru, Johor, Malaysia

D. Muhammad · S. R. W. Alwi
Faculty of Chemical and Energy Engineering, Universiti Teknologi Malaysia (UTM), Bahru, Johor, Malaysia

N. F. A. Kassim
Chemical Engineering Studies, College of Engineering, Universiti Teknologi MARA, Cawangan Pulau Pinang, Permatang Pauh, Pulau Pinang, Malaysia

M. Nasr et al. (eds.), *Artificial Intelligence Applications for a Sustainable Environment*, Green Chemistry and Sustainable Technology,
https://doi.org/10.1007/978-3-031-91199-6_6

caffeine concentration, and 20 mM H_2O_2. These findings confirm the effectiveness of bootstrap-enhanced machine learning in accurately predicting and optimizing the sonocatalytic processes.

Keywords CeO_2 NRs · Sonocatalytic degradation · Caffeine · Machine learning

6.1 Introduction

Emerging micropollutants, such as pharmaceuticals and personal care products (PPCPs), have raised significant environmental concerns over the past two decades. The high demand in various industries has resulted in the production of approximately 20 million tonnes/year of commercial PPCPs, with an expected continuous increase. PPCPs encompass a diverse range of organic compounds, including antibiotics, hormones, liquid regulators, antidepressants, nonsteroidal anti-inflammatory drugs, disinfectants, fragrances, preservatives, sunscreens, and winter protection agents [1]. Notably, caffeine, a natural component in coffee, tea, and chocolate, is also a common PPCP used as a stimulant. Due to its widespread use and persistence in the environment, caffeine has become a frequently detected pollutant, raising concerns over its accumulation and adverse impacts on aquatic ecosystems. Conventional water treatment methods are limited in removing such recalcitrant compounds [2].

To address the challenges posed by persistent organic pollutants like caffeine, advanced oxidation processes (AOPs) have emerged as promising technologies. Among these, sonocatalysis has gained attention for decomposing micropollutants in water and wastewater. Sonocatalysis employs ultrasonic irradiation in conjunction with semiconductor catalysts, leveraging the phenomenon of acoustic cavitation—nucleation, growth, and implosive collapse of microbubbles. This process generates highly reactive hydroxyl radicals (•OH) capable of oxidizing organic contaminants into harmless byproducts like CO_2 and H_2O. Ultrasonic waves within the frequency range of 20–100 kHz can initiate acoustic cavitation [3].

Cerium (IV) oxide (CeO_2), commonly referred to as ceria, is a widely studied rare-earth oxide known for its redox properties, specifically the reversible $Ce^{4+} \rightleftharpoons Ce^{3+}$ transition. This feature creates oxygen vacancies on its surface and bulk, enhancing its catalytic oxidation capabilities and making it a highly effective material for AOPs. However, the sonocatalytic degradation mechanism is complex and influenced by several factors, such as catalyst dosage, initial pH, temperature, reaction time, and contaminant concentration [4]. Traditional univariate studies, such as one factor at a time (OFAT) analysis, often fail to capture the intricate interactions among these parameters. Consequently, there is a need for advanced modeling techniques to better understand and optimize these processes [5].

Machine learning (ML) offers a powerful approach to modeling and optimizing AOPs like sonocatalysis. ML algorithms, such as support vector machines (SVM), regression trees, artificial neural networks (ANN), and Gaussian process regression (GPR), can analyze experimental data to predict degradation efficiency under various conditions [6]. By modeling complex parameter interactions, ML enables the identification of optimal conditions for caffeine degradation in wastewater. While ML has been applied to other pollutants, such as isopropylxanthic acid, diclofenac, ciprofloxacin, and carbamazepine, its application to PPCPs in sonocatalysis remains underexplored [7].

Bootstrap is a statistical resampling technique used to estimate the accuracy and variability of a model by generating multiple datasets from the original dataset. In bootstrap, random samples are drawn with replacement, meaning some data points may appear multiple times while others may be excluded in any given sample. Each bootstrap sample is used to train and validate a model, helping to evaluate its performance across different variations of the data. Bootstrap enhances machine learning by improving evaluation, reducing overfitting, and increasing model reliability, especially when data is scarce or variability is high [8].

In this study, CeO_2 nanorods (NRs) was synthesized via the hydrothermal method and characterized to evaluate its structural properties, morphology and functional groups. The sonocatalytic degradation of caffeine was then modeled using various ML algorithms. Experimental parameters, including initial pH, catalyst dosage, caffeine concentration, and reaction time, were used as inputs to train and test the ML models. The performance of ANN, SVM, regression trees, and GPR with bootstrap resampling was compared to identify the most accurate model. As far as the authors are aware, this is the first study to employ ML for modeling the sonocatalytic degradation of caffeine using CeO_2 NRs.

The primary objective of this work is to develop a comprehensive framework for the sonocatalytic degradation of caffeine using CeO_2 nanorods (NRs) and to leverage machine learning (ML) models for process optimization and prediction. Specifically, this study aims to:

1. Synthesize and characterize CeO_2 NRs to evaluate their structural, morphological, and functional properties for use in sonocatalysis.
2. Investigate the influence of key operational parameters, such as initial pH, catalyst dosage, caffeine concentration, and reaction time, on the degradation efficiency.
3. Apply advanced ML algorithms, including ANN, SVM, regression trees, and GPR, to model the sonocatalytic process and predict its performance under various conditions.
4. Employ bootstrap resampling techniques to enhance model reliability and assess the variability in predictions.
5. Compare the accuracy and effectiveness of different ML models to identify the best-suited approach for optimizing the sonocatalytic degradation of caffeine.

This work aims to bridge the gap between experimental studies and advanced data-driven modeling, providing a robust methodology for addressing the challenges of micropollutant removal in wastewater treatment.

6.2 Materials and Methods

6.2.1 *Materials*

Cerium (III) nitrate hexahydrate ($Ce(NO_3)_3{\cdot}6H_2O$, 99%, metal basis, Sigma-Aldrich) served as the cerium precursor. Sodium hydroxide (NaOH, 98% purity), hydrogen peroxide (H_2O_2, 30% w/w solution), and caffeine powder (molecular weight: 194.19 g/mol) were purchased from Merck. All chemicals were of analytical grade and used without further purification.

6.2.2 *Characterization of CeO_2 Nanorods (NRs)*

The crystalline structure of the synthesized CeO_2 nanorods was determined using a Bruker D8 Advance X-ray diffractometer (XRD) equipped with Cu Kα radiation ($\lambda = 1.5406$ Å). The XRD analysis was performed under operating conditions of 40 kV and 40 mA, with a step size of 0.02°. The diffraction peaks were matched with standard patterns from the Joint Committee on Powder Diffraction Standards (JCPDS) database to identify the crystalline phases. Raman spectroscopy was conducted with a Renishaw in Via Raman microscope at a spectral resolution of 1 cm^{-1} to further investigate the structural properties.

The morphology of the synthesized CeO_2 NRs was examined using transmission electron microscopy (TEM) (ZEISS LIBRA 120) and a high-resolution transmission electron microscope (HRTEM, JEOL JEM 2100T) operated at accelerating voltages of 100 kV and 200 kV, respectively. The sample was dispersed in ethanol via sonication, and a small droplet of the suspension was deposited onto a carbon-coated copper grid for drying prior to imaging. The functional groups and chemical characteristics of the CeO_2 NRs were examined through Fourier-transform infrared spectroscopy (FTIR) using a Thermo Scientific Nicolet 6700 instrument.

6.2.3 *Sonocatalytic Degradation of Caffeine*

The sonocatalytic degradation of caffeine was carried out in a batch-mode setup using an ultrasonic bath (Elmasonic S 60 H, Germany) operating at 37 kHz frequency and 150 W power. The sonocatalytic setup also included a 250 mL reaction vessel, a pump, a holding tank and a thermocouple for precise temperature monitoring and control as illustrated in the schematic diagram in Fig. 6.1. Ice cubes were periodically added to the holding tank to maintain the water bath temperature at 25 ± 3 °C and prevent overheating during the sonocatalysis process.

Prior to the addition of the synthesized CeO_2 NRs, the initial pH of 100 mL of the caffeine solution at a specific concentration was adjusted using 0.1 M NaOH or 0.1 M

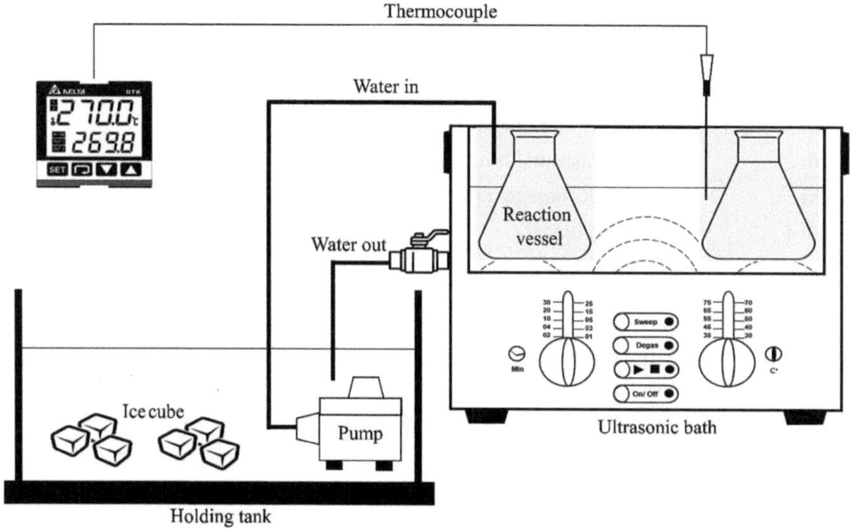

Fig. 6.1 Schematic diagram of the sonocatalytic experimental setup

HCl. The suspension was then subjected to magnetic stirring for 30 min to establish adsorption–desorption equilibrium. The suspension was added with 20 mM H_2O_2 as the oxidant and was sonicated in the ultrasonic bath for 150 min. Samples of 4 mL were withdrawn at 30-min intervals and filtered using a 0.22 μm polyvinyl difluoride (PVDF) syringe filter. The absorbance of the filtered solution was measured at the maximum wavelength ($\lambda_{\max}$) of 273.5 nm using a Lambda 25 UV–Vis spectrophotometer (PerkinElmer, USA). The sample concentration was determined using a standard calibration curve and the caffeine removal efficiency was calculated using the formula in Eq. 6.1:

$$\text{Degradation Efficiency}(\%) = \frac{C_i - C_t}{C_i} \times 100 \tag{6.1}$$

where C_i represents the initial caffeine concentration, and C_t is the caffeine concentration at a given time.

6.2.4 Development of Machine Learning

In this study, several ML algorithms, including support vector machines (SVM), regression trees, artificial neural networks (ANN), and Gaussian process regression (GPR), were employed to model and optimize the sonocatalytic degradation

Table 6.1 Input–output parameters and respective range for the sonocatalytic degradation of caffeine in water using CeO_2 nanorods (NRs)

Notation	Parameter	Range
Input 1	Initial pH of the solution (pH)	3.5–7.5
Input 2	Initial concentration of caffeine (C_o)	5.0–30.0 mg L^{-1}
Input 3	CeO_2 NRs dosage (D)	0.5–2.0 g L^{-1}
Input 4	Reaction time (t)	0–150 min
Target	Degradation of caffeine (DE)	0–100%

of caffeine. These algorithms are particularly well-suited for handling multidimensional data, allowing for accurate predictions and sensitivity analyses of factors such as pH, caffeine concentration, and CeO_2 nanorod dosage.

6.2.4.1 Input–Output Selection

The input and output data for the machine learning (ML) model were determined based on the experimental design, which involved varying key process parameters to observe their impact on caffeine's degradation efficiency (DE). Experimental data were collected by systematically altering four parameters: initial pH of the solution (pH), initial caffeine concentration (C_o), CeO_2 NRs dosage (D), and reaction time (t). The degradation efficiency served as the target output variable. Table 6.1 provides a summary of the input–output parameters along with their respective ranges.

To develop and validate the ML model, 90% of the experimental data was utilized for training, while the remaining 10% was reserved for testing.

6.2.4.2 Bootstrapping

Bootstrapping (BS) is a statistical resampling method widely used to enhance the robustness of data analysis, especially when the available dataset is limited. The technique involves generating multiple synthetic datasets, called bootstrap samples, by random sampling with replacements from the original dataset. This process preserves the inherent statistical characteristics of the original data while increasing the effective sample size. Bootstrapping is advantageous in ML applications, where larger datasets improve model training, validation, and generalization [8].

This study employed the bootstrapping method to expand the original 56 experimental data points to 560 data points to better train the machine learning (ML) model. Each bootstrap sample was created by randomly selecting data points from the original dataset, allowing repetition of individual points. The MATLAB programming environment was used to automate this process efficiently. By applying this resampling technique, a total of 10 bootstrap samples were generated, each containing 56 data points. These samples were concatenated to form an augmented dataset of 560 points, ensuring that the expanded data retained the variability and distribution of the original dataset.

6.2.4.3 Development of the ANN Model

Artificial neural network (ANN) models are computational systems inspired by the human brain, commonly used in machine learning to address complex regression. They are made up of layers of artificial neurons connected to each other. Each neuron processes input data by calculating a weighted sum, applying a nonlinear activation function, and sending the result to the next layer [9]. The structure includes an input layer (representing the features), one or more hidden layers (capturing complex patterns), and an output layer (producing predictions). ANNs are highly effective at modeling nonlinear systems, making them valuable for tasks with intricate and high-dimensional relationships between inputs and outputs [10].

The operation of a neural network can be mathematically described as follows. For a single-layer neuron, the output (y) is computed as:

$$y = f\left(\sum_{i=1}^{n} w_i x_i + b\right) \tag{6.2}$$

where (x_i) represents the inputs, (w_i) are the weights, (b) is the bias term, and $f(\cdot)$ is the activation function, such as a sigmoid, ReLU, or tanh function. For multi-layer networks, the process is extended through successive layers, with the output of each layer serving as the input for the next. The final prediction is determined after the network has processed the inputs through all layers.

ANNs are trained using an iterative optimization process, typically employing the backpropagation algorithm to minimize a cost function ($J(\theta)$. The cost function quantifies the error between predicted outputs and actual targets, commonly using mean squared error (MSE) for regression tasks:

$$J(\theta) = \frac{1}{m}\sum_{i=1}^{m} \left(y_i - \widehat{y_i}\right)^2 \tag{6.3}$$

where (m) is the number of training samples, y_i is the true output, and $\widehat{y_i}$ is the predicted output. Weights (w) and biases (b) are updated iteratively using gradient descent:

$$w^{(t+1)} = w^{(t)} - \eta \frac{\partial J(\theta)}{\partial w} \tag{6.4}$$

where η is the learning rate.

The key advantage of ANNs lies in their ability to approximate any continuous function given sufficient data and appropriate architecture, as stated by the Universal

Approximation Theorem [11]. They are highly flexible, capable of handling both structured and unstructured data, and can learn hierarchical feature representations, making them well-suited for tasks involving complex, nonlinear relationships [12].

6.2.4.4 Development of SVM Regression

Support vector machines (SVMs) are versatile supervised learning models used for tasks like classification, regression, and outlier detection. In regression, SVMs aim to predict continuous values by finding the optimal hyperplane that best fits the data. The key idea behind SVMs is to focus on the most important data points, called support vectors, which define the margin around the hyperplane. This approach helps SVMs handle high-dimensional feature spaces effectively and provides strong generalization, making them less prone to overfitting [13].

Mathematically, the SVM for a binary classification task aims to solve the following optimization problem:

$$\min_{w,b} \frac{1}{2}|\boldsymbol{w}|^2 \tag{6.5}$$

subject to:

$$y_i\left(\boldsymbol{w}^T\boldsymbol{x}_i + b\right) \geq 1, \quad i = 1,2,\ldots,N, \tag{6.6}$$

where ($\boldsymbol{w}$) is the weight vector defining the hyperplane, (b) is the bias term, ($\boldsymbol{x}_i$) represents the feature vector of the (i)-th data point, and ($y_i \in \{-1, +1\}$) denotes the corresponding class label. The optimization problem ensures that data points are correctly classified while maximizing the margin.

When the data is not linearly separable, the optimization is modified by introducing a slack variable (ξ_i) and a penalty parameter (C), resulting in a soft-margin SVM:

$$\min_{(w,b,\xi)} 1/2|\boldsymbol{w}|^2 + C\sum (i = 1)^N \xi_i \tag{6.7}$$

subject to:

$$y_i\left(\boldsymbol{w}^T\boldsymbol{x}_i + b\right) \geq 1 - \xi_i, \quad \xi_i \geq 0, \quad i = 1,2,\ldots,N. \tag{6.8}$$

SVMs use the kernel trick to handle nonlinear relationships. This method transforms the input data into a higher-dimensional space where a linear separation becomes possible. The kernel function ($K(\boldsymbol{x}_i, \boldsymbol{x}_j)$) calculates the similarity between data points in this transformed space. Common kernel functions include:

1. Quadratic Kernel: Incorporates quadratic interactions between features, capturing more complex relationships:

$$K(x_i, x_j) = \left(x_i^T x_j + 1\right)^2 \tag{6.9}$$

2. Cubic Kernel: Extends the feature interactions to cubic terms, allowing even more flexibility:

$$K(x_i, x_j) = \left(x_i^T x_j + 1\right)^3 \tag{6.10}$$

3. Fine Kernel: Refers to specific configurations with small kernel bandwidths, often tailored to provide high-resolution separation in small neighborhoods of the data.

In regression tasks (Support vector regression, SVR), SVMs minimize the error within an (ϵ)-insensitive zone, focusing only on deviations greater than (ϵ). The objective becomes:

$$\min_{w,b} \frac{1}{2}|w|^2 + C\sum_{i=1}^{N}\left(\xi_i^+ + \xi_i^-\right) \tag{6.11}$$

subject to:

$$y_i - \left(w^T x_i + b\right) \le \epsilon + \xi_i^+, \quad \left(w^T x_i + b\right) - y_i \le \epsilon + \xi_i^-, \quad \xi_i^+, \xi_i^- \ge 0. \tag{6.12}$$

SVMs are well-suited for high-dimensional data, resistant to overfitting, and adaptable to nonlinear problems through various kernel functions. They work especially well with small to medium-sized datasets that have complex patterns [14].

In MATLAB, SVM models can be developed using the 'fitcsvm' function for classification and 'fitrsvm' for regression. By specifying kernel functions (e.g., 'quadratic', 'cubic'), researchers can adapt the SVM to the problem's complexity. MATLAB's built-in tools also provide options to tune hyperparameters like (C) and (ϵ) for optimal performance, making SVM implementation straightforward and highly customizable [15].

6.2.4.5 Regression Trees

Decision trees and regression trees (collectively referred to as 'Tree' models) are non-parametric, interpretable models that segment the feature space into regions based on simple decision rules derived from the input data. A Tree works by recursively partitioning the dataset into subsets based on feature values, ultimately producing a hierarchical structure of nodes and branches. This structure allows for straightforward interpretation, as each path from the root to a leaf represents a set of rules leading to a prediction [16].

In a regression tree, the target variable is continuous, and the model predicts the output by averaging the target values of the samples within a given leaf node. The algorithm selects splits that minimize the error within the subsets by optimizing a metric such as mean squared error (MSE). At each split, the objective is to partition the data to achieve the greatest reduction in impurity, defined for regression as:

$$\text{MSE} = \frac{1}{N}\sum_{i=1}^{N}\left(y_i - \hat{y}\right)^2 \tag{6.13}$$

where (y_i) represents the true values, $(\hat{y})$ is the predicted value, and (N) is the number of data points in a node. The splitting criterion selects the feature and threshold that result in the largest decrease in MSE:

$$\Delta\text{MSE} = \text{MSE}_{\text{parent}} - \left(\frac{N_{\text{left}}}{N}\text{MSE}_{\text{left}} + \frac{N_{\text{right}}}{N}\text{MSE}_{\text{right}}\right) \tag{6.14}$$

where $(\text{MSE}_{\text{parent}})$, $(\text{MSE}_{\text{left}})$, and $(\text{MSE}_{\text{right}})$ represent the MSE of the parent, left child, and right child nodes, respectively, and (N_{left}) and (N_{right}) are the number of samples in the left and right child nodes.

Tree models can be classified based on the granularity of splits:

1. Fine Trees: These have a high level of granularity and allow deeper trees with more splits. Fine trees are capable of capturing complex relationships within the data, but they are more prone to overfitting, especially in noisy datasets.
2. Medium Trees: These represent a balance between complexity and generalization by limiting the maximum depth of the tree. Medium trees reduce the likelihood of overfitting while retaining sufficient flexibility to model the data effectively.
3. Coarse Trees: These have fewer splits and a shallow depth, making them less prone to overfitting but potentially underfitting the data. Coarse trees are suitable for datasets with simple underlying structures.

The advantages of Tree models include their interpretability, as the splits and rules are easy to visualize and understand. Additionally, Trees handle mixed data types (categorical and numerical) without requiring feature scaling. However, individual trees can be sensitive to data perturbations, a limitation often mitigated by ensemble methods such as random forests or Gradient Boosted Trees [17].

In MATLAB, Tree models can be implemented using functions like 'fitctree' for classification and 'fitrtree' for regression. The granularity of the Tree is controlled by parameters such as the maximum depth ('MaxNumSplits') and minimum leaf size ('MinLeafSize'). By tuning these parameters, practitioners can design Fine, Medium, or Coarse trees to balance accuracy and model complexity effectively [18].

6.2.4.6 Gaussian Process Regression

Gaussian process regression (GPR) is a powerful and flexible non-parametric Bayesian machine learning model used for regression tasks. GPR is particularly effective for modeling datasets with complex relationships and small to medium sample sizes, providing predictions along with an estimation of uncertainty. Unlike traditional regression models, which assume a specific functional form, GPR assumes that the data points are sampled from a multivariate Gaussian distribution. This approach makes it highly adaptable to a wide range of applications, including chemical engineering, where precise modeling of experimental data is crucial. In GPR, a Gaussian process (GP) is defined as a collection of random variables, any finite subset of which has a joint Gaussian distribution [19].

For a set of input features ($\boldsymbol{X} = [\boldsymbol{x}_1, \boldsymbol{x}_2, \ldots, \boldsymbol{x}_N]$) and corresponding output values ($\boldsymbol{y} = \left[y_1, y_2, \ldots, y_N\right]$), the GP is governed by a mean function ($m(\boldsymbol{x})$) and a covariance function (or kernel) ($k(\boldsymbol{x}, \mathrm{\dashv}b\{x'\})$), expressed as:

$$f(\boldsymbol{x}) \sim \mathcal{GP}(m(\boldsymbol{x}), k(\boldsymbol{x}, \boldsymbol{x}\prime)) \tag{6.15}$$

where:

($m(\boldsymbol{x}) = E\left[f(\boldsymbol{x})\right]$) is the mean function; ($k(\boldsymbol{x}, \boldsymbol{x}\prime) = E\left[(f(\boldsymbol{x}) - m(\boldsymbol{x}))(f(\boldsymbol{x}\prime) - m(\boldsymbol{x}\prime))\right]$) is the covariance function.

For regression, the predictions at a new input ($\boldsymbol{x}_*$) are obtained by conditioning the joint Gaussian distribution of the training outputs ($\mathrm{\dashv}b\{y\}$) and the predicted outputs ($f(\boldsymbol{x}_*)$). The posterior distribution is given by:

$$f_*|X, y, x_* \sim \mathrm{N}(\mu_*, \Sigma_*) \tag{6.16}$$

where:

$\mu_* = \boldsymbol{K}_*^{\boldsymbol{T}}\left(\boldsymbol{K} + \sigma_n^2\boldsymbol{I}\right)^{-1}\boldsymbol{y}$ is the mean prediction. $\Sigma_* = \boldsymbol{K}_{**} - \boldsymbol{K}_*^{\boldsymbol{T}}\left(\boldsymbol{K} + \sigma_n^2\boldsymbol{I}\right)^{-1}\boldsymbol{K}_*$ is the covariance prediction.

Here, $\boldsymbol{K}$ is the covariance matrix of the training data. $\boldsymbol{K}_*$ is the covariance vector between the training and test data. $\boldsymbol{K}_{**}$ is the covariance matrix of the test data. σ_n^2 represents the noise variance.

The choice of the kernel function ($k(\boldsymbol{x}, \boldsymbol{x}\prime)$) plays a crucial role in determining the flexibility and accuracy of the GPR model. The rational quadratic (RQ) kernel is a popular choice, as it can model data with varying degrees of smoothness. The RQ kernel is defined as:

$$k_{\mathrm{RQ}}\left(\boldsymbol{x}, \boldsymbol{x}'\right) = \sigma_f^2\left(1 + \frac{|\boldsymbol{x} - \boldsymbol{x}'|^2}{2\alpha \mathrm{l}^2}\right)^{-\alpha} \tag{6.17}$$

where, σ_f^2 controls the overall variance of the function. l is the length scale, determining the smoothness of the function. α is a scale-mixture parameter, allowing the kernel to behave as a weighted sum of squared exponential kernels [20].

GPR can be implemented in MATLAB using the 'fitrgp' function, which allows users to specify the kernel function and other hyperparameters. For instance, to use the rational quadratic kernel, one can set the 'KernelFunction' property to 'rationalquadratic'. The hyperparameters (σ_f^2), (l), and (α) can be optimized automatically using the built-in functionality for hyperparameter tuning [21].

6.3 Results and Discussions

6.3.1 Characterization of CeO_2

The XRD pattern of CeO_2 NRs, as shown in Fig. 6.2, reveals sharp and broad peaks corresponding to the fluorite cubic structure with a space group of *Fm3m*. The diffraction peaks indexed to the planes (111), (200), (220), (311), (222), (400), (331), and (420) align well with the standard data from the JCPDS database (Card No. 00-034-0394). The absence of additional impurity peaks confirms the successful hydrothermal synthesis and the formation of a pure CeO_2 phase.

HRTEM images (Fig. 6.3a–c) further confirm the nanorod morphology of the CeO_2 material. The nanorods have an irregular shape with an average diameter of 13.9 ± 3.8 nm. The selected area electron diffraction (SAED) pattern (Fig. 6.3d) validates the polycrystalline nature of the material and aligns with the fluorite cubic structure.

The FTIR spectrum of CeO_2 NRs with a spectral range of 4000–400 cm^{-1} is shown in Fig. 6.4. A broad absorption band at 3242 cm^{-1} corresponds to O–H stretching

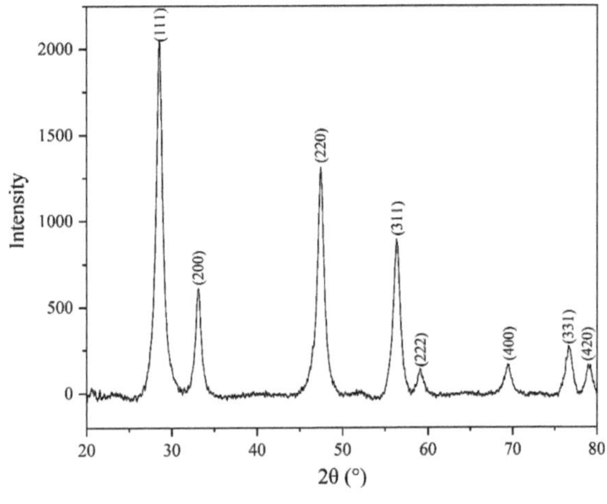

Fig. 6.2 XRD pattern of CeO_2 NRs

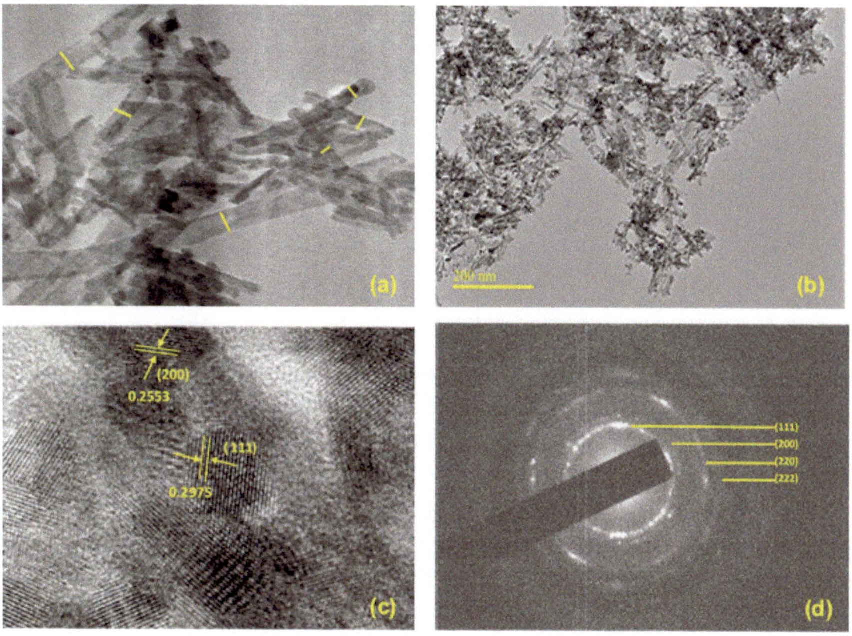

Fig. 6.3 **a** TEM image, **b**, **c** HRTEM image, and **d** SAED pattern of CeO_2 NRs

vibrations, attributed to adsorbed and hydrated water on the sample [22]. Another prominent band at 1634 cm^{-1} is associated with the O–H bending vibrations of adsorbed water molecules on the catalyst surface. The band observed between 750 and 400 cm^{-1} in the fingerprint region is attributed to CeO_2 stretching vibrations, specifically the symmetric Ce–O bond stretch [23].

Fig. 6.4 FTIR spectrum of CeO_2 NRs

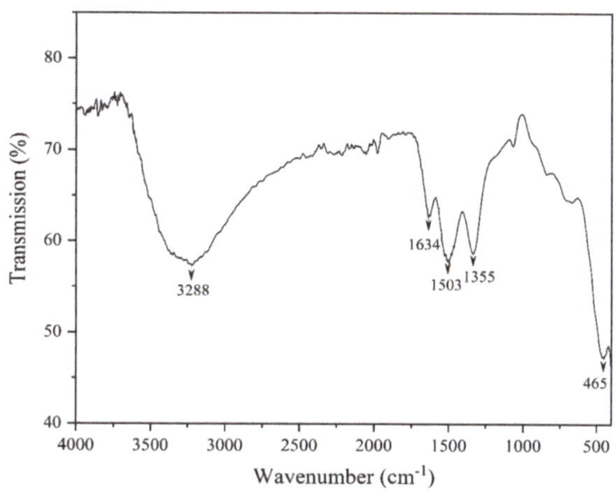

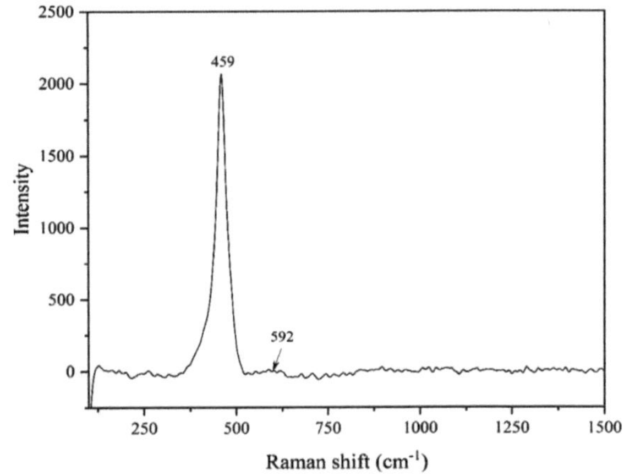

Fig. 6.5 Raman spectrum of CeO_2 NRs

The XRD findings were further supported by Raman spectroscopy's result. A prominent absorption peak at approximately 459 cm^{-1}, shown in Fig. 6.5, corresponds to the F_2g vibrational mode of fluorite-structured CeO_2 with a cubic symmetry (space group *Fm3m*). This peak is attributed to the symmetric stretching vibrations of oxygen atoms surrounding each $Ce4^+$ ion [24]. Additionally, a weaker peak at 592 cm^{-1} is likely associated with oxygen vacancies or defects due to the presence of $Ce3^+$, which are known to play a crucial role in enhancing catalytic activity [25].

6.3.2 Machine Learning Model

6.3.2.1 Bootstrap Results

The box plots compare the experimental dataset (top) and the resampled dataset obtained through bootstrapping (bottom) as shown in Fig. 6.6. The experimental dataset consists of 56 data points, whereas the bootstrapped dataset expands to 560, effectively increasing the dataset size while maintaining statistical characteristics. The spread of values across Inputs 1 to Input 4 and the target variable are similar between the two datasets, as evidenced by the comparable ranges and interquartile distances of the box plots. For Input 1, the standard deviation (SD) decreases slightly from 44.098 in the experimental data to 43.826 in the bootstrapped data, suggesting consistent variability across resampling. Similarly, Inputs 2 and 3 show a slight reduction in SD, from 1.492 to 1.429 and 0.386 to 0.358, respectively, indicating that the bootstrap maintains the variability but slightly reduces noise. For Input 4, the SD shows a slight increase from 7.503 to 7.593, while for the target, the SD remains effectively unchanged (29.437 in the original data versus 29.425 in the resampled data). Overall, the bootstrap resampling method successfully preserves

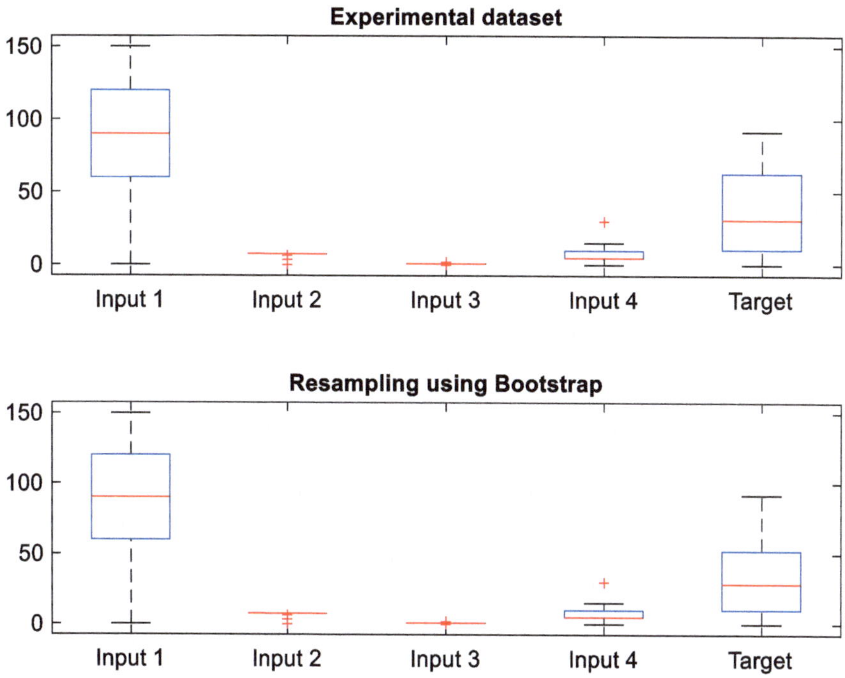

Fig. 6.6 Bootstrap box plot

the statistical characteristics of the original dataset while enhancing its robustness for further development.

6.3.2.2 Correlation Results

Linear correlation in descriptive statistics assesses the strength and direction of the relationship between two variables when the relationship is linear. One widely used measure for this purpose is Pearson's linear correlation coefficient (r). This coefficient quantifies the degree to which two variables are linearly related and takes on values ranging from -1 to $+1$.

A coefficient of $+1$ indicates a perfect positive linear relationship, meaning that as one variable increases, the other increases proportionally. Conversely, a coefficient of -1 represents a perfect negative linear relationship, where one variable increases as the other decreases proportionally. A value of 0 suggests no linear correlation, implying that the variables are not linearly related.

Pearson's correlation coefficient is calculated as the ratio of the covariance of two variables to the product of their standard deviations. This formula normalizes the correlation, making it dimensionless and comparable across different datasets [26]. However, the method assumes that the relationship between the variables is

Table 6.2 Correlation analysis

	Input 1	Input 2	Input 3	Input 4	Target
Input 1	1.000	0.190	0.117	0.087	0.675
Input 2	0.190	1.000	0.318	0.250	0.103
Input 3	0.117	0.318	1.000	−0.038	−0.132
Input 4	0.087	0.250	−0.038	1.000	−0.214
Target	0.675	0.103	−0.132	−0.214	1.000

inherently linear, that the data are continuous, and that they approximate a normal distribution. Violating these assumptions may lead to inaccurate interpretations of the correlation [27].

Table 6.2 shows Pearson's correlation coefficients for the variables involved in the sonocatalytic degradation process. The variables include four inputs (Input 1: initial pH, Input 2: initial caffeine concentration, Input 3: CeO_2 NRs dosage, Input 4: reaction time) and one target (degradation efficiency). Diagonal elements of Table 6.2 show perfect self-correlation ($r = 1.000$) for each variable. Off-diagonal elements represent the correlation between different variables. Based on Table 6.2, the correlation between Input 1 (initial pH) and the target (degradation efficiency) is ($r = 0.675$), indicating a moderate positive linear relationship. In contrast, Input 4 (reaction time) has a weak negative correlation with the target ($r = -0.214$), suggesting that increasing reaction time has a slight inverse relationship with degradation efficiency. Similarly, Input 3 (CeO_2 NRs dosage) shows almost no correlation with the target ($r = -0.132$). The correlations among the inputs are generally low (e.g., ($r = 0.190$) between Input 1 and Input 2), suggesting minimal linear dependence between process parameters. This analysis provides critical insights from a modeling perspective. Based on the correlation analysis, a nonlinear modeling method is required since the input variables have a low correlation or weak relationship (<0.9) with the target variables [28].

6.3.2.3 Comparative Performance of Machine Learning

The results presented in Fig. 6.7 compare the performance of four machine learning (ML) models, which are Fine tree, Gaussian process regression with rational quadratic kernel (GPR-RQ), ANN, and support vector machine with Fine function kernel (SVM Fine). These ML models are evaluated using root mean squared error (RMSE) and R-squared (R^2) metrics on test data. The models were tested with and without bootstrapped data, where models using bootstrapped data are denoted as (BS).

The Fine tree model without bootstrapping exhibits poor performance, with an RMSE of 15.40 and an R^2 value of 0.65, indicating its inability to capture the complex relationships in the data. However, the bootstrapped Fine tree model improves significantly, achieving near-perfect accuracy (with $R^2 = 0.999$) and a low RMSE of 0.18,

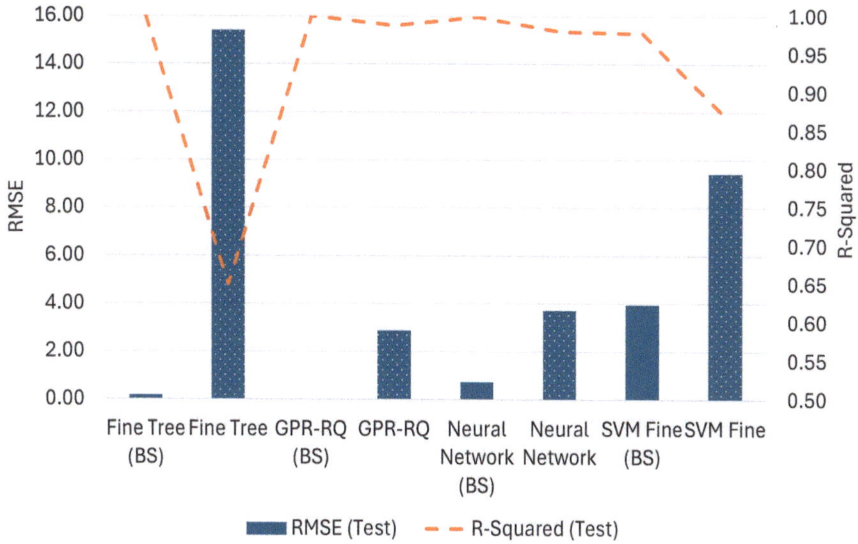

Fig. 6.7 Comparison of ML model using bootstrap data (BS) and not using bootstrap data based on RMSE and R^2 analysis

highlighting the effectiveness of bootstrapping in enhancing predictive performance for decision tree-based models. The GPR-RQ model also shows strong performance, with RMSE values of 2.86 and $R^2 = 0.99$ without bootstrapping, reflecting a high capacity to model nonlinear relationships in the data. When using bootstrapped data, the GPR-RQ model achieves almost perfect predictions of $R^2 = 1.00$ and lesser error (RMSE = 0.0013), making it the most accurate model overall.

The ANN model follows a similar trend, performing well without bootstrapping with $R^2 = 0.98$ and RMSE = 3.72. Bootstrapping further enhances its predictive ability, yielding R^2 nearly reach 1.00 and RMSE of 0.73, indicating its robustness and adaptability to new augmented data. In addition, the SVM Fine model demonstrates moderate performance on both with and without bootstrapping compared to other ML models. The RMSE values of 9.42 (with no bootstrapping) and 3.96 (with bootstrapped data) are relatively high, and the corresponding R^2 values (0.87 and 0.98, respectively) suggest that the model has less accuracy in generalization with the test data.

Based on these results, the GPR-RQ model using bootstrapped data emerges as the best-performing model. It near-perfect R^2 value and very small RMSE signify an exceptional ability to capture and predict the underlying relationships in the dataset. The superiority of GPR-RQ can be attributed to its inherent flexibility in modeling complex nonlinear patterns, which are critical in processes like sonocatalytic degradation involving multiple interacting variables. Furthermore, the results emphasize bootstrap data resampling in improving the performance of ML models, especially for datasets with limited size, by enhancing model variability and reducing overfitting. The full results of the ML model are tabulated in Table 6.3.

Table 6.3 ML model with RMSE and R^2 results

Model type	RMSE (test)	R-squared (test)
GPR-RQ (BS)	0.001	0.999
Fine tree (BS)	0.176	0.999
Neural network (BS)	0.734	0.999
GPR-RQ	2.858	0.988
Neural network	3.715	0.980
SVM fine (BS)	3.956	0.977
SVM fine	9.419	0.869
Fine tree	15.398	0.650

Figures 6.8, 6.9, and 6.10 illustrate the simulation results for the effects of the initial pH of the solution, initial caffeine concentration, and catalyst dosage on the sonocatalytic degradation of caffeine. The results indicate that the developed Gaussian process regression with rational quadratic (GPR-RQ) kernel model effectively reproduces the experimental data. This demonstrates the model's ability to accurately capture the complex interactions between process parameters and predict the degradation efficiency, validating its reliability for simulating sonocatalytic processes.

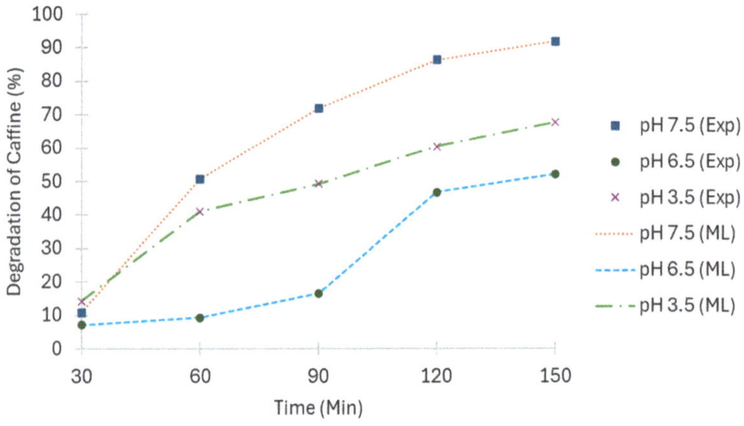

Fig. 6.8 Effect of solution pH on the sonocatalytic degradation of caffeine (Experimental conditions: [CAFF] = 5 mg L^{-1}, [CeO_2] = 1.0 g L^{-1}, [H_2O_2] = 20 mM and US frequency = 37 kHz)

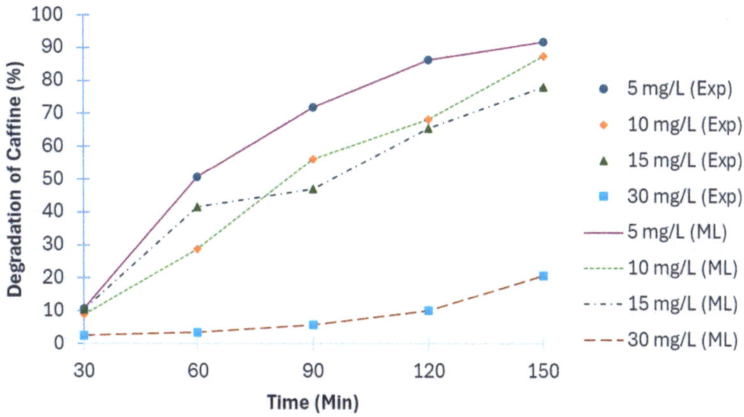

Fig. 6.9 Effect of initial concentration of caffeine on the sonocatalytic degradation of caffeine (experimental conditions: pH = 7.5, $[CeO_2] = 1.0\ g\ L^{-1}$, $[H_2O_2] = 20$ mM and US frequency = 37 kHz)

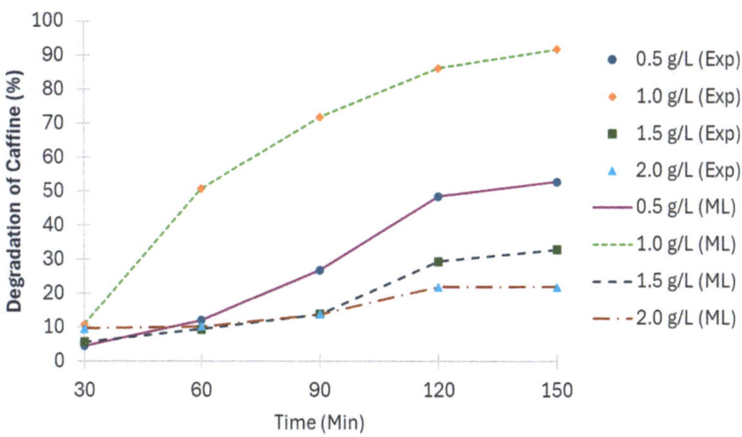

Fig. 6.10 Effect of CeO_2 NRs dosage on the sonocatalytic degradation of caffeine (experimental conditions: [CAFF] = 5 mg L^{-1}, pH = 7.5, $[H_2O_2] = 20$ mM and US frequency = 37 kHz)

6.3.3 Sonocatalytic Degradation of Caffeine

6.3.3.1 Effect of the Initial pH of the Solution

The pH of the reaction medium is known to influence the sonocatalytic degradation of caffeine significantly. The effect of the initial pH of the solution was evaluated at three different levels, as depicted in Fig. 6.8. The figure demonstrates a strong correlation between the predicted and experimental caffeine degradation values across the three pH levels, with a minimal root mean squared error (RMSE) of 0.00017 and a high

R-squared value of 0.9999. These results confirm the accuracy of the model and underscore the significance of pH in optimizing degradation efficiency.

The study found that caffeine degradation increased from 51.2 to 91.5% as the pH shifted from acidic (pH 3.5) to weakly basic (pH 7). This trend aligns with findings from earlier studies, which also reported reduced catalytic activity at low pH levels (e.g., pH 3.5) [29].

The observed pH dependence in the sonocatalytic degradation of caffeine can be attributed to its amphoteric nature and the surface properties of CeO_2, such as charge density [30]. Changes in pH can alter the surface charge of CeO_2 NRs, affecting the adsorption of caffeine molecules—a crucial step for effective sonocatalytic oxidation.

Additionally, the acid–base characteristics of surface sites on CeO_2 NRs provide further insights, particularly in relation to site availability, which can be explained by the point of zero charge (pH_{PZC}). For CeO_2 NRs, the pH_{PZC} was determined to be approximately 6.95. This means the catalyst surface is positively charged in solutions with a pH below 6.95 and negatively charged at higher pH levels. When the pH is below 6.95, the protonated CeO_2 surface enhances the adsorption of anionic molecules. Conversely, at pH levels above 6.95, deprotonation occurs, favoring the adsorption of cationic molecules. This dual behavior highlights the critical role of pH in optimizing the sonocatalytic degradation process. As reported in previous studies [32], CeO_2 NRs enhance the adsorption of caffeine molecules, particularly in the pH range of 6.95–10.4 ($pKa_{caffeine}$) [31], with optimal performance observed around pH 7.5. Therefore, this indicates that the pH of the solution significantly influences the efficiency of sonocatalytic degradation, with near-neutral pH conditions playing a crucial role in improving the process.

6.3.3.2 Effect of Initial Concentration of Caffeine

An additional experiment was conducted to investigate the influence of initial caffeine concentration on its sonocatalytic degradation, as depicted in Fig. 6.9. The data shown in the figure reveals a close alignment between the predicted and experimental values, with a small RMSE of 0.00028 and an almost perfect R-squared value of 0.9999. The findings revealed that as the initial caffeine concentration increased from 5 to 30 mg/L, the degradation efficiency significantly declined. Experimental data demonstrated a strong correlation with predicted values. At a lower concentration of 5 mg/L, approximately 91.5% of the caffeine was degraded after 150 min. In contrast, when the concentration was increased to 30 mg/L—representing a sixfold increase—the degradation efficiency dropped drastically to just 19.8%. It was also noted that the sonocatalytic degradation rate became notably less effective for caffeine concentrations exceeding 15 mg/L [33].

This decline in efficiency can be explained by the increased number of caffeine molecules and their intermediate compounds saturating the active sites on the CeO_2 NRs. As a result, the availability of reactive species such as hydroxyl radicals (•OH) decreased because the sonocatalyst particles could not effectively absorb the energy generated by acoustic cavitation. This limitation led to reduced production of •OH radicals, which are essential for the oxidation and breakdown of caffeine molecules. Consequently, the overall sonocatalytic degradation efficiency declined as the caffeine concentration increased [34].

6.3.3.3 Effect of CeO_2 NRs Dosage

Figure 6.10 demonstrates the effect of CeO_2 NRs dosages (0.5–2.0 g L^{-1}) on the sonocatalytic degradation efficiency of caffeine. As shown in the figure, there is minimal variation between the predicted and experimental values, with a very small RMSE of 0.00016 and an almost perfect R-squared value of 0.9999. This strong correlation suggests the high accuracy of the model in predicting the degradation efficiency based on the CeO_2 NRs dosage.

Increasing the CeO_2 NRs loading from 0.5 to 1.0 g L^{-1} significantly enhanced the degradation efficiency, achieving the highest performance of 91.5% at 1.0 g L^{-1}. However, further increases in catalyst dosage beyond this level resulted in a marked decline in efficiency, dropping below 50.0%. This can be associated with the turbulence that induced ultrasonic cavitation at this dosage, providing a better dispersing of CeO_2 NRs [35]. In contrast, when the dosage was insufficient, the cavitation effect generated a small number of nuclei, which led to a reduction in •OH radicals produced by CeO_2 NRs. However, it should be noted that the high dosage of CeO_2 NRs results in particle agglomeration and shielding effects occur. These effects reduce the surface area of CeO_2 NRs exposed to ultrasonic irradiation, which inhibits the formation of •OH radicals [33], thereby decreasing the degradation efficiency of caffeine.

The figure also indicates that the predicted caffeine degradation efficiencies are in close agreement with the experimentally measured values. This consistency confirms the predictive model's robustness and accuracy in capturing the system's behavior across varying CeO_2 NR dosages.

Conclusion

This study highlights the effectiveness of CeO_2 NRs in facilitating the sonocatalytic degradation of caffeine. The results demonstrated that key process parameters, such as the initial pH of the solution, initial caffeine concentration, and CeO_2 NRs dosage, could be accurately modeled using machine learning approaches. The Gaussian process regression with rational quadratic (GPR-RQ) kernel model, enhanced by bootstrap, successfully predicted the caffeine degradation percentage, achieving an R^2 value of 0.999 and a low RMSE of 0.001. The study also found that degradation efficiency was significantly influenced by these parameters. The highest caffeine degradation (91.5%) was achieved at the initial pH of the solution of 7.5, a CeO_2 NRs dosage of 1 g L^{-1}, an initial caffeine concentration of 5 mg L^{-1}, and an H_2O_2 concentration of 20 mM. Overall, the results confirmed that bootstrap-enhanced machine learning can effectively predict experimental outcomes in the sonocatalytic degradation of caffeine using CeO_2 NRs as the catalyst. While the proposed sonocatalytic degradation process using CeO_2 NRs demonstrated high efficiency and accurate predictive modeling, several limitations were denoted: (i) The study did not comprehensively explore the reusability of CeO_2 NRs. Reusability is crucial for evaluating the economic feasibility and sustainability of the process. (ii) Sonocatalysis involves ultrasonic irradiation, which may require significant energy input. The overall energy efficiency of the process compared to alternative wastewater treatment technologies has not been thoroughly assessed. Building on the findings of this study, several avenues for future work can be pursued to address these limitations and further advance the research: (a) Extend the study to include other PPCPs and emerging micropollutants to evaluate the versatility and applicability of the process across diverse contaminants. (b) Transition from laboratory-scale experiments to pilot-scale studies in real wastewater treatment facilities to validate the practicality and scalability of the proposed process. (c) Conduct a detailed energy consumption analysis to optimize the ultrasonic irradiation parameters and improve the overall energy efficiency of the sonocatalytic process.

Acknowledgements The authors gratefully acknowledge the Ministry of Higher Education Malaysia for the financial support in the form of the Long-Term Research Grant Scheme (LRGS/1/2018/USM/01/1/3). The authors also gratefully acknowledge Universiti Teknologi MARA (UiTM) for providing the necessary research facilities.

Data Availability Datasets used and/or analyzed in this study are available upon reasonable request.

Ethics Approval and Consent to Participate Not applicable.

Consent for Publication Not applicable.

Conflict of Interest The authors declare no competing interests.

References

1. Wang J, Wang S (2016) Removal of pharmaceuticals and personal care products (PPCPs) from wastewater: a review. J Environ Manag 182:620–640
2. de Almeida LNB, Josué TG, Fidelis MZ, Abreu E, Bechlin MA, dos Santos OAA, Lenzi GG (2021) Process comparison for caffeine degradation: fenton, photo-fenton, UV/H_2O_2 and UV/Fe^{3+}. Water Air Soil Pollut 232:147
3. Fernando Martínez R, Cravotto G, Cintas P (2021) Organic sonochemistry: a chemist's timely perspective on mechanisms and reactivity. J Org Chem 86(20):13833–13856
4. Montini T, Melchionna M, Monai M, Fornasiero P (2016) Fundamentals and catalytic applications of CeO_2-based materials. Chem Rev 116:5987–6041
5. Gupta B, Gupta AK, Tiwary CS, Ghosal PS (2021) A multivariate modeling and experimental realization of photocatalytic system of engineered S-C_3N_4/ZnO hybrid for ciprofloxacin removal: influencing factors and degradation pathways. Environ Res 196:110390
6. Li X, Zhang X, Zhang J, Gu J, Zhang S, Li G, Shao J, He Y, Yang H, Zhang S, Chen H (2023) Applied machine learning to analyze and predict CO_2 adsorption behavior of metal-organic frameworks. Carbon Capture Sci Technol 9:100146

7. Fadzeelah AKN, Bashah NAA, Rohman FS, Senin SF, Abdullah AZ (2024) Sonocatalytic degradation of caffeine using CeO_2 nanorods: modeling by artificial neural network. Desalin Water Treat 320:100721
8. Leong YK, Chang C-K, Arumugasamy SK, Lan JC-W, Loh H-S, Muhammad D, Show PL (2018) Statistical design of experimental and bootstrap neural network modelling approach for thermoseparating aqueous two-phase extraction of polyhydroxyalkanoates. Polymers 10(2):132
9. Rohman FS, Muhammad D, Aziz N (13 Nov 2023) Modeling of plug flow reactor using block-oriented model for ethylene glycol production. In: AIP conference proceedings, vol 2907(1), p 030001
10. Idris I, Ahmad Z, Othman MR, Rohman FS, Rushdan AI, Azmi A (2022) Application of artificial neural network to predict water flux from pre-treated palm oil mill effluent using direct contact membrane distillation. Mater Today: Proc 63:S411–S417
11. Shure L (2015) Artificial neural networks for beginners. The Mathwork Inc. https://blogs.mathworks.com/loren/2015/08/04/artificial-neural-networks-for-beginners/. Accessed 10 Nov 2024
12. Rohman FS, Muhammad D, Sudibyo, Murat MN, Azmi A (2022) Application of feed forward neural network for fouling thickness estimation in low density polyethylene tubular reactor. Mater Today: Proc 63:S95–S100
13. Rohman FS, Muhammad D, Azmi A, Murat MN (2023) Estimation of fouling thickness in low density polyethylene tubular reactor process using soft sensor model. In: AIP conference proceedings, vol 2785, p 020004
14. IBM (2023) What are support vector machines (SVMs)? https://www.ibm.com/topics/support-vector-machine. Accessed 15 Nov 2024
15. Mathwork (2024) Machine Learning with MATLAB. https://www.mathworks.com/campaigns/offers/next/machine-learning-with-matlab.html. Accessed 15 Nov 2024
16. Meghana P, Annepu V, Jweeg MJ, Bagadi K, Aljibori HSS, Mohammed MN (2024) Analysis of neural network algorithm in comparison to multiple linear regression and random forest algorithm. In: 2024 ASU international conference in emerging technologies for sustainability and intelligent systems (ICETSIS), Manama, Bahrain, pp 437–443
17. Hancock T, Smyth C (2009) 2.31—tree-based clustering and extensions. In: Brown SD, Tauler R, Walczak B (eds) Comprehensive chemometrics. Elsevier, pp 683–705
18. Mathwork (2011) RegressionTree. https://www.mathworks.com/help/stats/regressiontree.html. Accessed 15 Nov 2024
19. Schulz E, Speekenbrink M, Krause A (2018) A tutorial on Gaussian process regression: modelling, exploring, and exploiting functions. J Math Psychol 85:1–16
20. Hemachandran K, Tayal S, George PM, Singla P, Kose U (2022) Bayesian reasoning and Gaussian processes for machine learning applications. Chapman and Hall/CRC, New York
21. Mathwork (2015) Gaussian process regression models. https://www.mathworks.com/help/stats/gaussian-process-regression-models.html. Accessed 15 Nov 2024
22. Negi K, Kumar M, Singh G, Chauhan S, Chauhan MS (2018) Nanostructured CeO_2 for selective-sensing and smart photocatalytic applications. Ceram Int 44:15281–15289
23. Rani S, Verma S, Sharma B, Kumar S (2023) Tuning of structural, optical/spectral and thermal properties of cubic CeO_2 nanoparticles via incorporation of zirconia. Phys B Condens Matter 661:414922
24. Yao H, Wang Y, Jing Y, Luo G (2018) Ultrafast, continuous and shape-controlled preparation of CeO_2 nanostructures: nanorods and nanocubes in a microfluidic system. Ind Eng Chem Res 57:7525–7532
25. Xiao M, Zhang X, Yang Y, Cui X, Chen T, Wang Y (2022) M (M = Mn Co, Cu)-CeO_2 catalysts to enhance their CO catalytic oxidation at a low temperature: Synergistic effects of the interaction between Ce3+-Mx+-Ce4+ and the oxygen vacancy defects. Fuel 323:124379
26. Profillidis VA, Botzoris GN (2019) Chapter 5—statistical methods for transport demand modeling. In: Profillidis VA, Botzoris GN (eds) Modeling of transport demand. Elsevier, pp 163–224

27. Schober P, Boer C, Schwarte LA (2018) Correlation coefficients: appropriate use and interpretation. Anesth Analg 126(5):1763–1768
28. Hahs-Vaughn DL (2023) Foundational methods: descriptive statistics: bivariate and multivariate data (correlations, associations). Tierney RJ, Rizvi F, Ercikan K (eds) International encyclopedia of education, 4th ed. Elsevier, pp 734–750
29. Guo Y, Shen T, Wang C, Sun J, Wang X (2015) Rapid removal of caffeine in aqueous solutions by peroxymonosulfate oxidant activated with cobalt ion. Water Sci Technol 72:478–483
30. Panda D, Manickam S (2017) Recent advancements in the sonophotocatalysis (SPC) and doped-sonophotocatalysis (DSPC) for the treatment of recalcitrant hazardous organic water pollutants. Ultrason Sonochem 36:481–496
31. Elhalil A, Elmoubarki R, Machrouhi A, Sadiq M, Abdennouri M, Qourzal S, Barka N (2017) Photocatalytic degradation of caffeine by ZnO-ZnAl2O4 nanoparticles derived from LDH structure. J Environ Chem Eng 5:3719–3726
32. Khataee A, Karimi A, Arefi-Oskoui S, Darvishi Cheshmeh Soltani R, Hanifehpour Y, Soltani B, Joo SW (2015) Sonochemical synthesis of Pr-doped ZnO nanoparticles for sonocatalytic degradation of Acid Red 17. Ultrason Sonochem 22:371–381
33. Huang Y, Wang G, Zhang H, Li G, Fang D, Wang J, Song Y (2017) Hydrothermal-precipitation preparation of CdS@ (Er^{3+}: $Y_3Al_5O_{12}/ZrO_2$) coated composite and sonocatalytic degradation of caffeine. Ultrason Sonochem 37:222–234
34. Mirzaei A, Haghighat F, Chen Z, Yerushalmi L (2019) Sonocatalytic removal of ampicillin by Zn(OH)F: effect of operating parameters, toxicological evaluation and by-products identification. J Hazard Mater 375:86–95
35. Sancheti SV, Saini C, Ambati R, Gogate PR (2018) Synthesis of ultrasound assisted nanostuctured photocatalyst (NiO supported over CeO_2) and its application for photocatalytic as well as sonocatalytic dye degradation. Catal Today 300:50–57

Chapter 7
Structural Equation Modelling for Identifying the Determinants for Adoption of Household Water Treatment

Kunjal Badrakiya, Iqra Sharmeen, Saurab Patel, Yugmi S. Patel, and M. Mansoor Ahammed

Abstract In areas with known or potential water contamination, consumers often prefer to treat their tap water at the household level primarily due to concerns about water quality. Household water treatment (HWT) methods range from point-of-entry to point-of-use options. Examples include simple boiling or straining to more advanced systems that remove a broader range of contaminants. In this chapter, the factors influencing households' adoption of water treatment methods are first discussed. Their adoption depends on personal preferences, affordability, health considerations and existing water quality issues. This chapter investigates the intricate relationship between public perceptions regarding the adoption of HWT options and Artificial Intelligence (AI) specifically utilizing Structural Equation Modelling (SEM). By employing SEM to analyse data collected from surveys and interviews, this research aims to understand the complex interplay of factors influencing public acceptance of HWT technologies. SEM is an umbrella term which combines multiple regression, factor analysis and path analysis to model complex relationships between observed and latent variables, and it is better suited for cognition or psychological analysis. This chapter presents an overview on the use of SEM for different applications in water and wastewater management. A questionnaire was prepared after conducting a pilot-scale study, and a survey was carried out among 100 respondents in the city of Surat, India. Results of the study indicate that source of water, past experiences (health risks), water quality, affordability, efficiency and information source for method play a significant role in HWT adoption decisions. Key determinants identified using SEM include water quality, source of water supply and past experiences of the residents, with water quality having a stronger association with adoption decisions than the source of water supply and past experiences. Understanding these determinants is crucial for developing effective policies and interventions to ensure

K. Badrakiya · I. Sharmeen · S. Patel · Y. S. Patel · M. Mansoor Ahammed (✉)
Department of Civil Engineering, Sardar Vallabhbhai National Institute of Technology, Surat, India
e-mail: mansoorahammed@gmail.com

M. Nasr et al. (eds.), *Artificial Intelligence Applications for a Sustainable Environment*, Green Chemistry and Sustainable Technology,
https://doi.org/10.1007/978-3-031-91199-6_7

safe and affordable water supply, thereby improving public health and reducing burden of waterborne diseases.

Keywords Water source · Water quality · Point-of-use treatment · Drinking water · Water safety

7.1 Introduction

Providing access to safely managed drinking water is essential to achieve Sustainable Development Goals 6 (SDG 6). As of 2022, 2.2 billion people lacked its access, while 1.5 billion lack basic drinking water services [1]. Safe drinking water poses minimal health risks throughout a person's life, considering varying sensitivities at different stages and minimizing risks associated with long-term consumption [2]. The water quality is deteriorated during collection, transport and storage of water in the distribution system [3]. Water quality is primarily determined by physical, chemical and microbiological characteristics [4]. Pollutants include carcinogenic disinfection by-products (DBPs) such as trihalomethanes (THMs), haloacetic acids and halo aldehydes formed during disinfection of organic matter like humic and fulvic acids [5]. Even after disinfection, some microbes may still be present, exposing public to elevated risks of waterborne diseases such as cholera, diarrhoea, dysentery and typhoid. During transport and storage, the stream picks up sediment and debris as it flows from treatment plant or storage reservoir to the point of distribution. If the water remains stagnant in reservoirs or distribution pipes for extended periods, it can provide breeding ground for pathogens.

Waterborne diseases are one of the important public health problems especially in developing regions of the world. Household water treatment (HWT) has been shown to be an effective means of reducing this disease burden [6]. Various types of water purification devices are currently used for treatment of water at the household level. Such devices use a variety of basic processes such as filtration, adsorption, ion exchange, reverse osmosis, disinfection or distillation to achieve a desired contaminant removal. Activated carbon filter, ceramic candle filter, iodine resin gravity filter, polyester filter, ultra violet irradiation, reverse osmosis and hollow fibre membrane filters are different techniques employed either singly or in combination in these devices. In developed countries, granular activated carbon filter is the most commonly used device. It is used to remove the taste and odour associated with chlorine residuals, certain organic chemicals and chemical compounds produced by various microorganisms. In developing countries, where microbiological water quality is the major concern, water purifiers that can remove pathogenic microorganisms are more common.

In urban centres of the developing countries, people generally have access to drinking water which had already undergone treatment at a treatment facility. Yet, in these centres, people are showing increasing interest in installing household devices to further treat the water supplied to them as they are either not satisfied with the quality of the water supplied or as an additional barrier against contamination. In rural areas, however, people still depend on untreated source water and several low-cost treatment methods such as solar disinfection, biosand filters, chlorine tablets and natural coagulants are employed for household water purification [7].

While the market for commercial household devices has been showing a high growth rate, the effectiveness of these devices is seldom evaluated. A few laboratory studies have shown that these devices often fail to meet the requirements of a water treatment device. For example, Clasen and Menon [8] tested microbiological performance of three common types of water treatment devices. They found that the devices reduced bacteria by more than 4 logs. However, the reduction in viruses was lower than 3.4 logs. There were also high levels of residual iodine in treated water from the unit that used iodine resin. Verma and Arankalle [9] tested eight commercially available domestic water purification units for their virus removal efficiency and concluded that majority of water purifiers under use were inadequate for virus removal with virus removal ranging between 0.21 and 6.53 logs.

The selection and adoption of appropriate HWT depends on a multitude of factors, namely personal preferences, affordability, health considerations and existing water quality issues. The use of Structural Equation Modelling (SEM) to investigate the key determinants influencing the consumers' decision in selecting appropriate treatment option is investigated in this study. SEM is a statistical technique that identifies latent (unobserved) variables influencing a group of observed variables and specifies structural equations representing the relationships between these latent and observed variables [10]. SEM finds its application in water and wastewater management studies [11], correlating water parameters [12], assessing treatment efficiency [13], understand public perceptions [14], modelling impact of climate change on water resources [15], among others.

This chapter presents an overview on the use of SEM for different applications in water and wastewater management. A questionnaire was prepared after conducting a pilot-scale study, and a survey was carried out among 100 residents in the city of Surat, India, on their use of HWT. The key objectives of this chapter are to identify the factors influencing selection of appropriate HWT options/methods and understanding public attitudes towards HWT adoption and representing them via appropriate statistical modelling approaches.

7.2 Household Water Treatment

HWT practices improve water quality and reduces the burden of waterborne diseases due to re-contamination. The treatment methods can be classified into two categories:

a. For waters that have not undergone any treatment and whose microbial treatment is of greater concern.
b. For water that has already undergone treatment at a centralized treatment plant

There are various treatment options available, selection of which depends on multitude of factors. Some of the methods/devices are discussed below.

7.2.1 For Raw Water with No Prior Treatment

Solar Water Disinfection (SODIS):

The method uses PET or glass bottles filled with water and exposed to sunlight during which UV-A rays (315–400 nm) penetrate the container; altering DNA of pathogens followed by thermal inactivation [16]. For instance, *E. coli* is destroyed at 20 °C after an exposure of 60 min. Usual exposure time is 6–10 h, depending on the sunlight intensity, water turbidity and contaminant levels. Thus, SODIS safeguards users against waterborne diseases. However, low oxygen levels in water reduces the efficiency of SODIS for microbial destruction. SODIS is useful in areas with limited access to clean water and electricity but is not as efficiently in areas receiving intermittent sunlight.

Filtration:

Biosand filters: They rely on a combination of physical and biological processes to remove contaminants by passing water through layers of graded sand and gravel [17]. As time passes, a biofilm, called Schmutzdecke, develops on the media surface that aids in microbial removal.

Cloth filtration: It is a simple and affordable method for removing larger particles like dirt and sand but has limited efficiency for dissolved impurities. The efficiency depends on the fabric's weave and the size of the particles being removed. Cotton, muslin and polyester are some of the commonly used fabrics [18].

Ceramic filters: These filters use porous ceramic material to physically retain bacteria and protozoa by size exclusion and adsorption [18]. The filter's porous structure traps contaminants while allowing clean water to pass through, thereby safeguarding consumers against recontamination. Depending on the type of filtering media, removal of turbidity, sediments, bacteria and heavy metals can be achieved [19]. It is important to note that these filters need periodic cleaning to ensure optimal performance. Ceramic filters have a limited virus removal efficiency, and are therefore, suitable when the pathogen of concern is known.

Coagulation:

In addition to its nutritional and medicinal benefits, *Moringa oleifera*, commonly called 'drumstick', has gained attention as a natural coagulant for removing suspended solids and turbidity [20]. When added to turbid water, the coagulants neutralize the charges on suspended particles, causing them to clump together and settle to the bottom. The seeds contain proteins and tannins that act as natural coagulants and also generates lesser sludge than alum and other synthetic coagulants.

7.2.2 For Water that Has Undergone Treatment

Membrane Processes:

Membrane processes such as reverse osmosis (RO) remove dissolved impurities from water by passing it through a membrane with varying pore size ranges. Water is forced under high pressure through a semi-permeable membrane that retails salts, minerals and bacteria. Cellulose-based membranes such as cellulose acetate and polyamide are most commonly used [21]. RO removes hardness-causing cations, making permeate suitable for consumption. However, significant quantity of reject water is generated which requires further disposal. One of the operational issues is membrane fouling in which contaminants accumulate on or within the pores, that eventually leads to reduced flux and higher energy consumption. Probable reasons can be poor feed quality that can be improved by filtering through activated carbon prior to operation [22]. However, RO-based HWT devices are gaining popularity all over the world.

Activated Carbon Impregnated with Silver Ions:

This method combines the adsorption capabilities of activated carbon with the antimicrobial properties of silver ions to remove waterborne pathogens [23]. The silver oxides present in activated carbon release Ag^+ ions into water that interfere with microbial DNA replication, protein synthesis and enzyme activity. Silver ions are preferred as they ensure safe water storage by slowly releasing Ag^+ ions in water, thereby minimizing possibilities of recontamination and providing sustained antimicrobial protection [24].

7.2.3 Factors Affecting Household Water Treatment

HWT is a critical strategy for improving water quality and public health, especially in regions with limited access to safe drinking water. However, its implementation by consumers remains a significant challenge. This section explores the various factors that influence the adoption of HWT, providing a comprehensive understanding of the barriers and opportunities for promoting its use. Table 7.1 summarizes the various factors affecting the HWT adoption decisions and are discussed here.

Table 7.1 Summary of factors affecting HWT adoption decisions

Socioeconomic factors	Income, education, gender
Technological factors	Complexity, reliability, affordability, availability of the device/method
Behavioural factors	Beliefs, risk perceptions, trust
Institutional factors	Public health initiatives, government policies,
Environmental factors	Water quality, climate

Socioeconomic Factors:

Socioeconomic factors such as income and education play a significant role in HWT adoption. Lower-income households may face financial constraints, limiting their ability to invest in HWT technologies. Additionally, a lack of knowledge about waterborne diseases and the benefits of HWT can hinder adoption. Lower education levels may contribute to the preference for bottled water over HWT [25]. Gender also plays a role, with women often being more concerned about water quality and more likely to adopt HWT practices [26].

Technological Factors:

Technological factors, such as complexity, reliability and affordability of the method/device can significantly impact HWT adoption. Some technologies, such as ceramic and bio-sand filters, require specific maintenance practices, which may deter users. These filters are cleaned by periodic scrubbing plus washing and removal of Schmutzdecke, respectively, thereby making their operation cumbersome [16]. It is also important to note that consumers may be hesitant to invest in products that are prone to breakdowns or require frequent maintenance. Boiling, cloth filtration (straining) and SODIS are highly reliable as they do not involve any sophisticated instruments.

Additionally, the initial cost and ongoing maintenance expenses can be prohibitive for many households. For instance, RO purifiers are costly and the membranes need periodic replacement to compensate for wear and tear due to shear stress [27]. Some water treatment technologies, such as those that rely on electricity (boiling) or chemicals (chlorination), may incur additional operating costs. The availability of HWT technologies in local markets and the accessibility of information about these products can also influence adoption rates, with households having easy and continuous access showing higher willingness.

Behavioural Factors:

Behavioural factors, such as beliefs, risk perception and trust, shape HWT adoption. Cultural beliefs and practices can influence attitudes towards water and the use of HWT technologies. For instance, consumers prefer a certain taste in drinking water and disregard treatment options that may alter it; such as filters and RO [28]. However, it should be noted that, the human perception of taste and odour is highly subjective and influenced by various cultural factors. What one person perceives as unpleasant, another may find acceptable or even pleasant. The psychological concept of 'naturalness' among rural population discourages the use of HWT options [29]. They perceive water from streams, borewells, etc., to be natural and safe, and thus, any of the point-of-use treatment options interfere with the 'natural' characteristics of water. Individuals' perceptions of the risk of waterborne diseases can also impact their decision to adopt HWT. Their experiences and awareness shape their perceptions, with consumers infected with waterborne diseases in the past being more likely to practice HWT. Also, trust in the quality and safety of HWT technologies can be a significant factor in adoption decisions.

Additional Considerations:

Institutional Factors: Institutional factors such as government policies and public health initiatives can significantly influence HWT adoption. Government regulations and subsidies can promote the use of HWT, while public health campaigns can raise awareness about the importance of safe water and encourage the adoption of HWT practices. The availability of reliable water supply infrastructure can also impact HWT adoption.

Environmental Factors: Environmental factors, such as water quality and climate, can influence the need for and effectiveness of HWT. In regions with poor water quality, HWT can be essential to ensure safe drinking water. These treatment options are beneficial particularly in areas with intermittent water supply, as they allow for storage of treated water without compromising with its quality [4].

Social Factors: Social norms and peer pressure influence the adoption decision, particularly in communities where certain practices are considered more acceptable.

The adoption of HWT is influenced by a complex interplay of the above-mentioned factors. Addressing these factors through a combination of strategies, including education and awareness campaigns, subsidies and incentives and the development of affordable and user-friendly HWT technologies, is essential for promoting the widespread use of HWT and improving access to safe drinking water.

7.2.4 Role of IoT in HWT

IoT plays a vital role in assessing and optimizing water treatment stages. Some of the key functions include real-time monitoring, predictive maintenance, flow rate optimization, early detection of failures or breakdowns, data-driven decision-making and automated system control [30]. IoT allows for remote measurement of water quality indicators like pH, temperature, dissolved oxygen (DO), turbidity, TDS, conductivity and many more [31]. Thus, IoT offers numerous benefits including improved water quality, enhanced efficiency, reduced costs and improved safety. This section enlists some key types of IoT used for assessing different water treatment stages.

Coagulation: Sensors for turbidity and pH allow for real-time monitoring by continuously monitoring the water's turbidity and pH levels during the coagulation process. This data helps operators adjust the dosage of coagulants to achieve optimal results [32]. Flow meters are used to measure the flow rate of water and chemicals entering the coagulation tank, ensuring proper mixing and chemical dosing [31].

Filtration: Pressure transducers allow for predictive maintenance that monitor pressure drops across filter beds to detect clogging and the need for backwashing or filter replacement, thereby preventing filter failure and ensuring consistent water quality [30]. Additionally, flow meters help optimize filtration rates, maximizing the efficiency of the filtration process.

Disinfection: Residual chlorine sensors are used to measure the concentration of chlorine or other disinfectants to ensure adequate disinfection levels [33]. pH sensors are also used to check pH levels as fluctuations in pH can affect the efficiency of disinfectants. Biological sensors, for instance coliform bacteria sensors, aid in detecting the presence of specific bacteria or pathogens to confirm disinfection effectiveness [34]. If disinfectant levels drop below acceptable limits, IoT systems can trigger alerts, allowing for prompt corrective action.

7.3 Structural Equation Modelling (SEM) and Its Use in Water Quality Management

This section provides an overview of SEM and its applications in water quality management.

7.3.1 Fundamentals of SEM

SEM is a powerful statistical technique used to analyse complex relationships between variables. It combines multiple regression, factor analysis and path analysis to model relation between observed and unobserved variables [35]. SEM allows researchers to test theoretical models and assess the fit of observed data to these models. SEM finds its application in the fields of social sciences [36], psychology [37], education [38], business and engineering [39], where researchers need to model complex relations between variables.

SEM involves two primary components: the measurement model and the structural model. The measurement model specifies how latent variables (unobserved constructs) are measured by observed variables (indicators). The structural model, on the other hand, defines the relationships between latent variables, representing the theoretical causal relationships.

One of the key advantages of SEM is its ability to handle complex relationships between variables and also accounts for measurement error. Additionally, SEM allows researchers to test multiple hypotheses simultaneously, providing a comprehensive understanding of the underlying theoretical framework. However, the interpretation of the results can be challenging, especially for complex models. Additionally, the specification of the model, including the choice of latent variables and their relationships, is crucial and can impact the findings. Some of the key terms are explained here and summarized in Table 7.2.

a. *Latent variables*: A latent variable, also known as a hidden or unobserved variable, is a theoretical construct that cannot be directly measured. Instead, it is inferred from multiple observed variables. These variables are assumed to reflect or indicate the underlying latent construct. Think of it as a concept or idea

Table 7.2 Key terms of SEM

Sr. no.	Terms	Definitions
1	Latent variables	These are unobserved variables that are inferred from observed variables. They represent theoretical constructs or concepts that are not directly measurable, e.g. satisfaction, intelligence
2	Manifest variables	These are directly measured variables that serve as indicators for latent variables
3	Measurement model	This specifies the relationships between latent variables and indicators
4	Structural model	This specifies theoretical casual relationships between latent variables
5	Path model	It is a diagram that represents the relationships between variables as a series of directed arrows, or paths, connecting the variables (Fig. 7.1)
6	Endogenous variables	They are similar to dependent variables as they are influenced by other variables in the model. However, they can also influence other endogenous variables
7	Exogenous variables	They are similar to independent variables as they are thought to influence other variables in the model but they are not themselves influenced by any other variables

that exists, but its true form is not directly observable. The terms 'latent' and 'constructs' are used interchangeably. They are represented in Fig. 7.1 as circles or ovals (A and B). In SEM, latent variables are linked to observed variables through measurement models. These variables specify how they influence the observed variables. For instance, 'intelligence' is a complex concept that cannot be directly measured. However, it can be inferred from multiple indicators like IQ scores, academic performance and problem-solving abilities [39].

b. *Manifest variables*: A manifest variable, also known as an observed variable or indicator variable, is a directly measurable quantity. It represents a concrete, observable characteristic or attribute that can be quantified. They are used to measure latent variables and can be either quantitative (numerical) or qualitative (categorical). By using manifest variables, researchers can construct more robust and valid models in SEM, leading to a deeper understanding of complex psychological and social phenomena [39].
c. *Measurement model*: The measurement model specifies the relationship between latent variables and their observed indicators. It focuses on how well the observed variables measure the underlying latent constructs. It is used to assess the validity and reliability of the measurement instruments and to evaluate the degree to which the observed variables accurately represent the latent constructs [39]. It consists of three components:

   i. Latent variables.
   ii. Manifest variables.
   iii. Factor loadings.

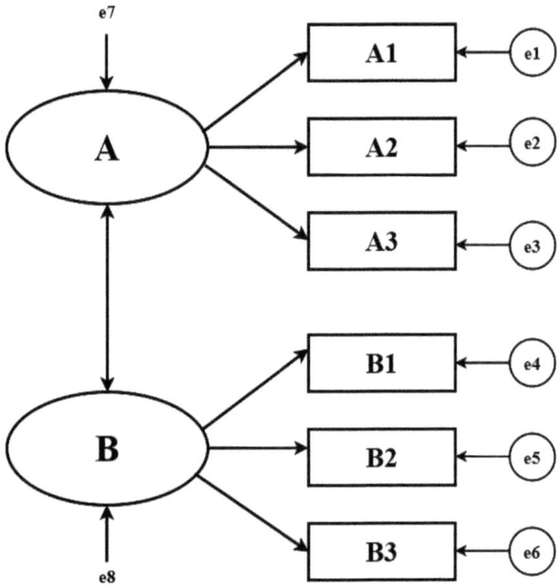

Fig. 7.1 Illustration of a path model

d. *Structural model*: The structural model specifies the relationships between latent variables. It focuses on the causal relationships among the constructs. Its primary purposes are to test hypotheses about the causal relationships between latent variables and to estimate the direct and indirect effects of latent variables on each other [39]. The two components are:

   i. Path Coefficients; representing the strength and direction of the relationships between latent variables.
   ii. Covariances, representing the shared variance between latent variables that is not explained by the direct effects.

e. *Exogenous variables*: Variables that are not influenced by other variables within the model. They are considered independent variables or causes. They serve as the starting point for causal relationships within the model, providing the initial input or stimulus that affects other variables [40].
f. *Endogenous variables*: Variables that are influenced by other variables within the model. They are considered dependent variables or effects. They are the variables that we are trying to explain or predict as they represent the outcomes or effects of causal relationships [40].

   The characteristics of exogenous and endogenous variable are compared in Table 7.3.
g. *Path model*: A path model is a specific type of structural equation model that focuses on the direct and indirect effects of variables on each other. Its primary purpose is to identify the causal pathways through which variables influence each other as well as to quantify the magnitude of these effects.

Table 7.3 Characteristics of exogenous and endogenous

Exogenous variables	Endogenous variables
They do not have any incoming paths from other variables in the model	They have at least one incoming path from another variable in the model
They can influence other variables but are not themselves influenced by them	They can be influenced by both exogenous and other endogenous variables
They are often measured variables, but they can also be latent variables	They are often latent variables, but they can also be measured variables

While, measurement model provides the foundation for the structural model by ensuring that the latent variables are accurately measured. Structural model builds upon the measurement model by specifying the relationships between the latent variables, and Path model is a specific type of structural model that focuses on the direct and indirect effects of variables on each other.

The traditional concept of dependent and independent variables often used doesn't directly translate to SEM due to the following reasons:

i. SEM often deals with latent variables, which are abstract constructs that cannot be directly measured. A variable can be both exogenous and endogenous in different models, depending on the specific research question and model specification [41].
ii. SEM allows for reciprocal relationships between variables, meaning that two variables can influence each other simultaneously [42].
iii. SEM can model complex relationships, including mediation and moderation effects. A mediator variable explains the relationship between two other variables, while a moderator variable influences the strength of that relationship [43] (Fig. 7.2).
iv. SEM can handle multiple dependent variables simultaneously, allowing for the analysis of complex systems of relationships.

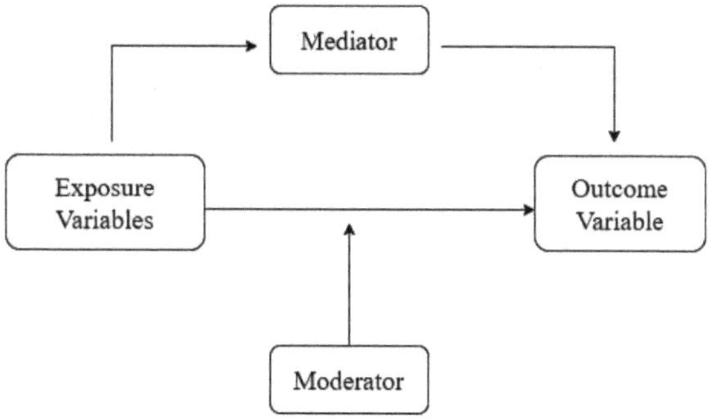

Fig. 7.2 Mediator and moderator variables (adapted from Moe et al. [44])

Table 7.4 Key differences between EFA and CFA

Exploratory Factor Analysis (EFA)	Confirmatory Factor Analysis (CFA)
EFA is used to explore the underlying structure of a set of variables without any prior hypotheses about the structure [40, 45]	CFA is a hypothesis-driven technique that is used to test a specific hypothesized factor structure [45]
EFA is often used in the early stages of research	CFA is a more rigorous technique than EFA and is applied at latter stages to test hypothesis
The primary goal of EFA is to reduce a large number of variables into a smaller number of factors	The primary goal of CFA is to assess the extent to which the observed data fit a specific factor structure
EFA is more flexible as it can accommodate different factor structures and require fewer assumptions about the data	While, CFA requires more specific assumptions about the data

7.3.2 Factor Analysis

There are two primary types of factor analysis are commonly used: Exploratory Factor Analysis (EFA) and Confirmatory Factor Analysis (CFA) [35]. While they are both used to understand the relationships between variables, they differ in their approach and purpose. Table 7.4 highlights some key differences between EFA and CFA that may aid in a better understanding of these terms.

7.3.3 Application of SEM in Water Quality Management Studies

SEM has emerged as a powerful tool for unravelling complex relationships within water management systems. It allows researchers to investigate intricate cause-and-effect linkages between various influencing factors. In the context of water and wastewater quality management, SEM offers several key applications, some of which are described below.

SEM can be employed to identify causal relationships between different factors influencing water quality. For instance, SEM can help determine how factors like land use patterns, climate change and pollution sources impact the quality of raw water [46]. This knowledge aids in optimizing treatment processes and ensuring safe drinking water. According to Wang et al. [47], fragmentation of watersheds intensifies the impact of urban pollution in water bodies. In a study conducted by da Silva et al. [48], SEM analysis showed that high metal concentration in water can cause genetic damage. Additionally, SEM can be used to explore the relationship between different treatment stages such as coagulation, filtration, disinfection, etc., to identify bottlenecks and inefficiencies, leading to improved treatment performance.

SEM is valuable for assessing the impact of policy interventions on water quality. By analysing the effects of regulatory measures such as discharge permits, water quality standards and pollution control technologies, SEM can help evaluate their effectiveness. Moreover, it can be used to simulate different policy scenarios, enabling policymakers to anticipate the potential consequences of future decisions. For instance, SEM was used to develop forest conservation strategies in Magdalena basin, Mexico [49]. Another study conducted by Gupta et al. [50], in India, the government should charge a small fee for Public Welfare Scheme to prevent people from taking them for granted.

SEM can aid in understanding how public perception of water quality influences water conservation behaviours and compliance with regulations, policymakers can develop more effective communication strategies. Neha and Kansal [51] identified various communication channels like TV, newspapers and workshops to promote use of reclaimed water. Similarly, behavioural change campaigns were shown to improve water recycling practices in Saudi Arabia [52].

SEM can be used to develop predictive models for water quality parameters. By identifying key factors influencing water quality and their interrelationships, SEM can help forecast future trends, enabling early warning systems and proactive management strategies. Ryberg et al. [53] predicted that heavy precipitation and anthropogenic interferences can elevate atrazine (a pesticide) runoff into surface waters.

SEM can assess the complex interactions between water, energy and food systems, identifying potential trade-offs and synergies to promote sustainable resource management. According to a study conducted by Daohan et al. [54], in China, the unsustainable groundwater withdrawal has a detrimental effect on the WEF nexus.

SEM can enhance the decision-making processes by considering the combined effects of climate change, urbanization and pollution. The study on the Kerch Peninsula revealed that vegetation has adapted to temperature changes due to climatic stress. However, the increasing influence of precipitation on vegetation recovery needs climate monitoring [55].

7.3.4 The SEM Process

The SEM process typically involves several steps. First, a theoretical model is specified, outlining the hypothesized relationships between variables. Next, it is necessary to ensure that the model is identifiable, meaning that the parameters can be uniquely estimated from the data. Next, the model is estimated using statistical software. The estimated parameters, including path coefficients and factor loadings, provide information about the strength and direction of the relationships between variables. Once the model is estimated, its fit to the data is evaluated using various fit indices, such as the chi-square test, root mean square error of approximation (RMSEA) and comparative fit index (CFI) [45]. If the model does not fit the data well, modifications may be made, such as adding or removing paths or latent variables. Figure 7.3 represents the sequence of steps involved in testing hypothesis using SEM and are described here.

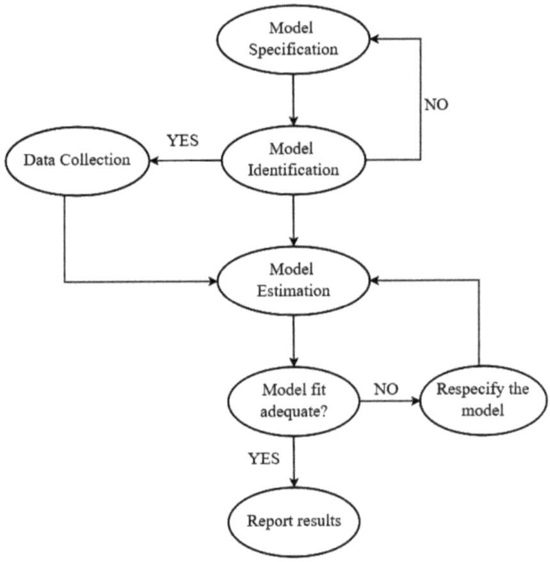

Fig. 7.3 Flow chart of SEM process

i. *Model specification*: It involves defining the theoretical relationships between latent and observed variables. The process typically involves conceptualizing the theoretical framework as well as specifying the measurement and structural model. It requires careful consideration of the research questions, underlying theory and the available data. However, it is an iterative process and researchers may need to refine their model based on several diagnostic information.
ii. *Model identification*: It is the ability to uniquely estimate the parameters of the model from the observed data. A model is identified if it has enough information to solve for all unknown parameters. If a model is not identified, it means that multiple sets of parameter estimates could fit the data equally well, making it impossible to draw definitive conclusions [56]. In such cases, re-start with the initial step, i.e. model specification, and once the model is identified, one can proceed with data collection.
iii. *Model estimation*: It involves assessing the fit of the specified model to the observed data and interpreting the estimated parameters [57]. This can be done by assessing absolute fit indices (e.g. chi-square, RMSEA), comparative fit indices (e.g. CFI, TLI) and incremental fit indices (e.g. standardized root mean residual (SRMR)). If the model fit is inadequate, the model needs to undergo a justifiable re-specification. Once the satisfactory fit of the respecified model is achieved, it is possible to report the path coefficients for a quantitative analysis of the latent variables. This step allows researchers to examine the significance and magnitude of parameter estimates, such as factor loadings and path coefficients, to understand the strength and direction of relationships between variables. By carefully evaluating the model fit and parameter estimates, one can assess the

theoretical adequacy of the model and draw meaningful conclusions from the analysis.

7.4 Case Study on Adoption of HWT

A study was conducted in Surat, India, to understand the factors shaping public opinions regarding use of HWT methods/devices. The objectives of the study were to understand which HWT methods are currently being used and the reasons behind their choices and to gauge understanding of HWT, its importance, risks and available methods. The study also explored residents' beliefs about HWT effectiveness, costs and inconveniences. How factors such as gender, education and income influence willingness to adopt HWT was also studied. Further, the study also tried to identify the HWT methods that consumers find most acceptable and feasible.

7.4.1 Study Area

Surat is located on the banks of Tapti River in Gujarat, India. The city faces challenges related to water quality, particularly due to industrial pollution and inadequate water treatment infrastructure [58]. The river serves as the primary drinking water source for the city. The physical and chemical characteristics of the river water undergoes continuous deterioration due to accelerating discharges of industrial and domestic effluents. Surat presents an ideal study area for the concerned survey due to the following reasons:

i. The city's diverse population spread across urban and rural areas offers a wide range of socioeconomic backgrounds.
ii. Poor water quality with respect to hardness, alkalinity, total dissolved solids, chlorides and sulphates highlights the need for effective water treatment.

7.4.2 Methodology

The consumers' adoption decisions are influenced by how they perceive the benefits and barriers for using a particular method/device. For this, a survey was conducted among 100 households from different localities of Surat to identify the influencing factors and develop strategies to promote their wider use. A draft questionnaire was developed and pilot-tested with a small group of representative individuals. Feedback from the pilot test was used to refine the questionnaire. The final questionnaire included demographic information, source of water, consumer satisfaction, awareness of waterborne diseases, current water treatment practices and their willingness to adopt HWT. The interviews were conducted at respondents' households, allowing

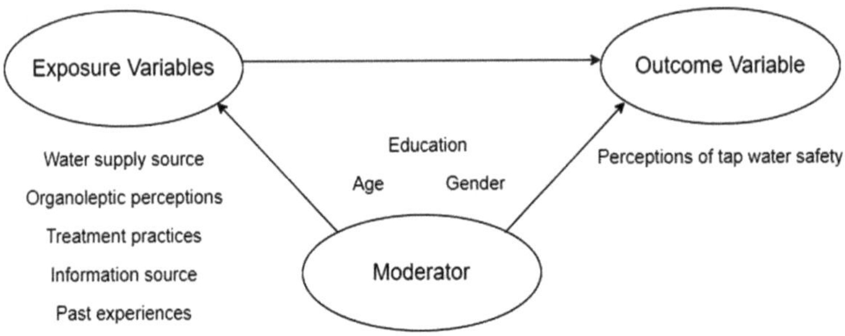

Fig. 7.4 Initial hypothesis of the study

interviewers for in-depth probing and clarification of responses. The collected data was analysed using SPSS, a statistical software, and modelled in AMOS (Analysis of Moment Structures). Descriptive analysis of the data gave percentages and means to describe the demographic characteristics of respondents. The next step was inferential analysis that helped identify significant relationships between variables and draw inferences about the targeted population. The primary purpose was to assess the consumer perceptions of their tap water supply, and thus, it was considered as the outcome variable. The exposure variables influencing the outcome variables were the following: source of water used/available, organoleptic perceptions, treatment practices, source of information for waterborne diseases, available HWT options as well as consumers' past experiences. As mentioned in Fig. 7.2, the moderator influence both exposure and outcome variables and are therefore considered in the analysis. In the concerned study, the gender, age and educational qualifications of the respondents were noted as moderating variables as shown in Fig. 7.4.

7.4.3 Questionnaire Survey

The questionnaire was meticulously crafted to ensure clarity, relevance and ease of understanding for respondents. The responses were collected on a 5-point Likert scale. The aim of the survey was to collect both quantitative and qualitative data. The questionnaire comprised of 23 questions, divided into 5 sections:

i. *Socio-demographic information*: Age, gender, education level, occupation and household size.
ii. *Quality perceptions*: This section focuses on how the respondents perceive the quality of their intended water supply in terms of taste and odour, mineral deposits and aesthetics.
iii. *Source of water*: The aim was to identify the respondents' preferred source of water from among piped connection, borewell and bottled water.
iv. *Experiences*: Past instances of waterborne diseases, objectionable taste and odour, irregular water supply.

v. *Preferred HWT methods*: The expected outcome was to understand the respondents' preference for boiling, RO and ceramic filters.

7.4.4 *Analysis of the Survey Results*

The age distribution of the respondents ranged from 19 to 63 years, with a mean age of 37 years (Fig. 7.5a). It can also be observed that the survey population consisted of a significant proportion in the prime working age groups of 25–44 years, comprising 65% of the total. This demographic is likely to be more aware of health risks and may be willing to invest in water treatment solutions. It is also evident that a substantial proportion of the sample (70%) had education till graduation degree and beyond, indicating a relatively educated population that is likely to be more receptive to water treatment technologies and practices (Fig. 7.5b). However, younger age groups, particularly 15–24 years, might require more targeted education and awareness while those with lower education levels may need simpler and more accessible information about the need and benefits of HWT. The sample population, consisting of 54% male and 46% female, was relatively balanced in terms of gender representation. Such overall diverse demographic profiles ensure that the results are not skewed towards a particular section.

Preferred Treatment Options

The study indicates minimal difference among households having positive and negative preferences for RO as shown in Fig. 7.6. The consumers use RO within their households as it provides a higher level of purification by removing dissolved solids and heavy metals [21]. The popularity of RO systems in Surat may be attributed to growing health concerns and consumer willingness to invest in RO systems. This acceptance is in line with the fact that high- and medium-income households predominate in the survey locations. It is also observed that 80% of respondents have had

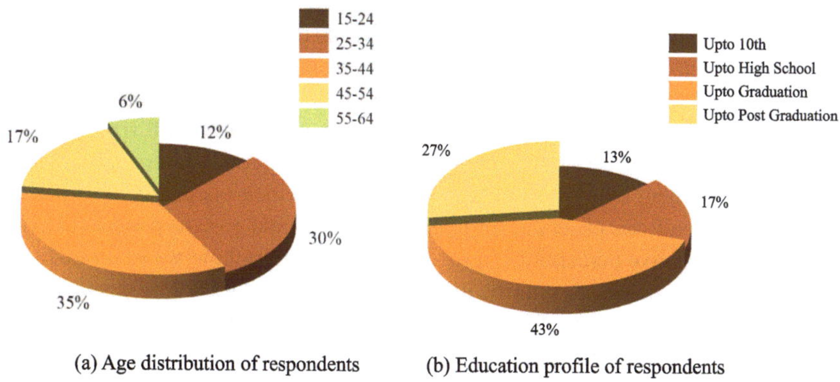

Fig. 7.5 **a** Age distribution of respondents, **b** Education profile of respondents

Fig. 7.6 Consumers' preferences of treatment options

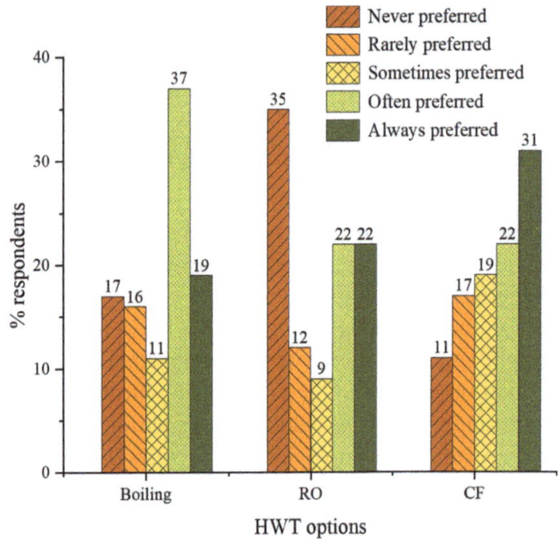

experiences of waterborne diseases that makes them suspicious of their existing water quality. On the contrary, the higher RO reject rates can be due to several reasons. They are more expensive to purchase and maintain and can lead to mineral depletion in the water [22]. Some individuals prefer a certain taste of water and disregard RO as it removes minerals that affect the taste of water. Some users believe that RO water lacks essential minerals, leading to a lower preference for RO systems. Some households still rely on traditional water purification methods like boiling or filtration, which they perceive as sufficient for their needs.

Ceramic filters (CF) are preferred by 53% respondents while 28% disregard their use. Several factors could contribute to a higher percentage of positive preferences among households, some of them being effective microbial removal, better affordability and easy maintenance. Unlike RO, CF retain essential minerals in water. This is valued by many consumers who believe in the health benefits of mineral-rich water [18]. They do not require electricity for operation and minimize water wastage, which can be an added benefit. Despite these advantages, a notable portion of households do not prefer CF, particularly due to limited filtration capacity against viruses and heavy metals which can be a point of concern for individuals with specific water quality concerns. In case of turbid water, the flow rate through filters can be lower than other treatment options, which can further hinder their operation.

A similar trend was observed for boiling with 56% and 33% respondents having positive and negative perceptions, respectively. Boiling drinking water is one of the traditional methods of water purification [2]. It inhibits the growth of pathogens by denaturing cell protoplasm of bacteria, viruses, protozoan spores and cysts that are resistant to chemical disinfection [22]. The waterborne pathogens are effectively destroyed at temperatures above 60 °C [59]. It is a simple and straight-forward method, requiring minimal equipment and giving consumers a sense of control over the purification process. The survey results also indicate a substantial rejection rate for boiling as it imposes certain limitations. Boiling requires significant energy that is derived from woods, gas or electricity, which can be costly and environmentally unsustainable especially in areas with limited access to electricity and fuel [18]. The method is time-consuming, inefficient in eradicating DBPs and other chemicals, as well as does not safeguard against future contamination, making it unsuitable for large households [60]. Figure 7.6 describes the consumers' views on different HWT alternatives.

Consumer Perceptions of Water Quality

Based on the survey results (Fig. 7.7), it can be concluded that a majority of the households (56%) are satisfied with the taste and odour of their source water. This indicates that the water quality in Surat has improved significantly over the years, likely due to advancements in water treatment technologies and improved water management practices. However, 16% of households are still dissatisfied with the taste and odour of their water. This suggests that there may be specific areas, particularly the low-income areas, where water quality issues persist. However, it is difficult to give conclusive evidence of factors shaping consumer opinions of T&O of their drinking water. A large proportion of the households (46%) sometimes experience high hardness in their source water. High levels of hardness-causing cations like calcium and magnesium lead to scale buildup in pipes, appliances and fixtures, affecting their efficiency and lifespan. Additionally, it can impact the taste and appearance of water which can be objectionable to the users. Hard water also has detrimental health impacts. Prolonged consumption of hard water increases the chances of cardiovascular diseases [61]. High calcium levels can increase the risk of kidney stone formation while excess magnesium can lead to diarrhoea and other digestive problems [61]. However, several minerals, including calcium and magnesium, are required in trace amounts for cardiovascular and nervous functions, bone and teeth health, among others [62].

Aesthetics, including factors like colour and turbidity, have the least impact on overall water quality. However, they significantly influence consumer perception and satisfaction. While not directly affecting health, the aesthetic qualities of water are crucial for consumer acceptance as they directly assess water quality based on organoleptic perceptions. A substantial proportion of people experience aesthetic-related issues in their source water.

Consumer Preference for Water Source

There is a strong preference among the consumers for piped connections (Fig. 7.8). These can be due to several reasons. Piped water connections offer a consistent and reliable supply of water, eliminating the need for manual pumping or water storage. This convenience is particularly valuable in urban and semi-urban areas with established water infrastructure. In low-income areas, irregular supply from municipal connections can be frustrating for households, leading to negative perceptions. In areas with adequate supply, households still disregard piped connections as they might not trust the municipal authorities' ability to provide water supply safe for

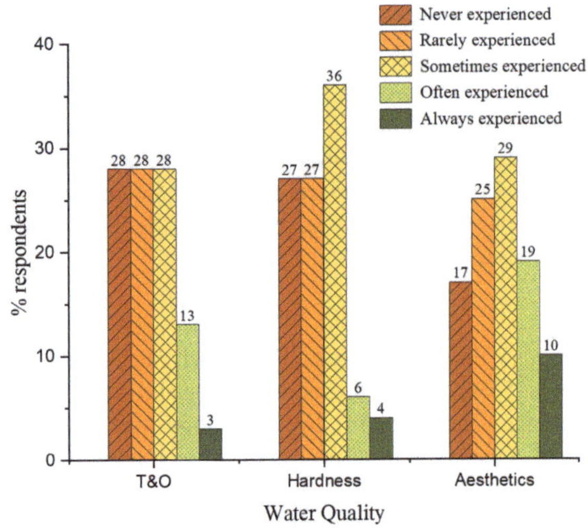

Fig. 7.7 Consumers' perceptions of water quality

consumption. Caputo et al. [63] stated that consumers' mistrust arises when the authorities do not regularly monitor the water quality to identify potential issues. With adequate treatment levels, bottled water can provide a safer and more hygienic water source compared to alternative options like borewells or surface water. This is especially important for households with vulnerable members, such as children and the elderly. The consideration is higher in households where majority of members are employed as purchasing packaged water can be costly and withdrawing water from borewells can get very cumbersome.

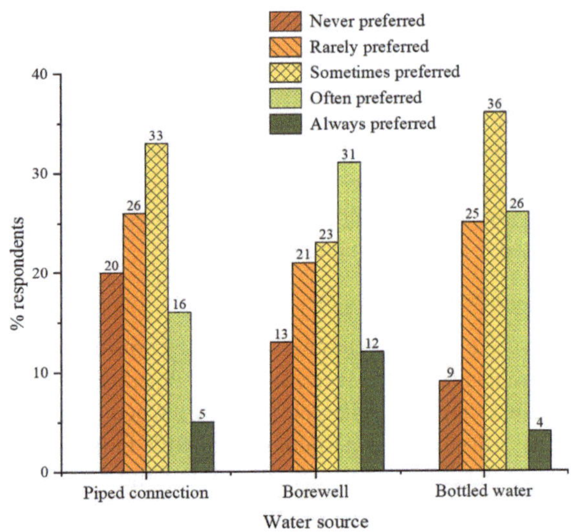

Fig. 7.8 Consumers' preference for water source

Still, a substantial proportion of households prefer borewells, mainly due to limited access to piped water. Another driving factor is the phenomenon of naturalness where consumers strongly perceive groundwater as safe. Although, borewells are self-reliant and allow households to manage their water supply without relying on external sources, it can still be contingent in non-rural areas. Thus, it is important to understand the characteristics of the water collected from a particular source, which in turn, help decide the level of treatment required. There can be two probable drivers of bottled water consumption. First, bottled water is often perceived as a high-quality and safe water source, particularly in urban areas where concerns about water quality may be prevalent. This perception can drive demand especially among affluent households. Second, using packaged water is considered a luxury especially among high-income localities.

Consumer Experiences

According to the survey, 80% of the respondents had experienced the debilitating effects of waterborne diseases and are more likely to adopt HWT. The first-hand experience of illness, coupled with the understanding of the potential health risks associated with contaminated water, motivates individuals to take proactive measures to protect themselves and their families. This increased awareness and heightened concern for water safety drive the adoption of HWT methods. Figure 7.9 represents the experiences and difficulties faced by consumers.

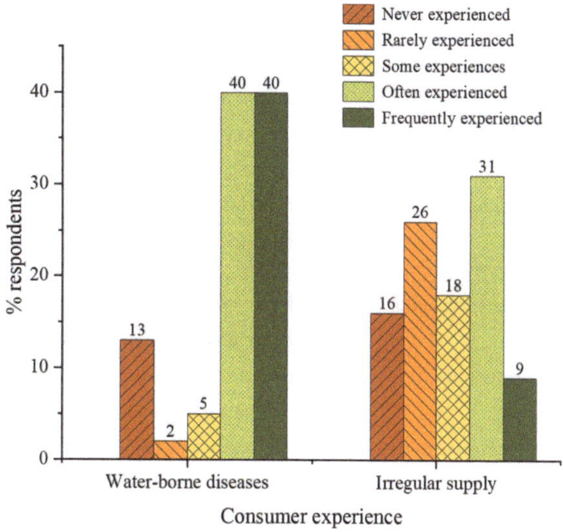

Fig. 7.9 Consumers' experiences with their water supply

7.4.5 The SEM Model

To achieve the objective of the study, SEM was performed using AMOS software. SEM allows for simultaneous analysis of multiple dependent variables and their underlying latent constructs. By utilizing SEM, it is possible to examine the complex relationship between various listed factors that are most likely to influence the HWT adoption decisions.

Initially, the model was prepared considering the 23 questions as observed/manifest variables that are divided into 8 sections as it is shown in Table 7.5. Figure 7.10 shows the model obtained after performing SEM in AMOS software. However, the final model obtained is quite different from the initial hypothesized model. This is because several variables were eliminated due to poor fit indices, model identification problem or simply due to lack of theoretical support. Also, if a variable is highly correlated with another variable, it may be redundant and can be removed without significantly affecting the model's explanatory power.

Table 7.5 Structure of survey questionnaire for assessing determinants of HWT adoption

Latent	Manifest			
Demographic	Age	Gender	Household size	Education
Water quality	Tap water has objectionable T&O (*QTY1*)			
	Tap water has objectionable colour (*QTY2*)			
	Tap water has excessive mineral content/is mineral deficit (*QTY3*)			
	Tap water is aesthetically unpleasant (*QTY4*)			
Water Source (WS)	Receive tap water supply from piped connection (*WS1*)			
	Withdraw drinking water from borewell (*WS2*)			
	Purchase bottled water for drinking (*WS3*)			
Past experiences	The existing water supply possess human health risks (*EXP1*)			
	The area/society/household receives intermittent water supply (*EXP2*)			
Perceived benefits	The treatment adopted is affordable (*ADV1*)			
	The treatment adopted is easy to use and reliable (*ADV2*)			
Intended use	Treated water is used for drinking only (*U1*)			
	Treated water is used for both drinking and cooking (*U2*)			
Information source	Newspaper and TV (*INFO1*)			
	Friends and Family (*INFO2*)			
	Advertisements in banners or hoardings (*INFO3*)			
Preferred treatment	Boiling (*TRT1*)			
	RO (*TRT2*)			
	Ceramic Filters (*TRT3*)			

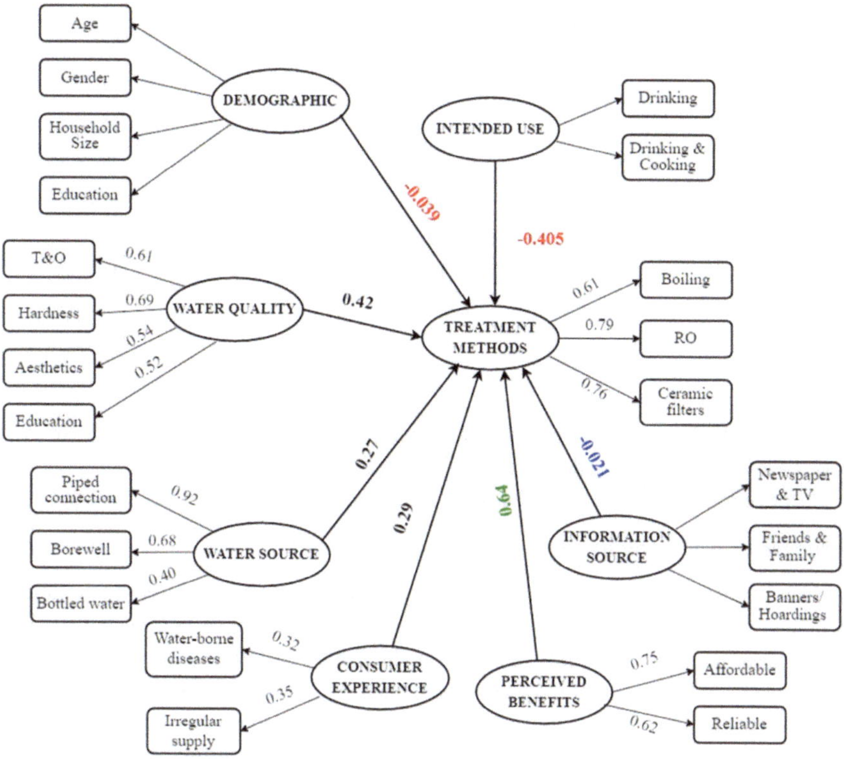

Fig. 7.10 Initial hypothesis of the model

The path coefficients, such as '*−0.405*', '*−0.039*', '*0.42*', '*0.27*', '*0.29*', '*0.64*' and '*−0.021*', shown in Fig. 7.10, represent the standard regression weights to quantify the strength and direction of relationships between variables and is calculated based on maximum likelihood estimation (MLE). A positive coefficient indicates a positive relationship and vice versa. The magnitude of the coefficient reflects the strength of the relationship. Path coefficients are typically standardized, meaning they represent the standard deviation change in the dependent variable for a one standard deviation change in the independent variable.

The latent components of 'Intended use' and 'Demographic' were eliminated as they carried a negative path coefficient of *−0.405* and *−0.039*, respectively, as shown in Fig. 7.10. Such negative coefficients suggest that an increase in the concerned variable lead to a decrease in the outcome variable 'consumer perceptions', which cannot be theoretically meaningful. During model re-specification, it was observed that the error term associated with *INFO 1* was also negative (−0.338), implying that *INFO 1* is perfectly predicted by its latent construct, which is statistically impossible. Later on, the latent variable 'Information source' was also removed as it gave negative path coefficient of *−0.021*. This re-specification gave positive path coefficients for the remaining variables but significantly worsened the model fit as shown in Table 7.6.

Table 7.6 Model fit indices after first modification

Type	Measure	Model Values	Recommended	Remarks
Absolute fit	Chi-square	277.931	–	–
	C_{min}/df	3.431	Between 2 and 5	Good fit
	RMSEA	0.157	<0.08	Poor fit
Incremental fit	CFI	0.595	>0.9	Poor fit
	TLI	0.545	>0.9	Poor fit
	NFI	0.518	>0.9	Poor fit
Parsimonious fit	PNFI	0.529	>0.05	Good fit
	PCFI	0.461	>0.05	Good fit

Further, the latent component of 'Perceived benefits' was omitted as it also gave poor fit indices for the model as shown in Table 7.7.

After redefining, the final SEM model obtained is shown in Fig. 7.11.

Influence of Water Quality on Adoption of HWT Devices

The path coefficient of *0.40* (Fig. 7.11) between water quality and treatment method indicates a positive and significant relationship between these two variables. This means that for every one-unit increase in the latent variable of water quality, the outcome variable of treatment method is expected to increase by 0.40 units, holding other factors constant. This finding suggests that as the perceived water quality deteriorates, households are more likely to adopt water treatment methods. This aligns with the theoretical expectation that individuals are more likely to seek solutions to problems when they perceive the severity of the problem to be high.

The path coefficients *of 0.55, 0.77* and *0.51*, respectively, for taste and odour, hardness and aesthetics on water quality indicate strong positive relationships between these variables. These findings imply that households with water sources that have higher levels of taste and odour, hardness and poor aesthetics are more likely to perceive their water quality as poor. This perception of poor water quality, in turn,

Table 7.7 Model fit indices after second modification

Type	Measure	Model values	Recommended	Remarks
Absolute fit	Chi-square	193.119	–	–
	C_{min}/df	3.388	Between 2 and 5	Good fit
	RMSEA	0.155	<0.08	Poor fit
Incremental fit	CFI	0.638	>0.9	Poor fit
	TLI	0.580	>0.9	Poor fit
	NFI	0.563	>0.9	Poor fit
Parsimonious fit	PNFI	0.486	>0.05	Good fit
	PCFI	0.551	>0.05	Good fit

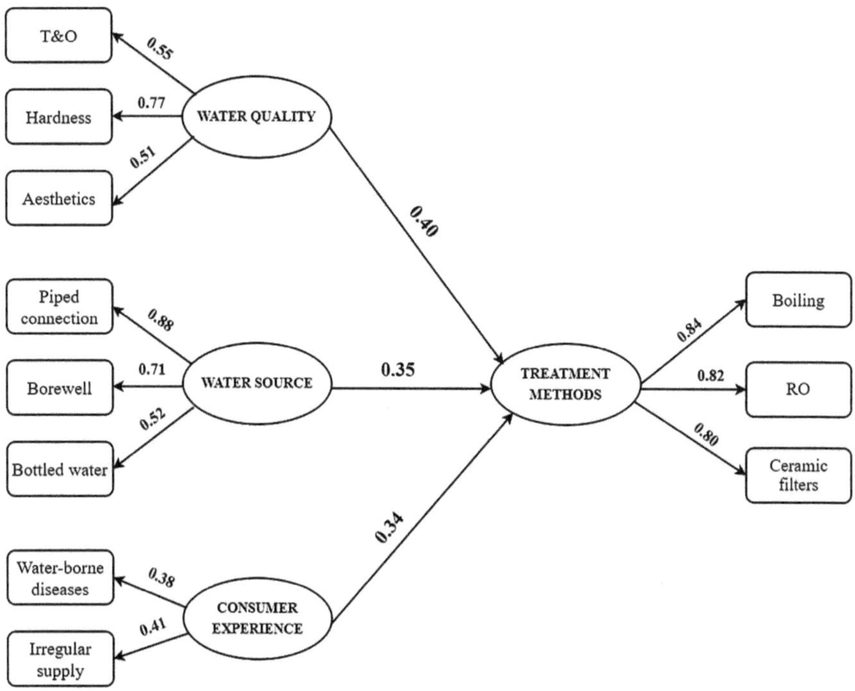

Fig. 7.11 Final SEM model

can motivate households to adopt water treatment methods to improve the quality of the water they consume.

Hardness (*0.77*) emerges as the primary factor affecting water quality. Surat is situated on the banks of Tapi River, which gives a logical explanation of excessive hardness in source water. High levels of hardness lead to scale buildup in pipes and appliances and alter the taste of water which can be objectionable for consumers. This also explains the influence of taste and odour (*0.55*) on water quality perceptions.

Aesthetics, being the least influencing factor (*0.51*), includes colour and turbidity of water. While not directly affecting health, the aesthetic qualities of water are crucial for consumer satisfaction.

Influence of Water Source on Adoption of HWT Devices

The path coefficient of *0.35* (Fig. 7.11) between water source and treatment methods indicates a moderate positive relationship between these two variables. This finding suggests that households with less reliable water sources, such as those relying on surface water or hand pumps, are more likely to adopt water treatment methods. This is likely due to the higher perceived risk of waterborne diseases associated with these water sources. By using water treatment methods, households can improve the quality of their water and reduce the risk of waterborne illnesses.

The path coefficient analysis reveals a strong relationship between water source and the three primary modes of water access: piped connection, borewell and bottled water. Piped connection exhibits the highest path coefficient (*0.88*), indicating its significant influence on overall water access. However, it's important to note that this doesn't necessarily imply that piped water is always of superior quality. Factors such as water treatment processes, distribution network conditions and potential contamination can significantly impact water quality.

While not as dominant as piped connections, borewells still hold considerable importance, as evidenced by their path coefficient of *0.71*. This suggests that borewells serve as a viable alternative, particularly in areas where piped water supply may be inadequate or intermittent. However, higher reliance on borewells can raise concerns about groundwater depletion.

Bottled water, with a path coefficient of *0.52*, emerges as a less significant source of drinking water. This lower coefficient indicates that bottled water is primarily used as a supplementary source or for specific purposes. Factors such as cost and environmental concerns may limit the widespread adoption of bottled water.

Influence of Consumer Experience on Adoption of HWT Devices

The path coefficient of *0.34* (Fig. 7.11) between consumer experiences and treatment methods indicates a moderate positive relationship between these two variables. This finding highlights the importance of consumer experiences in influencing the adoption of water treatment methods. Positive experiences, such as consistent water supply, good water taste and absence of odour, can lead to increased satisfaction and trust in the water source. However, negative experiences, such as frequent water shortages, poor water quality, or unpleasant taste and odour, can erode consumer trust and motivate them to seek additional water treatment solutions.

A path coefficient of *0.38* between waterborne diseases and consumer experience indicates a moderate but notable relationship. This suggests that the prevalence of waterborne diseases, such as diarrhoea, typhoid and cholera, has a direct impact on consumer satisfaction. When households experience these health issues, they are likely to perceive their water supply as unsafe and unreliable. This negative perception can lead to decreased trust in the water supply system and a reduced willingness to pay for water services.

The path coefficient of *0.41* between irregular water supply and consumer experience highlights the significant impact of water scarcity on consumer satisfaction. When households face frequent water shortages or low water pressure, it can disrupt daily routines, impact household productivity and lead to frustration and dissatisfaction. Moreover, irregular water supply can exacerbate the risk of waterborne diseases, as it may encourage the storage of water in unhygienic conditions.

7.5 Limitations and Recommendations of the Study

Although the adoption of HWT is crucial for ensuring access to safe drinking water, particularly in developing countries like India, social surveys aimed at identifying public perceptions towards HWT possess certain limitations.

One significant limitation is the potential for sampling bias [64]. The study relied on a small or unrepresentative sample of households, thus, may not accurately reflect the diversity of HWT practices across all socioeconomic groups in Surat. This limited the generalizability of findings to the broader population or to other cities in India with different socioeconomic, cultural and environmental contexts. Factors such as income level, education and access to information significantly influenced HWT adoption, and a skewed sample may have failed to capture the full range of experiences and needs. Another limitation is their limited depth [65]. The survey relied on closed-ended questions, which restricted the respondents' ability to express nuanced opinions and may have failed to capture the full depth of their emotions.

Another significant limitation is social desirability bias. Respondents may have felt pressure to provide socially acceptable answers, even if they don't reflect their true beliefs or behaviours. This can skew survey results, particularly on sensitive topics [66]. Moreover, the survey lacked contextual information about respondents and their situations [67]. This made it difficult to understand the underlying reasons for their responses and to interpret the findings accurately. Such studies on HWT adoption focus on short-term adoption rates without examining long-term sustainability. In Surat, as in other parts of India, even if a household initially adopts a treatment method, it may not continue using it if the costs, maintenance or perceived benefits do not align with their expectations.

However, it is possible to overcome the above-mentioned limitations by considering the following recommendations:

Diversified sampling: Future studies should aim to include a more diverse set of households from different income, education and geographic groups within the study area. Stratified sampling would ensure that the study captures variability across different demographic segments, providing a more accurate reflection of adoption trends. Additionally, considering both urban and rural areas would allow for comparison and contrast in adoption patterns [68].

Mixed-method approaches: Combining qualitative and quantitative methods can provide a richer understanding of the factors influencing HWT adoption [69]. Surveys can quantify the extent of adoption, while in-depth interviews and focus group discussions can explore the reasons behind the adoption or non-adoption of various water treatment methods. Understanding community perceptions, trust in technology and barriers to adoption will be crucial for identifying the most effective interventions.

Longitudinal studies: To better understand the dynamics of water treatment adoption over time, longitudinal studies should be conducted [70]. These studies could track the behaviour of households over several months or years, capturing changes in attitudes, practices and the sustainability of water treatment methods in response to environmental changes, economic conditions or policy interventions. Long-term data would provide insights into the durability of HWT practices and factors influencing their sustained use.

7.6 Conclusions and Future Scope

The study conducted among 100 households in Surat aimed to delve into the factors influencing the adoption of HWT methods/devices. The key determinants that shape consumer perceptions and decision-making processes were identified through SEM. Out of the seven sections considered initially as influencing the adoption decision, only three, namely water quality, water source and consumer experiences, were found to have considerable influence. Water quality emerged as the most significant factor, with a path coefficient of 0.40. This underscores the critical role of water quality in driving the demand for water treatment solutions. The consumers are most concerned regarding the hardness levels of their water supply. These concerns are likely to rise from past health-related experiences like as well as breakdown of pipes and other household appliances. Although excessive hardness is undesirable, soft water is also unfit for consumption. To mitigate these issues, Surat households often turn to water softeners like RO. It can also be noted that bottled water has least influence on water quality. This is justified as households preferring bottled water (30%) are least concerned about its quality as it is infused with added minerals and is convenient to access. Water source was identified as the second most influential factor, with a path coefficient of 0.35. Despite receiving water from piped connections, many consumers remain suspicious of its quality that may encourage them to practice HWT. This paradoxical behaviour highlights the need for increased transparency between water utilities and consumers. Consumer experience was found to be the least influential factor, with a path coefficient of 0.34. This indicates that while past experiences with water quality and treatment may play a role, it is not much significant in driving adoption decisions. Water source (0.35) and consumer experience (0.34) were identified as secondary factors influencing HWT adoption. While these factors play a role, their impact is less pronounced compared to water quality.

In summary, the study identified water quality, source of water and consumer experience as considerable factors influencing decisions of households to adopt HWT methods/devices. Water hardness emerged as a major point of concern for consumers. Also, consumers who have suffered from waterborne diseases and those who receive intermittent water supply are almost equally likely to practice HWT. Other factors such as demography, benefits of HWT, information sources and intended water use were not explicitly considered in the model due to their insignificant contribution and complex nature. However, they are likely to play a role in shaping consumer perceptions when a larger and more diverse population is considered or the study area is changed. To promote HWT adoption, policymakers and public health officials should focus on improving water quality and on raising awareness about the benefits of HWT.

The findings of the study demonstrate the potential of statistical modelling approaches in identifying determinants shaping consumer opinions for adopting HWT. While these findings contribute to the growing body of evidence supporting the adoption of HWT practices, further research shall aid in fully understanding the implications of our findings. One of the options is leveraging machine learning (ML) algorithms to enhance public awareness campaigns about the benefits of HWT [71]. ML techniques have shown promise in targeted advertising and personalized messaging [72], which could be adapted to tailor HWT awareness campaigns to specific demographics and regions. By analysing data on public attitudes, preferences and behaviours, ML models could identify the most effective communication channels to maximize HWT adoption. ML models can predict the likelihood of individuals adopting HWT based on various factors. This information can help prioritize outreach efforts and allocate resources effectively to communities most likely to benefit from HWT interventions [73]. With the help of ML models, policymakers and practitioners can make informed decisions about the most effective and equitable strategies for promoting the widespread adoption of this sustainable practice.

References

1. Global Sustainable Development Report (2023) United Nations
2. Guidelines for Drinking-water Quality, 4th ed, 1st addendum, World Health Organization, 2017
3. Shields K, Bain R, Cronk R, Wright J, Bartram J (2015) Association of supply type with fecal contamination of source water and household stored drinking water in developing countries: a bivariate meta-analysis. Environ Health Perspect 123:1222. https://doi.org/10.1289/ehp.1409002
4. Masanyiwa Z, Zilihona I, Kilobe B (2019) Users' perceptions on drinking water quality and household water treatment and storage in small towns in Northwestern Tanzania. Open J Soc Sci 7:28–42. https://doi.org/10.4236/jss.2019.71003
5. Kumar M, Shekhar S, Kumar R, Kumar P, Govarthanan M, Chaminda T (2023) Drinking water treatment and associated toxic byproducts: concurrence and urgence. Environ Pollut 320. https://doi.org/10.1016/j.envpol.2023.121009
6. Rosa G, Clasen T (2010) Estimating the scope of household water treatment in low-and medium-income countries. Am J Trop Med Hyg 82(2):289–300. https://doi.org/10.4269/ajtmh.2010.09-0382
7. WHO International Scheme to Evaluate Household Water Treatment Technologies (2016) World Health Organization
8. Clasen T, Menon S (2007) Microbiological performance of common water treatment devices for household use in India. Int J Environ Health Res 17(2):83–93. https://doi.org/10.1080/09603120701217695
9. Verma V, Arankalle V (2009) Virological evaluation of domestic water purification devices commonly used in India emphasizes inadequate quality and need for virological standards. Tropical Med Int Health 14(8):885–891. https://doi.org/10.1111/j.1365-3156.2009.02318.x
10. Tiwari V, Aditjandra P, Dissanayake D (2020) Public attitudes towards electric vehicle adoption using Structural Equation Modelling. Transp Res Procedia 48:1615–1634. https://doi.org/10.1016/j.trpro.2020.08.203
11. Wang W, Zhang F, Zhao Q, Liu C, Jim C, Johnson V, Tan M (2023) Determining the main contributing factors to nutrient concentration in rivers in arid northwest China using partial least squares structural equation modelling. J Environ Manag 343

12. Liu Y, Ma J, Chu J, Sun W, Wang Q, Liu Y, Zou P, Ma J (2024) Machine learning and structural equation modeling for revealing the influence factors and pathways of different water management regimes acting on brown rice cadmium. Sci Total Environ 954:176033. https://doi.org/10.1016/j.scitotenv.2024.176033
13. Roth G, Neu-Baker N, Brenner S (2015) SEM analysis of particle size during conventional treatment of CMP process wastewater. Sci Total Environ 508:1–6
14. Ebrahimi E, Asadi H, Rahmani M, Bayat H (2024) Tracing of phosphorus in river sediments using the structural equation modeling. Water Pract Technol 19(10):3941–3955. https://doi.org/10.2166/wpt.2024.245
15. Sohail M, Chen S (2022) A PLS-SEM analysis to check public willingness to use water from filtration plants installed by public own organizations; a study of psychological behavior toward sustainable development. Psychol Res Behav Manag 2671–2682. https://doi.org/10.2147/PRBM.S386097
16. Meera V, Ahammed MM (2008) Solar disinfection for household treatment of roof-harvested rainwater. Water Supply 8:153–160. https://doi.org/10.2166/ws.2008.054
17. Nair AT, Ahammed MM, Davra K (2014) Influence of operating parameters on the performance of a household slow sand filter. Water Supply 14:643–649. https://doi.org/10.2166/ws.2014.021
18. WHO International Scheme to Evaluate Household Water Treatment Technologies (2019) World Health Organization
19. Yang H, Xu S, Chitwood D, Wang Y (2020) Ceramic water filter for point-of-use water treatment in developing countries: principles, challenges and opportunities. Front Environ Sci Eng 14(5). https://doi.org/10.1007/s11783-020-1254-9
20. Sulaiman M, Zhigila D, Mohammed K, Umar D, Aliyu B, Manan F (2019) Moringa oleifera seed as alternative natural coagulant for potential application in water treatment: a review. J Adv Res Mater Sci 56(1):11–21
21. Yang Z, Zhou Y, Feng Z, Rui X, Zhang T, Zhang Z (2019) A review on reverse osmosis and nanofiltration membranes for water purification. Polymers 11:1252. https://doi.org/10.3390/polym11081252
22. Venkatesha R, Rao A, Kedare S (2020) Appropriate household point-of-use water purifier selection template considering a rural case study in western India. Appl Water Sci 10:124. https://doi.org/10.1007/s13201-020-01207-1
23. Shimabuku-Biadola Q, Arakawa F, Nishi L, Wernke G, Cusioli L, Bergamasco R (2024) Activated carbon impregnated with $Ag^{(0)}$ and Cu oxides for the removal of *Escherichia coli*, *Giardia* and *Cryptosporidium* (oo) cysts from water. Desalination Water Treat 318:100335. https://doi.org/10.1016/j.dwt.2024.100335
24. Laan H, Halem D, Smeets PWMH, Soppe AIA, Kroesbergen J, Wubbels G, Nederstigt J, Gensburger I, Heijman SGJ (2014) Bacteria and virus removal effectiveness of ceramic pot filters with different silver applications in a long term experiment. Water Res 51:47–54. https://doi.org/10.1016/j.watres.2013.11.010
25. Leveque J, Burns R (2017) Predicting water filter and bottled water use in Appalachia: a community-scale case study. J Water Health 15(3):451–461. https://doi.org/10.2166/wh.2017.219
26. Daniel D, Pande S, Rietveld L (2021) Socio-economic and psychological determinants for household water treatment practices in indigenous–rural Indonesia. Front Water 3:649445. https://doi.org/10.3389/frwa.2021.649445
27. March H, Garcia X, Domene E, Sauri D (2020) Tap water, bottled water or in-home water treatment systems: Insights on household perceptions and choices. Water 12:1310. https://doi.org/10.3390/W12051310
28. Lawrencia D, Maniam G, Chuah L, Poh P (2023) Critical review of household water treatment in southeast Asian countries. WIREs Wiley 10:20491948. https://doi.org/10.1002/wat2.1640
29. Garcia X, Pargament D (2015) Reusing wastewater to cope with water scarcity: economic, social and environmental considerations for decision-making. Resour Conserv Recycl 101:154–166. https://doi.org/10.1016/j.resconrec.2015.05.015

30. Channa A, Munir K, Hansen M, Tariq M (2024) Optimisation of small-scale aquaponics systems using artificial intelligence and the IoT: current status, challenges, and opportunities. Encyclopedia 4(1):313–336. https://doi.org/10.3390/encyclopedia4010023
31. Miller M, Kisiel A, Cembrowska-Lech D, Durlik I, Miller T (2023) IoT in water quality monitoring—are we really here? Sensors 23(2):960. https://doi.org/10.3390/s23020960
32. Arab M, Akbarian H, Gheibi M, Akrami M, Fathollahi-Fard A, Hajiaghaei-Keshteli M, Tian G (2022) A soft-sensor for sustainable operation of coagulation and flocculation units. Eng Appl Artif Intell 115:105315. https://doi.org/10.1016/j.engappai.2022.105315
33. Chinnappan C, William A, Nidamanuri S, Jayalakshmi S, Bogani R, Thanapal P, Syed S, Venkateswarlu B, Syed MJ (2023) IoT-enabled chlorine level assessment and prediction in water monitoring system using machine learning. Electronics 12(6):1458. https://doi.org/10.3390/electronics12061458
34. Saravanakumar K, Rajeswari R (2019) Microbial fuel cell–based self-powered biosensor for environment monitoring in IoT cloud framework. Concurr Comput: Pract Exp 31(15):e5165. https://doi.org/10.1002/cpe.5165
35. Fan Y, Chen J, Shirkey G, John R, Wu S, Park H, Shao C (2016) Applications of structural equation modeling (SEM) in ecological studies: an updated review. Ecol Process 5:19. https://doi.org/10.1186/s13717-016-0063-3
36. Schuberth F, Henseler J, Dijkstra TK (2018) Confirmatory composite analysis. Front Psychol 9:2541. https://doi.org/10.3389/fpsyg.2018.02541
37. Ghasemy M, Teeroovengadum V, Becker J, Ringle M (2020) This fast car can move faster: a review of PLS-SEM application in higher education research. High Educ 80:1121–1152. https://doi.org/10.1007/s10734-020-00534-1
38. Manley S, Hair J, Williams R, McDowell W (2021) Essential new PLS-SEM analysis methods for your entrepreneurship analytical toolbox. Int Entrep Manag J 17:1805–1825. https://doi.org/10.1007/s11365-020-00687-6
39. Partial Least Squares Structural Equation Modeling (PLS-SEM) using R, Hair J., 2021
40. Kline R (2016) Principles and practice of structural equation modeling, 4th ed, The Guilford Press
41. Bollen K, Noble M (2011) Structural equation models and the quantification of behavior. Proc Natl Acad Sci USA 108(3):15639–15646. https://doi.org/10.1073/pnas.1010661108
42. Hoffmann R, Kröger H, Pakpahan E (2018) The reciprocal relationship between material factors and health in the life course: evidence from SHARE and ELSA. Eur J Ageing 15:379–391. https://doi.org/10.1007/s10433-018-0458-3
43. Ballen C, Salehi S (2021) Mediation analysis in discipline-based education research using structural equation modeling: beyond "what works" to understand how it works, and for whom. J Microbiol Biol Educ 22(2). https://doi.org/10.1128/jmbe.00108-21
44. Moe V, von Soest T, Fredriksen E, Olafsen K, Smith L (2018) The multiple determinants of maternal parenting stress 12 months after birth: the contribution of antenatal attachment style, adverse childhood experiences, and infant temperament. Front Psychol 9:1987. https://doi.org/10.3389/fpsyg.2018.01987
45. Mumcu H, Aktas M (2020) Development of an attitude-towards-using-mathematics scale for high-school students and an analysis of student attitudes. Int J Math Educ Sci Technol 51(1):3–25. https://doi.org/10.1080/0020739X.2019.1679398
46. Lee J, Lee S, An K, Hwang S, Kim N (2020) An estimated structural equation model to assess the effects of land use on water quality and benthic macroinvertebrates in streams of the Nam-Han river system, South Korea. Int J Environ Res Public Health 17(6):2116. https://doi.org/10.3390/ijerph17062116
47. Wang Y, Liu X, Wang T, Zhang X, Feng Y, Yang G, Zhen W (2021) Relating land-use/land-cover patterns to water quality in watersheds based on the structural equation modelling. Catena 206:105566. https://doi.org/10.1016/j.catena.2021.105566
48. da Silva J, Chagas C, Souza T, Araujo M, Araujo L, Santos A, Oliveira M et al (2024) Using structural equation modeling to assess the genotoxic and mutagenic effects of heavy metal contamination in the freshwater ecosystems: a study involving Oreochromis niloticus in an urban river. Sci Total Environ 913:169529. https://doi.org/10.1016/j.scitotenv.2023.169529

49. Santibanez-Andrade G, Castillo-Arguero S, Vega-Pena E, Lindig-Cisneros R, Zavala-Hurtado J (2015) Structural equation modeling as a tool to develop conservation strategies using environmental indicators: the case of the forests of the Magdalena river basin in Mexico City. Ecol Ind 54:124–136. https://doi.org/10.1016/j.ecolind.2015.02.022
50. Gupta V, Gonzales H, Hungwe D, Kanokwan Y, Takahashi F (2024) Understanding global perspectives for the acceptance of community sanitation welfare schemes through a localised qualitative survey in Kho Nagorian, Jaipur, India. Sci Rep 14:16002. https://doi.org/10.1038/s41598-024-65191-1
51. Neha, Kansal A (2022) Acceptability of reclaimed municipal wastewater in cities: evidence from India's National Capital Region. Water Policy 24(1):212–228. https://doi.org/10.2166/WP.2021.197
52. Abubakar I, Mu'azu N (2022) Household attitudes toward wastewater recycling in Saudi Arabia. Util Policy 76:101372. https://doi.org/10.1016/j.jup.2022.101372
53. Ryberg K, Stone W, Baker N (2020) Causal factors for pesticide trends in streams of the United States: atrazine and deethylatrazine. J Environ Qual 49:152–162. https://doi.org/10.1002/jeq2.20045
54. Daohan H, Guijun L, Chengshuang S, Qian L (2020) Exploring interactions in the local water-energy-food nexus (WEF-Nexus) using a simultaneous equations model. Sci Total Environ 703:135034. https://doi.org/10.1016/j.scitotenv.2019.135034
55. Krivoguz D, Bespalova E, Zhilenkov A, Chernyi S, Kustov A, Degtyarev A, Zinchenko E (2024) Unveiling climate–land use and land cover interactions on the Kerch peninsula using structural equation modeling. Climate 12:120. https://doi.org/10.3390/cli12080120
56. Wang J, Wang X (2019) Structural equation modeling: applications using Mplus, 2nd ed. Wiley
57. Maydeu-Olivares A (2017) Maximum likelihood estimation of structural equation models for continuous data: standard errors and goodness of fit. Struct Equ Modeling 24(3):383–394. https://doi.org/10.1080/10705511.2016.1269606
58. Bansal N (2018) Industrial development and challenges of water pollution in coastal areas: the case of Surat, India. In: IOP conference series: earth and environmental science, vol 120, p 012001. https://doi.org/10.1088/1755-1315/120/1/012001
59. Malan A, Sharma H (2023) Assessment of drinking water quality and various household water treatment practices in rural areas of Northern India. Arab J Geosci 16:96. https://doi.org/10.1007/s12517-022-11153-8
60. Abubakar I (2021) Understanding the socioeconomic and environmental indicators of household water treatment in Nigeria. Util Policy 70:101209. https://doi.org/10.1016/j.jup.2021.101209
61. Sengupta P (2013) Potential health impacts of hard water. Int J Prev Med 4(8):866–875
62. Godswill C, Somtochukwu V, Ikechukwu O, Kate E (2020) Health benefits of micronutrients (vitamins and minerals) and their associated deficiency diseases: a systematic review. Int J Food Sci 3(1):1–32
63. Caputo A, Tomai M, Lai C, Desideri A, Pomoni E, Méndez HC, Langher V (2022) The perception of water contamination and risky consumption in El Salvador from a community clinical psychology perspective. Int J Environ Res Public Health 19:1109. https://doi.org/10.3390/ijerph19031109
64. Chen SW, Keglovits M, Devine M, Stark S (2022) Sociodemographic differences in respondent preferences for survey formats: sampling bias and potential threats to external validity. Arch Rehabil Res Clin Transl 4(1):100175. https://doi.org/10.1016/j.arrct.2021.100175
65. Social research methods, 6/E, Neuman LW (2007) Pearson education India
66. Tourangeau R, Yan T (2007) Sensitive questions in surveys. Psychol Bull 133(5):859–883. https://psycnet.apa.org/doi/10.1037/0033-2909.133.5.859
67. Andrade C (2020) The limitations of online surveys. Indian J Psychol Med 42(6):575–576. https://doi.org/10.1177/0253717620957496
68. Roy K, Zvonkovic A, Goldberg A, Sharp E, LaRossa R (2015) Sampling richness and qualitative integrity: challenges for research with families. J Marriage Fam 77(1):243–260. https://doi.org/10.1111/jomf.12147

69. Almeida F (2018) Strategies to perform a mixed methods study. Eur J Educ Stud 5(1):137–151. https://doi.org/10.5281/zenodo.1406214
70. Rindfleisch A, Malter A, Ganesan S, Moorman C (2008) Cross-sectional versus longitudinal survey research: concepts, findings, and guidelines. J Mark Res 45(3):261–279. https://doi.org/10.1509/jmkr.45.3.261
71. Sharma A, Patel N, Gupta R (2022) Enhancing consumer engagement through AI-driven personalized email campaigns: a comprehensive analysis using natural language processing and reinforcement learning algorithms. Eur Adv AI J 11(8)
72. Nayyar G, Suman S, Lakshmi TR (2024) Optimizing Ad campaigns with machine learning: data-driven approaches in modern media. Int J Multiphys 18(3):p1746
73. Ahmad F, Almuayqil S, Humayun M, Naseem S, Khan W, Junaid K (2020) Prediction of COVID-19 cases using machine learning for effective public health management. Comput Mater Contin 66:2265–2282

Chapter 8
Integrating Artificial Intelligence with Electrocoagulation for Sustainable Leachate Treatment: A Comparative Study of RSM and ANN for Pollutant Reduction

Chinenye Adaobi Igwegbe

Abstract The use of artificial intelligence (AI) techniques in conjunction with electrocoagulation-flocculation (ECF) for the sustainable treatment of landfill leachate (LCH) was examined The research focuses on reducing critical pollutants, including biochemical oxygen demand (BOD), chemical oxygen demand (COD), and color. Experimental data were modeled and optimized using Response Surface Methodology (RSM) and Artificial Neural Networks (ANN). While RSM was effective in identifying key parameter interactions and optimal conditions, ANN exhibited higher predictive accuracy, with R^2 values of 0.941, 0.9412, and 0.9723 for BOD, COD, and color reduction, respectively. The EC process, using aluminum electrodes, achieved notable pollutant reductions of up to 75.44% for BOD, 80.89% for COD, and 84.12% for color, closely aligning with the predicted values at an initial pH of 2.08, current density of 1.01 A, electrolysis time of 9.98 min, settling time of 30.34 min, and a temperature of 63.92 °C. Although this study primarily focused on integrating AI with ECF, the findings open opportunities for future incorporation of IoT technologies. IoT-enabled sensors could provide real-time monitoring and data acquisition for dynamic process optimization, further enhancing the efficiency and sustainability of ECF systems. By integrating EC with AI-driven tools, this study provided a reliable and efficient framework for improving pollutant removal and promoting environmentally sustainable water treatment solutions. The results highlighted the potential of AI-based optimization to enhance the design and performance of complex environmental processes.

C. A. Igwegbe (✉)
Department of Chemical Engineering, Nnamdi Azikiwe University, Awka, Nigeria
e-mail: ca.igwegbe@unizik.edu.ng; chinenyeigwegbe@upwr.edu.pl

Department of Applied Bioeconomy, Wrocław University of Environmental and Life Sciences, Wrocław, Poland

M. Nasr et al. (eds.), *Artificial Intelligence Applications for a Sustainable Environment*, Green Chemistry and Sustainable Technology,
https://doi.org/10.1007/978-3-031-91199-6_8

Keywords Electrocoagulation · Neural networks · Response surface methodology · Landfill leachate · BOD · COD · Environmental sustainability

8.1 Introduction

Landfill leachate is a byproduct of waste decomposition in landfills and a major environmental concern due to its complex and variable composition. It typically contains high levels of organic matter, ammonia, heavy metals, and persistent pollutants [1], posing significant risks to surface and groundwater systems if left untreated [2]. As global waste generation continues to rise, managing landfill leachate effectively has become a pressing environmental challenge. Current treatment methods for landfill leachate include physical, chemical, and biological processes [3]. While these methods can be effective, they often have limitations such as high operational costs, excessive sludge generation, or incomplete removal of contaminants. Additionally, many conventional methods require the use of external chemicals, which can introduce secondary pollutants and raise sustainability concerns [4]. These drawbacks necessitate the exploration of cleaner, more efficient, and sustainable treatment technologies [5].

Electrocoagulation (EC) has gained attention as an innovative water treatment technology due to its simplicity, minimal chemical requirements, and ability to remove a wide range of contaminants [6]. EC involves the use of sacrificial electrodes such as aluminum [7], iron [8], copper [9], or stainless steel to generate coagulants in situ through electrochemical reactions. These coagulants destabilize and aggregate colloidal particles, facilitating the removal of pollutants [10] such as biochemical oxygen demand (BOD), chemical oxygen demand (COD), and color-causing compounds [11]. Compared to conventional coagulation methods, EC offers advantages such as lower sludge production, reduced chemical usage, and easier operation, making it a promising solution for sustainable wastewater management [12].

Despite its potential, the performance of EC is highly dependent on operational parameters such as initial pH, current density, electrolysis time, and settling time. These parameters influence the generation of coagulants, the formation of flocs, and the overall efficiency of pollutant removal. pH is a critical parameter in EC as it affects the speciation of aluminum hydroxides [13]. Current density controls the rate of aluminum dissolution at the anode and, consequently, the generation of coagulants [14]. Higher current densities increase the coagulant production rate, enhancing pollutant removal efficiency [15]. However, excessively high current densities can lead to unnecessary energy consumption and electrode passivation [16]. Electrolysis time determines the cumulative amount of aluminum ions released into the solution [17]. Prolonged electrolysis ensures adequate coagulant availability but may lead to electrode degradation if extended excessively. After electrolysis, the treated water requires time for floc formation and settling. Adequate settling time ensures that the aggregated particles can effectively separate from the water, enhancing removal efficiency [18]. Achieving optimal performance requires careful tuning of these factors, but their interactions are often nonlinear and complex, making traditional optimization approaches inadequate. The conventional one-factor-at-a-time (OFAT) method, which involves varying one parameter while keeping others constant, is time-consuming and fails to capture the synergistic or antagonistic effects of parameter interactions [19, 20]. This limitation highlights the need for advanced optimization techniques that can model and analyze complex systems effectively.

Response Surface Methodology (RSM) and Artificial Neural Networks (ANN) stand out as two prominent approaches in environmental engineering. Artificial intelligence (AI) techniques, particularly those grounded in machine learning (ML), have shown immense potential in the optimization of environmental processes. These methods capitalize on the ability to analyze large datasets and uncover patterns that enable accurate predictions and actionable solutions for process improvement [21–23]. Among machine learning techniques, ANN are recognized as a highly effective method [24]. ANN represents a machine learning technique inspired by the structure and function of the human brain [25]. It excels in capturing highly nonlinear relationships and managing large, intricate datasets, providing enhanced predictive accuracy compared to conventional methods [23, 26]. However, implementing ANN requires meticulous tuning of hyperparameters and access to substantial training data [27]. RSM is a statistical tool, that employs polynomial equations to model and optimize processes, offering a valuable means of understanding variable interactions and determining optimal conditions within a specified experimental framework [28, 29]. It has found extensive application in environmental studies, although its reliance on second-order relationships can limit its ability to model complex nonlinear behaviors effectively. While both RSM and ANN have their respective strengths and limitations, they serve as complementary tools for advancing the optimization of environmental processes.

The integration of AI techniques with ECF has the potential to revolutionize the treatment of landfill leachate by enabling precise control and optimization of operational parameters. By combining the strengths of RSM and ANN, this study aims to (1) assess the effectiveness of EC using aluminum electrodes for reducing BOD, COD, and color in landfill leachate, (2) compare the predictive capabilities of RSM and ANN (an AI model) for modeling and optimizing the EC process, and (3) develop an integrated framework for enhancing pollutant removal efficiency while minimizing environmental and economic costs. This chapter explores how AI-driven tools, such as ANN, can enhance the efficiency and sustainability of the electrocoagulation process by uncovering nonlinear relationships between operational parameters and optimizing pollutant removal. By minimizing resource consumption and environmental impact, AI applications in EC demonstrate the potential to transform traditional wastewater treatment technologies into sustainable and scalable solutions for environmental management. This integration aligns with the principles of sustainability, as it promotes the use of resource-efficient technologies, minimizes waste generation, and ensures environmentally friendly water treatment solutions. This study contributes to advancing sustainable practices in environmental management by addressing the limitations of conventional methods and utilizing AI-driven optimization. Landfill leachate treatment is a critical component of waste management and environmental protection, especially in the context of increasing global waste production. This study not only provides a comparative analysis of RSM and ANN in the context of EC but also establishes a framework for applying AI in complex environmental processes. The findings have broader implications for the design, optimization, and implementation of sustainable water treatment technologies, demonstrating the transformative potential of AI in achieving environmental goals.

8.2 Aluminum Reaction Mechanism in Electrocoagulation

The success of electrocoagulation (EC) as a water treatment process hinges on the electrochemical reactions that occur at the electrodes. These reactions generate coagulants in situ, which play a critical role in destabilizing and aggregating pollutants, ultimately leading to their removal. Aluminum was selected as the electrode material for this study due to its high efficiency, cost-effectiveness, and ability to generate hydroxide species that act as robust coagulants [12, 30]. The underlying mechanisms and reaction pathways are detailed below.

8.2.1 Electrochemical Reactions and Coagulant Formation

At the core of the EC process are the anodic and cathodic reactions that facilitate the generation of aluminum ions (Al^{3+}) and hydrogen gas (H_2) in the aqueous medium. These reactions (Eqs. 8.1, 8.2, 8.3 and 8.4) can be described as follows [31, 32]:

Anode Reactions:

- Oxidation of aluminum at the anode:

$$Al_{(s)} \rightarrow Al^{3+}_{(aq)} + 3e^- \tag{8.1}$$

In this step, aluminum metal dissolves, releasing trivalent aluminum ions into the solution.

- Hydrolysis of aluminum ions to form hydroxides:

$$Al^{3+}_{(aq)} + 3H_2O \rightarrow Al(OH)_3 + 3H^+ \tag{8.2}$$

These hydroxides act as the primary coagulants, neutralizing the charges on suspended particles and promoting aggregation.

Cathode Reaction:

- Reduction of water at the cathode, producing hydrogen gas and hydroxide ions:

$$2H_2O_{(l)} + 2e^- \rightarrow H_{2(g)} + 2OH^-_{(aq)} \tag{8.3}$$

Overall Reaction:

- Combining the above reactions, the net equation is:

$$Al_{(s)} + 3H_2O_{(l)} \rightarrow Al(OH)_3 + H_2 \tag{8.4}$$

The hydrogen gas generated at the cathode can further enhance the separation of pollutants by aiding flotation.

These reactions are fundamental to the EC process, facilitating the removal of contaminants through coagulation and flotation mechanisms [32].

8.2.2 The Role of Aluminum Hydroxides

The aluminum hydroxides ($Al(OH)_3$) formed during the EC process are highly effective coagulants due to their amphoteric nature [33] and strong charge-neutralizing properties [12]. These hydroxides adsorb onto colloidal particles, neutralizing their surface charges and causing destabilization. Once destabilized, the particles aggregate into larger flocs, which can be removed through sedimentation or flotation. Aluminum hydroxides can also interact with dissolved organic compounds and heavy metals, forming insoluble complexes that are readily separated from the treated water. These multi-functional properties make aluminum electrodes particularly suitable for treating complex wastewater streams like landfill leachate.

8.3 Materials and Methods

8.3.1 Sample Collection and Preparation

Landfill leachate (LCH) samples were sourced from Agu-Awka, Anambra State, Nigeria and stored at 4 °C till usage. Samples were characterized for pH, COD and BOD following standard methods: APHA 4500-H [34], APHA 5220-D [35] and APHA 5210-B [34], respectively, with initial values of 7.6, 3972 mg L^{-1}, and 1413 mg L^{-1}. The LCH's absorbance was analyzed using a UV–visible spectrophotometer (APEL PD-3000UV Spectrophotometer) across wavelengths ranging from 200 to 600 nm. This process identified the maximum wavelength (λmax) for accurate assessment of absorbance or color in both untreated and treated samples. The color intensity was quantified by measuring absorbance at λmax = 245 nm, which corresponds to the brown color of the LCH. Color reduction was calculated based on the decrease in absorbance relative to the initial sample, with the starting absorbance recorded as 0.5845.

Aluminum electrodes (purity > 99.5%) with dimensions of 100 cm × 9 cm, obtained from reliable laboratory equipment and chemical suppliers in Onitsha, Nigeria, were thoroughly prepared prior to use. They were initially cleaned with acetone to remove surface impurities. Subsequently, they underwent meticulous cleaning with propane-2-one, an organic solvent effective in dissolving residual contaminants. This step was critical to eliminating any residual metal ions or organic substances that could interfere with the EC process. This thorough cleaning protocol ensured minimal cross-contamination, thereby enhancing the accuracy and reproducibility of experimental outcomes.

8.3.2 Experimental Setup

This study examined the influence of various factors on the EC process. The parameters considered were initial pH, current intensity, electrolysis time, temperature, and settling time to determine their effects on treatment efficiency. Using high-purity Al rods (purity > 99.5%) as sacrificial electrodes, a series of experiments were meticulously conducted. The selection of operating variable ranges for the EC experiments was driven by the need to thoroughly investigate the influence of different factors on the EC process. A wide initial pH range from 2 to 10 was chosen to assess the process's sensitivity across acidic to alkaline conditions. The current intensity was systematically adjusted from 0.5 to 3 A to capture a spectrum of intensities reflecting practical operational scenarios. Electrolysis times ranging from 2 to 10 min were selected to study the temporal effects of the current application on effluent treatment. The settling time, which ranged from 0 to 40 min, was explored to evaluate the effectiveness of the flocculation process over time.

The experimental setup (Fig. 8.1) involved vertically positioning the electrodes, with an the effective surface area of 900 cm^2 vertically in a reactor with dimensions of 36 × 15 × 23 cm. These electrodes were connected in parallel monopolar mode with an interelectrode distance of 5 cm. This specific distance was selected to minimize the IR drop, thereby decreasing ohmic resistance and reducing power consumption [36, 37]. The electrodes were immersed to a depth of 8.4 cm into the wastewater for optimal exposure. Before each experiment, the pH of the wastewater was meticulously adjusted within the range of pH 2–10 using 1 M HCl or NaOH solutions and set at a constant temperature (30–60 °C). Maintaining a constant voltage of 220 V, the current intensity was systematically regulated at specific levels: 0.5, 1.0, 1.5, 2.0, 2.5, and 3 A. These were conducted at various electrolysis times (2, 4, 6, 8, and 10 min), transmitting different current intensities into 500 ml of each effluent. The electrocoagulation studies were executed using a regulated DC power supply of 220 V, 0-3A (HUPE Model LLLN003C), with the voltage and current meticulously controlled through a voltage and current multimeter (LDA DT-830D). To ensure a consistent experimental environment, a magnetic stirrer was used to immerse the samples in each beaker. This setup, which involved placing the mixture on a hot plate/stirrer, facilitated agitation of the electrolyte with a constant 3-min fast mixing at 150 rpm, followed by a 15-min flocculation time (slow mixing) at 30 rpm. Subsequently, 20 mL of the treated samples were extracted from each beaker at different sedimentation times (0–40 min), and the respective efficiencies were evaluated using Eq. 8.5:

$$\%E = \frac{E_0 - E_f}{E_0} \times 100 \tag{8.5}$$

Here, E_0 and E_f represent the initial and resulting values of BOD, COD, and color. These values obtained from these experiments were used for subsequent analyses (RSM and ANN modeling).

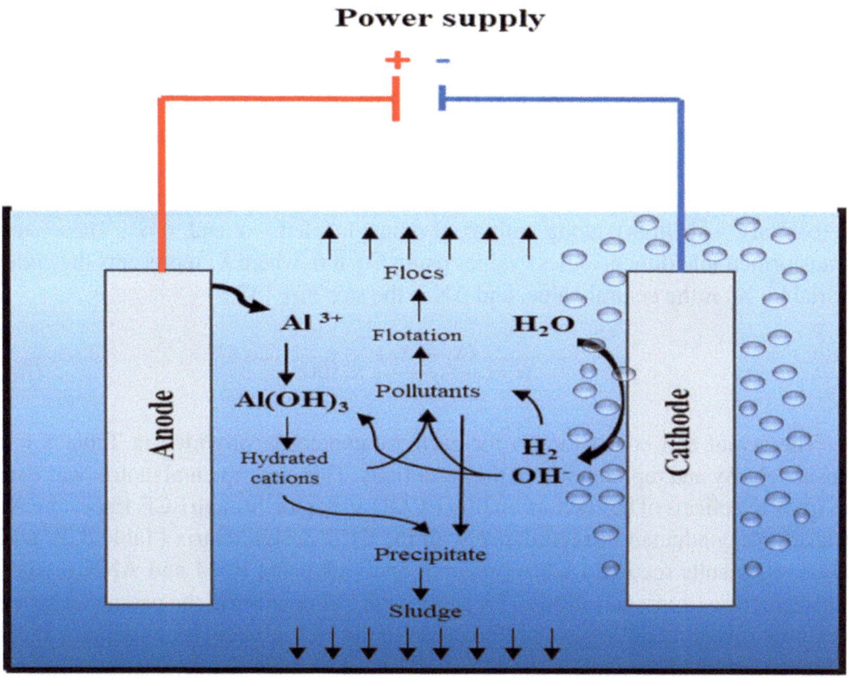

Fig. 8.1 Schematic diagram of the electrocoagulation process using aluminum electrodes. Reprinted with permission from [38]. © 2021 MDP1. https://creativecommons.org/licenses/by/4.0/

Table 8.1 Process parameters and their ranges for the ECF experiment using central composite design (CCD)

Variables	Range of actual and coded variables				
	$-\alpha$	-1	0	$+1$	$+\alpha$
A: initial pH	2	4	6	8	10
B: current intensity (A)	1	1.5	2	2.5	3
C: electrolysis time (min)	2	4	6	8	10
D: settling time (min)	0	10	20	30	40
E Temperature ($^{\circ}C$)	30	40	50	60	70

8.3.3 Design of the Modeling Experiment

A central composite design (CCD) was utilized in this study to establish an experimental framework with five-level factors. This design comprised 50 experimental trials, developed using Design Expert software (Stat-Ease, version 13.0.1.0). The CCD configuration included 10 axial points, 32 factorial points, and 8 repetitions at the center to ensure optimization of BOD, COD and color via the ECF process. The primary input variables (initial pH, current density, electrolysis time, settling time, and temperature) were carefully selected and coded for the experimental design, as shown in Table 8.1. The variables were assigned the following coded levels: -1 (low), 0 (central), + 1 (high), along with axial points labeled $-\alpha$ and $+\alpha$. These were transformed into dimensionless values using Eq. 8.6, where X_i represents the coded variable, X_0 is the central value, and ΔX is the step size [37]:

$$Xi = \frac{(X_i - X_0)}{\Delta X} \times 100 \tag{8.6}$$

The actual and coded ranges for each parameter are provided in Table 8.1 to ensure clarity and replicability of the conditions. This experimental matrix was used to study the effects of key factors on BOD, COD and color through ECF. Experimental trials were conducted in accordance with the CCD design matrix (Table 8.2), with observed results recorded for subsequent analyses using RSM and ANN models. This structured approach allowed for a thorough examination of the parameter space, enabling robust modeling and optimization of the ECF process for maximum BOD, COD and color reduction.

8.3.4 ANN Modeling

In this study, an ANN model was developed based on the CCD framework, comprising 50 experimental runs (refer to Table 8.2). A multilayer perceptron (MLP)-based feed-forward ANN was constructed using MATLAB's "nftool," with the

Table 8.2 RSM and ANN predicted and observed response values for the ECF process using aluminum electrodes

Run	pH	Current dosage (A)	EC time (min)	Settling time (min)	Temperature	BOD (%)			COD (%)			COL (%)		
						Obs	RSM Pred	ANN Pred	Obs	RSM Pred	ANN Pred	Obs	RSM Pred	ANN Pred
1	4	1.5	4	10	40	67.28	67.23	65.66	72.65	72.36	72.32	60.05	58.01	62.36
2	4	2.5	8	10	60	72.43	72.77	72.46	76.61	76.71	77.33	71.89	73.62	71.93
3	4	1.5	4	10	60	68.51	68.69	68.50	73.63	73.83	73.72	62.81	60.63	61.62
4	6	2	6	20	50	70.54	70.92	70.87	74.69	74.49	74.29	62.14	63.25	63.71
5	4	1.5	8	30	60	72.57	72.76	72.42	77.50	77.59	78.16	72.96	73.48	72.17
6	4	2.5	4	30	40	71.05	70.82	71.08	74.79	75.19	75.11	72.12	71.47	71.99
7	8	2.5	4	30	40	70.88	70.99	70.78	76.30	76.17	75.63	66.38	66.24	65.75
8	4	1.5	8	30	40	71.69	71.83	71.73	76.53	76.67	76.06	69.65	69.71	70.37
9	4	2.5	4	10	40	68.22	68.24	68.11	73.42	72.74	72.39	58.98	61.02	59.23
10	8	1.5	4	30	60	69.11	68.54	69.37	75.34	74.44	74.88	52.22	51.03	52.11
11	10	2	6	20	50	68.71	68.72	68.72	74.03	74.26	73.87	54.73	55.05	52.98
12	8	1.5	4	10	60	67.43	67.08	67.57	73.90	73.87	73.92	48.99	48.30	47.59
13	8	1.5	8	10	40	69.61	69.05	69.66	74.12	73.68	73.71	56.84	55.60	55.95
14	6	2	6	20	50	71.18	70.92	70.87	73.85	74.49	74.29	63.34	63.25	63.71
15	6	1	6	20	50	69.09	69.36	68.99	75.20	74.57	75.01	54.70	59.11	54.59
16	4	1.5	4	30	40	69.87	69.97	69.77	74.05	74.33	74.28	59.79	60.73	60.70
17	6	2	2	20	50	69.71	70.02	69.68	75.73	75.9	75.76	59.98	60.52	60.50
18	4	1.5	4	30	60	70.02	70.08	70.03	74.47	74.87	75.08	60.97	61.42	59.79
19	4	2.5	8	30	40	71.16	71.82	71.18	76.82	77.01	77.27	78.53	78.66	79.58
20	4	2.5	4	10	60	70.23	69.71	70.23	74.00	73.74	73.79	63.34	62.87	61.46

(continued)

Table 8.2 (continued)

Run	pH	Current dosage (A)	EC time (min)	Settling time (min)	Temperature	BOD (%)			COD (%)			COL (%)		
						Obs	RSM Pred	ANN Pred	Obs	RSM Pred	ANN Pred	Obs	RSM Pred	ANN Pred
21	6	2	10	20	50	74.90	74.38	74.86	79.31	78.83	79.56	79.42	79.50	79.33
22	6	2	6	0	50	67.42	68.15	67.51	69.43	70.18	69.65	39.74	42.36	41.36
23	4	2.5	8	30	60	72.87	72.77	72.83	77.69	77.47	78.12	82.47	81.66	81.86
24	6	2	6	40	50	71.91	70.96	71.89	73.49	72.43	73.40	57.05	55.06	61.92
25	6	2	6	20	50	70.28	70.92	70.87	74.68	74.49	74.29	63.97	63.25	63.71
26	8	2.5	8	10	60	71.28	71.42	71.24	75.68	75.23	75.71	63.97	64.50	64.08
27	6	2	6	20	50	71.13	70.92	70.87	74.55	74.49	74.29	63.45	63.25	63.71
28	6	2	6	20	50	71.20	70.92	70.87	74.34	74.49	74.29	64.67	63.25	63.71
29	8	2.5	4	10	60	69.61	68.91	70.20	74.78	74.64	74.46	55.10	54.50	55.21
30	8	2.5	8	30	60	71.75	71.49	71.70	75.43	75.52	76.40	75.19	74.47	71.24
31	6	2	6	20	70	70.18	70.68	70.20	76.17	76.54	76.23	71.15	71.04	68.81
32	2	2	6	20	50	70.95	70.72	72.88	76.14	75.61	75.85	73.06	73.36	71.87
33	4	1.5	8	10	60	73.38	72.61	73.37	77.77	77.31	78.03	73.66	73.17	72.13
34	8	1.5	8	30	40	69.97	70.63	70.08	74.42	74.42	74.62	60.41	59.78	58.75
35	8	1.5	4	30	40	69.25	69.32	69.49	73.81	74.46	73.76	53.74	51.55	53.37
36	6	2	6	20	30	69.87	69.16	69.82	75.86	75.18	75.71	65.90	66.63	62.66
37	4	2.5	4	30	60	70.59	70.94	70.59	74.94	75.26	74.85	70.34	71.39	71.45
38	4	1.5	8	10	40	70.23	70.33	70.26	74.64	75.46	74.47	69.45	67.47	70.54
39	6	2	6	20	50	70.80	70.92	70.87	74.60	74.49	74.29	63.57	63.25	63.71

(continued)

Table 8.2 (continued)

Run	pH	Current dosage (A)	EC time (min)	Settling time (min)	Temperature	BOD (%)			COD (%)			COL (%)		
						Obs	RSM Pred	ANN Pred	Obs	RSM Pred	ANN Pred	Obs	RSM Pred	ANN Pred
40	8	2.5	8	10	40	70.13	70.02	70.09	74.55	74.41	74.27	62.13	60.77	61.64
41	6	2	6	20	50	71.44	70.92	70.87	74.20	74.49	74.29	63.34	63.25	63.71
42	8	2.5	4	30	60	69.48	70.22	69.86	75.82	75.69	75.26	61.78	64.96	63.01
43	8	2.5	8	30	40	71.29	71.44	71.39	75.40	75.62	75.70	68.93	72.68	69.34
44	8	1.5	8	30	60	70.64	70.67	70.70	73.73	74.77	74.65	62.72	62.34	64.24
45	8	2.5	4	10	40	67.92	68.33	68.00	74.07	74.2	74.17	55.02	53.85	55.02
46	4	2.5	8	10	40	70.29	70.48	70.33	74.95	75.32	74.98	68.24	68.70	67.97
47	6	2	6	20	50	70.54	70.92	70.87	74.69	74.49	74.29	62.14	63.25	63.71
48	6	3	6	20	50	71.67	71.19	71.66	75.37	75.69	74.93	78.04	74.25	78.00
49	8	1.5	8	10	60	69.98	70.44	69.81	75.04	74.97	75.30	60.44	60.09	61.63
50	8	1.5	4	10	40	66.53	66.51	66.51	72.80	72.96	72.40	44.95	46.89	46.41

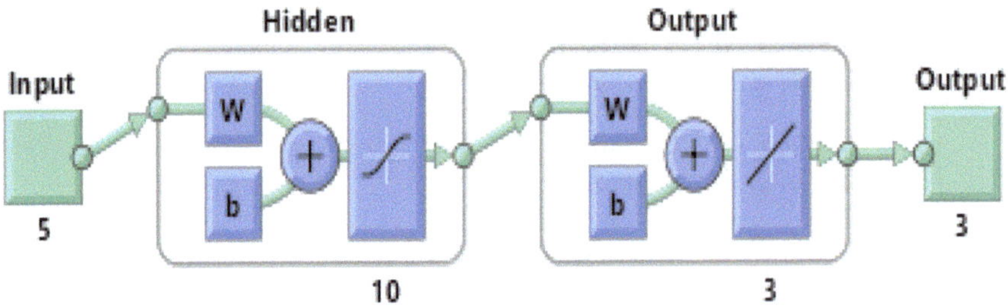

Fig. 8.2 Architecture of the multilayer perceptron neural network used in the study

Levenberg–Marquardt (trainlm) backpropagation algorithm applied as the learning technique. This algorithm iteratively adjusted the network weights by minimizing the sum of squared errors, ensuring efficient model training.

The ANN architecture consisted of three layers (Fig. 8.2):

- Input Layer: Included five parameters—initial pH, current density, electrolysis time, settling time, and temperature.
- Hidden Layer: Designed through a trial-and-error approach to determine the optimal number of neurons, systematically adjusted between 1 and 20 nodes.
- Output Layer: Represented the response variable, BOD, COD and color reduction (%).

The dataset was divided into three subsets: 80% for training, 10% for validation, and 10% for testing using the random method (dividerand). This partitioning ensured a robust evaluation of the model's performance. The training process terminated automatically when the generalization ceased to improve, as indicated by an increase in the mean square error (MSE) of the validation samples.

The optimal number of neurons in the hidden layer was determined through iterative testing. This systematic approach ensured that the network effectively captured the nonlinear relationships between the input parameters and the output while avoiding overfitting. Although the Levenberg–Marquardt backpropagation algorithm method required additional computational resources, it significantly reduced processing time and enhanced the model's performance [39].

The normalized values enabled the ANN to process the data effectively, improving model performance and predictive accuracy. To standardize the input values and facilitate ANN modeling, the process variables were normalized to a range between 0 and 1 using Eq. 8.7 [40]:

$$x_{ik} = \frac{y_{ik} - min\,y(k)}{\max y(k) - \min y(k)} \tag{8.7}$$

where x_{ik} is is the normalized value of the k-th response in the i-th trial, y_{ik} is the actual response value for the k-th parameter in the i-th trial, $\max y(k)$ and $\min y(k)$

represent the maximum and minimum values of the response k, respectively, and *i* denotes the trial index in the experimental design.

8.4 Results and Discussion

8.4.1 RSM Modeling and Parameter Optimization

The RSM effectively modeled the ECF process, providing insights into the interactions and impacts of various operational parameters on pollutant removal efficiency. The study revealed significant interactions among key parameters, including initial pH, current density, and electrolysis time, demonstrating their combined influence on optimizing the treatment process. The statistical analysis summarized in Tables 8.3 and 8.4 highlight the significance of individual and interactive effects, as evaluated using Analysis of Variance (ANOVA). ANOVA employs F-values and p-values to assess the validity and accuracy of the RSM model. Based on the threshold of p-value ≤ 0.05 [41], terms with significant impacts were identified, while non-significant terms were excluded from the reduced quadratic models to enhance model precision.

The reduced quadratic models for BOD, COD, and color reduction are presented in Table 8.5. These models demonstrated high predictive accuracy, with R^2 values of 0.9318, 0.9205, and 0.9684, respectively (Table 8.4). The high coefficients of determination indicate the models' ability to explain the variability in the experimental data. Key model terms such as initial pH, current density, and electrolysis time, along with higher-order interactions, were statistically validated as significant contributors to the pollutant removal process. The optimization results derived from RSM showed that pollutant removal efficiency was sensitive to the synergistic effects of process parameters (Eqs. 8.8, 8.9 and 8.10). The predictive capabilities of the RSM-derived models were validated by comparing the observed values (experimental results) and predicted response values for BOD, COD, and color reduction, as shown in Table 8.2. The close agreement between experimental and predicted values reinforces the reliability of the models.

The ANOVA results in Table 8.3 further confirm the statistical significance of the models and their terms. For BOD reduction, parameters such as pH, current density, and electrolysis time showed highly significant contributions with p-values < 0.0001. COD and color reduction followed a similar trend, indicating the robust applicability of the quadratic models in predicting pollutant removal efficiencies. Moreover, Table 8.4 provides a comprehensive comparison of the predicted and observed results, showcasing minimal error margins, which underscores the efficacy of the RSM approach in optimizing the ECF process. The reduced quadratic equations (Table 8.5) for each pollutant highlight the importance of both first-order and interaction terms in driving the optimization process. For example, the equation for COD reduction indicates the significant role of the interaction between current density and

Table 8.3 ANOVA results for the quadratic model used in the ECF process on landfill leachate using aluminum electrodes

Parameter	Source	Sum of squares	df	Mean square	F-value	p-value
BOD	**Model**	117.20	20	5.86	19.82	<0.0001
	A-pH	10.01	1	10.01	33.85	<0.0001
	B-current dosage	8.34	1	8.34	28.22	<0.0001
	C-EC time	47.68	1	47.68	161.24	<0.0001
	D-settling time	19.75	1	19.75	66.81	<0.0001
	E-temperature	5.72	1	5.72	19.36	0.0001
	AB	1.33	1	1.33	4.51	0.0424
	AC	0.6078	1	0.6078	2.06	0.1624
	AD	0.0124	1	0.0124	0.0419	0.8392
	AE	1.58	1	1.58	5.34	0.0281
	BC	1.47	1	1.47	4.96	0.0339
	BD	0.0458	1	0.0458	0.1547	0.6969
	BE	0.0003	1	0.0003	0.0009	0.9769
	CD	3.07	1	3.07	10.38	0.0031
	CE	1.34	1	1.34	4.53	0.0418
	DE	3.62	1	3.62	12.26	0.0015
	A^2	2.84	1	2.84	9.59	0.0043
	B^2	0.8211	1	0.8211	2.78	0.1064
	C^2	3.30	1	3.30	11.16	0.0023
	D^2	3.68	1	3.68	12.43	0.0014
	E^2	1.98	1	1.98	6.71	0.0149
	Residual	8.57	29	0.2957		
	Lack of fit	7.41	22	0.3368	2.02	0.1716
	Pure error	1.17	7	0.1665		
	Cor total	125.78	49			
COD	**Model**	106.71	20	5.34	16.79	<0.0001
	A-pH	4.55	1	4.55	14.31	0.0007
	B-current dosage	3.13	1	3.13	9.85	0.0039
	C-EC time	21.42	1	21.42	67.38	<0.0001
	D-settling time	12.71	1	12.71	39.99	<0.0001
	E-temperature	4.64	1	4.64	14.61	0.0006
	AB	1.48	1	1.48	4.67	0.0391
	AC	11.29	1	11.29	35.53	<0.0001
	AD	0.4536	1	0.4536	1.43	0.2419
	AE	0.6356	1	0.6356	2.00	0.1680

(continued)

Table 8.3 (continued)

Parameter	Source	Sum of squares	df	Mean square	F-value	p-value
	BC	0.5228	1	0.5228	1.64	0.2099
	BD	0.4584	1	0.4584	1.44	0.2395
	BE	0.4301	1	0.4301	1.35	0.2542
	CD	1.17	1	1.17	3.67	0.0653
	CE	0.2869	1	0.2869	0.9026	0.3499
	DE	1.72	1	1.72	5.40	0.0274
	A^2	0.3929	1	0.3929	1.24	0.2753
	B^2	0.8275	1	0.8275	2.60	0.1175
	C^2	16.57	1	16.57	52.12	<0.0001
	D^2	20.25	1	20.25	63.70	<0.0001
	E^2	3.77	1	3.77	11.87	0.0018
	Residual	9.22	29	0.3179		
	Lack of Fit	8.58	22	0.3901	4.30	0.0273
	Pure Error	0.6352	7	0.0907		
	Cor total	115.93	49			
Color	**Model**	3562.41	20	178.12	44.45	<0.0001
	A-pH	838.14	1	838.14	209.17	<0.0001
	B-current dosage	573.35	1	573.35	143.09	<0.0001
	C-EC time	900.41	1	900.41	224.71	<0.0001
	D-settling time	402.97	1	402.97	100.57	<0.0001
	E-temperature	48.71	1	48.71	12.16	0.0016
	AB	31.24	1	31.24	7.80	0.0092
	AC	1.13	1	1.13	0.2808	0.6002
	AD	7.53	1	7.53	1.88	0.1810
	AE	2.89	1	2.89	0.7217	0.4025
	BC	6.41	1	6.41	1.60	0.2161
	BD	119.35	1	119.35	29.79	<0.0001
	BE	1.18	1	1.18	0.2940	0.5918
	CD	0.4656	1	0.4656	0.1162	0.7357
	CE	18.91	1	18.91	4.72	0.0381
	DE	7.45	1	7.45	1.86	0.1832
	A^2	1.83	1	1.83	0.4576	0.5041
	B^2	23.56	1	23.56	5.88	0.0218
	C^2	91.46	1	91.46	22.83	<0.0001
	D^2	422.97	1	422.97	105.56	<0.0001

(continued)

Table 8.3 (continued)

Parameter	Source	Sum of squares	df	Mean square	F-value	p-value
	E^2	62.44	1	62.44	15.58	0.0005
	Residual	116.20	29	4.01		
	Lack of fit	111.09	22	5.05	6.92	0.0069
	Pure error	5.11	7	0.7299		
	Cor total	3678.61	49			

Table 8.4 Summary of ANOVA response metrics for ECF process modeling using RSM

Response	R^2	Adj. R^2	Pred. R^2	Adequate precision	CV%
BOD reduction (%)	0.9318	0.8848	0.7634	22.3511	0.7731
COD reduction (%)	0.9205	0.8656	0.7106	23.6765	0.7525
Color reduction (%)	0.9684	0.9466	0.8825	30.2924	3.14

Table 8.5 Reduced quadratic models for pollutant removal responses (BOD, COD, and color) in the ECF process

Quadratic models for each response	Equation
$\boldsymbol{BOD} = 70.92 - 1.00\boldsymbol{A} + 0.9135\boldsymbol{B} + 2.18\boldsymbol{C} + 1.41\boldsymbol{D} + 0.757\boldsymbol{E} + 0.816\boldsymbol{AB} - 0.89\boldsymbol{AE} - 0.86\boldsymbol{BC} - 1.24\boldsymbol{CD} + 0.82\boldsymbol{CE} - 1.35\boldsymbol{DE} - 1.19\boldsymbol{A}^2 + 1.28\boldsymbol{C}^2 - 1.36\boldsymbol{D}^2 - 0.10\boldsymbol{E}^2$	8
$\boldsymbol{COD} = 74.49 - 0.67\boldsymbol{A} + 0.56\boldsymbol{B} + 1.46\boldsymbol{C} + 1.13\boldsymbol{D} + 0.68\boldsymbol{E} + 0.86\boldsymbol{AB} - 2.38\boldsymbol{AC} - 0.93\boldsymbol{DE} + 2.88\boldsymbol{C}^2 - 3.18\boldsymbol{D}^2 + 1.37\boldsymbol{E}^2$	9
$\boldsymbol{Color} = 63.25 - 9.16\boldsymbol{A} + 7.57\boldsymbol{B} + 9.49\boldsymbol{C} + 6.35\boldsymbol{D} + 2.21\boldsymbol{E} + 3.95\boldsymbol{AB} + 7.73\boldsymbol{BD} + 3.08\boldsymbol{CE} + 3.43\boldsymbol{B}^2 + 6.76\boldsymbol{C}^2 - 14.54\boldsymbol{D}^2 + 5.59\boldsymbol{E}^2$	10

electrolysis time, reflecting the interplay of operational variables in enhancing pollutant removal. These findings demonstrated the utility of RSM not only in process optimization but also in understanding the intricate relationships between process parameters and their effects on water quality improvement.

Table 8.6 presents the optimal operating conditions for the ECF process as predicted by the RSM model, alongside the experimentally validated results for BOD, COD, and color reduction. Optimal operating conditions identified through RSM were an initial pH of 2.08, a current density of 1.01 A, an electrolysis (EC) time of 9.98 min, a settling time of 30.34 min, and a temperature of 63.92 °C. Under these conditions, the RSM model predicted reductions of 75.33% for BOD, 84.72% for COD, and 92.71% for color. These values are compared with the experimental outcomes to confirm the reliability and accuracy of the RSM model in achieving significant pollutant reductions. Experimental trials conducted at these conditions validated the predictions, achieving reductions of 75.44% for BOD, 80.89% for

COD, and 84.12% for color. The experimental results closely align with the predicted values, demonstrating the model's effectiveness in optimizing the ECF process for sustainable landfill leachate treatment.

8.4.2 ANN Modeling and Predictive Accuracy

The feed-forward multilayer perceptron architecture employed Levenberg–Marquardt backpropagation and was optimized for five input parameters (pH, current density, electrolysis time, settling time, and temperature). Table 8.7 highlights the performance metrics of the ANN model across the training, validation, and testing phases, emphasizing its predictive accuracy. The training phase achieved the lowest MSE of 0.1872 and the highest coefficient of determination (R^2) of 0.9983, indicating an excellent fit to the training data. During the validation phase, an MSE of 7.8194 and an R^2 of 0.9056 were observed, reflecting the model's ability to generalize to unseen data. The testing phase recorded an MSE of 1.7433 and an R^2 of 0.9705, confirming the model's robust predictive accuracy. These metrics demonstrate the ANN's effective learning and generalization across all phases. The consistent performance across training, validation, and testing phases underscores the reliability of the modeling framework used in this study.

Figure 8.3 displays regression plots illustrating the correlation between experimental data (target) and ANN-predicted outputs for training, validation, and testing phases of the model. Each subplot represents a phase, with the diagonal "Y = T" line showing an ideal fit where predicted values equal observed data. The close clustering of data points around the "Y = T" line across all phases signifies strong predictive accuracy. The slopes and intercepts of the regression lines further affirm the model's capacity to generalize and accurately predict outcomes for BOD, COD, and color reductions in the electrocoagulation-flocculation process.

Figure 8.4 shows the MSE progression during ANN training, validation, and testing across 19 epochs. The training curve (blue) demonstrates a rapid decline in MSE, indicating effective learning. The validation curve (green) decreases initially, stabilizing after 13 epochs, with the best performance marked, showing the optimal stopping point. The testing curve (red) follows a similar trend, confirming the model's generalization capability. The low MSE values and convergence of the curves highlight the ANN's accuracy and robustness in predicting outcomes for the ECF process.

The training state plot (Fig. 8.5) provides insights into the optimization process during the ANN training phase. The gradient plot demonstrates a consistent decline in gradient values over the training epochs, indicating that the optimization algorithm effectively minimized the error function. The steady nature of the gradient toward the later epochs highlights the convergence of the model. The Mu plot, representing the learning rate adaptation, shows an initial increase followed by stabilization, illustrating the Levenberg–Marquardt algorithm's ability to balance exploration and exploitation in the parameter space. Validation failure points are marked to ensure

Table 8.6 Optimal operating conditions and model validation results for the ECF process

Type of value	Operating variables					Responses (%)			Desirability
	Initial pH	Current intensity (A)	Electrolysis time (min)	Settling time	Temperature (°C)	BOD (%)	COD (%)	Color reduction (%)	
Predicted optimum values	2.08	1.01	9.98	30.34	63.92	75.33	84.72	92.71	1.00
Experimental value						75.44	80.89	84.12	

Table 8.7 Performance metrics of the ANN model during training, validation, and testing phases

Phase	Number of samples	MSE	R^2
Training	34	0.1872	0.9983
Validation	8	7.8194	0.9056
Testing	8	1.7433	0.9705

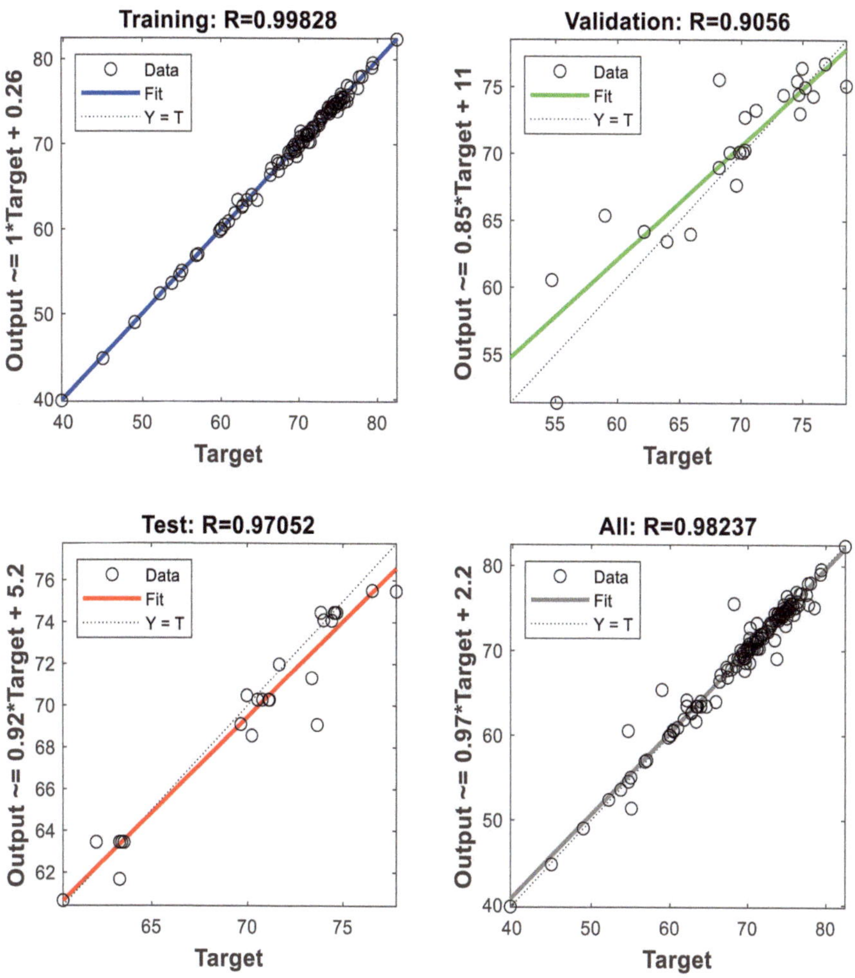

Figures 8.3 Regression plots for training, validation, testing, and overall performance of the ANN model

Fig. 8.4 Performance plots (MSE evolution) during training, validation, and testing of the an model over 19 epochs

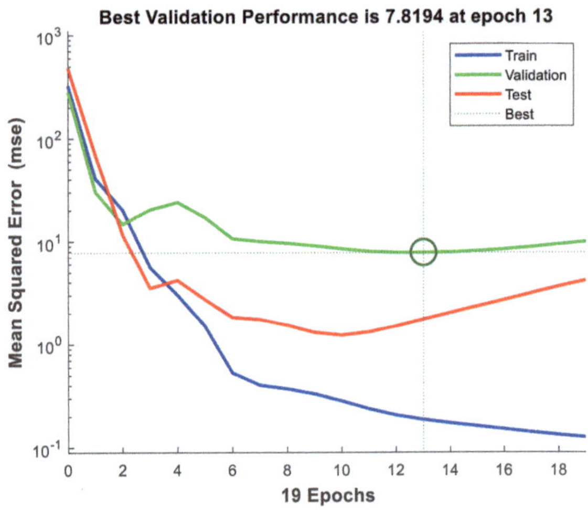

early stopping if the model begins to overfit. In this case, the lack of significant validation failures reflects effective generalization by the ANN model, ensuring robust performance across training and validation datasets.

The error histogram (Fig. 8.6) illustrates the distribution of prediction errors across the training, validation, and testing datasets. The central peak near zero error indicates the model's ability to accurately predict target values for the majority of instances; it highlights the model's high accuracy and robust performance across all phases. A narrow error distribution with minimal outliers confirms that the ANN achieved high predictive accuracy during training and testing phases. The consistent distribution of errors across datasets reflects the model's generalization ability and resistance to overfitting. The presence of slightly higher errors in the validation and test phases

Fig. 8.5 Training state plot showing gradient, Mu, and validation fail during ANN training

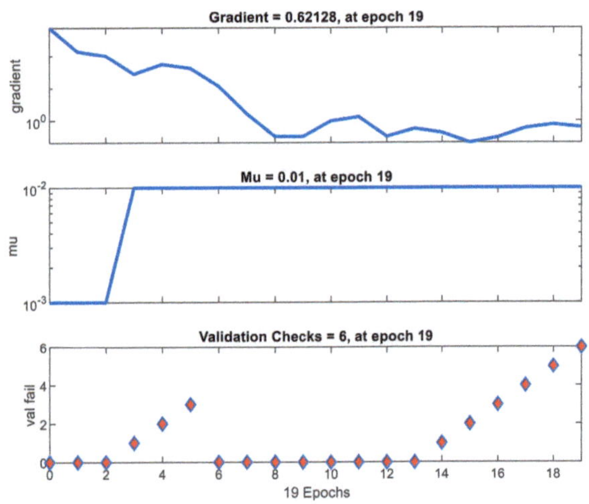

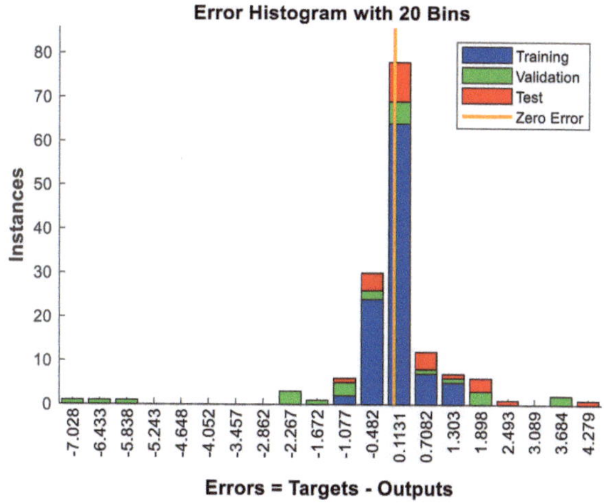

Fig. 8.6 Error histogram displaying the distribution of absolute errors for training, validation, and testing phases in ANN modeling, as generated by MATLAB's ANN fitting tool

compared to training is expected and indicates a reliable balance between model complexity and performance. The predicted ANN values for BOD, COD, and color reduction (Table 8.2) are in close agreement with the observed values (experimental results); this reinforces the reliability of the models.

8.4.3 Comparative Analysis of RSM and ANN

The comparative analysis between the RSM and ANN models demonstrates distinct strengths in modeling the ECF process, as reflected in the results for BOD, COD, and color reductions. Observed and predicted values indicate that while both models effectively represent the process, ANN consistently delivers higher predictive accuracy across all responses (Fig. 8.7). For BOD reduction, RSM achieved an R^2 value of 0.9317, while ANN attained a superior R^2 value of 0.941, indicating improved precision in capturing the observed trends. Similarly, for COD reduction, RSM recorded an R^2 of 0.9205, which was outperformed by ANN with an R^2 of 0.9412. In the case of color reduction, ANN further demonstrated its strength, with an R^2 of 0.9723 compared to RSM's R^2 of 0.9684.

The performance metrics in Table 8.8 also highlight ANN's enhanced reliability. ANN yielded lower MSE and higher R^2 values for training, validation, and testing phases, underscoring its robustness in modeling the ECF process. This consistency in ANN predictions reflects its ability to model nonlinear interactions more effectively than RSM. Overall, the results from the figures and table reinforce the complementary nature of the two methods. While RSM provides useful insights into parameter interactions, ANN excels in predictive accuracy, making it the preferred choice for optimizing complex environmental processes like ECF.

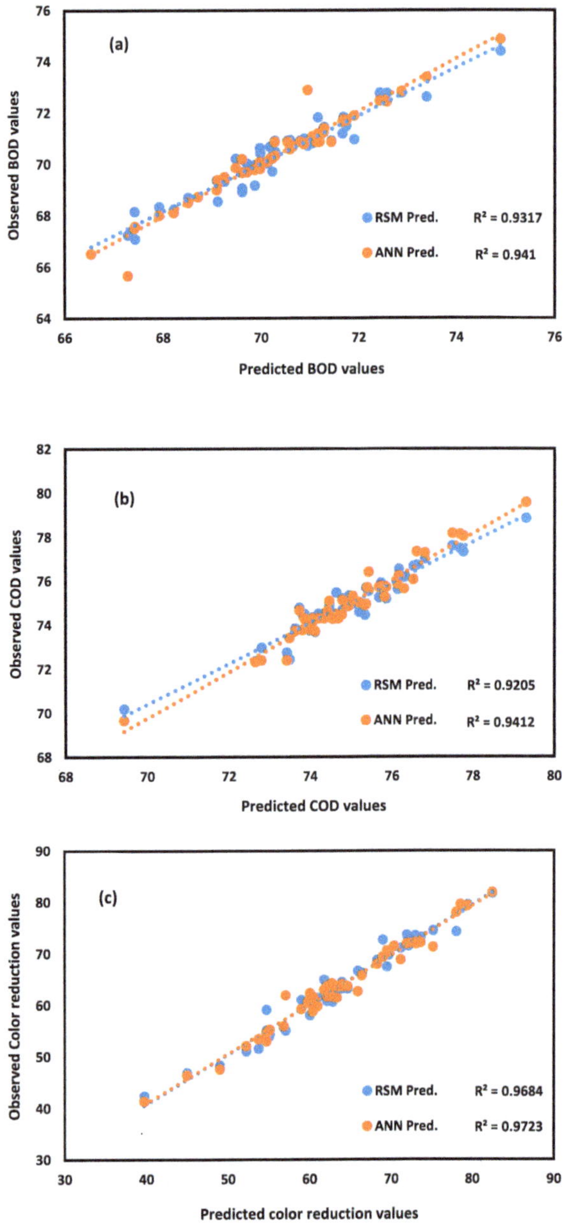

Fig. 8.7 The observed values versus the predicted values for **a** BOD, **b** COD and **c** color reduction from LCH

Table 8.8 Comparison of predicted responses for landfill leachate treatment using RSM and ANN

Error function equation	BOD		COD		Color	
	RSM	ANN	RSM	ANN	RSM	ANN
$AAD = [\frac{1}{n}\sum_{i=1}^{n}(\frac{\%R_{i,pred}-\%R_{i,exp}}{\%Ri,_{exp}})]X\,100$	0.0067	0.0023	0.0046	0.0043	0.0186	0.0168
$RMSE = \left(\frac{1}{n}\sum_{i=1}^{n}(\%R_{i,pred} - \%R_{i,\text{exp}})^2\right)^{1/2}$	0.7764	0.4388	0.4292	0.4056	1.5250	1.4343

Where n is the number of data points or samples, $\%R_{i,pred}$ is the predicted value, and $\%R_{i,exp}$ is the experimental value

8.4.4 Impact of Operational Parameters on Pollutant Removal

The surface plots (Fig. 8.8) provide a detailed visualization of the influence of key operational parameters on pollutant removal, specifically for BOD, COD, and color reduction. The interaction of current density and electrolysis time reveals that increasing the current density from 1.5 A to 2.5 A improves BOD reduction from 68% to approximately 75%, as observed in the surface plots. However, further increases in current density show diminishing returns, likely due to electrode passivation and higher energy consumption [16]. Similarly, extending electrolysis time up to 8 min enhances pollutant removal, with BOD reduction reaching 74%, but beyond this point, the efficiency plateaus as sufficient coagulants have been generated.

Settling time also plays a significant role in the process. An increase in settling time from 10 to 30 min improved color reduction from 70 to 84%, providing sufficient time for floc formation and sedimentation. On the other hand, temperature negatively impacts pollutant removal, as higher temperatures (above 50 °C) reduce color reduction efficiency from 84% to around 70%. This decline is attributed to the reduced solubility and stability of aluminum hydroxides at elevated temperatures [37]. The initial pH of the solution emerges as a critical factor in pollutant reduction. Under slightly acidic to neutral conditions (pH ~ 6), COD reduction peaks at 82%, as observed in the surface plots. At lower pH values (pH ~ 2–4), BOD and COD reduction rates are enhanced due to the amphoteric behavior of aluminum hydroxides, which form effective coagulants [33]. However, highly acidic or basic conditions reduce efficiency due to suboptimal coagulant formation.

The statistical significance of variable interactions is depicted by the shape of the surface plots, with elliptical shapes indicating significant interactions, such as between pH and electrolysis time. For instance, at an electrolysis time of 8 min and pH 6, COD reduction exceeds 80%, underscoring the importance of these parameters. Similarly, the interaction between settling time and current density reveals that a settling time of 30 min combined with a current density of 2.5 A maximizes color reduction, achieving over 90%. Overall, the results underscore the importance of

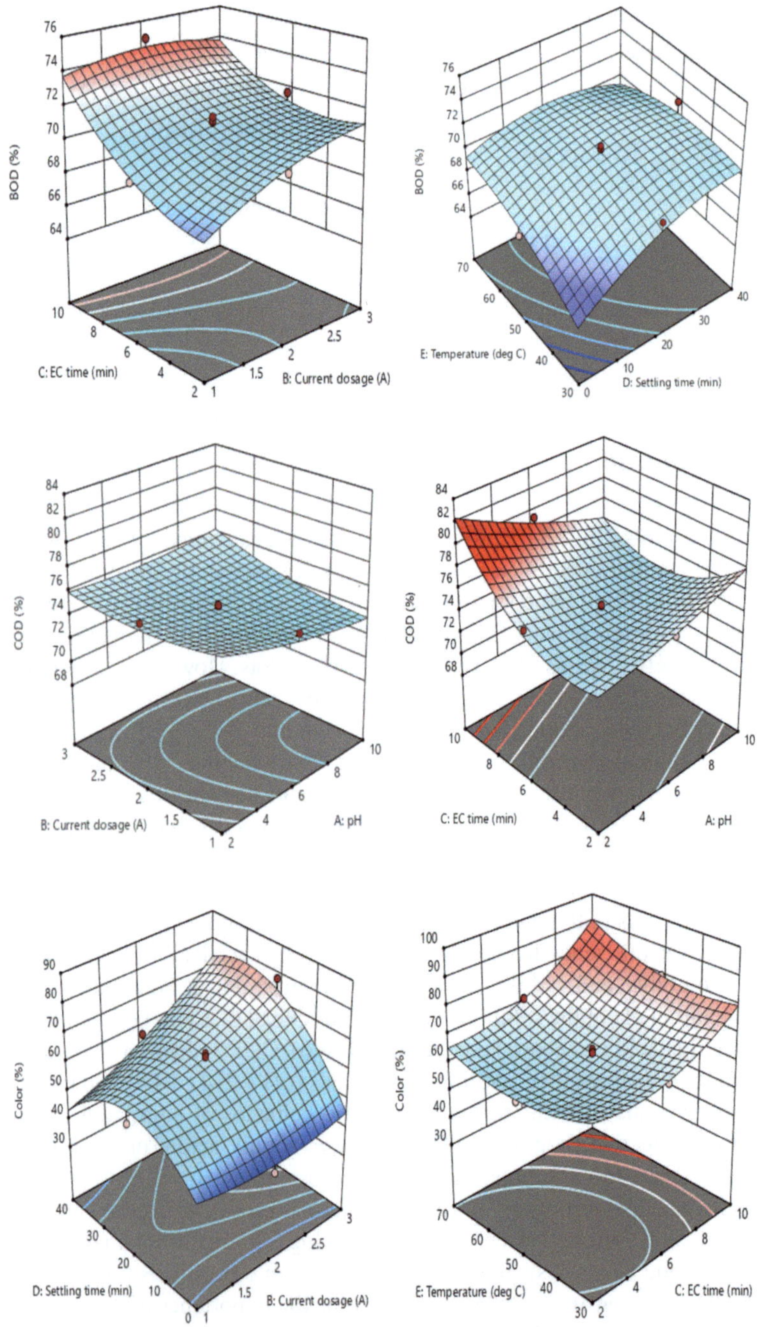

Fig. 8.8 Surface plots depicting the effects of process variables on BOD, COD and color

balanced operational conditions. Key parameters such as current density, electrolysis time, settling time, temperature, and pH interact synergistically to optimize pollutant removal, with significant improvements observed under optimal conditions. These findings emphasize the necessity of fine-tuning these variables to achieve efficient wastewater treatment outcomes.

8.4.5 Implications for Sustainable Water Treatment

The combination of ANN and ECF presents a powerful approach to optimizing complex water treatment processes. By leveraging the superior predictive capabilities of ANN, operational parameters can be fine-tuned with precision, reducing resource consumption and cutting operational costs. Additionally, the use of aluminum electrodes in ECF supports sustainability efforts by minimizing the need for external chemicals and reducing sludge generation. This study highlights how AI-driven optimization can transform water treatment technologies, offering a more efficient and environmentally friendly solution with the potential for broader applications in environmental engineering.

8.4.6 Limitations of the Proposed Models to Predict Optimum Values of Landfill Leachate Treatment

While the integration of RSM and ANN has proven to be highly effective in optimizing and predicting landfill leachate treatment processes, the proposed models have certain limitations. Both RSM and ANN depend heavily on the quality and quantity of experimental data [42]. Limited datasets or noisy data can negatively impact the accuracy and reliability of the predictions [43]. For ANN, in particular, the need for a substantial dataset is critical to avoid issues such as overfitting or underfitting during the training process [44].

Another limitation is the generalizability of the models. Since the current study focuses on specific experimental conditions such as pH range, electrode materials, and operational parameters, the models may require recalibration or retraining to be effectively applied to other landfill leachates with different chemical compositions or treatment methods. Additionally, ANN models, while highly accurate, are computationally intensive. The process of hyperparameter tuning and training can demand significant computational resources [45], which could pose challenges in environments with limited access to such resources.

A further drawback is the interpretability of model outputs. While RSM provides straightforward equations that explain variable interactions, ANN operates as a "black box," making it difficult to understand the underlying relationships between the input parameters and the outcomes [46]. This lack of transparency may hinder practical decision-making in some cases. Moreover, the scalability of the models is another concern. Predictions are based on small-scale experimental setups, and scaling up the electrocoagulation-flocculation process to industrial levels may introduce variables such as electrode wear, energy consumption, and flow dynamics, which are not accounted for in the current models.

Finally, the assumptions underlying the models can be limiting. For instance, RSM assumes that the system can be adequately represented by a second-order polynomial equation, which may not always capture highly nonlinear interactions [47]. Despite these limitations, the combination of RSM and ANN remains a robust tool for addressing complex environmental challenges. Future work could focus on expanding datasets, applying transfer learning techniques for ANN, and validating the models under more diverse and real-world operational conditions to enhance their applicability and reliability.

8.4.7 Study Limitations and Recommendations

This study successfully demonstrated the integration of ANN and RSM for optimizing the ECF process in the treatment of landfill leachate; however, there are certain limitations that need to be addressed for broader application and future advancements. One significant limitation is the lack of an in-depth analysis of the economic feasibility of electricity consumption and electrode utilization during the ECF process. While process optimization enhanced pollutant removal efficiency, further research is necessary to evaluate the costs associated with electricity usage and electrode wear in large-scale operations. The incorporation of AI-driven tools for predictive maintenance of electrodes and adaptive energy management systems could improve the cost-effectiveness of the process and make it more feasible for industrial applications.

Additionally, this study was conducted at a laboratory scale, which may not capture the challenges and variability encountered in real-world conditions. Future research should focus on scaling up the process and examining its performance under different leachate compositions and environmental conditions. The integration of renewable energy sources, such as solar or wind power, could also offset the energy demands of the ECF process, aligning it with sustainability goals. AI can play a critical role in optimizing energy usage by managing inputs and balancing renewable and non-renewable energy sources, thereby reducing operational costs and enhancing environmental benefits.

Another key recommendation is to leverage AI not only for parameter optimization but also for real-time monitoring and predictive maintenance. Artificial intelligence has the potential to enhance the economic feasibility of the ECF process by predicting electrode degradation, enabling timely maintenance, and optimizing energy consumption. These advancements would reduce downtime and operating costs, making the process more efficient and cost-effective. Lastly, the environmental impact of residual sludge generated during the process needs further exploration. Future studies could investigate sustainable sludge management strategies that comply with environmental regulations and align with circular economy principles, with AI assisting in designing innovative reuse or disposal methods. Addressing these limitations and incorporating these recommendations would significantly enhance the feasibility, scalability, and sustainability of the ECF process, enabling its broader adoption as a reliable solution for landfill leachate treatment.

8.5 Conclusion

This study highlights the effective integration of Artificial Neural Networks (ANN) and Response Surface Methodology (RSM) with Electrocoagulation-Flocculation (ECF) for the sustainable treatment of landfill leachate (LCH). Both modeling techniques demonstrated their strengths in optimizing operational parameters to achieve significant reductions in BOD, COD, and color. The ANN model consistently outperformed RSM in predictive accuracy, with superior R^2 values of 0.941, 0.9412, and 0.9723 for BOD, COD, and color reduction, respectively. Optimal conditions identified through RSM included an initial pH of 2.08, current density of 1.01 A, electrolysis time of 9.98 min, settling time of 30.34 min, and a temperature of 63.92 °C. Experimental validation confirmed pollutant reductions of 75.44% for BOD, 80.89% for COD, and 84.12% for color, closely aligning with the predicted values.

The surface plots elucidated key interactions among operational parameters, such as the synergistic effects of pH and electrolysis time on COD reduction, where optimal removal exceeded 80% at pH 6. Similarly, increasing settling time to 30 min maximized color reduction to over 90%. However, elevated temperatures (>50 °C) negatively affected pollutant removal, underscoring the need for controlled operational conditions. These findings demonstrated the importance of optimizing current density, electrolysis time, and settling time for efficient pollutant removal. The ANN model's superior accuracy and ability to handle complex nonlinear relationships emphasized its potential for advancing water treatment technologies. By reducing resource consumption and operational inefficiencies, ANN-driven optimization aligns with sustainability goals.

Future work should explore the socio-economic feasibility of integrating machine learning tools, like ANN-based algorithms, into broader environmental management strategies. This includes raising awareness about the benefits of using these advanced optimization techniques to mitigate the movement of landfill leachate into rivers and oceans, ultimately protecting aquatic ecosystems and public health. Additionally, studies should evaluate the scalability of this approach for large-scale applications and its potential to integrate with renewable energy sources for improved energy efficiency. Incorporating Internet of Things (IoT) technologies into the ECF process could enable real-time monitoring and dynamic optimization of critical parameters like pH, current density, and temperature. The integration of IoT-enabled sensors with AI models can further enhance pollutant removal precision, contributing to a circular economy.

The combination of AI tools with ECF technology not only improves pollutant removal efficiency but also provides a scalable and eco-friendly solution for managing landfill leachate. This study underscored the transformative potential of integrating artificial intelligence (AI) with environmental engineering processes to promote sustainable and efficient wastewater treatment.

Disclosure Statements

Acknowledgements I wish to acknowledge Nnamdi Azikiwe University, Nigeria; Wroclaw University of Environmental and Life Sciences, Poland; and the Polish National Agency for Academic Exchange (NAWA)—Ulam NAWA for International Scientists 2022

Compliance with Ethical Standards: This article does not contain any studies involving human or animal subjects.

Conflict of Interest: The author declare that there are no conflicts of interest.

Funding Statement: The author solemnly declare that no external funding was received for this research.

Consent to Publish: Not applicable.

Data Availability Statement: All relevant data are included in this article.

References

1. Igwegbe CA, López-Maldonado EA, Landázuri AC, Ovuoraye PE, Ogbu AI, Vela-García N et al (2024) Sustainable municipal landfill leachate management: current practices, challenges, and future directions. Desalin Water Treat 320:100709
2. Parvin F, Tareq SM (2021) Impact of landfill leachate contamination on surface and groundwater of Bangladesh: a systematic review and possible public health risks assessment. Appl Water Sci 11:100
3. Mojiri A, Zhou JL, Ratnaweera H, Ohashi A, Ozaki N, Kindaichi T et al (2021) Treatment of landfill leachate with different techniques: an overview. Water Reuse 11:66–96
4. Oba SN, Ighalo JO, Aniagor CO, Igwegbe CA (2021) Removal of ibuprofen from aqueous media by adsorption: a comprehensive review. Sci Total Environ 780:146608
5. Kuppan N, Padman M, Mahadeva M, Srinivasan S, Devarajan R (2024) A comprehensive review of sustainable bioremediation techniques: eco friendly solutions for waste and pollution management. Waste Manag Bull
6. Othmani A, Kadier A, Singh R, Igwegbe CA, Bouzid M, Aquatar MO, et al (2022) A comprehensive review on green perspectives of electrocoagulation integrated with advanced processes for effective pollutants removal from water environment. Environ Res 114294
7. Idusuyi N, Ajide O, Abu R, Okewole O, Ibiyemi O (2022) Low cost electrocoagulation process for treatment of contaminated water using aluminium electrodes from recycled cans. Mater Today Proc 56:1712–1716
8. Obi CC, Nwabanne JT, Igwegbe CA, Ohale PE, Okpala COR (2022) Multi-characteristic optimization and modeling analysis of electrocoagulation treatment of abattoir wastewater using iron electrode pairs. J Water Proc Eng 49:103136
9. Afsharnia M, Biglari H, Rasouli SS, Karimi A, Kianmehr M (2018) Sono-electrocoagulation of fresh leachate from municipal solid waste; simultaneous applying of iron and copper electrodes. Int J Electrochem Sci 13:472–484

10. Tegladza ID, Xu Q, Xu K, Lv G, Lu J (2021) Electrocoagulation processes: a general review about role of electro-generated flocs in pollutant removal. Process Saf Environ Prot 146:169–189
11. Igwegbe CA, Onukwuli OD, Ighalo JO, Umembamalu CJ (2021) Electrocoagulation-flocculation of aquaculture effluent using hybrid iron and aluminium electrodes: a comparative study. Chem Eng J Adv 6:100107
12. Mao Y, Zhao Y, Cotterill S (2023) Examining current and future applications of electrocoagulation in wastewater treatment. Water 15:1455
13. Aityoub A, Abouelfida A, Benyaich A, Romane A, Elmeray M, Idouhli R et al (2020) Optimization of physicochemical parameters during the electrocoagulation cadmium elimination. J Chem 2020:3071420
14. Al-Marri JS, Abouedwan AB, Ahmad MI, Bensalah N (2023) Electrocoagulation using aluminum electrodes as a sustainable and economic method for the removal of kinetic hydrate inhibitor (polyvinyl pyrrolidone) from produced wastewaters. Front Water 5:1305347
15. Bassyouni D, Ali S, Abdel-Aziz M, Elashtoukhy E (2023) Electrocoagulation technique and statistical analysis for treatment of real effluent from the pulp and paper industry. Int J Electrochem Sci 18:100389
16. Ingelsson M, Yasri N, Roberts EP (2020) Electrode passivation, faradaic efficiency, and performance enhancement strategies in electrocoagulation—a review. Water Res 187:116433
17. Igwegbe CA, Onukwuli OD, Ighalo JO, Umembamalu CJ, Adeniyi AG (2021) Comparative analysis on the electrochemical reduction of colour, COD and turbidity from municipal solid waste leachate using aluminium, iron and hybrid electrodes. Sustain Water Resourc Manag 7:39
18. Iwuozor KO (2019) Prospects and challenges of using coagulation-flocculation method in the treatment of effluents. Adv J Chem-Sect A 2:105–127
19. Chettri D, Verma AK, Chirania M, Verma AK (2024) Optimization strategies for enhanced production of microbial products. In: Industrial microbiology and biotechnology: a new horizon of the microbial world. Springer, pp 465–487
20. Ghosh V, Lonhare A (2024) Response surface methodology: a statistical tool to optimize process parameters (quantitative data) to maximize the microbial biomass and their bioactive metabolites. In: Industrial and manufacturing designs: quantitative and qualitative analysis. pp 139–171
21. Kumar Y, Marchena J, Awlla AH, Li JJ, Abdalla HB (2024) The AI-powered evolution of big data. Appl Sci 14:10176
22. Nabeel MZ (2024) AI-enhanced project management systems for optimizing resource allocation and risk mitigation. Asian J Multidiscipl Res Rev 5:53–91
23. Ukoba K, Olatunji KO, Adeoye E, Jen T-C, Madyira DM (2024) Optimizing renewable energy systems through artificial intelligence: Review and future prospects. Energy Environ 0958305X241256293
24. El Alaoui M, Rougui M (2024) Examining the application of artificial neural networks (ANNs) for advancing energy efficiency in building: a comprehensive reviews. J Sustain Res 6
25. Wani OA, Mahdi SS, Yeasin M, Kumar SS, Gagnon AS, Danish F et al (2024) Predicting rainfall using machine learning, deep learning, and time series models across an altitudinal gradient in the North-Western Himalayas. Sci Rep 14:27876
26. Igwegbe CA, Onukwuli OD, Ighalo JO, Ezeliora CD, Onyechi PC (2022) ANN prognostication and GA optimization of municipal solid waste leachate treatment using aluminum electrodes via electrocoagulation-flocculation method. In: Current trends and advances in computer-aided intelligent environmental data engineering. Elsevier, pp 161–183
27. Jin M, Liao Q, Patil S, Abdulraheem A, Al-Shehri D, Glatz G (2022) Hyperparameter tuning of artificial neural networks for well production estimation considering the uncertainty in initialized parameters. ACS Omega 7:24145–24156
28. Reji M, Kumar R (2022) Response surface methodology (RSM): an overview to analyze multivariate data. Indian J Microbiol Res 9:241–248

29. Rajewski J, Dobrzyńska-Inger A (2021) Application of response surface methodology (RSM) for the optimization of chromium (III) synergistic extraction by supported liquid membrane. Membranes 11:854
30. Sahu O (2019) Suitability of aluminum material on sugar industry wastewater with chemical and electrochemical treatment processes. Int J Indus Chem 10:335–347
31. Igwegbe CA (2019) Evaluation of bio- and electro- coagulants' activities on Fish pond wastewater and Solid waste leachate. Ph.D. dissertation. Department of Chemical Engineering
32. Morales-Figueroa A, Teutli-Sequeira EA, Linares-Hernández I, Martínez-Miranda V, García-Morales MA, Roa-Morales G (2022) Optimization of the electrocoagulation process with aluminum electrodes for rainwater treatment. Front Environ Sci 10:860011
33. Bagastyo AY, Ayu AP, Barakwan RA, Trihadiningrum Y (2020) Recovery of alum sludge by using membrane-based electrochemical process. J Ecol Eng 21:237–247
34. APHA (1992) Standard methods for the examination of water and wastewater, 18th edn. Washington, DC
35. APHA (1998) Standard methods for the examination of water and wastewater
36. Nariyan E, Aghababaei A, Sillanpää M (2017) Removal of pharmaceutical from water with an electrocoagulation process; effect of various parameters and studies of isotherm and kinetic. Sep Purif Technol 188:266–281
37. Igwegbe CA, Obi CC, Onyechi CC, Davoud B, Białowiec A, Onukwuli OD (2024) Integration of experimental and intelligent modeling for optimizing iron electrocoagulation-flocculation recovery of aquafarm effluent. Desalination Water Treat 320:100832
38. Almukdad A, Hawari AH, Hafiz M (2021) An enhanced electrocoagulation process for the removal of fe and mn from municipal wastewater using dielectrophoresis (Dep). Water 13:485
39. Yadav A, Chithaluru P, Singh A, Joshi D, Elkamchouchi DH, Pérez-Oleaga CM et al (2022) An enhanced feed-forward back propagation Levenberg–Marquardt algorithm for suspended sediment yield modeling. Water 14:3714
40. Igwegbe CA, Onukwuli OD, Ighalo JO, Menkiti MC (2021) Bio-coagulation-flocculation (BCF) of municipal solid waste leachate using picralima nitida extract: RSM and ANN modelling. Curr Res Green Sustain Chem 4:100078
41. Igwegbe CA, Onyechi CC, Białowiec A, Onukwuli OD (2024) Enhancing municipal solid waste leachate treatment efficiency: AI-based prediction of electrocoagulation/flocculation recovery using iron electrodes. Environ Technol 1–16
42. Reza A, Chen L, Mao X (2024) Response surface methodology for process optimization in livestock wastewater treatment: a review. Heliyon 10:e30326
43. Gupta S, Gupta A (2019) Dealing with noise problem in machine learning data-sets: a systematic review. Procedia Comput Sci 161:466–474
44. Alzubaidi L, Zhang J, Humaidi AJ, Al-Dujaili A, Duan Y, Al-Shamma O et al (2021) Review of deep learning: concepts, CNN architectures, challenges, applications, future directions. J Big Data 8:1–74
45. Raiaan MAK, Sakib S, Fahad NM, Mamun AA, Rahman MA, Shatabda S et al (2024) A systematic review of hyperparameter optimization techniques in convolutional neural networks. Decis Anal J 11:100470
46. Zhang Z, Beck MW, Winkler DA, Huang B, Sibanda W, Goyal H (2018) Opening the black box of neural networks: methods for interpreting neural network models in clinical applications. Ann Transl Med 6
47. Banza M, Seodigeng T, Rutto H (2023) Comparison study of ANFIS, ANN, and RSM and mechanistic modeling for chromium (VI) removal using modified cellulose nanocrystals–sodium alginate (CNC–alg). Arab J Sci Eng 48:16067–16085

Part III
AI in Sustainable Energy Solutions

Chapter 9
Integrating Artificial Intelligence in CO_2 Capture and Conversion Processes for Sustainable Energy Solutions

Oscar E. Medina, Pedro Narváez, Andrés A. Amell, Diana López, and Alexander Santamaría

Abstract Artificial intelligence (AI) is revolutionizing the landscape of Carbon Capture and Utilization (CCU) technologies, providing innovative tools to address climate change and achieve sustainable energy transitions. This chapter elucidates how AI aligns with and enhances the key objectives of CCU: increasing operational efficiency, scaling industrial applications, and promoting environmentally sustainable practices. By employing advanced machine-learning models and predictive algorithms, AI enables significant breakthroughs in optimizing CO_2 capture processes, such as enhancing material performance, improving separation precision, and reducing energy consumption. Furthermore, AI transforms CO_2 utilization by discovering and refining pathways to convert carbon dioxide into high-value products, including fuels, polymers, and industrial chemicals. Through detailed analysis, the chapter demonstrates how AI facilitates the design of more effective catalysts, predicts reaction outcomes with unprecedented accuracy, and accelerates the deployment of economically viable CCU systems. In addition, the role of AI in addressing data integration challenges, improving process automation, and enhancing decision-making frameworks is thoroughly examined. These capabilities position AI as a tool for operational improvements and a transformative enabler for achieving global carbon neutrality. By clearly linking AI's potential to the overarching goals of CCU, the chapter provides a forward-looking perspective on leveraging AI as a strategic driver for innovation and sustainability in the energy sector.

Keywords Artificial intelligence · CO_2 hydrogenation · Machine learning · Process prediction · Valuable products

---

O. E. Medina (✉) · D. López · A. Santamaría
Instituto de Química, Química de Recursos Energéticos y Medio Ambiente, Universidad de Antioquia UdeA, Medellín, Colombia
e-mail: oemedinae@unal.edu.co

O. E. Medina · P. Narváez · A. A. Amell
Facultad de Ingeniería, Grupo de Ciencia y Tecnología de Gases y Uso Racional de la Energía, Universidad de Antioquia, Medellín, Colombia

M. Nasr et al. (eds.), *Artificial Intelligence Applications for a Sustainable Environment*, Green Chemistry and Sustainable Technology,
https://doi.org/10.1007/978-3-031-91199-6_9

Nomenclature

AARD	Average absolute relative deviation
AdaBoost	Adaptive boosting
AI	Artificial intelligence
AL	Active learning
AMP	2-Amino-2-methyl-1-propanol
ANFIS	Adaptive neuro-fuzzy inference systems
ANN	Artificial neural networks
BPNN	Back-propagation neural networks
CART	Classification and regression tree
CCS	Carbon capture and storage
CCU	Carbon capture and utilization
CNNs	Convolutional neural networks
CNT	Carbon nanotubes
COSMO-RS	Conductor-like Screening Model for Real Solvents
DCFCs	Biochar in direct carbon fuel cells
DEA	Diethanolamine
DEEA	(Diethylamino)ethanol
DEIM	Discrete empirical interpolation
DESs	Deep eutectic solvents
DFT	Density functional theory
DL	Deep learning
DNN	Deep neural network
DT	Decision tree
DTR	Decision tree regression
eCO_2RR	Electrochemical CO_2 reduction
EDA	Ethylenediamine
E-AHF	Extended adaptive hybrid functions
GNNs	Graph neural networks
GN-SINDy	Greedy sampling neural network in sindy
GP	Gaussian processes
GPBO	Gaussian process Bayesian optimization
GPR	Gaussian process regression
GRNN	Regression neural networks
HBAs	Hydrogen bond acceptors
HBDs	Hydrogen bond donors
IFC	Ionic fragment contributions
IGCC	Integrated coal gasification combined cycle
ILs	Ionic liquids
LCA	Life cycle assessment
LSSVM	Least-squares support vector machines
MAPA	(Methylamino)propylamine
MD	Molecular dynamics

MDEA	Methyl diethanolamine
MEA	Monoethanolamine
ML	Machine learning
MLIPs	Machine-learning interatomic potentials
MLR	Multiple linear regression
MLP	Multi-layer perceptron
MNPCS	Nitrogen-groups modified porous carbon microspheres materials
MOFs	Metal–organic frameworks
MP-MCTS	Multi-player Monte Carlo tree search
MWCNT	Multi-walled carbon nanotubes
NLP	Natural language processing
NSGA-II	Nondominated sorting genetic algorithm II
OCP	Open catalyst project
PZ	Piperazine
QSPR	Quantitative structure–property relationships
Q-DEIM	Q-discrete empirical interpolation
RBF	Radial basis function
RBFNNs	Radial basis function neural networks
RF	Random forest
RL	Reinforcement learning
RMSE	Root mean square error
RNN	Recurrent neural networks
SHAP	Shapley Additive explanations
SINDy	Sparce Identification of Nonlinear Dynamics
SMILES	Simplified molecular input line entry systems
SVM	Support vector machines
SWIR	Short-wave infrared
TEA	Triethanolamine
WDPC	Biomass-pyrolysis derived char
XAI	Explainable AI
XGBoost	Extreme gradient boosting

9.1 Introduction

The ever-increasing concentration of carbon dioxide (CO_2) in the atmosphere is one of the most pressing challenges of our time [1, 2]. As a primary greenhouse gas, CO_2 contributes significantly to global warming and climate change, prompting an urgent need for practical solutions to mitigate its impact [3]. Since the onset of the industrial revolution, economic growth and technological advancements have come at the cost of increased CO_2 emissions, mainly from fossil fuel combustion, deforestation, and various industrial processes [4, 5]. With the atmospheric CO_2 concentration exceeding 400 parts per million (ppm), a level not seen for millions of years, there is a critical need for sustainable energy solutions that address this challenge [6]. As the international community strives to meet the targets set by the Paris Agreement—to limit global warming to well below 2 °C above pre-industrial levels—reducing CO_2 emissions has become a central focus of climate policy [7].

Among the myriad strategies to combat rising CO_2 levels, CCU emerges as a promising approach. Unlike traditional carbon capture and storage (CCS) methods, which focus on capturing CO_2 and storing it underground, CCU aims to convert captured CO_2 into valuable products, such as fuels, chemicals, and building materials [8]. This approach reduces the amount of CO_2 released into the atmosphere and provides economic incentives by creating marketable products. However, the implementation of CCU technologies globally faces significant challenges [8]. These include the high energy demands of capture processes, the need for advanced materials and catalysts for efficient conversion, and the economic feasibility of the overall process [9, 10]. Alongside traditional CO_2 sources, biogenic CO_2—produced through biological processes such as fermentation and biomass combustion—has gained attention as a critical area in pursuing low-carbon solutions [11]. Biogenic CO_2, as part of the natural carbon cycle, offers a sustainable and renewable option for CO_2 utilization. Incorporating biogenic CO_2 capture and conversion into CCU systems is particularly relevant for producing low-carbon PTX (Power-to-X) products, which include synthetic fuels, chemicals, and other carbon-based materials generated using renewable energy. This approach aligns with creating carbon–neutral or even carbon-negative processes by integrating renewable feedstocks and clean energy sources.

Capturing CO_2 from industrial and biogenic emissions involves several methods, including pre-, post-, and oxy-combustion, and refining processes like absorption, adsorption, membrane separation, and cryogenic distillation [12]. Each method has advantages and limitations, often balancing energy efficiency, cost, and scalability. For instance, chemical absorption using amines is one of the most mature technologies but is energy-intensive and requires significant heat input for solvent regeneration [13]. Similarly, adsorption processes, while potentially more energy-efficient, require developing advanced adsorbent materials with high selectivity and capacity for CO_2 [14]. Once captured, converting CO_2 into valuable products is another challenge. Catalysis plays a crucial role in this process, as it facilitates the transformation of CO_2 into chemicals and fuels. However, CO_2 is a highly stable molecule, and its activation requires both highly efficient and stable catalysts under reaction conditions [15–17]. Developing such catalysts often involves complex trial-and-error approaches, extensive experimentation, and significant resource investment. For these technologies to be viable on a commercial scale, the cost of CO_2 capture and conversion must be competitive with traditional methods of producing fuels and chemicals [18, 19]. This requires optimizing every step of the process, from the energy required for capture to conversion efficiency and the marketability of the final products.

Artificial intelligence (AI) has recently emerged as a transformative area in various fields, offering unprecedented data analysis, optimization, and predictive modeling capabilities [20]. In the context of CO_2 capture and conversion, AI presents a unique opportunity to address many existing challenges. By leveraging AI-driven techniques, such as machine learning (ML), deep learning (DL), and reinforcement learning (RL), researchers and engineers can optimize capture processes, enhance catalyst design, predict system behaviors under various conditions, and improve the overall efficiency and scalability of CCU technologies.

AI can enhance the efficiency of CO_2 capture methods by optimizing operational parameters in real-time. For example, ML algorithms can analyze data from ongoing capture processes to identify optimal conditions for maximum CO_2 absorption or adsorption, such as temperature, pressure, and solvent concentration [12]. Additionally, AI can predict the behavior of capture materials under different conditions, aiding in the design of more efficient and cost-effective capture systems.

The design of catalysts is critical for CO_2 conversion, as their efficiency and selectivity directly influence the feasibility of the process. AI-driven techniques, such as generative models and RL, can be used to discover new catalysts with enhanced properties [21]. By simulating the interactions between CO_2 molecules and potential catalyst materials, AI can predict the effectiveness of catalysts before synthesizing them in the laboratory, significantly reducing the time and cost associated with experimental trials. AI also enables the development of predictive models that simulate entire CO_2 capture and conversion processes in a variety of scenarios [22]. These models can be used to optimize process parameters, minimize energy consumption, and maximize product yield. Furthermore, AI algorithms can learn from historical data to predict the long-term behavior of systems, identify potential failure points, and suggest proactive maintenance schedules, thus enhancing the reliability and longevity of CCU technologies.

Integrating AI in CO_2 capture and conversion processes represents a promising frontier for sustainable energy solutions. By addressing key challenges in process optimization, catalyst design, predictive modeling, and economic feasibility, AI has the potential to transform the way CO_2 emissions are managed.

This review is structured to provide a comprehensive exploration of the integration of AI in CO_2 capture and conversion processes. It is organized into six sections. The first section introduces the fundamental principles of CO_2 capture and conversion technologies and describes the chemical and engineering challenges these processes face. The second section delves into the different types of AI techniques—such as ML, DL, and RL—that can be applied to this field, providing an overview of their capabilities and limitations.

The third section focuses on the current state of AI applications in optimizing CO_2 capture processes, including case studies that illustrate the practical benefits of AI-driven optimization. The fourth section explores the role of AI in catalyst discovery and design for CO_2 conversion, highlighting recent breakthroughs achieved through AI methodologies. This section also discusses the role of AI in some CO_2 conversion technologies. The final section provides a critical analysis of the future directions and opportunities for further research in this interdisciplinary field. This review contributes to the literature by summarizing current advances, identifying gaps, and proposing a roadmap for future developments. The novelty of this review lies in its interdisciplinary approach, combining insights from chemical engineering, data science, and environmental economics to present a holistic view of how AI can be leveraged for sustainable CO_2 management.

The objectives of this chapter are threefold. First, it seeks to establish a clear understanding of how AI can address critical challenges in CCU technologies, such as improving energy efficiency, enhancing catalyst performance, and enabling scalable deployment. Second, the chapter aims to explore integrating AI-driven techniques into specific aspects of CCU processes, including real-time optimization, predictive modeling, and catalyst discovery, highlighting their transformative potential. Third, it aspires to provide a forward-looking perspective by identifying key research gaps and proposing strategic directions for advancing AI applications in this interdisciplinary field. By aligning these objectives with the broader goals of reducing carbon emissions and fostering sustainable energy practices, this chapter underscores the pivotal role of AI in creating economically viable and environmentally sustainable solutions for CO_2 management.

9.2 Fundamental Principles of CO_2 Capture and Conversion Technologies

9.2.1 Overview of CO_2 Capture Technologies

CO_2 capture technologies play a pivotal role in mitigating greenhouse gas emissions by separating carbon dioxide from various industrial gas streams, such as flue gas, ambient air, and biogenic carbon from fermentation and biomass combustion. These technologies differ in their approaches, offering unique advantages and facing specific challenges. This section provides an overview of the primary CO_2 capture technologies, including adsorption, absorption, oxy-combustion, cryogenic separation, algae systems, and membrane separation, highlighting their strengths and limitations in real-world applications [13].

Adsorption is a widely used method for CO_2 capture, where solid adsorbents, such as activated carbon, zeolites, or metal–organic frameworks (MOFs), are employed to capture CO_2 from gas mixtures [23]. The CO_2 molecules adhere to the surface of the adsorbent, effectively separating them from the other gases. One of the key benefits of adsorption is that it requires less energy for regeneration compared to absorption-based systems [24]. Adsorption systems are also flexible and adaptable, operating through temperature or pressure swing operations, depending on the specific requirements. However, developing efficient and cost-effective adsorbents remains challenging, as advanced materials can be expensive, and their performance can degrade over time due to humidity and impurities. Moreover, adsorbents require periodic regeneration, which can add complexity to the process [25].

Absorption-based CO_2 capture, on the other hand, is a mature and versatile technology commonly used in industrial processes. In this method, a liquid solvent absorbs CO_2 from the gas mixture, and the solvent is subsequently regenerated to release the captured CO_2 [26]. Absorption technologies, particularly chemical solvents like amines, offer high CO_2 capture efficiency and can be adapted to various processes. However, the energy-intensive nature of solvent regeneration poses a significant limitation [27]. Additionally, solvent degradation over time can lead to environmental emissions, reducing the overall efficiency of the system. Research efforts continue to focus on the development of more stable and energy-efficient solvents that can be regenerated with lower energy consumption [28].

Cryogenic separation is another CO_2 capture method that involves cooling the gas stream to extremely low temperatures to condense CO_2 from the mixture. This technology can achieve very high-purity CO_2 and has been commercially proven in specific applications where such purity is essential [29]. However, cryogenic separation is highly energy-intensive due to the extreme cooling requirements, making it costly and generally less viable for widespread use, except in cases where high-purity CO_2 is critical [30].

Algae-based systems represent a more sustainable CO_2 capture technology that utilizes photosynthesis to convert CO_2 into biomass or biofuels. This approach is particularly suited for integrating nutrient-rich wastewater treatment systems, providing a cost-effective solution in specific cases [31]. While algae systems present a promising and environmentally friendly option, they require large areas of land for cultivation, and harvesting and dewatering algae can be labor-intensive and complex. Furthermore, the efficiency of algae systems is highly dependent on environmental factors such as sunlight, temperature, and nutrient availability, which can pose challenges in scalability [32].

Membrane separation offers an energy-efficient alternative to thermal-based CO_2 capture technologies, as most membrane systems do not require heat. Instead, selective membranes are used to separate CO_2 from other gases [33]. Membrane systems are compact and modular, making them suitable for continuous operation and easy scalability in industrial applications. However, these systems are vulnerable to membrane fouling and degradation, and developing high-selectivity, high-permeability membranes are essential for making the process cost-effective. Ongoing research focuses on improving membrane materials and performance to extend their operational lifespan and reduce associated costs [34].

Each CO_2 capture technology has advantages and limitations, making them suitable for different industrial applications based on energy consumption, CO_2 purity requirements, and scalability [35]. Adsorption and absorption technologies are widely used in large-scale CO_2 capture processes, offering high efficiency and adaptability, while cryogenic separation is more appropriate for applications requiring exceptionally pure CO_2. Algae systems and membrane technologies present sustainable and energy-efficient alternatives but face scalability and material performance challenges [36]. The continued development of new materials, solvents, and process optimizations will be crucial for enhancing efficiency and reducing the costs of CO_2 capture technologies. These advances will support broader adoption of carbon capture technologies, which is key in addressing global climate challenges.

9.2.2 CO_2 Capture by Absorption

CO_2 capture via absorption is one of the most widely employed methods for removing carbon dioxide from gas streams, particularly in industries and power plants [37]. A liquid solvent (absorbent) selectively dissolves CO_2 from a gas mixture within a gas–liquid contactor, such as an absorber column [38]. As the CO_2-rich gas flows through the absorber, it interacts with the absorbent, leading to the selective absorption of CO_2 molecules into the liquid phase, leaving behind the other gases in the stream (see Fig. 9.1). The CO_2-laden solvent is then transferred to a regeneration unit, which is heated to release the captured CO_2. This regenerated solvent is recycled back to the absorption process. At the same time, the captured CO_2 can be compressed and stored or used for various industrial applications such as enhanced oil recovery or the production of chemicals [39].

Absorption processes offer high selectivity for CO_2 and large capacity, making them ideal for advanced CO_2 capture applications. However, certain challenges hinder the widespread utilization of this method, including high energy consumption, solvent degradation, corrosion, toxicity, and costs associated with the final disposal of degraded absorbents. Additionally, the high cost and mass loss of absorbents due to instability remain major obstacles. Table 9.1 summarizes a comparison of different absorbents used for CO_2 capture. Absorbents such as ionic liquids and certain modified porous silicas that primarily rely on the solubility of CO_2 in the absorbent without a chemical reaction are classified here. Examples include systems operating at high

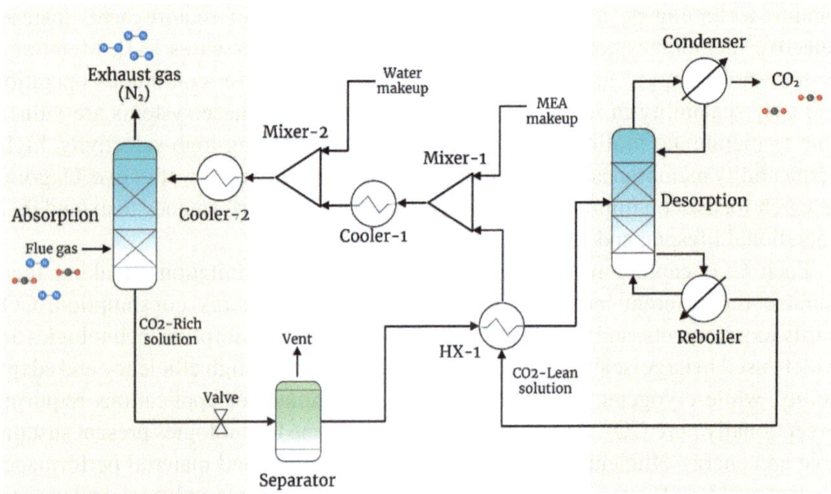

Fig. 9.1 Schematic representation of a conventional chemical CO_2 absorption process. The system includes an absorber column, where the gas–liquid contact facilitates the selective absorption of CO_2 into the solvent, and a desorber column, where the solvent is regenerated by releasing the captured CO_2. HX-1 (heat exchanger) preheats the solvent before entering the absorber and recovers heat from the lean solvent returning from the desorber, enhancing energy efficiency

Table 9.1 Studies on CO_2 capture via absorption method

Absorbents	Conditions	Diffusion coefficient D (CO_2) (m^2 s^{-1})	Amount of CO_2 capture (Absorption capacity)	Temperature (°C)	Absorption type	Absorption equilibrium (min)	References
2EAE-TMPAD	–	1.01×10^{-9}	–	60	Chemical	–	[40]
2EAE-DEAB	–	0.98×10^{-9}	–	60	Chemical	–	[40]
2EAE-1DMA2P	–	0.92×10^{-9}	–	60	Chemical	–	[40]
2MAE-2DMAE	–	0.92×10^{-9}	–	60	Chemical	–	[40]
2EAE-2DMAE	–	0.85×10^{-9}	–	60	Chemical	–	[40]
DMCA-MCA	–	0.61×10^{-9}	–	60	Chemical	–	[40]
Deep eutectic solvents (DESs): 1,5-diazabicyclo [4.3.0] non-5-ene (DBN) and 1,2,4-triazole (Tz) DES [2DBN:Tz]	100 kPa	–	0.19 g CO_2 g^{-1} DES	25	Chemical	60	[41]
[3DBN:Tz]	100 kPa	–	1.89 mol CO_2 mol^{-1} DES	25	Chemical	30	[41]
[DBN:3Tz]	100 kPa	–	–	25	Chemical	180	[41]
Methyldiethanolamine (MDEA)	300 kPa recovered CO_2 pressure	–	–	–	Chemical	–	[42]

(continued)

Table 9.1 (continued)

Absorbents	Conditions	Diffusion coefficient D (CO_2) ($m^2\ s^{-1}$)	Amount of CO_2 capture (Absorption capacity)	Temperature (°C)	Absorption type	Absorption equilibrium (min)	References
Tertiary amine TEMDA (N, N, N′, N′-Tetramethylethylenediamine) and space hindered amine DACH (trans-1,4-Diaminocyclohexane) with Chlorella sp. L166	300 kPa recovered CO_2 pressure	–	Carbon sequestration capacity: 123.27 mg L^{-1} d^{-1}	–	Chemical	–	[43]
Anhydrous multi-hybrid absorbent from tetraethylenepentamine-modified porous silica	Specific surface area: 669.54 $m^2\ g^{-1}$	–	0.419 mmol g^{-1} at 20% CO_2 or 0.377 mmol g^{-1} @ 0.15 bar and 27 °C	57	Physical	95% regeneration efficiency after 5 cycles	[44]
I-liquid-D-I NOHM	50 bar	–	3.890 mmol g^{-1}	20	Physical	–	[45]
H-ZSM-5-liquid/[P66614][Br]	10 bar	–	0.460 mmol g^{-1}	25	Physical	–	[46]

where 2EAE-TMPAD is 2-(ethylamino) ethanol, N, N, N′, N′-tetramethylpropane-1,3-diamine; 2EAE-DEAB is 2-(ethylamino) ethanol, 4-diethylamino-2-butanol; 2EAE-1DMA2P is 2-(ethylamino) ethanol, 1-dimethylamino-2-propanol; 2MAE-2DMAE is 2-(methylamino)ethanol, 2-dimethylamino ethanol; 2EAE-2DMAE is 2-(ethylamino) ethanol, 2-dimethylamino ethanol; DMCA-MCA is N, N, dimethylcyclohexylamine, N-methylcyclohexylamine; 3DBN:Tz is 3(1,5-diazabicyclo[4.3.0]non-5-ene):1,2,4-triazole; DBN:3Tz is 1,5-diazabicyclo[4.3.0]non-5-ene:3(1,2,4-triazole) pressures like I-liquid-D TNOHM and H-ZSM-5-liquid, which utilize pressure to enhance the physical solubility of CO_2. Systems involving amines or deep eutectic solvents that chemically react with CO_2 to form bonds are classified under chemical absorption. These include DES systems like [2DBN-Tz] and amine-based systems like TEMDA and MDEA.

Absorption processes generally utilize two primary strategies for capturing CO_2: physical absorption and chemical absorption [47]. Physical absorption relies on Henry's law, which governs the solubility of gases based on partial pressure, making it ideal for capturing CO_2 at low and high temperatures. The regeneration of solvents in physical absorption is achieved through either an increase in temperature or a decrease in pressure—or a combination of both [48]. This method benefits high-pressure gas streams, such as those encountered in the pre-combustion stage of integrated coal gasification combined cycle (IGCC) plants. In contrast, chemical absorption relies on forming weak bonds between CO_2 and the solvent. When the CO_2-laden solvent is heated, the CO_2 is released, allowing for solvent regeneration [12]. Chemical absorption is widely applied for CO_2 capture at lower partial pressures, with amine and carbonate solutions being the most common solvents.

Significant progress has been made in absorption-based CO_2 capture technologies in recent years. Research efforts have explored various solvents, including physical, chemical, and mixed solvents, to improve capture efficiency and minimize energy consumption. Physical solvents such as rectisol, propylene carbonate, and Selexol capture CO_2 by dissolving it via weak van der Waals forces, making them less energy-intensive for regeneration [49]. However, they are limited by their lower CO_2 capture capacity and sensitivity to partial pressure. In contrast, chemical solvents such as monoethanolamine (MEA), diethanolamine (DEA), and methyl diethanolamine (MDEA) form stronger chemical bonds with CO_2, which enhances their capture capacity but requires more energy for solvent regeneration [50].

The combination of physical and chemical solvents, known as solvent blends, has emerged as a promising approach to improve absorption efficiency. Mixed solvents can enhance CO_2 absorption capacity, selectivity, and regeneration energy efficiency [51]. For example, Nwaoha et al. [52] demonstrated that a tri-solvent blend of AMP (2-amino-2-methyl-1-propanol), PZ (piperazine), and MEA significantly outperformed conventional MEA solutions in terms of lower energy consumption and higher absorption rates. In addition to solvent blends, several studies have explored the potential of adding surfactants to the absorption process to enhance CO_2 capture efficiency. Valluri and Kawatra [53] demonstrated that adding frothers as surfactants to Na_2CO_3 solutions increased CO_2 absorption efficiency from 55.6% to 99.9%. The surfactant enhanced the surface area for CO_2 transport and improved mass transfer within the packed bed. Additionally, frothers reduced surface tension, preventing bubble coalescence and creating smaller, more uniform bubbles that accelerated CO_2 absorption.

Another recent development is the use of phase change solvents for CO_2 capture. These solvents exist as a single-phase solution under normal conditions but undergo phase separation when subjected to temperature, pressure, or other factors such as ionic strength changes. Phase change solvents, such as those studied by Zhang et al.

[54], form two distinct phases during CO_2 absorption—one rich in CO_2 and the other lean. This property significantly reduces energy consumption during solvent regeneration, as only the CO_2-rich phase requires treatment. The iCap process, which uses a blend of 3-(methylamino)propylamine (MAPA) at 2 M and 2-(diethylamino)ethanol (DEEA) at 5 M, exemplifies this approach by selectively regenerating only the CO_2-rich phase.

Recent research has proposed a variety of novel solvents for CO_2 capture, including deep eutectic solvents (DESs), ionic liquids (ILs), and non-aqueous amine-based solvents. In recent studies, DESs, formed by mixing hydrogen bond donors and acceptors, have demonstrated high CO_2 solubility and selectivity. For instance, Pishro et al. [55] developed DESs using ethylenediamine (EDA), monoethanolamine (MEA), and other amines, reporting high CO_2 solubility at 40 °C and 0.8 MPa. Similarly, ILs such as $[Bmim]^+[BF4]^-$ and their blends with amines have shown promise as alternatives to conventional solvents, with higher CO_2 loading and improved stability.

Introducing water-lean amino acid salt-based biphasic absorbents has also shown potential for reducing energy consumption during solvent regeneration [56]. Potassium prolinate and potassium sarcosinate, combined with weak polar solvents like 2-methoxyethanol, demonstrated 40–50% less regeneration energy than conventional MEA solutions while maintaining high cyclic capacity.

While advances in absorption technologies have significantly improved CO_2 capture efficiency, several challenges remain for large-scale industrial implementation. The high viscosity and insolubility of some novel solvent systems can cause operational issues such as fouling and blockages in equipment [57]. Moreover, the environmental impact of these solvents, particularly in terms of their life cycle (from production to disposal), must be thoroughly evaluated to ensure sustainable carbon capture solutions [58]. Future research should focus on optimizing solvent formulations to balance high CO_2 capture efficiency with reduced energy consumption and environmental impact. Comparative studies of different solvent systems and process configurations will also be crucial to identify the most suitable solutions for specific industrial applications.

9.2.3 CO_2 Capture by Adsorption

Adsorption-based CO_2 capture is a widely employed method for separating and concentrating carbon dioxide from various gas mixtures, such as flue gas or ambient air [59]. This process involves using solid adsorbents—materials with a high affinity for CO_2 molecules. When the gas mixture encounters the adsorbent, CO_2 molecules adhere to the surface of the material, effectively separating them from other gas components. Figure 9.2 illustrates a typical adsorption process, where the efficiency of CO_2 capture is highly dependent on the properties of the adsorbent material [39].

The choice of adsorbent is critical to the success of the adsorption process. Common adsorbents used for CO_2 capture include activated carbon [60], zeolites

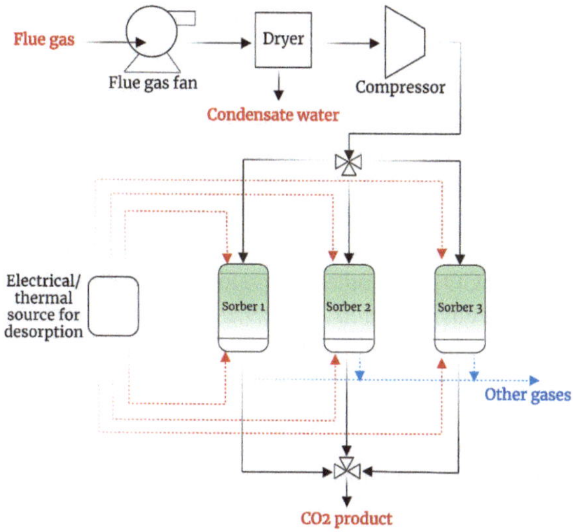

Fig. 9.2 Schematic representation of a typical CO_2 adsorption process. The system utilizes solid adsorbents, such as activated carbon, zeolites, or metal–organic frameworks, to capture CO_2 molecules selectively. The adsorption process involves the gas stream passing through three packed columns containing adsorbent material, where CO_2 adheres to the surface of the adsorbent. Upon saturation, the adsorbent is regenerated via temperature or pressure changes, releasing the captured CO_2 for storage or utilization

[61], metal–organic frameworks (MOFs) [62], aluminosilicates [63], and some rare earths [64]. These materials are favored due to their high surface areas, excellent porosity, and tunable chemical properties, which enhance their capacity to capture CO_2. Once the adsorbent material becomes saturated with CO_2, it can be regenerated by applying heat, pressure, or a combination of both, which releases the captured CO_2 for storage or further utilization [65]. This process can be repeated multiple times, making adsorption-based capture a flexible and efficient method for mitigating greenhouse gas emissions from industrial sources.

One key advantage of adsorption-based CO_2 capture is its relatively low energy requirement [66]. CO_2 can form bonds with the surface of the adsorbent material, forming surface carbonates or bicarbonates. While these are coordination bonds rather than covalent bonds, they are more easily decomposed, significantly reducing the energy required for regeneration compared to chemical absorption methods. This lower energy demand is particularly beneficial for large-scale applications, where minimizing energy consumption is crucial for cost-effective operation [67].

The efficiency of adsorption-based CO_2 capture depends on various mechanisms by which CO_2 interacts with the adsorbent [66]. These mechanisms include *(i)* Heterogeneous and surface adsorption: CO_2 molecules are adsorbed onto the surface of the adsorbent, where they adhere to active sites; *(ii)* Electrostatic interactions: The interaction between CO_2 molecules and charged adsorbent surfaces can enhance the adsorption process; *(iii)* Acid–base interactions: Adsorbents with acidic or basic properties can form temporary interactions with CO_2 molecules; (*iv*) Physisorption: Weak van der Waals forces enable CO_2 molecules to adhere to the adsorbent surface without forming strong chemical bonds and *v*) H-bonding: Some adsorbents may form weak hydrogen bonds with CO_2, further increasing adsorption capacity.

In addition to these mechanisms, the design of the absorbent plays a significant role in its performance. Narrow-pored materials doped with CO_2-philic heteroatoms, such as nitrogen or oxygen, are often used to improve CO_2 capture efficiency [15, 68]. The ideal adsorbent material should have high mechanical strength, thermal stability, selectivity for CO_2 over other gases, low preparation costs, and resistance to degradation in humid environments. Table 9.2 summarizes various adsorbents and their synthesis methods, highlighting materials that meet these criteria. The table provides a comparative overview of various adsorbents used in CO_2 capture, emphasizing how differences in fabrication methods, surface areas, and adsorption mechanisms impact their performance. Generally, adsorbents derived from biomass and carbon-based materials offer moderate to high specific surface areas and rely on physical interactions, such as electrostatic attraction or acid–base interactions, for CO_2 capture. These materials are often selected for being eco-friendly and cost-effective. Metal–organic frameworks (MOFs), like UTSA-16, which utilize temperature swing adsorption, are designed for efficient cyclic use. MOFs are particularly advantageous in applications where temperature-based CO_2 release is desirable, allowing for effective regeneration with minimal energy input. Innovative synthetic materials, such as nitrogen-doped carbon microspheres and amine-functionalized silicas, offer specialized functionalities that enhance CO_2 selectivity and capacity. These materials often employ physical and chemical adsorption mechanisms, enabling them to capture CO_2 selectively even in mixed gas streams, such as those containing nitrogen. Magnetic and biodegradable adsorbents bring additional benefits through unique properties like ease of separation and compatibility with sustainable disposal processes. However, they may exhibit lower surface areas than other types, relying more on chemical and specific molecular interactions to achieve effective CO_2 capture. Lastly, some synthetic and chemically activated adsorbents, such as CaO-based nanoadsorbents and modified resins, are specifically tailored to offer high durability and stability over multiple regeneration cycles, making them suitable for industrial applications with high throughput and repeated use requirements.

Adsorption-based CO_2 capture has several advantages over other CO_2 capture methods, such as chemical absorption. In addition to lower energy requirements, adsorption processes can be easily tailored to specific industrial applications. Adsorbents can be modified to capture CO_2 from different gas streams, including those with varying concentrations of CO_2 or contaminants. Moreover, the modular nature of adsorption systems makes them suitable for integration into existing industrial infrastructures. However, the selection of an appropriate adsorbent material remains a challenge. Although materials such as activated carbon, zeolites, and MOFs exhibit excellent CO_2 capture performance, they can degrade in the presence of water vapor or other impurities common in industrial gas streams. Humidity, in particular, can reduce the efficiency of CO_2 adsorption by competing for active sites on the adsorbent surface. As such, future research should focus on the development of adsorbents

Table 9.2 Studies on carbon dioxide capture via adsorption method

Adsorbents	Fabrication method/process	Specific surface area ($m^2\ g^{-1}$)	Diffusion coefficient $D(CO_2)$ ($m^2\ s^{-1}$)	Amount of CO_2 capture (CO_2 adsorption uptake)	Experimental conditions/regeneration cycles	Adsorption mechanism	References
Biomass-pyrolysis derived char (WDPC)-KOH—HNO_3	–	755.7	–	~1–1.6 mol kg^{-1}	30 °C, 1 atm, and 100 min	Electrostatic attraction and acid–base interaction	[69]
WDPC-KOH—CO_2	–	1618.9	–	~1–1.6 mol kg^{-1}	30 °C, 1 atm, and 100 min	Electrostatic attraction and acid–base interaction	[69]
WDPC-KOH	–	970.3	–	~1–1.6 mol kg^{-1}	30 °C, 1 atm, and 100 min	Electrostatic attraction and acid–base interaction	[69]
Temperature swing adsorption (TSA) process operated on a metal–organic framework (MOF) adsorbent UTSA-16	Temperature Swing adsorption	–	–	–	25–40 °C	–	[65]
Carbon microspheres materials (PCS)	Ethanol-assisted hydrothermal technology	–	CO_2/N_2 (15:85) separation coefficient: 56.9	3.02 mmol g^{-1}	313 K, 1 bar, and 20 min/30 cycles	Physical adsorption, chemical adsorption	[70]

(continued)

Table 9.2 (continued)

Adsorbents	Fabrication method/ process	Specific surface area ($m^2\ g^{-1}$)	Diffusion coefficient $D(CO_2)$ ($m^2\ s^{-1}$)	Amount of CO_2 capture (CO_2 adsorption uptake)	Experimental conditions/ regeneration cycles	Adsorption mechanism	References
Nitrogen-groups modified porous carbon microspheres materials (MNPCS)	Ethanol-assisted hydrothermal technology		CO_2/N_2 (15:85) separation coefficient: 56.9	4.77 mmol g^{-1}	313 K, 1 bar, and 20 min/30 cycles	Physical adsorption, chemical adsorption	[70]
Covalently tethered amines were in-situ polymerized into mesoporous silica support, namely PE-SBA-15	In-situ polymerization	235.0	–	0.27 mmol g^{-1} for 400 ppm CO_2 concentration 0.86 mmol g^{-1} for 5% ppm CO_2 concentration 0.9 mmol g^{-1} for 10% ppm CO_2 concentration 0.94 mmol g^{-1} for 15% ppm CO_2 concentration	25 °C/10 cycles	Diffusion, chemical interactions, weak non-covalent complexes formation, and physisorption	[71]

(continued)

Table 9.2 (continued)

Adsorbents	Fabrication method/process	Specific surface area ($m^2\ g^{-1}$)	Diffusion coefficient $D(CO_2)$ ($m^2\ s^{-1}$)	Amount of CO_2 capture (CO_2 adsorption uptake)	Experimental conditions/regeneration cycles	Adsorption mechanism	References
Magnetic biodegradable adsorbent (NaOH@Chitosan-Fe_3O_4)	Alkali post-treatment	3.3	Diffusion coefficient: 2.850 $m^2\ s^{-1}$ Mass flux: 0.0042 mol m^{-2} s Mass transfer coefficient: 10.324 m s^{-1}	3.21 mmol g^{-1}	25 °C and 9 bar/10 cycles	Heterogeneous and surface adsorption, adsorbent carbonation, electrostatic excretion changes, chemical interactions, molecular penetration, and H-bonding	[72]
Corn starch	Chemical activation	711.0 1742.0	–	3.805 mmol g^{-1} 4.278 mmol g^{-1}	–	–	[73]
Sugarcane bagasse	Physical and chemical activation	622.0 873.0 1149.0	–	2.25 2.68 4.28	–	–	[74]
A400 Purolite as a solid resin base (solid adsorbent)	Impregnation method	–	PCO_2 (bar): 0.004 bar	58 mg g^{-1}	20 °C, 24 h, and 0.5 g loading/3 cycles	Diffusion of CO_2 molecules into the adsorbent pores	[75]

(continued)

Table 9.2 (continued)

Adsorbents	Fabrication method/ process	Specific surface area ($m^2\ g^{-1}$)	Diffusion coefficient $D(CO_2)$ ($m^2\ s^{-1}$)	Amount of CO_2 capture (CO_2 adsorption uptake)	Experimental conditions/ regeneration cycles	Adsorption mechanism	References
10% surfactants and 40% amines (T40/S10) sample	Impregnation method	1.1	PCO_2 (bar): 0.004 bar	84 mg g^{-1}	20 °C, 24 h, and 0.5 g loading/3 cycles	Diffusion of CO_2 molecules into the adsorbent pores	[75]
ZrO_2-stabilized CaO nanoadsorbent CaZr	One-pot MWCNT-template method	1.7	–	0.1 g $CO_2\ g^{-1}$ adsorbent	–/15 cycles	Physical adsorption	[76]
CaZr-CNT2.5	One-pot MWCNT-template method	2.5	–	0.1 g $CO_2\ g^{-1}$ adsorbent	–/15 cycles	Physical adsorption	[76]
CaZr-CNT5	One-pot MWCNT-template method	6.0	–	0.09 g $CO_2\ g^{-1}$ adsorbent	–/15 cycles	Physical adsorption	[76]
CaZr-CNT10	One-pot MWCNT-template method	4.9	–	0.18 g $CO_2\ g^{-1}$ adsorbent	–/15 cycles	Physical adsorption	[76]

Where CNT is carbon nanotubes; WDPC is biomass-pyrolysis derived char; MNPCS is nitrogen-groups modified porous carbon microspheres materials, MWCNT is multi-walled carbon nanotubes that maintain their performance in humid environments while being cost-effective and scalable for industrial applications.

9.2.4 CO_2 Capture by Oxy-Combustion

CO_2 can also be captured in pre-combustion, post-combustion, and oxy-fuel combustion. Each method captures CO_2 at a different stage in the combustion process, with unique operational advantages and challenges [77]. Oxy-fuel combustion is especially promising for near-zero CO_2 emissions, making it ideal for integration with large primary CO_2 emitters, such as power plants and the cement industry. Unlike pre- and post-combustion techniques, which capture CO_2 before or after combustion, oxy-fuel combustion uses nearly pure oxygen (around 95 mol %) mixed with recycled flue gases (RFG) instead of air [78]. This approach eliminates most nitrogen from the combustion process, which is crucial for minimizing nitrogen oxide (NO_x) emissions—a primary environmental pollutant. Nitrogen's absence reduces NO_x formation through the oxidation of N_2, which typically intensifies at high flame temperatures, especially those exceeding 1426 °C [79]. In oxy-fuel combustion, the low nitrogen content and the higher heat capacity of CO_2 compared to N_2 help stabilize flame temperatures and control laminar flame speed, reducing the need for dilution and additional fuel to maintain efficient combustion characteristics [80].

The oxy-fuel combustion setup generates a flue gas with high CO_2 concentrations, often around 90% on a dry basis, simplifying CO_2 capture and reducing reliance on energy-intensive separation methods. Additionally, the combustion process benefits from the reverse Zeldovich mechanism, where low nitrogen levels support NO reduction back to N_2, further lowering NO_x emissions. This CO_2-rich exhaust is suitable for purification through mechanical separation methods, which use temperature-based condensation rather than traditional absorption processes, making it economically and operationally efficient [81].

Flame temperature control is achieved by diluting pure oxygen with recycled flue gases, either in a wet or dry form. Wet recycling involves recirculating hot, humid flue gases before cleanup, retaining more energy but risking corrosion and erosion due to acidic gases and particulate matter. In contrast, dry recycling, which occurs after filtration and desulfurization, is generally preferred to mitigate these risks [82].

Oxy-fuel combustion's adaptability, particularly its capacity for retrofitting existing power plants, further enhances its economic viability. By leveraging existing air-based combustion infrastructure, oxy-fuel technology facilitates CO_2 capture with a high-purity stream and offers reduced capture costs [83]. Table 9.3 provides a detailed comparison of pre-combustion, post-combustion, and oxy-fuel combustion methods, illustrating the distinct benefits, challenges, and operational efficiencies each CCS approach offers.

Table 9.3 Comparison of combustion technologies

Combustion technology	Pre-combustion	Post-combustion	Oxy-fuel combustion
Efficiency penalty (%)	7–10	8–12	7–11
Separation task	CO_2/H_2	CO_2/N_2	O_2/N_2
Capture technology	Chemical absorption, such as alkanol amines Physical absorption, such as ionic liquids Membrane process using polymeric and ceramic membranes Physical adsorption including zeolites, activated carbon, alumina, and metal-organic frameworks Cryogenic distillation Gas hydrate crystallization Hybrid process	Chemical absorption, such as the amine scrubbing process Physical absorption, such as ionic liquids Physical adsorption, such as zeolites, activated carbon, and metal-organic frameworks Membrane process using polymeric and ceramic membranes Gas hydrate crystallization Cryogenic distillation Hybrid process	CO_2CPU process CLC technology Cryogenic distillation Physical adsorption, such as zeolites and activated carbon Membrane process using polymeric and ceramic membranes Ion transport membrane reactor Hybrid process
Advantage	Producing a carbon-free H_2-based fuel	Causing minimum changes in the original configuration of the existing power plant Its maintenance does not stop the operation and can be controlled Mature technology	Producing a highly concentrated CO_2 stream Low NO_X production Increasing both the convective and radiative heat transfer Capturing can be applied without additional material or solvent
Disadvantage/ Limitation	Required additional cost for the syngas generation More costly in comparison with the other methods	High NO_X production Low CO_2 purity in the flue gas leads to treating a high volume of gas to separate CO_2	Requiring approximate pure O_2

9.3 Introduction of Machine-Learning Approaches

In the rapidly evolving CCU field, ML approaches have emerged as transformative tools, enabling more efficient materials discovery and optimization [84]. This section delves into the data sources, databases, and various ML techniques that are becoming fundamental to the advancement of CO_2 capture and conversion technologies. [85] This section explores the foundational elements of these approaches, the latest computational tools available, and how these methodologies are being applied in innovative ways to address complex challenges in CCU.

9.3.1 Data Sources and Computational Tools: Building the Foundation for ML Models

The success of ML models in materials science heavily depends on the availability of high-quality, comprehensive datasets that provide detailed insights into material properties and behaviors. These datasets are fundamental for training models that predict material performance and facilitate high-throughput screening for advanced materials design [86]. One of the pioneering efforts in this space was the "Materials Genome Project," which laid the groundwork for the "Materials Project," a widely used database that collates density functional theory (DFT)-calculated properties of materials [87]. Other significant databases include the Automatic FLOW for Materials Discovery, the Open Quantum Materials Database, and the Materials Cloud, each contributing to the rapidly growing repository of material data available to researchers [87]. A noteworthy development is the Open Catalyst Project (OCP), a large-scale dataset created by Fundamental AI Research at Meta AI and Carnegie Mellon University [88]. This project offers over 260 million DFT calculation results, specifically designed to accelerate the discovery of new catalysts for renewable energy storage through AI-driven research. OCP's unique ability to predict relaxed energy states marks a significant advancement over traditional databases focusing mainly on structure–property relationships.

As computational power and methodologies evolve, tools such as xTB and DeepDFT are becoming popular alternatives to traditional DFT calculations [89]. These tools offer less computationally intensive ways to estimate electronic structures, opening new possibilities for large-scale simulations and the generation of extensive datasets. Additionally, ML interatomic potentials (MLIPs) have emerged as powerful tools, enabling simulations at a scale previously infeasible with ab initio molecular dynamics (MD), thus significantly lowering the computational barriers in quantum chemistry [90].

Despite these advances, much of the catalyst design research relies heavily on experimental data extracted from the literature, given the complexities associated with the computational modeling of behaviors. The integration of natural language processing (NLP) techniques is beginning to address these challenges, allowing for automated extraction of information from unstructured academic texts, thus enhancing the organization and standardization of data in this field.

9.3.2 Supervised Learning

Supervised learning remains a cornerstone in ML, where models are trained on labeled datasets to predict outcomes for new, unseen data [91]. This approach is particularly valuable in CO_2 capture and conversion, where accurately predicting material behavior under different conditions is crucial.

9.3.2.1 Traditional Machine Learning

Traditional ML algorithms, such as multiple linear regression (MLR), random forest (RF), gradient boosting, support vector machines (SVM), and Gaussian process regression (GPR), have been extensively applied in CCU research [92]. Feature engineering is pivotal in these models, as carefully crafted input features can significantly enhance model performance and reduce computational demands. Notably, radial basis function neural networks (RBFNNs) and classification and regression trees (CART) have demonstrated their utility in CCU applications [93], with the former excelling in approximation tasks and the latter in decision-making processes.

9.3.2.2 Deep Learning

Deep learning (DL), a subset of ML that uses deep neural networks (DNNs), has been rapidly adopted in materials discovery. Unlike traditional algorithms, DL methods reduce the need for extensive feature engineering by automatically extracting meaningful patterns from data during training [94]. For example, convolutional neural networks (CNNs) have been successfully employed in interpreting X-ray diffraction patterns, infrared spectra, Raman spectroscopy, and mass spectroscopy data, streamlining the analysis of complex materials [95]. Graph neural networks (GNNs), another DL innovation, are particularly adept at modeling the intricate relationships between atoms and molecules, making them highly effective in predicting chemical interactions and structure–property relationships [96]. Their application has shown promise in reducing the computational cost associated with traditional DFT calculations, making them invaluable in catalyst design and discovery.

9.3.3 Feature Selection and Explainability: Enhancing Model Insights

Feature selection is a critical step in developing efficient ML models to identify the most relevant features that improve prediction accuracy. Methods such as wrapper algorithms employ greedy strategies to optimize feature sets, though they may overlook complex feature interactions [97]. As ML models are often perceived as "black boxes," the push for explainable AI (XAI) has gained attention. XAI focuses on making model outputs more interpretable and providing insights into decision-making.

Tree-based models like decision trees (DT), RF, and XGBoost offer built-in mechanisms to assess feature importance, typically by measuring the reduction in impurity when features are used for splitting decisions [98]. Additionally, methods such as SHAP (SHapley Additive exPlanations) provide a game-theoretic approach to interpreting model predictions, assigning contribution scores to each feature [99]. However, it is crucial to balance algorithmic explanations with scientific domain knowledge to derive meaningful insights from the data.

9.3.4 Active Learning: Maximizing Efficiency in Data-Driven Research

Active learning (AL) has emerged as a valuable ML strategy to optimize model training while minimizing the number of labeled samples required [100]. By selectively querying the most informative samples from an unlabeled dataset, AL frameworks can significantly enhance model accuracy with fewer data points. This approach is particularly advantageous in materials science, where data collection and annotation can be costly and time-consuming [101]. Most AL strategies are pool-based, operating on an initial labeled dataset and iteratively selecting new samples for labeling based on model uncertainty. However, care must be taken to avoid overconfidence in model predictions, particularly when combining DL models with AL, as this can lead to a lack of diversity in queried samples.

9.3.5 Gaussian Process Bayesian Optimization: Enhancing Model Performance

Bayesian optimization, often coupled with Gaussian processes (GP), is a robust method for optimizing complex functions with minimal evaluations [102]. Like AL, Bayesian optimization can leverage uncertainty sampling to guide the selection of new samples, but its primary focus is refining model performance rather than merely labeling data [103]. Gaussian process Bayesian optimization (GPBO) specifically uses GPs to model the objective function, making it a powerful tool for exploring high-dimensional spaces efficiently and effectively [104]. ML approaches, from traditional algorithms to DL and AL strategies, are revolutionizing the landscape of CO_2 capture and conversion. By leveraging vast datasets, advanced computational tools, and innovative algorithms, researchers can accelerate the discovery and optimization of materials and catalysts, driving the field toward more sustainable and efficient solutions [104]. As these technologies evolve, their integration with domain-specific knowledge and experimental data will be key to unlocking their full potential in CCU applications [105].

9.4 AI Applied to CO_2 Capture

9.4.1 Machine-Learning-Driven Advancements in Amine-Mediated CO_2 Capture

Amine-mediated CO_2 capture is a cornerstone technology for mitigating emissions due to its significant CO_2 absorption capacity, approximately 4000 mmol L^{-1}, compared to water (1200 mmol L^{-1}) [106]. Despite the effectiveness of traditional amine-based systems, recent innovations in ML are revolutionizing how to predict and optimize CO_2 absorption processes, particularly by tailoring amine molecules for improved performance. This section explores the growing role of ML in enhancing our understanding and development of amine-based CO_2 capture systems, detailing various ML methodologies and their applications in predicting physicochemical properties and optimizing solvent design.

ML models, especially those based on supervised learning, have shown promising results in predicting the thermodynamic properties of CO_2 in amine solvents. For instance, back-propagation neural networks (BPNN) and RBFNN have been employed to model CO_2 solubility in various amine solvents like monoethanolamine (MEA), diethanolamine (DEA), and methyldiethanolamine (MDEA) [107]. These models have demonstrated excellent alignment with experimental data across various conditions, such as different amine concentrations, temperatures, and CO_2 partial pressures. The comparative analysis of BPNN and RBFNN models reveals that while both are effective, BPNNs tend to yield slightly lower prediction errors, showcasing their robustness in handling diverse input parameters.

Expanding on these foundations, additional studies have employed ML techniques like general regression neural networks (GRNN) to forecast solvent solubility, viscosity, and density [108]. By integrating extensive datasets from the literature, these models provide more comprehensive predictions, enhancing our ability to design effective and practical solvents in real-world applications. Zhang et al. [108] used 433 datasets to predict CO_2 solubility and viscosity by BPNN and CO_2 density by GRNN. The results indicate that BPNN with a single hidden layer and a limited number of hidden neurons effectively predicts CO_2 solubility and aqueous solution viscosity. Meanwhile, GRNN demonstrates superior performance in predicting the density of aqueous solutions. This ML-based predictive framework shows strong potential to provide fast and accurate predictions for CO_2-related thermodynamic properties in solution environments. This framework could be highly valuable in applications that require fast and reliable predictions, enhancing the efficiency of CO_2 capture and storage processes.

To further refine the predictive accuracy of CO_2 solubility in amine systems, more sophisticated ML approaches, such as adaptive neuro-fuzzy inference systems (ANFIS), least-squares support vector machines (LSSVM), and adaptive boosting with classification and regression trees (AdaBoost-CART), have been developed [109]. These models are trained using a broad spectrum of data points, including variations in solvent concentration, temperature, and pressure, to ensure their versatility. Yarveicy et al. [109] predicted the CO_2 solubility of piperazine in an aqueous solvent by artificial neural networks (ANN), ANFIS, LSSVM, and AdaBoost-CART. The AdaBoost-CART model, in particular, has emerged as a superior predictive tool, demonstrating enhanced accuracy over other neural-based models, especially when forecasting CO_2 solubility in solvents like piperazine and ethanolamine [110]. Moreover, an analysis of the importance of features within these models highlights that the dominant factors influencing CO_2 solubility differ among solvents. For instance, Pouryousefi et al. [110] showed that while temperature plays a crucial role for MEA and triethanolamine (TEA), DEA's solubility is more sensitive to pressure variations. Such insights are invaluable for developing tailored solutions for CO_2 capture, underscoring the importance of ML in advancing solvent design. Chen et al. [107] reported in-house models based on BPNN and RBFNN schemes to predict CO_2 density, viscosity, refractive index, heat capacity, thermal conductivity, and thermal diffusivity in quaternary systems, such as 4-(diethyl amino)-2-butanol and MDEA blended with MEA. BPNN and RBFNN models achieved reliable performance over wide ranges of input parameters, showing acceptable average absolute relative errors. However, The BPNN model slightly outperformed the RBFNN model, with lower average absolute relative and root-mean-square errors. Comparisons with eight other numerical models emphasized the accuracy of both neural networks, with the BPNN's effectiveness being particularly noteworthy. These results confirm the suitability of BPNN and RBFNN models for accurately capturing complex patterns in diverse parameter settings. Dong et al. [111] used 3935 datasets to predict the mass transfer coefficient of CO_2 in amine-mediated absorption considering 23 amines, demonstrating the potential of ML models to evaluate the physicochemical properties of these systems. The authors demonstrated that the BPNN models accurately predicted CO_2 absorption, with average absolute relative errors of less than 4%, closely matching experimental data. By adding parameters for amine properties and column characteristics, a generalized model was developed that accounts for all influencing factors, making it a practical tool for evaluating amine performance in CO_2 absorbers.

ML is not confined to predictive modeling but extends to the innovative design of new amine solvents. By integrating ML with kinetic experiments, MD simulations, and large-scale virtual screening, researchers can systematically discover and optimize amine molecules for enhanced CO_2 capture. For example, Orlov et al. [112] leveraged RF and XGBoost algorithms to predict quantitative structure–property relationships (QSPR) of amine solvents, ultimately identifying tertiary amines with high CO_2 capture capacities, as depicted in Fig. 9.3. The use of ML models to evaluate commercially available amines has led to the discovery of promising candidates, such as 1-ethyl-3-pyrrolidinol, which, when combined with piperazine, demonstrated a significant increase in CO_2 capture performance.

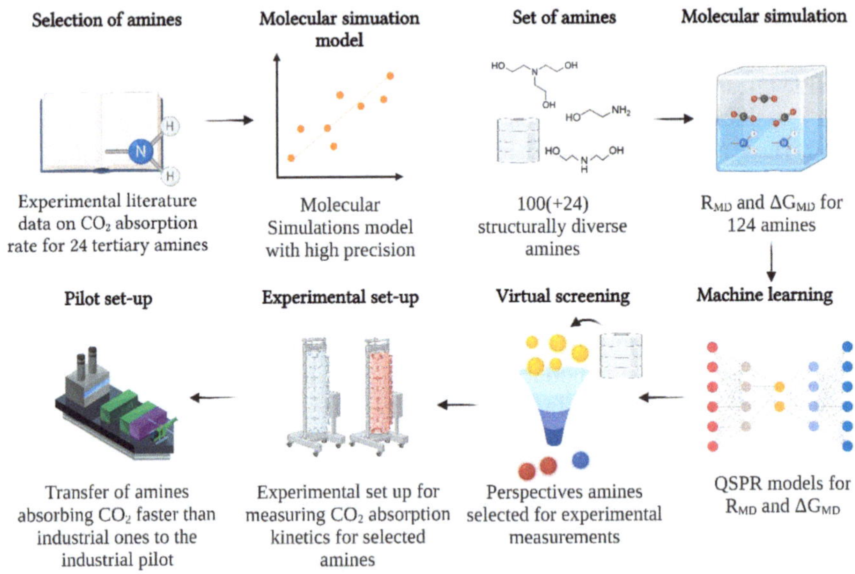

Fig. 9.3 Workflow illustrating the methodology to predict amine properties through the integration of machine learning (ML) and molecular dynamics (MD) simulations. The process begins with input data on molecular structures, which are subjected to MD simulations to extract physical and chemical properties. ML models are then trained on these properties to predict the performance of amines in CO_2 capture applications. This workflow streamlines the identification of high-performing amines by reducing the reliance on experimental screening. Reproduced with permission from Orlov et al. [112]. Copyright 2022 Springer Nature Publication

Additionally, ML applications have benefited the optimization of solvent regeneration energy, a critical parameter in CO_2 capture efficiency. GPBO has been effectively used to predict regeneration energies, bypassing the need for time-intensive first-principles calculations [113]. This approach has shown that certain solvents, like the novel water-lean amine solvent K_2Sol (2.8 GJ tCO_2^{-1}), exhibit significantly lower regeneration energies compared to conventional solvents like MEA (4.3 GJ tCO_2^{-1}), thus offering more energy-efficient solutions for CO_2 capture.

While the advancements in ML have significantly enhanced our ability to predict and optimize CO_2 capture processes, the complexity of amine chemistry presents ongoing challenges. The kinetics and capacity for CO_2 absorption are influenced by multiple factors, including the molecular structure of the amines, steric hindrance, temperature, and CO_2 pressure. For instance, the number of carbon atoms bonded to the nitrogen atom in an amine affects its pKa value, thereby influencing CO_2 binding strength. Furthermore, the proportions of carbamate anions, ammonium cations, and (bi)carbonate species vary depending on several conditions, complicating the development of precise predictive models.

Despite these challenges, ML continues to evolve, offering increasingly sophisticated algorithms that can capture these nonlinear relationships. Integrating ML with domain-specific knowledge and experimental validation remains key to advancing the field, driving the discovery of new amines, and optimizing CO_2 capture processes. ML is rapidly transforming the landscape of amine-mediated CO_2 capture, providing powerful tools for predicting solvent properties, optimizing design, and improving energy efficiency.

9.4.2 Machine Learning for CO_2 Capture in Ionic Liquids

Ionic liquids (ILs), as non-volatile molten salts composed of organic cations and organic/inorganic anions, have emerged as promising solvents for CO_2 capture due to their high CO_2 absorption capacity, thermal stability, and ease of regeneration. The tunable properties of ILs, depending on the specific structure and composition of their ions, make them versatile for various applications in CO_2 capture. This section explores the integration of ML techniques in predicting and optimizing the CO_2 capture performance of ILs, highlighting the advantages of ML in enhancing predictive accuracy over traditional models and advancing the design of high-performing ILs.

The application of ML to ILs for CO_2 capture began with efforts to demonstrate the potential of ML models in achieving superior predictive performance compared to traditional methods. Zhao et al. [114] utilized MLR and support vector machine (SVM) algorithms to predict the viscosity of ILs using a dataset of 1502 experimental viscosity measurements from the IL Thermo Database, incorporating features such as temperature, pressure, and quantum-chemical descriptors. The SVM model demonstrated significantly better predictive accuracy than the MLR model, as evidenced by a lower average absolute relative deviation.

Building on these findings, Venkatraman et al. [115] employed DT and RF models to elucidate the relationship between the structural characteristics of ILs and their CO_2 absorption capabilities. Using a comprehensive dataset of over 10,000 CO_2 solubility measurements for 185 ILs, the ML models outperformed the COSMOtherm model. COSMOtherm is a universal tool for predictive property calculation of liquids. It combines quantum chemistry and thermodynamics to compute various equilibrium thermodynamic properties or derived quantities. Similarly, Ghiasi et al. [116] developed a CART model to predict CO_2 solubility in various ILs based on an extensive dataset of over 5330 experimental data points from 66 ILs. The model incorporated key IL properties, such as critical temperature, critical pressure, and acentric factor, along with temperature and pressure conditions. The results showed that the CART model provided highly accurate predictions, with an average absolute relative deviation of just 0.04%. It outperformed other ML models, including Radial Basis Function Artificial Neural Networks (RBF-ANN), Multi-Layer Perceptron (MLP) ANN, ANFIS, and LSSVM, highlighting its effectiveness in modeling CO_2 solubility in ILs.

Furthermore, Aghaie et al. [117] categorized literature datasets into thermodynamic and molecular structural variables to develop ML models for predicting CO_2

solubility in ILs. Using techniques such as LSSVM, DT, RF, and MLR, they identified pressure, temperature, and the fraction of the highest occupied and lowest unoccupied molecular orbitals as critical factors influencing CO_2 capture capacity in ILs. These insights emphasize the capability of ML models to discern complex relationships and prioritize influential variables in CO_2 capture processes.

Advanced ML techniques have been employed further to refine the predictive models for CO_2 solubility in ILs. Song et al. [118] developed an ANN-based model combined with a group contribution (GC) approach to predict CO_2 solubility, utilizing a dataset of 10,116 measurements. The ANNGC model, which leverages chemical thermodynamics and engineering principles to estimate thermodynamic properties from smaller chemical groups, showed superior predictive performance, highlighting the effectiveness of integrating domain-specific knowledge with ML techniques.

In addition, the prediction of Henry's law constants, which are fundamental to understanding gas solubility in liquids, has been enhanced through ML models. Wu et al. [119] employed an MLP model trained on 160 experimental Henry's law constants for CO_2 in 32 ILs, finding that the MLP model outperformed RF and MLR models in predictive accuracy. Similarly, integrating ionic fragment contributions (IFC) with GC, SVM, and ANN models facilitated the development of QSPR models, successfully predicting CO_2 and N_2 solubility using large datasets [120]. These approaches offer valuable guidance for identifying promising ILs candidates for CO_2 capture.

The Conductor-like Screening Model for Real Solvents (COSMO-RS), a quantum chemistry-based thermodynamic method, has been used alongside ML to predict the solubility and selectivity of ILs, particularly in CO_2 and H_2S capture. Taheri et al. [121] developed a screening method that integrates COSMO-RS with ANN to predict solubility and selectivity, demonstrating the synergy between traditional computational chemistry methods and ML techniques in optimizing IL properties. Further advancing this integration, Kuroki et al. [122] used Henry's law constant datasets derived from COSMO-RS theory and DFT calculations to train a GPR model. The study found that electronic states of ionic species were the most significant factors influencing CO_2 capture capacity. Using this model, the authors screened 402,114 IL candidates, identifying those with superior CO_2 absorption properties, which were subsequently validated through synthesis and experimental analysis. The authors emphasize the importance of precisely designing the electronic states of ionic species through quantum chemistry calculations to optimize gas absorption properties. The systematic introduction of functional groups, elemental substitutions, and other ionic liquids (ILs) modifications can accelerate research by efficiently exploring the vast chemical space. Additionally, the proposed method for developing functional liquids is not limited to ILs but can also be extended to other gas-absorbing liquids, such as deep eutectic solvents or multicomponent systems with weak intermolecular interactions, offering a broader application for advancing gas absorption technologies.

Recent advancements in DL, particularly using graph neural network (GNN) architectures, have shown promising results in predicting CO_2 absorption in ILs. Jian et al. [123] compared traditional ML models with GNN architectures, including graph convolution networks, graph attention networks, and graph isomorphism networks. The GNN models, which directly process molecular structures as inputs, bypass the need for computationally expensive first-principles calculations, making them highly suitable for large-scale applications. This development marks a significant leap forward in the predictive modeling of IL properties, offering new avenues for research and discovery. To expedite the discovery of high-performance ILs, ML frameworks have been developed to model complex workflows. Zhang et al. [124] designed a framework combining multi-player Monte Carlo tree search (MP-MCTS) and recurrent neural networks (RNN) to generate and evaluate ILs in a parallelized manner. Figure 9.4 shows the schematic representation of the designed algorithm that combines MP-MCTS and RNN. The MP-MCTS algorithm searches the chemical space of ILs by iteratively generating and testing new structures, with tree nodes representing symbols of SMILES (simplified molecular input line entry systems) strings, while the RNN, trained on a dataset of 414,972 SMILES strings, predicts relevant properties. This framework effectively integrates CO_2 solubility, selectivity, and desorption characteristics, aiding in the discovery of novel ILs with high selectivity for CO_2 over other gases in flue gas and syngas applications.

Integrating ML into the study of ionic liquids for CO_2 capture has significantly enhanced the ability to predict and optimize IL properties, guiding the discovery of new, high-performing candidates. From employing traditional ML models to leveraging advanced DL architectures, ML techniques have provided robust frameworks for understanding and improving ILs' CO_2 capture capabilities.

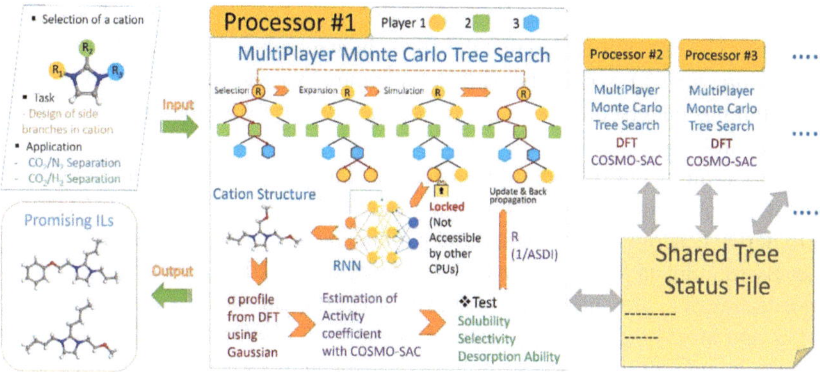

Fig. 9.4 A schematic of the algorithm to design an application-specific ionic liquid. With given inputs about the target application and cation, side branches of the cation core are generated by the algorithm combining the multi-player Monte Carlo tree search (MP-MCTS) and recurrent neural network (RNN). A new IL (a preset anion + cation with newly designed branches) undergoes a performance check for the target application, and then internal parameters in MP-MCTS are updated. With a structured parallel strategy, multiple attempts to design ILs are executed simultaneously, and the corresponding processors share the same tree state file. When a certain node is under atomistic simulation to evaluate the reward (R) value, the node is locked and not accessible by other processors. Reproduced with permission from Zhang et al. [124]. Copyright 2021 Elsevier

9.4.3 Machine Learning for CO_2 Capture in Deep Eutectic Solvents

DESs, which are composed of hydrogen bond acceptors (HBAs) and hydrogen bond donors (HBDs), have attracted considerable attention as potential solvents for CO_2 capture due to their unique properties, including low cost, simple preparation, non-volatility, and non-flammability. Unlike ionic liquids (ILs), where ionic interactions dominate, the primary interaction mechanism in DESs is hydrogen bonding. This characteristic gives DESs distinct advantages in CO_2 capture, such as tunable solubility and selectivity. This section explores the application of ML techniques to predict and optimize CO_2 capture performance in DESs, focusing on enhancing solubility, absorption rates, selectivity, and reducing energy requirements.

The effectiveness of DESs in CO_2 capture can be optimized by modulating their physical and chemical properties, such as gas solubility, viscosity, and absorption enthalpy. Theoretical approaches like QSPR models and the COSMO-RS have been widely used to predict the solubility and Henry's constant of CO_2 in DESs [125, 126]. These models help in understanding physical CO_2 absorption, where factors like Henry's constant and ionic structure play a critical role, and chemical CO_2 absorption, where the functional groups of DESs are pivotal. However, chemical absorption increases regeneration energy demand due to strong covalent bonds formed with CO_2.

ML models are increasingly being employed to predict CO_2 solubility in DESs. For example, Sagar et al. [127] developed a BPNN model to predict CO_2 solubility using structural properties, experimental conditions, and thermodynamic behavior as input data. The model revealed that the alkyl chain length and pressure have a linear relationship with CO_2 solubility, while temperature exhibited an inverse correlation. These findings underscore the complex interplay of variables affecting CO_2 capture efficiency in DESs.

Incorporating COSMO-RS-derived features has proven to be a highly effective strategy in enhancing ML model accuracy [128]. Wang et al. [129] used COSMO-RS to derive descriptors from structural parameters generated by Gaussian 09 software. By training an RF model on CO_2 solubility using 1011 datasets across various temperatures and pressures, they achieved a coefficient of determination (R^2) of 0.9758 and an average absolute relative deviation (AARD) of 7.76%. The representative scheme of the processes followed by the authors is shown in Fig. 9.5. By combining QSPR validation with comparisons to models reported in the literature, such as the COSMO-RS model, traditional thermodynamic models, and equations of state methods, the developed QSPR model using COSMO-RS-derived parameters as molecular descriptors is suggested to provide reliable predictions of CO_2 solubility in DESs. This model could be valuable for selecting DESs in CO_2 capture processes. Building on this work, Lemaoui et al. [130] emphasize the critical role of understanding CO_2 solubility in cations for optimizing CO_2 capture using DESs. The CO_2 solubility of DESs was accurately predicted by an MLP model, which utilized molecular descriptors from the COSMO-RS. A comprehensive dataset of 2327 data

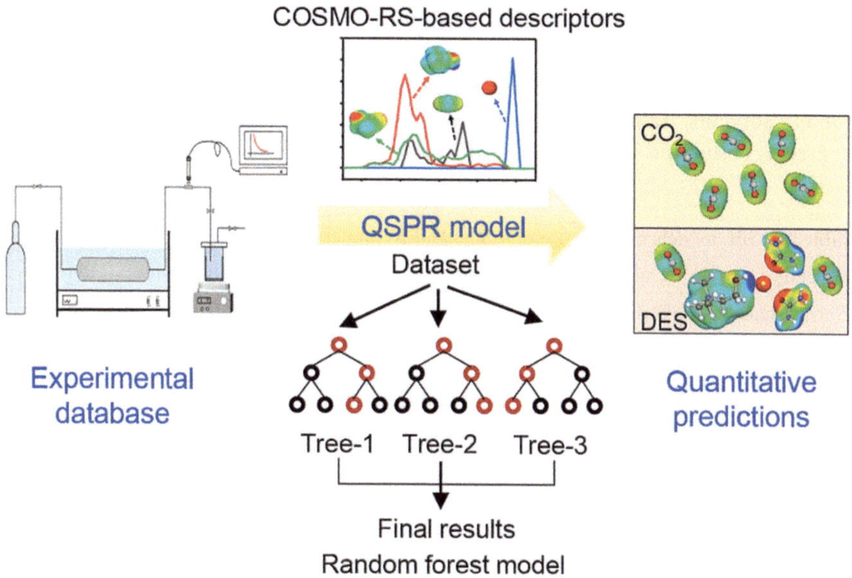

Fig. 9.5 Framework for establishing QSPR model based on machine-learning method. Reproduced with permission from Wang et al. [129]. Copyright 2021 KeAi

points from 94 unique DES mixtures was used, resulting in an excellent performance model with an R^2 of 0.986 ± 0.002 and an AARD of 4.504 ± 0.507. To facilitate the design of DESs with high CO_2 solubility, the model was applied to screen 1320 DES combinations, offering a valuable tool for the efficient development of new solvents. This approach reduces the need for costly and time-intensive experimental processes, promoting more rapid advancements in CO_2 capture technology.

Further advancements were demonstrated by Mohan et al. [131], who combined COSMO-RS with an ANN to predict CO_2 solubility in DESs. Their model, trained on 1973 data points for 132 DESs, achieved a superior AARD of 2.72%, significantly outperforming the COSMO-RS model alone, which had an AARD of 23.4%. These results highlight the efficacy of integrating ML models with computational chemistry approaches to design effective DESs for CO_2 capture.

ML techniques have also been employed to identify promising DES candidates for CO_2 capture. Makarov et al. [132] proposed a two-step screening method combining ML and thermodynamic modeling to discover DESs with high CO_2 absorption capacity. Initially, ML models were trained on a dataset of 162 DESs and 232 ILs, using transformer convolutional neural fingerprints and RF models with extended connectivity fingerprints to predict CO_2 absorption capacity. The SHAP method was then used to identify key descriptors, revealing that temperature and pressure significantly influence CO_2 absorption. In the second step, the ML predictions were integrated with the Redlich–Kister thermodynamic model to estimate the melting temperatures of the DESs. This approach enabled the identification of 1447 DESs as potential candidates with high CO_2 absorption capacity at room temperature. This two-step method demonstrates the power of combining ML models with classical thermodynamic approaches to streamline the discovery process for effective CO_2 capture solvents. Table 9.4 summarizes the studies in which ML models were used to predict certain properties of ILs/DESs.

Table 9.4 ML models used to predict ILS and DESs properties

Property	ML algorithm	Input	R^2 score	References
Surface tension	GENETIC-KNN FOA-KNN PSO-KNN	Temperature Quantitative structure	0.962 0.941 0.827	[133]
Melting point Density Viscosity Electrical conductivity Refractive index Surface tension	RFR DNN TransCNN TransCNF ASNN XGBoost	Quantitative structure	0.6 0.95 0.8 0.73 0.86 0.75	[134]
Electrical conductivity	KNN XGB Ada-DT Ada-RFR	Quantitative structure/ Thermodynamic properties Temperature	>0.9	[135]
CO_2 solubility in ILs	ANN-GC SVM-GC	ILs structure Temperature Pressure	0.9836 0.9783	[118]
CO_2 solubility in DESs	RFR	DESs molecular structure parameters Temperature Pressure	0.9500	[129]
Toxicity	ELM MLR SVM	ILs structure Electrostatic potential surface area Charge distribution area	0.940 0.849 0.874	[136]
Miscibility of CO_2 in ILs Supercritical CO_2 in ILs	MLP-ANN	Critical pressure of ILs Critical temperature of ILs Molecular weight of ILs System temperature System pressure	0.9987	[137]
Viscosity	MLR SVM	Quantum molecular description ($S_{\sigma\text{-profile}}$)	0.803 0.944	[114]

(continued)

Table 9.4 (continued)

Property	ML algorithm	Input	R^2 score	References
Viscosity	FFANN	Group contributions Temperature Pressure	0.972 (testing)	[138]
Viscosity	XGBoost	Morgan fingerprint that characterizes the quantitative structure of DESs Basic properties Water content		[139]

9.4.4 Integration of Machine-Learning and Thermodynamic Models in ILs/DESs Design for Carbon Capture

The development of ILs and DESs for carbon capture processes increasingly relies on the integration of ML with thermodynamic models. This combined approach is essential for accelerating the discovery and optimization of new solvents with high CO_2 capture efficiency. Generally, studies employing both ML and thermodynamic models follow a systematic process, as illustrated in Fig. 9.6.

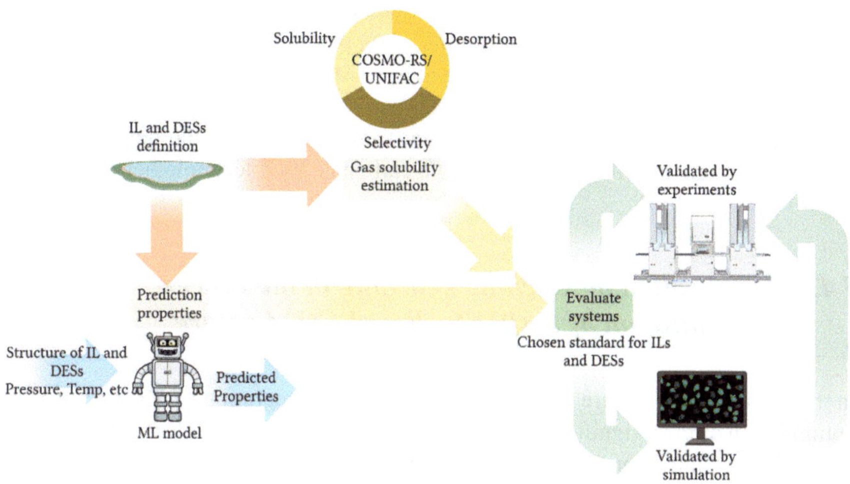

Fig. 9.6 Framework for the design and evaluation of ionic liquids (ILs) and deep eutectic solvents (DESs) for CO_2 capture. The process begins with defining the structure and properties of ILs and DESs, including parameters such as pressure and temperature. Using COSMO-RS/UNIFAC models, gas solubility, selectivity, and desorption properties are estimated. Machine-learning (ML) models predict physical and chemical properties of the solvents based on these inputs. The predicted properties are then evaluated using simulations and validated through experiments, ensuring the chosen ILs and DESs meet the required standards for CO_2 capture applications. This iterative process enables the efficient optimization of solvent systems, combining computational and experimental methodologies to achieve high-performance outcomes

The design process begins with the quantum-chemical optimization of the various cations and anions that constitute ILs and DESs. Quantum calculations provide detailed quantitative structural information about these ions, which is vital for understanding their behavior in carbon capture applications [140]. Next, experimental data and thermodynamic calculations are used to derive fundamental properties of these solvents, such as solubility and absorption enthalpy, which are then employed to train ML models [141].

ML-assisted search algorithms enable the exploration of a wide range of possible ion combinations, including novel ILs and DESs that have not yet been tested experimentally [142]. These models efficiently screen candidate solvents based on their potential for CO_2 capture. The most promising candidates can be experimentally verified or evaluated with advanced simulation software to confirm their performance in real carbon capture processes [121].

Two critical areas for future research may further enhance the accuracy and efficiency of this approach: (*i*) Improving solubility predictions: The current methods for predicting gas solubility in ILs and DESs, based on activity coefficient models, often lack precision. ML, with its capacity to model complex nonlinear relationships, holds the potential to refine these predictions, providing more accurate assessments of solubility and enhancing the overall design process and (*ii*) tailoring solvent design for specific applications: The requirements for ILs and DESs can vary significantly depending on the composition of the gas stream and the specific design of the carbon capture process. Future work should focus on applying inverse design techniques, supported by ML, to create solvents optimized for particular industrial conditions and gas compositions. By combining the strengths of ML and thermodynamic modeling, researchers can more quickly identify high-performance solvents. This reduces the time and cost associated with experimental trials and enables the development of tailored solutions for efficient and scalable carbon capture technologies.

9.4.5 Application of Machine Learning in the Process Optimization of ILs/DESs-Based CO_2 Capture

In the field of CO_2 capture using ILs and DESs, the integration of ML into process optimization offers a significant opportunity to enhance efficiency and reduce energy costs. The two most commonly used techniques for CO_2 capture in ILs/DESs-based processes are pressure swing and temperature swing technologies [143]. While the temperature swing method typically involves chemical absorption, the pressure swing method relies on physical absorption. ILs and DESs, due to their unique combination of functional groups, can perform both chemical and physical absorption of CO_2, meaning that their capture performance is highly dependent on the temperature and pressure conditions during the absorption and desorption phases of the process [144]. A typical ILs/DESs-based carbon capture system consists of an absorber and a desorber operating in a closed loop. The CO_2-rich gas stream enters the absorber, where the IL or DES captures CO_2 as it flows countercurrent to the gas. The CO_2-lean stream exits from the top, and the CO_2-rich stream is sent to the desorber for regeneration, either through high-temperature or low-pressure desorption [145]. The primary challenge faced by large-scale CO_2 capture plants is the high energy demand associated with these processes, making it essential to develop advanced tools for modeling and optimizing the process to reduce energy consumption and maximize CO_2 capture efficiency.

Traditional process optimization methods, such as Aspen Plus, Aspen HYSYS, and PRO II, rely on chemistry, hydrodynamics, and thermodynamics principles to simulate and optimize carbon capture processes [146]. However, these methods are computationally expensive and require an in-depth understanding of the underlying principles. The iterative nature of these solutions makes them complex and time-consuming, especially for large-scale optimizations [147]. As a data-driven approach, ML offers a promising alternative by reducing computational costs and time while maintaining comparable levels of detail. ML models can be trained using data from traditional simulation software or actual plant data, allowing for accurate process simulations without extensive real-world experiments [148]. For example, an artificial neural network (ANN) model was developed using simulated data from the CO_2SIM simulator to replicate the complex behavior of the CO_2 capture process with monoethanolamine (MEA) [149]. The ANN model was highly consistent with the physical model while requiring only a fraction of the computation time due to its non-iterative nature. This efficiency makes ML models well-suited for real-time optimization and control of CO_2 capture systems.

Another ML technique, RL models, has also been applied to optimize operating conditions for amine-based CO_2 capture systems. By training on simulated and plant data, RL models have demonstrated significant cost savings by optimizing key parameters such as reboiler duty, flow rates, and temperature. Combining ML models with optimization algorithms, such as sequential quadratic programming and genetic algorithms, has proven highly effective in identifying optimal operating conditions for CO_2 capture processes [150]. ML models can explore the effects of various operating parameters—such as absorption and desorption pressures, temperatures, flow rates, and gas compositions—on the overall performance of ILs/DESs in CO_2 capture [151]. These models can generate various performance metrics under different conditions, providing a powerful tool for optimizing the process. For instance, process optimization studies have shown that aprotic heterocyclic anion ILs can be economically competitive with advanced amine-based absorbents when subjected to higher regeneration temperatures and smaller absorption columns [152].

As the chemical and physical properties of different ILs/DESs vary, it is crucial to optimize the process parameters specific to each solvent. Studies have indicated that the optimal process conditions for each solvent can vary significantly, affecting both the capture efficiency and energy consumption. Future research should systematically compare the effects of different ILs/DESs on key process parameters to identify solvents that offer the best trade-offs between performance and energy demand. In addition to process optimization, the environmental impact of ILs/DESs throughout their entire life cycle—including production, usage, and recycling—should be carefully evaluated. While ILs and DESs offer promising CO_2 capture potential, few studies have addressed their life cycle environmental impact, particularly for mission-specific applications [153]. Life cycle assessment (LCA) studies should be conducted to ensure that ILs/DESs are effective in capturing CO_2 and environmentally sustainable throughout their use.

When modeling CO_2 separation processes involving ILs/DESs, various physical and transport properties must be considered, including density, viscosity, heat capacity, thermal conductivity, and mass transfer coefficients. ILs and DESs are often treated as pseudo-components in commercial process simulators, requiring users to define their properties and interactions with gas components (e.g., CO_2, N_2, O_2, H_2S) in phase equilibrium [154]. This introduces challenges in accurately simulating the behavior of ILs/DESs under different operating conditions. To address these challenges, ML models can be combined with simulation tools to develop comprehensive models that evaluate the impact of ILs/DESs properties on the entire CO_2 capture process [155]. However, it is important to note that no studies have been published that specifically evaluate the effect of feed gas variation on CO_2 capture performance using ML approaches. This gap in research represents an opportunity to improve the accuracy and predictive capability of models by considering the influence of different gas mixtures on the absorption process, which could lead to more effective and adaptable CO_2 capture systems.

9.4.6 A Framework for Carbon Capture Solvent Design Using Machine Learning

Based on a comprehensive literature review, this structured framework for designing carbon capture solvents using ML has been developed to address key challenges and enhance the efficiency of CO_2 capture processes. The framework consists of four essential stages:

1. **Data Acquisition and Feature Engineering (Stage 1):** This initial stage focuses on gathering diverse data on the molecular structures of solvents and key environmental parameters such as temperature, pressure, and solvent-specific characteristics that influence CO_2 capture efficiency. Selecting relevant features, such as solvent composition, molecular weight, and functional groups, is critical to creating accurate predictive models.
2. **Model Development (Stage 2):** In this stage, an ML model is developed to capture the nonlinear relationships between the input features and the desired solvent properties. Various ML algorithms, including BPNN, RF, and Artificial Neural Networks (ANN), are utilized based on their ability to model complex, high-dimensional data. Feature engineering is a crucial part of this stage, as it enhances the model's ability to accurately predict solvent properties by optimizing the input data representation.

3. **Property Prediction (Stage 3):** Once the model is trained, it is applied to predict key properties of the solvents, such as CO_2 solubility, viscosity, refractive index, heat capacity, and thermal conductivity. These predictions allow researchers to efficiently screen a wide range of candidate solvents, reducing the need for time-consuming experimental tests and focusing on the most promising candidates for CO_2 capture.
4. **Solvent Optimization (Stage 4):** The final stage involves selecting the optimal solvent from the pool of candidates based on the predicted properties. This process takes full advantage of ML's ability to analyze large datasets and uncover complex relationships, enabling rapid identification of high-performance solvents without needing first-principles-based simulations that are computationally expensive and slow.

By adhering to this framework, researchers can expedite discovering and optimizing ILs and DESs for CO_2 capture. This approach accelerates the development of more efficient carbon capture materials and provides a systematic methodology for overcoming the challenges of solvent design in a sustainable and resource-efficient manner, ultimately contributing to the advancement of carbon capture technologies.

9.5 AI for CO_2 Conversion

The urgency of climate change has driven researchers to seek innovative approaches to mitigate greenhouse gas emissions. One promising area is the conversion of carbon dioxide (CO_2) into valuable products, transforming a significant waste gas into a resource for sustainable energy. AI has opened new possibilities in optimizing CO_2 conversion processes, enhancing their efficiency and accuracy. Studies have shown that ML models, including ANFIS and various neural network architectures, are effective for predictive modeling in CO_2 solubility and conversion rates. These advances allow for more effective catalyst design and operational optimization, key factors in reducing fossil fuel dependence [156, 157]. Furthermore, AI frameworks excel at handling large datasets, identifying optimal conditions for conversion, and enabling significant strides toward sustainable energy solutions [6]. This section explores AI's transformative role in CO_2 conversion, underscoring its potential to support a carbon–neutral economy.

Addressing climate change requires innovative strategies to mitigate greenhouse gas emissions, especially CO_2, which is a primary contributor. CO_2 conversion addresses this challenge by turning waste CO_2 into valuable products, thus balancing emissions. AI plays an essential role by enhancing the efficiency and precision of these conversion processes. Studies have demonstrated AI's potential, such as using feed-forward neural networks and long short-term memory networks to predict future CO_2 emissions, helping guide policy decisions aimed at reducing emissions targets [14]. Another study highlighted the effectiveness of AI in optimizing catalytic processes for CO_2 conversion to light olefins, which offers a pathway to sustainable energy solutions [6]. Integrating AI in CO_2 conversion is thus a critical convergence of technology and environmental responsibility, accelerating the shift toward carbon neutrality.

The application of AI to CO_2 management represents a significant advancement in sustainable energy. Various AI techniques show strong potential for optimizing CO_2 capture and conversion processes. For example, Nassef et al. [156] demonstrated how ML models, such as ANFIS, could predict CO_2 solubility, achieving a 13.4% increase in absorption efficiency. Integrating AI with traditional chemical processes has improved catalyst design and operational efficiency, as seen in the successful use of Cascade Forward Neural Networks and Generalized Regression Neural Networks for CO_2 conversion into light olefins [157]. These advanced AI frameworks represent a transformative shift in carbon management strategies, addressing global challenges posed by climate change and providing real-time data insights that support informed decision-making.

The application of AI to CO_2 management is advancing sustainable energy by optimizing capture and conversion processes. ML techniques, such as ANFIS, Cascade Forward Neural Networks (CFNN), and GRNN, demonstrated significant potential in enhancing CO_2 solubility and conversion efficiency. For instance, ANFIS has successfully predicted CO_2 solubility, improving absorption efficiency by 13.4%, while CFNN and GRNN have optimized CO_2 conversion into light olefins [156]. These AI frameworks enhance catalytic performance, refine operational efficiency, and provide valuable real-time data insights, supporting informed decision-making in carbon management strategies. By integrating ML with traditional chemical processes, these advances contribute to climate change mitigation efforts and position AI as a critical tool for improving industrial processes and sustainability [2]. ANFIS has also proven valuable for optimizing solubility in capture solvents, uncovering relationships between operational parameters that enhance catalyst performance [157]. These AI-driven advances contribute to efficiency and align with broader sustainability goals, highlighting AI's potential to accelerate material discovery and optimize carbon management strategies.

GNNs have recently emerged as powerful tools in catalysis research, aiding in the precise discovery and optimization of CO_2 conversion catalysts. These networks excel in modeling complex molecular structures, accelerating the identification of promising catalytic materials. Through large datasets like those from the Open Catalyst Project, GNNs streamline the development of catalysts suited for electrochemical processes, reducing the need for trial-and-error experimentation [156]. Furthermore, AI enhances electrochemical systems by optimizing parameters like temperature and pressure, which improves catalyst efficiency. Integrating AI with traditional methods enables synergies that address the challenges of CO_2 emissions, establishing a foundation for sustainable practices and continuous improvement in catalyst design [4].

AI applications in CO_2 conversion represent a transformative approach to climate change and sustainability initiatives. With advanced algorithms like feed-forward neural networks (FFNN), ANFIS, and long short-term memory (LSTM) networks, researchers have achieved remarkable accuracy in forecasting CO_2 emissions, significantly lowering error rates in predictive models [157]. AI-driven ML frameworks have also reduced the time and cost of material experimentation, optimizing CO_2 conversion processes for practical use. Combining AI with current carbon capture technologies can enhance operational efficiencies, aligning with global climate objectives. Continued exploration into AI applications will refine CO_2 conversion techniques, support policymakers in creating effective emission reduction strategies, and lay a foundation for a more sustainable future [156].

AI's integration with CO_2 conversion technology offers unprecedented possibilities for innovation and efficiency, which is critical for combating climate change. For example, AI models like feed-forward neural networks and LSTM networks have substantially improved predictive accuracy in forecasting CO_2 emissions and conversion. However, data availability, experimental validation, and operational scalability remain. Addressing these issues will require ongoing research and collaboration among interdisciplinary stakeholders. Future initiatives should focus on unifying diverse AI methods to refine carbon capture and conversion processes, advancing sustainability efforts, and expediting the path to a carbon–neutral economy. Through interdisciplinary cooperation, we can fully leverage AI's potential to tackle environmental concerns [158].

9.5.1 AI in Catalysts Design

Integrating AI and ML transforms catalyst design by overcoming the limitations of traditional trial-and-error methods, which are time-consuming and costly. By combining ML with high-throughput experimentation, researchers can quickly identify and optimize catalysts. For example, the use of Bayesian optimization alongside DFT enables nuanced predictions and enhances the understanding of structure–activity relationships, accelerating the development of catalysts for essential reactions such as ammonia synthesis [159]. AI and ML also facilitate the analysis of large datasets, improving the accuracy of predictions related to catalytic performance and stability [160]. By accelerating discovery and enhancing the understanding of structure–activity relationships, AI and catalytic research synergize to create a data-driven approach that holds great promise for sustainable chemical production and efficiency improvements across various industrial processes. Furthermore, by refining synthesis processes through real-time data adaptation, AI contributes to more sustainable chemical manufacturing, minimizing resource waste and boosting reaction efficiency. This approach not only speeds up discovery but also supports the global demand for greener technologies and efficient industrial processes [161, 162].

The integration of ML into catalyst optimization represents a significant paradigm shift in materials science, introducing efficiencies in catalyst discovery and performance enhancement that were previously unattainable. With the help of advanced algorithms, researchers can analyze and predict catalytic behavior across extensive datasets, outperforming traditional methodologies. For example, high-throughput screening combined with ML has recently identified optimized Cu-Ag catalysts for ethylene epoxidation, highlighting how co-promoter loading enhances product selectivity [159]. Further, Bayesian optimization and Large Language Models have streamlined experimental design by efficiently iterating through potential catalyst compositions and reaction conditions, thus enabling a more thorough exploration of reaction parameters [160]. This combined use of ML and systematic experimentation accelerates development timelines and improves our understanding of structure–activity relationships, establishing a solid foundation for the rational design of catalysts aligned with sustainable chemical production goals.

ML techniques, such as Bayesian optimization and DL, have significantly transformed the process of identifying optimal catalyst compositions and reaction conditions, marking a substantial improvement over traditional method. Unlike conventional methods like the One Factor at A Time (OFAT) and Design of Experiments (DoE), which can overlook complex interactions between parameters, Bayesian optimization allows for systematic exploration of the compositional landscape through probabilistic modeling. This approach incorporates real-time experimental data into decision-making processes, yielding more dynamic and efficient research [159]. Additionally, DL algorithms can process extensive datasets from high-throughput experimentation and computational simulations, unveiling intricate structure–activity relationships that are beyond the reach of manual analysis [163]. Employing these advanced methods enables researchers to expedite catalyst discovery, improve the accuracy of performance predictions, and contribute to innovations in sustainable chemical production. This integration of AI into experimental chemistry highlights the need for ongoing methodological advancements in catalyst design.

Figure 9.7 illustrates an AI-driven workflow for catalyst discovery and optimization. It begins with high-throughput experimentation and computational methods (stage a), such as DFT, which generate extensive data on catalyst properties and reactions. This data is then processed by ML algorithms (stage b), such as Bayesian optimization and DL, which identify patterns and predict optimal catalyst configurations. As the AI models are refined through iterative analyses, the insights gained are used to synthesize and test improved catalyst compositions (stages c and d). This integrated approach enhances predictive accuracy and accelerates the development of efficient and sustainable catalytic materials for industrial applications (stages e and f).

The convergence of high-throughput experimentation (HTE) with AI marks a transformative development in catalyst design. HTE systematically screens vast libraries of materials under controlled conditions, generating extensive datasets that are crucial for optimization. When integrated with ML algorithms, this synergy allows researchers to quickly identify patterns and correlations within the data, significantly accelerating the discovery of novel catalysts with improved performance. This approach has proven to be much more efficient than traditional trial-and-error methods. Additionally, AI-driven predictive models derived from HTE datasets enhance the understanding structure–activity relationships, enabling more informed catalyst development decisions and streamlining the optimization process.

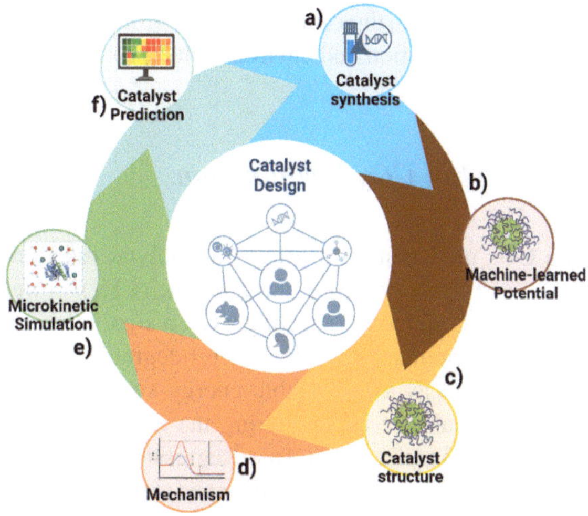

Fig. 9.7 Framework for AI-driven catalyst design. The process includes **a** catalyst synthesis, **b** application of machine-learned potentials, **c** structural optimization, **d** mechanistic studies, **e** microkinetic simulations, and **f** catalyst prediction. This iterative approach integrates experimental and computational methods to accelerate the discovery of high-performance catalysts

This integration accelerates catalyst discovery and fosters innovation in sustainable chemical production and energy conversion technologies [20].

For example, ML algorithms such as SVM and RF Regression can detect key structure–activity relationships, thus informing catalyst optimization [162]. Furthermore, high-throughput techniques, combined with AI, support the iterative refinement of models through AL approaches, enhancing predictive capabilities by leveraging real-time experimental data [160]. The resulting methodologies offer a comprehensive framework for addressing the inherent challenges in heterogeneous catalysis, leading to significant advancements in sustainable chemical production.

Integrating AI into catalyst design represents a pivotal evolution in materials science, enhancing catalyst development's efficiency and effectiveness. Another method used is high-throughput screening (HTS), coupled with ML algorithms, which enables researchers to navigate the complexities of catalyst properties and performance more proficiently than traditional methods. For instance, the ML combination with DFT has shown immense potential in identifying optimal catalyst compositions and configurations [162]. Innovative AI frameworks, such as those developed by Kazuki and Yuta, which employ Bayesian optimization alongside large language models, demonstrate how AI can streamline the design process while refining experimental outcomes [160]. As catalyst design increasingly relies on collaboration between AI, experimental validation, and theoretical insights, the pathway opens for sustainable solutions to global energy and environmental challenges [162].

Future research should aim to develop automated systems incorporating advanced ML techniques, such as large language models combined with Bayesian optimization, to streamline catalyst optimization workflows [160]. Automated systems promise to address the increasing demands for clean energy technologies by optimizing reactions like ammonia synthesis, improving efficiency and cost-effectiveness. Additionally, research to create universal descriptors may further enhance the accuracy of predictive models, thus contributing to groundbreaking advancements in sustainable chemical production and environmental protection [161].

9.5.2 AI in CO_2 Methanation

The CO_2 methanation process—a catalytic reaction that converts carbon dioxide (CO_2) and hydrogen (H_2) into methane (CH_4)—offers promise as a CCU strategy. This process supports recycling CO_2, a significant greenhouse gas while producing renewable methane, which can be integrated into existing natural gas infrastructures or used as a sustainable energy source. AI, specifically ML, has begun to play a critical role in optimizing the methanation process, with notable methods such as RF and Decision Tree Regression (DTR) showing exceptional predictive power for gas concentrations in experimental methanation setups. These techniques enable data-driven insights that enhance the efficiency and accuracy of catalytic reactions and have implications for improving methane detection and monitoring, further supporting carbon management objectives [164, 165].

CO_2 methanation represents a crucial pillar of CCU efforts in reducing atmospheric CO_2 while generating renewable energy. The thermochemical process ($CO_2 + 4H_2 \leftrightarrows CH_4 + 2H_2O$) typically occurs in fixed-bed reactors at elevated temperatures (250 °C to 400 °C). This reaction enables CO_2 recycling while facilitating the production of sustainable methane, which can be injected into existing energy infrastructures. The shift toward ML in predicting reaction conditions and catalyst performance has enhanced the optimization potential of methanation processes [164].

Applying ML in CO_2 methanation has transformed the optimization of catalytic processes, offering the ability to analyze complex datasets and predict reaction outcomes with remarkable precision. Studies have demonstrated that ML techniques, such as RF and DTR, effectively model gas outlet concentrations in methanation reactions, showing high predictive accuracy with a high correlation to experimental data ($R > 0.98$) and providing insights into catalyst performance. These models utilize empirical data to forecast performance metrics, such as gas concentrations, under varying conditions and provide insights into catalytic reactions, accounting for variables such as coke formation and temperature effects. Integrating data-driven methodologies with mechanistic models enhances the understanding of the complex relationships inherent in chemical processes and helps optimize methanation reactions [164, 165]. By applying ML techniques, researchers have also improved the accuracy of methane detection and monitoring. For instance, ML can process extensive datasets in methane monitoring systems, including multi-band retrieval techniques, to detect methane emissions more accurately, which is critical to comply with emerging environmental regulations [156]. ML facilitates the optimization of catalytic reactions and strengthens the data foundation needed for accurate carbon management and sustainable energy strategies [164].

Moreover, advanced ML techniques can integrate diverse datasets and forecast catalytic behavior in complex environments. Multi-band retrieval techniques, often used in methane monitoring, can be applied to catalytic methanation, where synthesizing experimental and predictive data enhances catalyst formulation and CO_2 conversion. These predictive methodologies are critical in advancing research on catalytic efficiency, a crucial element in CO_2 methanation as the field moves toward more sustainable energy solutions [165].

Reactor optimization for CO_2 methanation has significantly benefited from advanced AI techniques that streamline the identification of optimal operational parameters. RF and DTR, with high predictive precision, have demonstrated success in predicting gas outlet concentrations and facilitating the selection of ideal catalysts and reaction conditions. Additionally, GNNs have emerged as an effective tool for modeling the spatial–temporal dependencies within reactor systems, offering improved data-driven insights and mechanistic understanding. This integration of AI models enhances predictive accuracy while reducing computational demands, paving the way for real-time methanation process control and optimization. Deploying AI-based models fosters a more efficient reactor operation and control approach, leading to tighter process management and superior methanation outcomes. This shift allows researchers and engineers to manage complex reaction dynamics with enhanced agility and precision, directly supporting advancements in CO_2 capture and methane production [164].

By optimizing CO_2 methanation reactors, surrogate modeling offers a simplified yet effective approach to real-time control and performance monitoring. These models retain predictive accuracy while being less computationally demanding than traditional mechanistic approaches. Studies demonstrate the effectiveness of ML-based surrogate models, such as GNNs and Sparse Identification of Nonlinear Dynamics (SINDy), which capture essential data patterns and predict reaction outcomes in real-time [164]. Surrogate models (GNNs and SINDy) simplify reaction complexity and facilitate interpretability, providing insights into critical parameters that influence catalytic reactions and CO_2 conversion. By merging ML techniques, such as RF, with surrogate modeling, researchers can accurately predict gas concentrations, as demonstrated in methanation processes. This hybrid approach of data-driven and mechanistic models significantly advances the goal of sustainable CO_2 management and energy generation, positioning surrogate models as a cornerstone of future CO_2 methanation research and development [156].

The process flow diagram of CO_2 methanation integrated with ML visually represents the catalytic conversion of CO_2 and H_2 into renewable methane (CH_4) within a fixed-bed reactor. The critical components of the process include the input of CO_2 and H_2, followed by the reaction occurring in the reactor under elevated temperatures and pressures. ML algorithms, specifically RF and DTR, are integrated to analyze real-time data on gas concentrations and reactor conditions, enhancing predictive accuracy and optimizing catalytic performance. The output section illustrates the production of methane and water, while a feedback loop emphasizes how insights from ML inform real-time adjustments in reactor operation.

On the other hand, Fig. 9.8 shows the Greedy sampling neural network in sparse identification of nonlinear dynamical systems (GN-SINDy), an advanced adaptation of the classical SINDy framework, with additional steps that enhance model

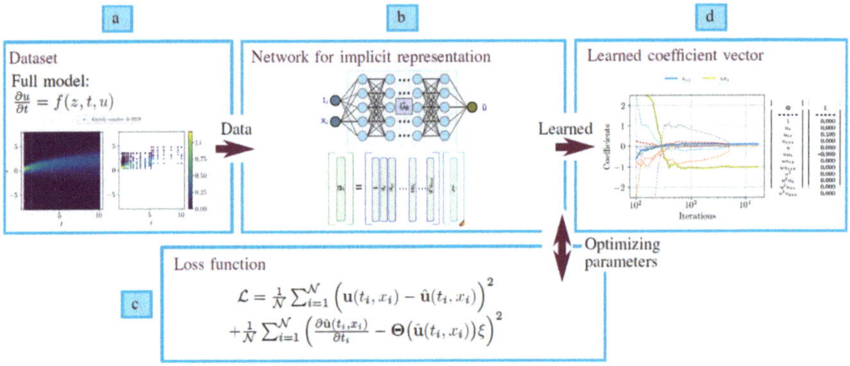

Fig. 9.8 Schematic diagram of Greedy Sampling Neural Network in SINDy (GN-SINDy) for PDE sparse identification. The figure illustrates **a** dataset generation from the full model, **b** the neural network for implicit representation, **c** the loss function for optimization, and **d** the learned coefficient vector, showing the evolution of parameters during training. Adapted from Peterson et al. [166], Copyright 2024 Authorea Preprints

accuracy and interpretability. This structured workflow is particularly valuable as an option for Power-to-X systems, specifically in CO_2 methanation reactors, where it aids in optimizing reactor performance by capturing key system dynamics. The first stage, step (a), uses the Q-discrete empirical interpolation (Q-DEIM) technique to sample the most informative data points from the dataset strategically. This method builds upon the earlier DEIM approach, refining it through a QR factorization with column pivoting to select critical data points carefully. In the context of CO_2 methanation, this focused "sensor placement" strategy is essential, as it ensures efficient sampling within the reactor, especially in two- or three-dimensional configurations. Capturing such critical data supports enhanced model accuracy, which is necessary for optimizing reactor processes in Power-to-X applications. In step (b), the sampled pairs representing time and position are input into a deep neural network (DNN). The DNN is trained to recognize and learn the intricate relationships between these spatiotemporal pairs and the observed values, such as temperature or gas concentration, within the reactor. By understanding these internal dynamics, the DNN generates a highly accurate approximation of the reactor environment, supporting analyses that are vital for fine-tuning CO_2 conversion processes. Step (c) continues to refine the model by optimizing its core parameters using stochastic gradient descent. This process minimizes discrepancies between predicted and actual values, with a loss function that balances accuracy in mapping data points while enforcing consistency in the DNN's predictions. This optimization is critical in CO_2 methanation, where small variations in reaction conditions can substantially impact efficiency and product yield. Finally, in step (d), the coefficient vector undergoes fine-tuning to produce a model that highly fidelity predicts and explains the original dataset. This capability is especially beneficial in CO_2 methanation applications, as it provides operators with a robust model that can interpret real-time data and offer insights into the system's behavior, facilitating more sustainable Power-to-X operations. The

GN-SINDy framework offers a powerful approach for Power—to—X systems with its systematic sampling, DL insights, and iterative adjustments. It produces a model that balances predictive accuracy with explanatory depth, making it ideal for optimizing complex processes in CO_2 methanation reactors and enhancing overall reactor performance.

These AI-driven methodologies optimize catalytic reactions and advance methane monitoring and emissions tracking, which are essential for regulatory compliance, such as in the EU Methane Act [164]. Additionally, advanced detection techniques, such as those used in methane plume identification via satellites, underscore the potential of ML to bolster methane emissions management. Combining AI with traditional mechanistic insights fosters a more profound understanding of catalytic reactions and operational efficiencies, supporting efforts to reduce emissions and promote renewable energy [165].

Future advancements in CO_2 methanation may focus on the synergistic application of ML and digital twins for real-time optimization and monitoring. These digital frameworks could leverage predictive ML models to monitor reactor performance under diverse operational conditions, providing real-time data insights for enhanced process control. Moreover, ongoing AI research may improve methane detection accuracy by integrating Short-Wave Infrared (SWIR) systems with ML, enabling comprehensive methane tracking and adherence to environmental standards [164]. The progression of AI in this field holds promise for both optimizing methanation efficiency and supporting global carbon reduction goals.

9.5.3 AI in CO_2 Transformation into Methanol

Converting CO_2 into methanol presents a promising strategy for carbon recycling and renewable energy production. Methanol serves as a clean-burning fuel, a key component in chemical production, and a potential medium for renewable energy storage, making it a versatile and sustainable alternative. Recent advances in AI, particularly through ML algorithms such as GPR and ANNs, have greatly enhanced our ability to optimize the CO_2-to-methanol conversion process [167, 168]. Furthermore, the integration of computer simulations with data-driven approaches is accelerating the identification of new materials for more efficient and sustainable CO_2 conversion [169, 170]. By using hydrogen generated from renewable sources, the methanol synthesis process can further reduce its carbon footprint, contributing to a cleaner and more sustainable energy system.

Integrating AI in predicting methanol production from CO_2 hydrogenation represents a significant advancement in catalysis. Using ML, researchers can analyze complex datasets from various catalytic processes, improving the accuracy and efficiency of methanol production predictions. For instance, a GPR model was developed using data from Aspen Hysys simulations, achieving an R^2 of 0.98078, which illustrates the model's predictive accuracy under different operating conditions [168]. Ensemble-tree models also show high predictive accuracy for methanol synthesis, underscoring AI's potential to optimize catalyst performance by adjusting variables such as pressure and temperature [170]. AI allows real-time monitoring of production processes and enhances decision-making, paving the way for sustainable industrial practices in methanol synthesis.

ML models like GPR and Artificial Neural Networks (ANN) are increasingly applied to optimize methanol conversion rates from CO_2 hydrogenation. These models can analyze complex data patterns and make high-accuracy predictions. GPR, for instance, has shown robust predictive capabilities, achieving a root mean square error (RMSE) of 0.83127 and an R^2 of 0.98078 even under uncertain conditions, as demonstrated by Samad et al. [167]. ANN also offers promise in refining methanol production by managing variables such as temperature and pressure, which are essential for conversion efficiency. A study by Daniel Chuquin-Vasco et al. [168] employed an ANN model with a dataset of 133 samples, achieving an RMSE of 0.0085, further confirming ANNs potential to optimize methanol synthesis. These machine-learning techniques underline the transformative impact of AI-driven methods in advancing renewable energy strategies and addressing environmental issues related to CO_2 emissions.

Optimizing catalytic processes through AI has become essential in efficiently transforming CO_2 into methanol. ML algorithms improve predictive accuracy in methanol synthesis by adapting to variable conditions in real-time. For example, studies using GPR models have demonstrated success, achieving an RMSE of 0.83127 and an R^2 of 0.98078 in predicting conversion rates, showcasing AI's adaptability in real-time applications [167]. Additionally, ensemble-tree models analyzed 1425 data points, revealing critical factors such as space velocity and pressure that impact methanol yield [170]. Integrating AI with catalytic processes enhances the understanding of reaction mechanisms and aids in developing responsive, intelligent systems that support sustainable methanol production.

Thermocatalytic reduction of CO_2 to methanol is crucial in reducing greenhouse gas emissions, where advanced catalysts play a pivotal role in enhancing reaction efficiency and yield. Copper-based catalysts are particularly effective for CO_2 hydrogenation [171]. AI also plays a transformative role in optimizing reaction conditions and catalyst performance by analyzing complex datasets generated in methanol synthesis. ML techniques identify the most promising catalysts quickly and accurately, accelerating the discovery process. AI-driven methods improve catalytic systems' efficiency, enabling better methanol yield prediction under varied conditions [168].

The integration of ML has deepened our understanding of catalytic mechanisms, enabling the identification of efficient pathways for CO_2 reduction, especially with catalysts like copper [169]. As AI and catalytic research continue to advance, the potential for optimizing methanol production increases, providing an impactful means of reducing CO_2 emissions. The synergetic use of AI in methanol production represents a cornerstone in sustainable energy and environmental conservation efforts, enabling us to recycle CO_2 into a valuable resource that supports a sustainable energy future.

Figure 9.9 presents a diagram outlining the key stages of transforming CO_2 into methanol, highlighting the integration of AI-driven optimization. Starting with CO_2 capture and renewable hydrogen production, the process moves to methanol synthesis, where catalysts facilitate the reaction between CO_2 and H_2. AI models continuously optimize conditions such as temperature and pressure to maximize yield.

Transforming CO_2 into methanol demonstrates the potential of AI to promote sustainable energy practices and contribute to climate change mitigation. ML techniques, such as Gaussian Process Regression and neural networks, help optimize methanol production by predicting conversion rates across varied conditions, as shown in research by Samad et al. [167]. Furthermore, AI in catalyst discovery is pivotal; data-driven approaches enable efficient catalytic pathway identification, enhancing process efficiency [169]. Innovative technologies will continue positioning AI as essential in advancing renewable methanol production, providing frameworks for optimizing catalyst performance and real-time process monitoring. AI's role in CO_2 transformation into methanol is thus a keystone strategy for reducing CO_2 emissions and utilizing resources sustainably.

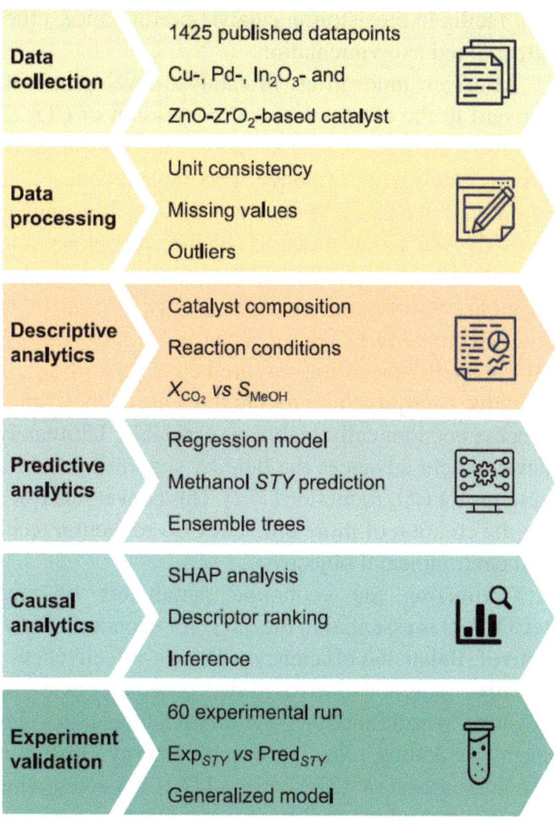

Fig. 9.9 AI-enhanced CO_2-to-methanol conversion process. The abbreviations included in the schematic are as follows: CO_2 conversion (X_{CO_2}), methanol selectivity (S_{MeOH}), methanol space–time yield (STY), SHapley Additive exPlanations (SHAP), experimental methanol space–time yield (Exp_{STY}), and predicted methanol space–time yield ($Pred_{STY}$). Adapted from Survana et al. [170], Copyright 2022 Elsevier

9.5.4 AI in Electrochemical Conversion of CO_2

Through advanced electrochemical processes, carbon dioxide (CO_2) can be turned into valuable products such as methanol. Researchers have begun exploring AI and ML to optimize critical parameters, such as catalyst performance and energy consumption, directly impacting CO_2 reduction efficiency. For example, studies have shown that AI models can significantly enhance the electrochemical CO_2 reduction process, improving product yield and reducing energy usage, as demonstrated in recent research [172]. Using biochar in direct carbon fuel cells (DCFCs) highlights how AI can address the complex, nonlinear behaviors of new materials, further expanding the potential of electrochemical systems [173]. Setting clear research objectives is crucial as the field progresses to leverage AI's potential and maximize its benefits in CO_2 conversion technologies.

Integrating AI in the electrochemical field has accelerated the development of CO_2 reduction technologies, driving down costs and enhancing operational efficiency. For instance, recent studies emphasize how AI can improve catalyst selection and optimize operational parameters, leading to higher conversion rates and more cost-effective processes [174]. AI's role is pivotal, as it enables data-driven insights that facilitate precision in catalyst performance, efficient process optimization, and streamlined experimentation.

AI-driven innovations in catalyst development have marked a significant step forward in the electrochemical conversion of CO_2. Researchers can use ML algorithms to analyze extensive datasets to uncover patterns that reveal the key factors driving catalyst performance. This capability accelerates the screening and selection of optimal materials, as shown in studies that apply extended adaptive hybrid functions (E-AHF) algorithms to enhance model accuracy and efficiency [172]. These algorithms, achieving $R^2 > 0.96$ in some instances, have optimized various catalysts, including N-doped porous carbons and Cu-Sn alloys, demonstrating high CO production selectivity (90%) that can be used in further processes as Fischer Tropsch [175]. By enhancing modeling and predictive capabilities, AI supports the design of more effective catalysts that yield high CO selectivity and energy efficiency, making the process economically viable and scalable. Ultimately, AI's contribution to catalyst development advances the field by supporting more sustainable, high-performance systems for CO_2 reduction [172]. This convergence of AI and electrochemistry allows for the creation of more efficient CO_2 reduction technologies, advancing economic and environmental objectives.

Optimizing the economic parameters of electrochemical CO_2 reduction (eCO_2RR) is essential in the quest for economically viable, sustainable energy alternatives. Balancing efficiency with cost-effectiveness is necessary to make eCO_2RR a realistic alternative to fossil fuel-based processes. Recent work using the extended adaptive hybrid functions algorithm demonstrates significant performance improvements, including reducing specific energy consumption to 9.95 kWh kg^{-1} and a production cost of $378 per ton [172]. By integrating ML models, researchers can monitor and adjust real-time parameters, maximizing yield and energy efficiency while maintaining economic feasibility [173]. AI's multi-objective optimization techniques are especially valuable in this context, as they allow for simultaneous enhancement of yield, conversion rates, and energy consumption, making eCO_2RR a viable solution for sustainable carbon conversion.

Multi-objective optimization approaches enable researchers to fine-tune process parameters to maximize CO_2 conversion rates and minimize energy requirements [172]. Techniques like the extended adaptive hybrid functions (E-AHF) algorithm have shown remarkable results in reducing energy usage and increasing the selectivity of CO_2 conversion reactions. Moreover, AI-driven optimization frameworks enable more effective management of the complex catalyst and operational parameters required to meet sustainability and economic targets [174]. This innovative use of ML reduces costs and energy demands and positions CO_2 electrochemical conversion as a feasible and efficient pathway to a lower-carbon future.

Figure 9.10 illustrates the electrochemical CO_2 conversion process, highlighting AI's role in optimizing each stage. CO_2 is captured and then enters an electrochemical cell where AI-driven models optimize catalyst selection and operating conditions to enhance conversion efficiency. Advanced algorithms like E-AHF are applied to improve catalyst performance and energy use, resulting in a high-yield, cost-effective product output.

Applying algorithms such as extended adaptive hybrid functions (E-AHF) and nondominated sorting genetic algorithm II (NSGA-II) has greatly optimized CO_2 reduction processes, achieving improvements in selectivity and economic viability [172]. In addition to performance enhancements, AI's predictive capabilities streamline catalyst development and operational adjustments, leading to cost-effective solutions [174]. As this field evolves, future research must prioritize developing robust AI frameworks that can handle the complexities of non-traditional fuels. Establishing

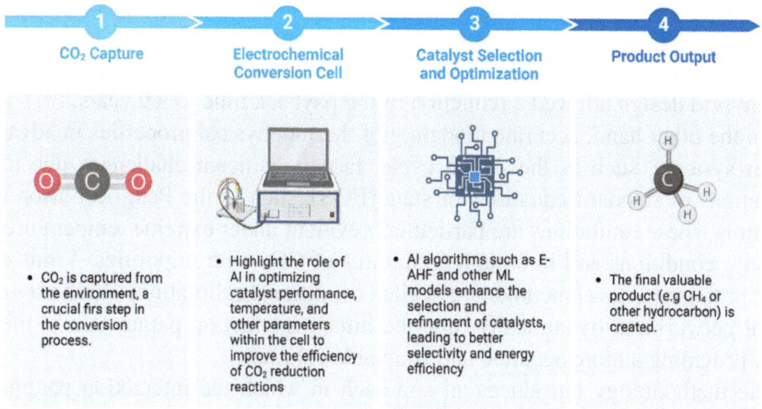

Fig. 9.10 AI-enhanced electrochemical CO_2 reduction process. The flowchart illustrates the steps involved in the conversion of CO_2 into valuable products, starting with CO_2 capture, followed by optimization of electrochemical conversion cells, catalyst selection and refinement using AI algorithms, and culminating in the production of valuable outputs such as CO or hydrocarbon

clear goals and fostering interdisciplinary collaboration will maximize AI's impact, allowing for more advanced and sustainable carbon conversion solutions that support a low-carbon, sustainable future [175].

9.5.5 Application of AI in the Allam Cycle for Sustainable Energy Generation

The Allam cycle has emerged as a promising solution in efficient and low-carbon power generation, leveraging supercritical carbon dioxide as the working fluid. However, the design, modeling and operation of these systems face significant challenges due to the extreme temperature and pressure conditions and the complex thermophysical interactions involved. This section explores how AI can transform Allam cycle systems, from improving their key components to efficient integration with other energy generation and storage technologies [176]. The presented case studies highlight the versatility and effectiveness of ML in the design, modeling, and operation of these systems, opening new opportunities to advance toward a sustainable and efficient energy transition.

In the study of Zou et al. [177], the authors presented the integration of ML in a hybrid system of the Allam cycle with the Goswami cycle that allowed the optimization of the combined power generation, heating and cooling, showing significant improvements in efficiency and economic profitability. The analysis focused on evaluating the economic and energy implications of this hybrid system through a comprehensive approach that included parametric analysis and an advanced optimization strategy based on AI and genetic algorithms. First, the performance of the system without the Goswami cycle was evaluated, achieving a power generation of 13.5 MW and 8.5 MW of heating, with a payback time of 16.32 years. Subsequently, the impact of integrating the Goswami cycle was analyzed, obtaining an increase in power generation to 15 MW, together with 5 MW of cooling and 6.7 MW of heating. This hybrid design allowed a reduction of the payback time to 9.9 years.

On the other hand, accurate modeling of thermophysical properties in advanced power systems, such as the Allam cycle, faces significant challenges due to the limitations of standard equations of state (EOS), such as the Peng-Robinson (PR) equation. These limitations are particularly evident under extreme temperature and pressure conditions and in complex mixtures of CO_2 with impurities. Vilnis et al. [178] proposed a novel methodology called conditional calibration, which uses deep neural networks to dynamically tune the binary interaction parameters in the PR EOS, providing a more accurate and adaptable solution.

The methodology introduces an approach in which the interaction parameters are not treated as fixed values nor exclusively temperature-dependent but as functions conditioned on multiple operating variables such as pressure, temperature, and composition. This is achieved through a multi-layer neural network that uses the

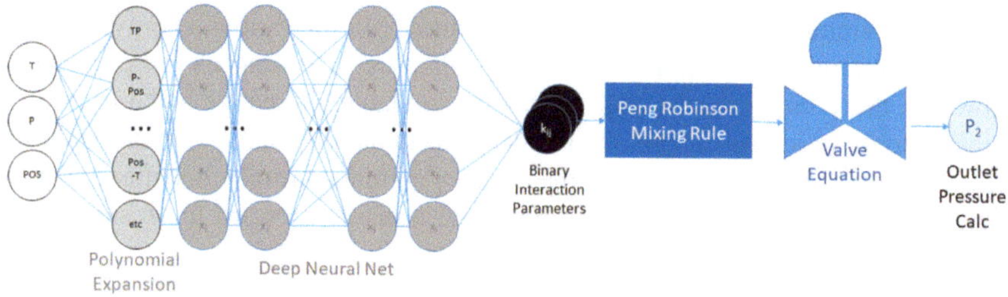

Fig. 9.11 Conditionally calibrated EOS predicting valve outlet pressure. Adapted from Vilnis et al. [178], Copyright 2018, The 6th International Supercritical CO_2 Power Cycles

differentiability of the process model to perform a gradient-based holistic calibration [179].

A diagram of the proposed method is shown in Fig. 9.11. The model uses three input variables: pressure, temperature, and valve position. These three input variables are expanded as a fourth-degree polynomial, giving 35 input variables for the forward propagation neural network. These variables are passed to a four-layer neural network with 64 hidden units per layer, and a final affine transformation yields the necessary $n(n - 1)$ k_{ij}'s, which are then reshaped and symmetrized into the Peng-Robinson EOS interaction matrix.

Finally, efficient storage and integration of renewable energy represent fundamental challenges in the transition toward sustainable energy systems. In this context, Power-to-X (PtX) technologies emerge as a strategic solution by converting excess renewable electricity into storable energy carriers, facilitating coupling across sectors. Wand et al. [180] introduced a novel Power-to-Gas-to-Power (PtG-PtP) system powered by the Allam cycle, which integrates carbon capture and water desalination into PtG processes. By combining exergy analysis with ML techniques such as artificial neural networks (ANN) and multi-objective optimization, their system maximizes energy and water production efficiency while reducing emissions, highlighting a novel approach to multi-energy system design in PtG technologies.

The system performance was evaluated using advanced simulation and optimization tools. Engineering Equation Solver (EES) was employed to perform detailed exergy and economic analysis, while MATLAB was used to implement an artificial neural network along with the moth-and-flame algorithm for multi-objective optimization. The study results showed that the integrated system offers an optimal freshwater flow rate of 378.5 kg s^{-1}, an energy efficiency of 39.03% and an operating cost rate of \$11,021 h^{-1}. Integrating waste heat from the Allam cycle into the Multiple Effect Desalination Unit enabled the maximization of thermal utilization and the reduction of the system's energy consumption. This study confirms that the combination of the Allam cycle with power-to-gas technology, and the use of neural networks as an optimization method, not only improves energy efficiency and economic viability but also simultaneously addresses carbon emissions and the intermittency of renewable energy [181].

Despite significant advances in implementing ML in various energy systems, specific literature on its application in the Allam cycle still needs to be explored. However, these initial results suggest that it is a viable alternative and a promising strategy to drive improvements in efficiency, sustainability and economic viability. This emerging approach underlines the need for continued research to maximize the benefits of ML and consolidate its role in the transition toward more sustainable energy systems [182].

9.6 Study Limitations and Recommendations

The integration of AI with CCU technologies represents a promising frontier in addressing climate change by enhancing the efficiency of CO_2 capture and conversion processes. While AI provides the transformative potential to optimize each stage of these processes—from improving capture methods to predicting catalytic efficiencies—numerous challenges must be overcome to fully realize its benefits at scale. Among the most pressing issues are data quality, model complexity, computational demands, and catalyst optimization, all impacting AI-driven CCU systems' reliability, scalability, and feasibility. Addressing these challenges will advance current CCU technologies and set the foundation for future developments in sustainable energy. For example, real-world CO_2 capture involves highly variable operating conditions across sectors, from power generation to chemical manufacturing. Datasets that are limited in representing these diverse conditions lead to AI models that lack generalizability, limiting their applicability in complex, dynamic environments. In addition, inconsistencies in data collection methods and differences in laboratory and field conditions can further reduce the accuracy of AI-driven models. Developing standardized protocols for data generation and enhancing data sharing among research institutions and industry stakeholders is essential to support AI-based optimizations in CCU technologies.

Model complexity and computational intensity also present significant barriers to the deployment of AI in CO_2 capture and conversion, especially when real-time decision-making is required. Advanced AI techniques, such as DL and RL, offer substantial improvements in prediction accuracy but at the cost of increased computational demand. These high processing power and memory requirements can be prohibitive, particularly for applications requiring on-site deployment where resources may be limited. Techniques such as transfer learning and model simplification have been introduced to mitigate these demands by enabling pre-trained models to adapt to new datasets without extensive retraining. However, these adaptations are not always straightforward, and significant trade-offs exist between model simplicity and performance accuracy. Achieving a balance that allows efficient, real-time application while maintaining high predictive accuracy is a critical challenge in making AI-driven CO_2 capture feasible at an industrial scale.

Catalyst design and optimization, a core aspect of CO_2 conversion, benefits significantly from AI, yet presents challenges. Catalysts facilitate the transformation of captured CO_2 into valuable products, such as fuels or chemicals, and the effectiveness of this transformation depends on the catalytic material's composition, structure, and stability under operating conditions. AI-driven models can accelerate the discovery of new catalysts by predicting promising compositions and reaction outcomes, yet real-world validation of these predictions remains a bottleneck. Catalyst design requires experimental validation to confirm AI predictions, which is often costly and resource-intensive. Moreover, catalytic performance is influenced by multiple factors, including the molecular structure and operating conditions, which add layers of complexity to predictive models. To address these challenges, AI must evolve to account for these intricate interactions and predict catalytically viable materials accurately and consistently. Ensuring that AI-generated predictions are experimentally reliable is essential to advancing catalyst research and improving CO_2 conversion technologies.

Integrating ML with high-throughput experimentation and advanced simulation techniques is a promising avenue for addressing these challenges. High-throughput methods allow researchers to quickly test numerous catalyst compositions and conditions, generating valuable datasets that can enhance AI models' training processes. For example, coupling ML with quantum mechanical approaches, such as DFT, enables more efficient simulations of catalyst behavior under various conditions. These combined techniques help predict key performance indicators such as catalytic efficiency, reaction yield, and durability, providing insights that guide experimental validation and ultimately streamline the catalyst discovery process. Developing such integrated workflows, where AI guides experimental design, marks a significant advancement in CCU research by enabling faster, more cost-effective discovery of viable catalysts for CO_2 conversion.

Another area with substantial potential for future advancements is using AI to improve the energy efficiency of CO_2 capture processes. Different methods of CO_2 capture, such as adsorption, absorption, and membrane separation, each have unique advantages and challenges related to energy consumption, scalability, and CO_2 purity levels. Absorption, for example, is widely used in industrial applications but often requires high energy input for solvent regeneration. AI can optimize these processes by analyzing data on operating conditions to identify the optimal parameters, such as temperature, pressure, solvent blends and solvent concentration that minimize energy requirements while maximizing CO_2 absorption. ML algorithms can further enhance process efficiency by predicting the behavior of various capture materials under different conditions, aiding in selecting materials that offer improved performance and lower energy costs. By enabling real-time adjustments based on data-driven insights, AI has the potential to significantly reduce the energy demands associated with CO_2 capture, which is crucial for scaling these technologies sustainably.

Future perspectives on AI-driven CO_2 capture and conversion focus on enhancing model interpretability and addressing scalability for industrial applications. Many AI algorithms, particularly those based on DL, are often seen as "black boxes," making it challenging to interpret their decision-making processes. As AI is central in optimizing CCU systems, developing XAI frameworks that provide transparent, interpretable insights will be essential. XAI techniques, such as SHAP and feature importance analysis, allow researchers to understand better the factors influencing model predictions, which is critical for gaining trust and facilitating the widespread adoption of AI in the industry. Furthermore, incorporating domain knowledge into AI models, such as chemical engineering and materials science insights, can help make AI models more interpretable and reliable in real-world applications.

For future developments in AI-driven CCU to be impactful globally, economic feasibility and policy support will be key. While AI enhances the efficiency of CCU processes, the initial costs associated with AI model development, deployment, and ongoing operation can be significant. Governments and industry stakeholders must collaborate to create policies and incentives that promote investment in AI-enhanced CCU systems. For instance, policies that provide subsidies or tax breaks for industries adopting AI in their carbon capture and conversion processes could encourage broader implementation. Additionally, establishing partnerships between academia, industry, and policymakers would foster innovation and streamline the commercialization of AI-driven CCU technologies, making them more accessible and economically viable.

9.7 Conclusions

This chapter has demonstrated the transformative potential of AI in revolutionizing CCU technologies, addressing critical challenges in efficiency, scalability, and sustainability. Through the integration of AI-driven methodologies, significant advancements have been made across various aspects of CCU processes, including absorption, adsorption, oxy-combustion, catalyst design, and the evaluation of CO_2 transformation into value-added products. In absorption processes, AI has significantly enhanced the design and performance of solvent systems by enabling the prediction of thermodynamic and kinetic properties, facilitating the development of novel formulations with higher CO_2 absorption capacities and reduced energy requirements for regeneration. ML models have provided insights into the complex interplay of factors affecting absorption efficiency, such as solvent composition, temperature, and pressure. Furthermore, real-time optimization powered by AI has minimized energy consumption and operational costs, making absorption technologies more viable for large-scale applications. In the realm of adsorption, AI has driven the discovery and optimization of advanced adsorbents, including activated carbons, zeolites, and metal–organic frameworks (MOFs). By leveraging predictive modeling, researchers have tailored adsorbent materials to specific operating conditions, enhancing their CO_2 capture capacities and resistance to degradation. AI has also enabled the precise tuning of material properties, such as surface chemistry and pore structure, ensuring higher selectivity and stability even in challenging environments like humid or high-pressure systems. Oxy-combustion technologies have also benefited from AI-driven advancements, particularly in optimizing the combustion process and the associated CO_2 capture. AI models have been instrumental in designing systems that efficiently recycle flue gases, stabilize flame temperatures, and minimize the production of NOx emissions. These improvements have increased the efficiency of oxy-combustion systems and reduced the need for energy-intensive post-combustion separation processes.

Catalyst design and evaluation represent another critical area where AI has made profound contributions. The inherent complexity of catalytic processes, especially in CO_2 conversion to value-added products such as fuels, polymers, and chemicals, has historically required extensive trial-and-error experimentation. AI has revolutionized this process by employing generative models, RL, and advanced simulations to predict catalyst performance and identify promising candidates. These methodologies have significantly reduced the time and resources required for catalyst development, enabling the discovery of materials with improved activity, selectivity, and stability. AI has also enhanced the evaluation and optimization of catalytic processes for CO_2 transformation. Predictive models have been employed to simulate reaction pathways, optimize reaction conditions, and assess the long-term performance of catalysts under various operating scenarios. These insights have facilitated the scaling of CO_2 conversion technologies, bridging the gap between laboratory-scale innovations and industrial applications. Additionally, integrating AI into catalyst evaluation has provided a deeper understanding of reaction mechanisms and catalyst deactivation phenomena, guiding the development of more robust and efficient catalytic systems.

Overall, this chapter underscores the role of AI in advancing CCU technologies, demonstrating its potential to address the pressing challenges of climate change and energy sustainability. By integrating AI-driven techniques into every stage of the CCU process, from material discovery to system optimization, researchers have paved the way for innovative, cost-effective, and environmentally sustainable solutions. The continued fusion of AI with experimental and industrial practices promises to unlock new opportunities for reducing carbon emissions and achieving global carbon neutrality. Combining insights from data science, chemical engineering, and environmental research, this interdisciplinary approach provides a comprehensive roadmap for the future of CCU technologies and their role in the sustainable energy transition.

Acknowledgements The authors gratefully acknowledge the financial support provided by Ministerio de Ciencia, Tecnología e Innovación—MinCiencias through the fiduciary FIDUCOLDEX within the framework of the program "Incremento del Grado de Madurez Tecnológico (TRL) de Sistemas Energéticos Sostenibles y Eficientes para la Transición Energética y la Reindustrialización del País" (Contract No. RC–393-2023). The authors also acknowledge the Universidad de Antioquia (UdeA) for the economic support.

Author Contribution **Oscar E. Medina involved in** conceptualization, data curation, formal analysis, investigation, methodology, validation, visualization, writing of the original draft**; Pedro Narvaez contributed to** formal analysis, investigation, writing of the original draft**; Andres A. Amell involved in** conceptualization, formal analysis, investigation, writing of reviewing and editing**; Diana López participated in** conceptualization, formal analysis, writing of reviewing and editing**; Alexander Santamaría involved in** conceptualization, formal analysis, investigation, writing of reviewing and editing.

Declaration of Competing Interest The authors declare that they have no known competing financial interests or personal relationships that could have appeared to influence the work reported in this paper.

References

1. Medina OE, Gallego J, Cespedes S, Nassar NN, Montoya T, Corteś FB et al (2022) Effect of pressure on thermo-oxidative reactions of saturates, aromatics, and resins (S-Ar-R) from extra-heavy crude oil. Fuel 311:122596
2. Medina OE, Gallego J, Acevedo S, Riazi M, Ocampo-Pérez R, Cortés FB et al (2021) Catalytic conversion of n-C7 asphaltenes and resins II into hydrogen using CeO2-based nanocatalysts. Nanomaterials 11(5):1301
3. Styring P (2019) Carbon dioxide utilization as a mitigation tool. Managing Global Warming. Elsevier; pp 527–51
4. Renda S, Martino M, Palma V (2024) CO_2 methanation over open cell foams prepared via chemical conversion coating. J Clean Prod 434:140221
5. Duyar MS, Treviño MAA, Farrauto RJ (2015) Dual function materials for CO_2 capture and conversion using renewable H2. Appl Catal B 168:370–376
6. Nassef AM, Olabi AG, Rezk H, Abdelkareem MA (2023) Application of artificial intelligence to predict CO_2 emissions: critical step towards sustainable environment. Sustainability 15(9):7648
7. Wirth DA (2017) The paris agreement as a new component of the UN climate regime. Int Organ Res J 12(4)
8. Medina OE, Amell AA, López D, Santamaría A (2025) Comprehensive review of nickel-based catalysts advancements for CO_2 methanation. Renew Sustain Energy Rev 207:114926
9. Ghaib K, Nitz K, Ben-Fares FZ (2016) Chemical methanation of CO_2: a review. ChemBioEng Rev 3(6):266–275
10. Mebrahtu C, Krebs F, Abate S, Perathoner S, Centi G, Palkovits R (2019) CO_2 methanation: principles and challenges. Studies in surface science and catalysis. Elsevier, pp 85–103
11. Medina OE, Moncayo-Riascos I, Franco CA, Cortés FB (2023) Molecular dynamic simulation and experiments on n-C7 Asphaltene pyrolysis under high-pressure conditions using ceria-based Nanocatalysts. ACS Appl Nano Mater 6(14):12805–12815
12. Olajire AA (2010) CO_2 capture and separation technologies for end-of-pipe applications–a review. Energy 35(6):2610–2628
13. Aaron D, Tsouris C (2005) Separation of CO_2 from flue gas: a review. Sep Sci Technol 40(1–3):321–348
14. Jerng SE, Park YJ, Li J (2024) Machine learning for CO_2 capture and conversion: a review. Energy and AI, 100361
15. Mateus F, Teixeira P, Lopes JM, Henriques C, Bacariza C (2023) CO_2 methanation on Ni catalysts supported over activated carbons derived from cork waste. Energy Fuels 37(12):8552–8562
16. Vita A, Italiano C, Pino L, Frontera P, Ferraro M, Antonucci V (2018) Activity and stability of powder and monolith-coated Ni/GDC catalysts for CO_2 methanation. Appl Catal B 226:384–395
17. Fu H, Lian H, Al doping enhances Ni/Ceo2 low-temperature CO_2 Methanation: revealing how to adjust catalyst reaction pathway and metal-support strong interaction to improve catalyst performance. Available at SSRN: 4502292
18. Abate S, Mebrahtu C, Giglio E, Deorsola F, Bensaid S, Perathoner S et al (2016) Catalytic performance of γ-Al_2O_3–ZrO_2–TiO_2–CeO_2 composite oxide supported Ni-based catalysts for CO_2 methanation. Ind Eng Chem Res 55(16):4451–4460

19. Ye R, Ma L, Hong X, Reina TR, Luo W, Kang L et al (2024) Boosting Low-Temperature CO_2 Hydrogenation over Ni-based Catalysts by Tuning Strong Metal-Support Interactions. Angew Chem 136(3):e202317669
20. Eko AS, Catalyst design: a narrative study on artificial intelligence
21. Zhang C, Sun S, Xu S, Wu C (2022) CO_2 capture over steam and KOH activated biochar: effect of relative humidity. Biomass Bioenerg 166:106608
22. Song C (2006) Global challenges and strategies for control, conversion and utilization of CO_2 for sustainable development involving energy, catalysis, adsorption and chemical processing. Catal Today 115(1–4):2–32
23. Wang X, Liu H, Li Y, Yang X, Gao F, Wang X et al (2023) Metal-organic frameworks for C2H2/CO_2 separation: recent development. Coord Chem Rev 482:215093
24. Podder J, Patra BR, Pattnaik F, Nanda S, Dalai AK (2023) A review of carbon capture and valorization technologies. Energies 16(6):2589
25. Yadav S, Mondal SS (2019) A complete review based on various aspects of pulverized coal combustion. Int J Energy Res 43(8):3134–3165
26. Dai Z, Deng L (2023) Membranes for CO_2 capture and separation: Progress in research and development for industrial applications. Separation and Purification Technology, 126022
27. Cichowska-Kopczyńska I, Nowosielski B, Warmińska D (2023) Deep eutectic solvents: properties and applications in CO_2 separation. Molecules 28(14):5293
28. Mao J, Li C, Yun Y, Liu J, Yang W, Li M et al (2024) Biphasic solvents based on dual-functionalized ionic liquid for enhanced post-combustion CO_2 capture and corrosion inhibition during the absorption process. Chem Eng J 481:148691
29. Lin M, Zhang Y, Yan G, Yu P, Duan S, Chen H et al (2023) Design and analysis of cryogenic CO_2 separation from a CO_2-rich mixture. Energy Sci Eng 11(7):2253–2266
30. Yerumbu N, Sahoo RK, Sivalingam M (2023) Multiobjective optimization of membrane in hybrid cryogenic CO_2 separation process for coal-fired power plants. Environ Sci Pollut Res 30(50):108783–108801
31. Cole GM, Greene JM, Quinn JC, McDaniel B, Kemp L, Simmons D et al (2023) Integrated techno-economic and life cycle assessment of a novel algae-based coating for direct air carbon capture and sequestration. J CO_2 Utilization 69:102421
32. Rupesh K, Prahaaladhan V, Matheshwaran S, Rahul S, Sudalai S, Arumugam A (2024) Carbon dioxide capture using algae. Advances and technology development in greenhouse gases: emission, capture and conversion. Elsevier; pp 331–50
33. Chen B, Xie H, Shen L, Xu Y, Zhang M, Zhou M et al (2023) Covalent organic frameworks: the rising-star platforms for the design of CO_2 separation membranes. Small 19(17):2207313
34. Song H, Peng Y, Wang C, Shu L, Zhu C, Wang Y et al (2023) Structure regulation of MOF nanosheet membrane for accurate H2/CO_2 separation. Angew Chem 135(17):e202218472
35. Dai Y, Niu Z, Luo W, Wang Y, Mu P, Li J (2023) A review on the recent advances in composite membranes for CO_2 capture processes. Sep Purif Technol 307:122752
36. Rath GK, Pandey G, Singh S, Molokitina N, Kumar A, Joshi S et al (2023) Carbon dioxide separation technologies: applicable to net zero. Energies 16(10):4100
37. Lu G, Wang Z, Bhatti UH, Fan X (2023) Recent progress in carbon dioxide capture technologies: a review. Clean Energy Sci Technol 1(1):32
38. Allangawi A, Alzaimoor EF, Shanaah HH, Mohammed HA, Saqer H, El-Fattah AA et al (2023) Carbon capture materials in post-combustion: adsorption and absorption-based processes. C 9(1):17
39. Song C, Liu Q, Deng S, Li H, Kitamura Y (2019) Cryogenic-based CO_2 capture technologies: state-of-the-art developments and current challenges. Renew Sustain Energy Rev 101:265–278
40. Sharif M, Fan H, Wu X, Yu Y, Zhang T, Zhang Z (2023) Assessment of novel solvent system for CO_2 capture applications. Fuel 337:127218
41. Ruan J, Ye X, Wang R, Chen L, Deng L, Qi Z (2023) Experimental and theoretical study on efficient CO_2 absorption coordinated by molecules and ions of DBN and 1, 2, 4-triazole formed deep eutectic solvents. Fuel 334:126709

42. Hara N, Taniguchi S, Yamaki T, Nguyen TT, Kataoka S (2023) Bi-objective optimization of post-combustion CO_2 capture using methyldiethanolamine. Int J Greenhouse Gas Control 122:103815
43. Yin Q, Mao W, Chen D, Song C (2023) Effect of adding tertiary amine TMEDA and space hindered amine DACH on the CO^2 chemical absorption-microalgae conversion system. Energy 263:125726
44. Ding Y, Ma L, Yang X, Zhu X, Wang H, Cheng M et al (2023) Anhydrous multi-hybrid absorbent with low viscosity and high regeneration efficiency for post-combustion CO_2 capture. Energy 263:125742
45. Bai H, Zheng Y, Yang R (2016) Recyclable liquid-like POSS derivatives with designed structures and their potential for CO_2 capture. Mater Des 99:145–154
46. Li P, Chen H, Schott JA, Li B, Zheng Y, Mahurin SM et al (2019) Porous liquid zeolites: hydrogen bonding-stabilized H-ZSM-5 in branched ionic liquids. Nanoscale 11(4):1515–1519
47. Ochedi FO, Yu J, Yu H, Liu Y, Hussain A (2021) Carbon dioxide capture using liquid absorption methods: a review. Environ Chem Lett 19:77–109
48. Borhani TN, Wang M (2019) Role of solvents in CO_2 capture processes: The review of selection and design methods. Renew Sustain Energy Rev 114:109299
49. Ghodile SB, Chapter-5 recent advances in carbon capture technologies: absorption and adsorption. Chief Editor, 77
50. Sanni SE, Oni BA, Okoro EE (2024) Physical and hybrid solvents for natural gas sweetening: Ethers, pyrrolidone, methanol and other sorbents. Advances in natural gas: formation, processing, and applications. Volume 2: Natural Gas Sweetening. Elsevier; pp 115–33
51. Gautam A, Mondal MK (2023) Review of recent trends and various techniques for CO_2 capture: Special emphasis on biphasic amine solvents. Fuel 334:126616
52. Nwaoha C, Saiwan C, Tontiwachwuthikul P, Supap T, Rongwong W, Idem R et al (2016) Carbon dioxide (CO_2) capture: Absorption-desorption capabilities of 2-amino-2-methyl-1-propanol (AMP), piperazine (PZ) and monoethanolamine (MEA) tri-solvent blends. J Natural Gas Sci Eng 33:742–750
53. Valluri S, Kawatra SK (2021) Use of frothers to improve the absorption efficiency of dilute sodium carbonate slurry for post combustion CO_2 capture. Fuel Process Technol 212:106620
54. Zhang X, Zhu Z, Sun X, Yang J, Gao H, Huang Y et al (2019) Reducing energy penalty of CO_2 capture using Fe promoted SO42–/ZrO2/MCM-41 catalyst. Environ Sci Technol 53(10):6094–6102
55. Pishro KA, Murshid G, Mjalli FS, Naser J (2020) Investigation of CO_2 solubility in monoethanolamine hydrochloride based deep eutectic solvents and physical properties measurements. Chin J Chem Eng 28(11):2848–2856
56. Li H, Guo H, Shen S (2020) Low-energy-consumption CO_2 capture by liquid–solid phase change absorption using water-lean blends of amino acid salts and 2-alkoxyethanols. ACS Sustain Chem Eng 8(34):12956–12967
57. Sema T, Naami A, Fu K, Edali M, Liu H, Shi H et al (2012) Comprehensive mass transfer and reaction kinetics studies of CO_2 absorption into aqueous solutions of blended MDEA–MEA. Chem Eng J 209:501–512
58. Chen G, Chen G, Peruzzini M, Zhang R, Barzagli F (2022) Understanding the potential benefits of blended ternary amine systems for CO_2 capture processes through 13C NMR speciation study and energy cost analysis. Sep Purif Technol 291:120939
59. Raganati F, Miccio F, Ammendola P (2021) Adsorption of carbon dioxide for post-combustion capture: a review. Energy Fuels 35(16):12845–68
60. Yin G, Liu Z, Liu Q, Wu W (2013) The role of different properties of activated carbon in CO_2 adsorption. Chem Eng J 230:133–140
61. Kumar S, Srivastava R, Koh J (2020) Utilization of zeolites as CO_2 capturing agents: advances and future perspectives. J CO_2 Utilization 41:101251
62. Ghanbari T, Abnisa F, Daud WMAW (2020) A review on production of metal organic frameworks (MOF) for CO_2 adsorption. Sci Total Environ 707:135090

63. Giraldo LJ, Medina OE, Ortiz-Perez V, Franco CA, Cortés FB (2023) Enhanced carbon storage process from flue gas streams using rice husk silica nanoparticles: an approach in shallow coal bed methane reservoirs. Energy Fuels 37(4):2945–2959
64. Medina OE, Galeano-Caro D, Brattekås B, Pérez-Cadenas AF, Carrasco-Marín F, Cortés FB et al (2023) Simultaneous CO_2 adsorption and conversion over Ni–Pd supported CeO_2 nanoparticles during catalytic n-C7 asphaltene gasification. Fuel 342:127733
65. Peh SB, Farooq S, Zhao D (2023) Techno-economic analysis of MOF-based adsorption cycles for postcombustion CO_2 capture from wet flue gas. Chem Eng Sci 268:118390
66. Dhoke C, Zaabout A, Cloete S, Amini S (2021) Review on reactor configurations for adsorption-based CO_2 capture. Ind Eng Chem Res 60(10):3779–3798
67. Bramsiepe C, Sievers S, Seifert T, Stefanidis G, Vlachos DG, Schnitzer H et al (2012) Low-cost small scale processing technologies for production applications in various environments—mass produced factories. Chem Eng Process 51:32–52
68. Shahkarami S, Azargohar R, Dalai AK, Soltan J (2015) Breakthrough CO_2 adsorption in bio-based activated carbons. J Environ Sci 34:68–76
69. Jiang D, Li H, Wang S, Cheng X, Bartocci P, Fantozzi F (2023) Insight the CO_2 adsorption onto biomass-pyrolysis derived char via experimental analysis coupled with DFT calculation. Fuel 332:125948
70. An X, Yang J, Qi G, Li T, Zhao K, Fu D (2023) Functional molecular engineering hierarchical pore-interface based on thermodynamic-kinetic synergy strategy for efficient CO_2 capture and separation. Fuel 332:126243
71. Al-Absi AA, Domin A, Mohamedali M, Benneker AM, Mahinpey N (2023) CO_2 capture using in-situ polymerized amines into pore-expanded-SBA-15: Performance evaluation, kinetics, and adsorption isotherms. Fuel 333:126401
72. Helmi M, Moazami F, Ghaemi A, Hemmati A (2023) Synthesis, characterization and performance evaluation of NaOH@ Chitosan-Fe_3O_4 as an adsorbent for CO_2 capture. Fuel 338:127300
73. Nazir G, Rehman A, Park S-J (2021) Role of heteroatoms (nitrogen and sulfur)-dual doped corn-starch based porous carbons for selective CO_2 adsorption and separation. J CO_2 Utilization 51:101641
74. Guo Y, Tan C, Sun J, Li W, Zhang J, Zhao C (2020) Porous activated carbons derived from waste sugarcane bagasse for CO_2 adsorption. Chem Eng J 381:122736
75. Samaddoost L, Soltani M, Fatehifar E, Asl EA (2023) Design of amine-functionalized resin via a facial method with efficient CO_2 capture from air. Process Saf Environ Prot 171:18–27
76. Mousavi SB, Heidari M, Rahmani F, Sene RA, Clough PT, Ozmen S (2023) Highly robust ZrO_2-stabilized CaO nanoadsorbent prepared via a facile one-pot MWCNT-template method for CO^2 capture under realistic calcium looping conditions. J Clean Prod 384:135579
77. Yadav S, Mondal S (2022) A review on the progress and prospects of oxy-fuel carbon capture and sequestration (CCS) technology. Fuel 308:122057
78. Koohestanian E, Shahraki F (2021) Review on principles, recent progress, and future challenges for oxy-fuel combustion CO_2 capture using compression and purification unit. J Environ Chem Eng 9(4):105777
79. Li X, Peng Z, Pei Y, Ajmal T, Rana KJ, Aitouche A et al (2022) Oxy-fuel combustion for carbon capture and storage in internal combustion engines–a review. Int J Energy Res 46(2):505–522
80. Ong CW, Chen C-L (2021) Intensification, optimization and economic evaluations of the CO_2-capturing oxy-combustion CO_2 power system integrated with the utilization of liquefied natural gas cold energy. Energy 234:121255
81. Okeke IJ, Ghantous T, Adams TA (2023) Design strategies for oxy-combustion power plant captured CO_2 purification. Chem Prod Process Model 18(1):135–154
82. Talei S, Fozer D, Varbanov PS, Szanyi A, Mizsey P (2024) Oxyfuel combustion makes carbon capture more efficient. ACS Omega 9(3):3250–3261
83. Escudero AI, Espatolero S, Romeo LM (2016) Oxy-combustion power plant integration in an oil refinery to reduce CO_2 emissions. Int J Greenhouse Gas Control 45:118–129

84. Al-Sakkari EG, Ragab A, Dagdougui H, Boffito DC, Amazouz M (2024) Carbon capture, utilization and sequestration systems design and operation optimization: Assessment and perspectives of artificial intelligence opportunities. Sci Total Environ, 170085
85. Chen H, Zheng Y, Li J, Li L, Wang X (2023) AI for nanomaterials development in clean energy and carbon capture, utilization and storage (CCUS). ACS Nano 17(11):9763–9792
86. Chennakrishnan B, Kumar A, Singh R (2024) Using artificial intelligence to evaluate uncertainty for optimizing CCUS. Fifth EAGE global energy transition conference & exhibition (GET 2024). European association of geoscientists & engineers, 1–5
87. Han L, Xiang Z (2023) Intelligent design and synthesis of energy catalytic materials. Fund Res
88. Tran R, Lan J, Shuaibi M, Wood BM, Goyal S, Das A et al (2023) The Open Catalyst 2022 (OC22) dataset and challenges for oxide electrocatalysts. ACS Catal 13(5):3066–3084
89. Jørgensen PB, Bhowmik A (2022) Equivariant graph neural networks for fast electron density estimation of molecules, liquids, and solids. npj Comput Mater 8(1):183
90. Yang Y, Zhao L, Han C-X, Ding X-D, Lookman T, Sun J et al (2021) Taking materials dynamics to new extremes using machine learning interatomic potentials. J Mater Inf 1(2):10
91. Raju CN, Murthy P, VidyaSagar VV, Unleashing the potential of supervised learning: exploring data variations to shape ML techniques
92. Du X, Salasakar S, Thakur GC (2024) A comprehensive summary of the application of machine learning techniques for C_{O2}-EOR Projects
93. Yan Y, Borhani TN, Subraveti SG, Pai KN, Prasad V, Rajendran A et al (2021) Harnessing the power of machine learning for carbon capture, utilisation, and storage (CCUS)–a state-of-the-art review. Energy Environ Sci 14(12):6122–6157
94. Shen B, Yang S, Hu J, Zhang Y, Zhang L, Ye S et al (2024) Interpretable causal-based temporal graph convolutional network framework in complex spatio-temporal systems for CCUS-EOR. Energy 309:133129
95. Jayaraman A, Lu S (2024) Machine learning for analysis of structural characterization. the digital transformation of product formulation. CRC Press, pp 106–36
96. Jiali L (2021) Deep learning based materials discovery and informatics. National University of Singapore (Singapore).
97. Karlupia N, Abrol P (2023) Wrapper-based optimized feature selection using nature-inspired algorithms. Neural Comput Appl 35(17):12675–12689
98. Alhams A, Abdelhadi A, Badri Y, Sassi S, Renno J (2024) Enhanced bearing fault diagnosis through trees ensemble method and feature importance analysis. J Vib Eng Technol, 1–17
99. Stier J, Gianini G, Granitzer M, Ziegler K (2018) Analysing neural network topologies: a game theoretic approach. Procedia Comput Sci 126:234–243
100. Fan C, Wu Q, Zhao Y, Mo L (2024) Integrating active learning and semi-supervised learning for improved data-driven HVAC fault diagnosis performance. Appl Energy 356:122356
101. Dimitriadou K, Papaemmanouil O, Diao Y (2016) AIDE: an active learning-based approach for interactive data exploration. IEEE Trans Knowl Data Eng 28(11):2842–2856
102. Wang X, Jin Y, Schmitt S, Olhofer M (2023) Recent advances in Bayesian optimization. ACM Comput Surv 55(13s):1–36
103. Wang Z, Dahl GE, Swersky K, Lee C, Nado Z, Gilmer J et al (2024) Pre-trained Gaussian processes for Bayesian optimization. J Mach Learn Res 25(212):1–83
104. Lu Q, Polyzos KD, Li B, Giannakis GB (2023) Surrogate modeling for Bayesian optimization beyond a single Gaussian process. IEEE Trans Pattern Anal Mach Intell 45(9):11283–11296
105. Fotias SP, Ismail I, Gaganis V (2024) Optimization of well placement in carbon capture and storage (CCS): Bayesian optimization framework under permutation invariance. Appl Sci 14(8):3528
106. Jerng SE, Gallant BM (2022) Electrochemical reduction of CO_2 in the captured state using aqueous or nonaqueous amines. Iscience 25(7)
107. Chen G, Luo X, Zhang H, Fu K, Liang Z, Rongwong W et al (2015) Artificial neural network models for the prediction of CO_2 solubility in aqueous amine solutions. Int J Greenhouse Gas Control 39:174–184

108. Zhang Z, Li H, Chang H, Pan Z, Luo X (2018) Machine learning predictive framework for CO^2 thermodynamic properties in solution. J CO_2 Utilization 26:152–9
109. Yarveicy H, Ghiasi MM, Mohammadi AH (2018) Performance evaluation of the machine learning approaches in modeling of CO_2 equilibrium absorption in Piperazine aqueous solution. J Mol Liq 255:375–383
110. Ghiasi MM, Abedi-Farizhendi S, Mohammadi AH (2019) Modeling equilibrium systems of amine-based CO_2 capture by implementing machine learning approaches. Environ Prog Sustain Energy 38(5):13160
111. Dong S, Quan H, Zhao D, Li H, Geng J, Liu H (2022) Generic AI models for mass transfer coefficient prediction in amine-based CO_2 absorber, Part I: BPNN model. Chem Eng Sci 264:118165
112. Orlov AA, Valtz A, Coquelet C, Rozanska X, Wimmer E, Marcou G et al (2022) Computational screening methodology identifies effective solvents for CO_2 capture. Commun Chem 5(1):37
113. Hwang J, Kim J, Lee HW, Na J, Ahn BS, Lee SD et al (2019) An experimental based optimization of a novel water lean amine solvent for post combustion CO_2 capture process. Appl Energy 248:174–184
114. Zhao Y, Huang Y, Zhang X, Zhang S (2015) A quantitative prediction of the viscosity of ionic liquids using S σ-profile molecular descriptors. Phys Chem Chem Phys 17(5):3761–3767
115. Venkatraman V, Alsberg BK (2017) Predicting CO_2 capture of ionic liquids using machine learning. J CO^2 Utilization 21:162–8
116. Ghiasi MM, Mohammadi AH (2017) Application of decision tree learning in modelling CO_2 equilibrium absorption in ionic liquids. J Mol Liq 242:594–605
117. Aghaie M, Zendehboudi S (2020) Estimation of CO_2 solubility in ionic liquids using connectionist tools based on thermodynamic and structural characteristics. Fuel 279:117984
118. Song Z, Shi H, Zhang X, Zhou T (2020) Prediction of CO_2 solubility in ionic liquids using machine learning methods. Chem Eng Sci 223:115752
119. Wu T, Li W-L, Chen M-Y, Zhou Y-M, Zhang Q-Y (2021) Prediction of Henry's law constants of CO_2 in imidazole ionic liquids using machine learning methods based on empirical descriptors. Chem Pap 75:1619–1628
120. Tian Y, Wang X, Liu Y, Hu W (2023) Prediction of CO_2 and N2 solubility in ionic liquids using a combination of ionic fragments contribution and machine learning methods. J Mol Liq 383:122066
121. Taheri M, Zhu R, Yu G, Lei Z (2021) Ionic liquid screening for CO_2 capture and H2S removal from gases: the syngas purification case. Chem Eng Sci 230:116199
122. Kuroki N, Suzuki Y, Kodama D, Chowdhury FA, Yamada H, Mori H (2023) Machine learning-boosted design of ionic liquids for CO_2 absorption and experimental verification. J Phys Chem B 127(9):2022–2027
123. Jian Y, Wang Y, Barati FA (2022) Predicting CO_2 absorption in ionic liquids with molecular descriptors and explainable graph neural networks. ACS Sustain Chem Eng 10(50):16681–16691
124. Zhang K, Wu J, Yoo H, Lee Y (2021) Machine learning-based approach for tailor-made design of ionic liquids: application to CO_2 capture. Sep Purif Technol 275:119117
125. Liu Y, Yu H, Sun Y, Zeng S, Zhang X, Nie Y et al (2020) Screening deep eutectic solvents for CO_2 capture with COSMO-RS. Front Chem 8:82
126. Song Z, Hu X, Wu H, Mei M, Linke S, Zhou T et al (2020) Systematic screening of deep eutectic solvents as sustainable separation media exemplified by the CO_2 capture process. ACS Sustain Chem Eng 8(23):8741–8751
127. Sagar A, Upadhyayula S (2022) Implementation of Artificial Neural Networks in the assessment of CO_2 solubility in deep eutectic and ionic liquid solvents–Performance and cost comparison. Sustain Chem Climate Action 1:100007
128. Klamt A (1995) Conductor-like screening model for real solvents: a new approach to the quantitative calculation of solvation phenomena. J Phys Chem 99(7):2224–2235
129. Wang J, Song Z, Chen L, Xu T, Deng L, Qi Z (2021) Prediction of CO_2 solubility in deep eutectic solvents using random forest model based on COSMO-RS-derived descriptors. Green Chem Eng 2(4):431–440

130. Lemaoui T, Boublia A, Lemaoui S, Darwish AS, Ernst B, Alam M et al (2023) Predicting the CO_2 capture capability of deep eutectic solvents and screening over 1000 of their combinations using machine learning. ACS Sustain Chem Eng 11(26):9564–9580
131. Mohan M, Demerdash O, Simmons BA, Smith JC, Kidder MK, Singh S (2023) Accurate prediction of carbon dioxide capture by deep eutectic solvents using quantum chemistry and a neural network. Green Chem 25(9):3475–3492
132. Makarov DM, Fadeeva YA, Golubev VA, Kolker AM (2023) Designing deep eutectic solvents for efficient CO_2 capture: A data-driven screening approach. Sep Purif Technol 325:124614
133. Abourehab MA, Shawky AM, Venkatesan K, Yasmin S, Alsubaiyel AM, AboRas KM (2022) Efficiency development of surface tension for different ionic liquids through novel model of machine learning technique: application of in-thermal engineering. J Mol Liq 367:120391
134. Baskin I, Epshtein A, Ein-Eli Y (2022) Benchmarking machine learning methods for modeling physical properties of ionic liquids. J Mol Liq 351:118616
135. Nakhaei-Kohani R, Madani SA, Mousavi S-P, Atashrouz S, Abedi A, Hemmati-Sarapardeh A et al (2022) Machine learning assisted structure-based models for predicting electrical conductivity of ionic liquids. J Mol Liq 362:119509
136. Cao L, Zhu P, Zhao Y, Zhao J (2018) Using machine learning and quantum chemistry descriptors to predict the toxicity of ionic liquids. J Hazard Mater 352:17–26
137. Mesbah M, Shahsavari S, Soroush E, Rahaei N, Rezakazemi M (2018) Accurate prediction of miscibility of CO_2 and supercritical CO_2 in ionic liquids using machine learning. J CO_2 Utilization 25:99–107
138. Paduszynski K, Domanska U (2014) Viscosity of ionic liquids: an extensive database and a new group contribution model based on a feed-forward artificial neural network. J Chem Inf Model 54(5):1311–1324
139. Shi D, Zhou F, Mu W, Ling C, Mu T, Yu G et al (2022) Deep insights into the viscosity of deep eutectic solvents by an XGBoost-based model plus SHapley Additive exPlanation. Phys Chem Chem Phys 24(42):26029–26036
140. Liu Q, Zhang X (2022) Multiscale screening of deep eutectic solvents for efficient extraction of m-cresol from model coal tar. ACS Omega 7(38):34485–34494
141. Nagulapati VM, Rehman HMRU, Haider J, Qyyum MA, Choi GS, Lim H (2022) Hybrid machine learning-based model for solubilities prediction of various gases in deep eutectic solvent for rigorous process design of hydrogen purification. Sep Purif Technol 298:121651
142. Liu Y, Dai Z, Zhang Z, Zeng S, Li F, Zhang X et al (2021) Ionic liquids/deep eutectic solvents for CO_2 capture: reviewing and evaluating. Green Energy Environ 6(3):314–328
143. Haider J, Saeed S, Qyyum MA, Kazmi B, Ahmad R, Muhammad A et al (2020) Simultaneous capture of acid gases from natural gas adopting ionic liquids: challenges, recent developments, and prospects. Renew Sustain Energy Rev 123:109771
144. Rahimi M, Moosavi SM, Smit B, Hatton TA (2021) Toward smart carbon capture with machine learning. Cell Reports Phys Sci 2(4)
145. Sun J, Sakai Y, Sato Y, Kansha Y (2022) Improving CO_2 capture process with latent heat reuse system and artificial neural network model. Chem Eng Trans 94:79–84
146. Wu X, Shen J, Wang M, Lee KY (2020) Intelligent predictive control of large-scale solvent-based CO_2 capture plant using artificial neural network and particle swarm optimization. Energy 196:117070
147. Li F, Zhang J, Oko E, Wang M (2015) Modelling of a post-combustion CO_2 capture process using neural networks. Fuel 151:156–163
148. Chen J, Wang F (2014) Cost reduction of CO_2 capture processes using reinforcement learning based iterative design: a pilot-scale absorption–stripping system. Sep Purif Technol 122:149–158
149. Liu X, Wu J, Lei Y, Wu X, Man Y, Luo H et al (2023) Data-driven surrogate optimized and intensified extractive distillation process for clean separation of isopropanol from water: a sustainable alternative. J Clean Prod 383:135475
150. Shalaby A, Elkamel A, Douglas PL, Zhu Q, Zheng QP (2021) A machine learning approach for modeling and optimization of a CO_2 post-combustion capture unit. Energy 215:119113

151. Subraveti SG, Li Z, Prasad V, Rajendran A (2019) Machine learning-based multiobjective optimization of pressure swing adsorption. Ind Eng Chem Res 58(44):20412–20422
152. Leperi KT, Yancy-Caballero D, Snurr RQ, You F (2019) 110th anniversary: surrogate models based on artificial neural networks to simulate and optimize pressure swing adsorption cycles for CO_2 capture. Ind Eng Chem Res 58(39):18241–18252
153. Magina S, Barros-Timmons A, Ventura SP, Evtuguin DV (2021) Evaluating the hazardous impact of ionic liquids–challenges and opportunities. J Hazard Mater 412:125215
154. Carminati HB, de Medeiros JL, Moure GT, Barbosa LC, Ofélia de Queiroz FA (2020) Low-emission pre-combustion gas-to-wire via ionic-liquid [Bmim][NTf2] absorption with high-pressure stripping. Renew Sustain Energy Rev 131:109995
155. Ma C, Xie Y, Ji X, Liu C, Lu X (2018) Modeling, simulation and evaluation of biogas upgrading using aqueous choline chloride/urea. Appl Energy 229:1269–1283
156. Nassef AM (2023) Improving CO_2 absorption using artificial intelligence and modern optimization for a sustainable environment. Sustainability 15(12):9512
157. Sedighi M, Mohammadi M, Ameli F, Amiri-Ramsheh B, Hemmati-Sarapardeh A (2025) A comparative study of machine learning frameworks for predicting CO_2 conversion into light olefins. Fuel 379:133017
158. Lv X, Yang Y, Qin D, Cao X, Xu H (2022) Artificial intelligence service recovery: the role of empathic response in hospitality customers' continuous usage intention. Comput Hum Behav 126:106993
159. Andrews N, Fan J, Forward R, Chen M, Loock H-P (2017) Determination of the thermal, oxidative and photochemical degradation rates of scintillator liquid by fluorescence EEM spectroscopy. Phys Chem Chem Phys 19(1):73–81
160. Lai NS, Tew YS, Zhong X, Yin J, Li J, Yan B et al (2023) Artificial intelligence (AI) workflow for catalyst design and optimization. Ind Eng Chem Res 62(43):17835–17848
161. Han Z-K, Sarker D, Ouyang R, Mazheika A, Gao Y, Levchenko SV (2021) Single-atom alloy catalysts designed by first-principles calculations and artificial intelligence. Nat Commun 12(1):1833
162. Goldsmith BR, Esterhuizen J, Liu JX, Bartel CJ, Sutton C (2018) Machine learning for heterogeneous catalyst design and discovery
163. Tamtaji M, Gao H, Hossain MD, Galligan PR, Wong H, Liu Z et al (2022) Machine learning for design principles for single atom catalysts towards electrochemical reactions. J Mater Chem A 10(29):15309–15331
164. Okur O (2023) Predictive modeling of CO_2 methanation reaction for carbon capture
165. Tiemann E, Zhou S, Kläser A, Heidler K, Schneider R, Zhu XX (2024) Machine learning for methane detection and quantification from space--a survey. arXiv preprint arXiv:240815122
166. Peterson L, Forootani A, Medina EIS, Gosea IV, Sundmacher K, Benner P (2024) Towards digital twins for power-to-X: comparing surrogate models for a catalytic CO_2 Methanation reactor. Authorea Preprints
167. Samad A, Saghir H, Musawwir A, Zulkefal M (2024) An intelligent system for predicting the methanol conversion rate from the direct Hydrogenation of CO_2 under uncertainty. Mater Proc 17(1):3
168. Chuquin-Vasco D, Parra F, Chuquin-Vasco N, Chuquin-Vasco J, Lo-Iacono-Ferreira V (2021) Prediction of methanol production in a carbon dioxide hydrogenation plant using neural networks. Energies 14(13):3965
169. Manae MA, Dheer L, Waghmare UV (2022) CO_2 utilization through its reduction to methanol: design of catalysts using quantum mechanics and machine learning. Trans Indian Nat Acad Eng 7(1):1–11
170. Suvarna M, Araujo TP, Pérez-Ramírez J (2022) A generalized machine learning framework to predict the space-time yield of methanol from thermocatalytic CO_2 hydrogenation. Appl Catal B 315:121530
171. Fulham GJ, Wu X, Liu W, Marek EJ (2024) Mechanistic insights into the role of zinc oxide, zirconia and ceria supports in Cu-based catalysts for CO_2 hydrogenation to methanol. Chem Eng J 480:147732

172. Xing L, Jiang H, Tian X, Yin H, Shi W, Yu E et al (2023) Combining machine learning with multi-physics modelling for multi-objective optimisation and techno-economic analysis of electrochemical CO^2 reduction process. Carbon Capture Sci Technol 9:100138
173. Cherni A, Halouani K (2024) Artificial intelligence for electrochemical prediction and optimization of direct carbon fuel cells fueled with biochar. Electrochem 5(1):29–44
174. Park J, Lee J (2024) Electrochemical energy conversion and storage processes with machine learning. Trends Chem
175. Choi J, Bang K, Jang S, Choi J, Ordonez J, Buttler D et al (2023) Deep learning of electrochemical CO_2 conversion literature reveals research trends and directions. J Mater Chem A 11(33):17628–17643
176. Chan W, Morosuk T, Li X, Li H (2023) Allam cycle: review of research and development. Energy Convers Manage 294:117607
177. Zou Y, Alghassab MA, Abdulwahab A, Sharma A, Ghandour R, Alkhalaf S et al (2024) Heat recovery from oxy-supercritical carbon dioxide cycle incorporating Goswami cycle for zero emission power/heat/cooling production scheme; techno-economic study and artificial intelligence-based optimization. Case Studies Thermal Eng 54:104084
178. Vilnis L, McCallum A, Freed D, Rafati N, Durham N, Camilo J, Supercritical CO_2 power cycles
179. Neises T, Turchi CS (2014) Supercritical CO_2 power cycles: design considerations for concentrating solar power. 4th international symposium—supercritical CO_2 power cycles, Pittsburgh, PA, pp 9–10
180. Wang L, Alirahmi SM, Yu H (2024) Development and analysis of a novel power-to-gas-to-power system driven by the Allam cycle for simultaneous electricity and water production. Energy Convers Manage 319:118934
181. Aghamolaei R (2024) Recent advancements in applying machine learning in power-to-X processes: a literature review. Sustainability 16:9555
182. Rao M (2021) Machine learning in estimating CO_2 emissions from electricity generation. engineering problems-uncertainties, constraints and optimization techniques. IntechOpen

Chapter 10
Integrating Hybrid AI Models for Improved PV Power Forecasting: A Path Toward Sustainable Energy

Mawloud Guermoui, Khaled Ferkous, Abdelaziz Rabehi, and Abdelfattah Belaid

Abstract This chapter emphasizes that more attention has to be given to applying hybrid deep learning techniques to the PV power forecast in order to optimize renewable energy systems toward a greener environment. Proposed models integrate advanced pre-processing methods, and decomposition algorithms for time series data are developed based on fusing decomposition algorithms along with deep learning models in constructing a robust and accurate framework for the forecast. In this work, we further propose incorporating a transformer model along with variances of VMD to take the performance a step ahead with regard to the estimation of PV power. Results about three grid-connected PV plants located in the South-Saharian region of Algeria, obtained after releasing the locality of the proposed model, show its performance: in fact, the normalized RMSE values obtained for the studied photovoltaic plants range from 2 to 2.4%. In this respect, the chapter has developed a complete solution for the estimation of PV energy output under different environmental conditions, capitalizing on strengths from these methods and trying to advance the reliability of PV power forecast. The contribution shall, in turn, support efficient renewable energy resource management and add toward global clean energy goals. The dependence on this, of course, is an essentiality with which one looks at the context.

Keywords PV power forecasting · Deep learning models · Variational mode decomposition · Hybrid models

M. Guermoui (✉) · A. Belaid
Centre de DÈveloppement Des Energies Renouvelables, CDER, UnitÈ de Recherche AppliquÈe en Energies Renouvelables, URAER, GhardaÔa, Algeria
e-mail: mawloud.guermoui@gmail.com

M. Guermoui · A. Rabehi
Telecommunications and Smart Systems Laboratory, University of ZianeAchour, Djelfa, Algeria

K. Ferkous
Materials, Energy Systems Technology and Environment Laboratory, Ghardaia University, Ghardaia, Algeria

M. Nasr et al. (eds.), *Artificial Intelligence Applications for a Sustainable Environment*, Green Chemistry and Sustainable Technology,
https://doi.org/10.1007/978-3-031-91199-6_10

10.1 Introduction

Photovoltaic (PV) power is an essential renewable energy source due to its clean, versatile, and decentralized characteristics [1]. Integrating PV power into energy grids brings both economic and environmental benefits but also presents challenges stemming from the variable and unpredictable nature of PV power generation [2]. To address these challenges and ensure the efficient operation of power systems, accurate forecasting of PV power output is vital. However, forecasting is complicated by the non-linear and volatile behavior of PV power time series, which are heavily influenced by continuously changing weather conditions [3].

A range of methods has been developed for PV power forecasting, including physical modeling and data-driven approaches [4]. Physical modeling often involves intricate calculations and significant data requirements, while data-driven methods have gained traction with the progress of machine learning (ML) and artificial intelligence (AI) [5]. These data-driven approaches leverage historical PV data and meteorological forecasts to predict future power output. Among these, short-term forecasting (STF) is particularly critical for grid operations, aiding in scheduling maintenance and optimizing power supply. Furthermore, accurate STF has notable financial implications for power companies, as even slight improvements in forecast precision can lead to significant cost savings [6].

Recent advancements in deep learning have introduced models such as Long Short-Term Memory (LSTM), Bidirectional LSTM (BiLSTM), and Gated Recurrent Units (GRU), which excel at modeling complex dependencies in time series data [7]. While traditional ML methods, including Support Vector Machines (SVM) and Extreme Learning Machines (ELM), have shown promising results, standalone ML models often face challenges like overfitting and hyperparameter optimization [8]. To overcome these limitations, hybrid models combining decomposition techniques with deep learning have emerged. Decomposition methods such as Variational Mode Decomposition (VMD) break PV power time series into distinct frequency components, isolating patterns and trends for improved forecasting accuracy [9]. This approach allows deep learning models to focus on specific patterns, enhancing overall prediction performance.

In this study, we propose an advanced hybrid deep learning model that integrates Variational Mode Decomposition (VMD) is integrated with state-of-the-art deep learning architectures to enhance PV power forecasting accuracy. VMD is employed to decompose PV power time series into multiple sub-series, isolating distinct patterns and trends. Each sub-series is then predicted using advanced deep learning models, specifically Long Short-Term Memory (LSTM), Gated Recurrent Units (GRU), Convolutional Neural Networks (CNN), and the Transformer model. These models were selected for their complementary strengths: LSTM's ability to capture long-term dependencies, GRU's efficiency in handling sequential data with reduced computational complexity, CNN's strong feature extraction and pattern recognition capabilities, and the Transformer model's exceptional proficiency in capturing global dependencies and handling long-range temporal relationships. The

Transformer's attention mechanism allows it to focus on the most relevant parts of the input sequence, enabling improved performance on complex time series tasks. After forecasting individual sub-series, the predictions are aggregated to reconstruct the final PV power output, ensuring robust and accurate forecasting results.

The proposed hybrid model was validated using real-world data from a PV station in Algeria, demonstrating significant improvements in forecast accuracy compared to traditional models. The integration of Variational Mode Decomposition (VMD) with LSTM, GRU, CNN, and the Transformer model enables a more comprehensive analysis of the time series data, effectively capturing the complex and variable nature of PV power generation. Specifically, LSTM captures long-term dependencies, GRU efficiently processes sequential data with lower computational overhead, CNN extracts critical features and identifies patterns within the data, and the Transformer model [10] excels at capturing global dependencies and long-range temporal relationships through its attention mechanism. This robust combination of models enhances the reliability of PV power predictions, contributing to more efficient grid management and advancing sustainable energy production.

This chapter is organized as follows: Sect. 10.2 outlines the methodology, detailing the VMD-based decomposition and the employed deep learning models. Section 10.3 provides a site description and explores the data used in the study. Section 10.4 presents the experimental results and discusses the findings and their implications. Finally, Sect. 10.5 concludes the paper, summarizing the contributions and outlining potential directions for future research.

10.2 Methodology

10.2.1 Variational Mode Decomposition

Variational Mode Decomposition (VMD) [9] is a powerful signal processing technique widely used for decomposing complex signals into simpler, well-defined modes, making it particularly valuable for applications such as forecasting and noise reduction. Unlike its predecessor, Empirical Mode Decomposition (EMD), which suffers from mode mixing and lacks a solid theoretical foundation, VMD offers a more robust and mathematically grounded approach.

VMD operates by breaking down a signal into a set of band-limited modes, each represented as an amplitude-modulated–frequency-modulated (AM-FM) component. The decomposition is achieved through an iterative process that optimizes a cost function to minimize the bandwidth of each mode. This process is conducted in the frequency domain using Hilbert transforms and analytic signals to ensure precision.

The main steps in VMD include:

Initialization: The center frequencies and initial modes are initialized.

Decomposition: For each mode, the Hilbert transform is applied to compute an analytic signal. The spectrum is shifted to the baseband by demodulating the signal with a pre-set frequency. The mode is then computed using Wiener filtering.

Update: The center frequency of each mode is updated by calculating the weighted average of its frequency spectrum.

Iteration: These steps are repeated iteratively until convergence, ensuring that the cost function, which minimizes the bandwidth of the modes, is optimized:

$$min_{\{uk\},\{wk\}}\left\{\sum\nolimits_k \|\delta_t\left[\left(\delta(t)+\frac{j}{\pi t}\right)*u_k(t)\right]e^{-jw_k t}\|_2^2\right\}s.t.\sum_k u_k(t)=f(t) \quad (10.1)$$

where $u_{k(t)}$ are the decomposed modes, and ω_k are their respective center frequencies.

One notable advantage of Variational Mode Decomposition (VMD) compared to techniques like Empirical Mode Decomposition (EMD) is its ability to avoid mode mixing and its well-defined convergence criteria. VMD has been successfully applied in various domains, including photovoltaic (PV) power forecasting, where it aids in extracting relevant features from noisy time series data, thereby enhancing prediction accuracy.

By decomposing complex, non-stationary signals into simpler components, VMD effectively isolates meaningful trends within time series data, enabling the development of more accurate forecasting models. Studies such as those by Dragomiretskiy and Zosso [9], the original developers of VMD, highlight its superiority in efficiently handling non-linear and non-stationary data.

10.2.2 Long Short-Term Memory

Long Short-Term Memory (LSTM) [11] is a specialized variant of recurrent neural networks (RNNs) designed to effectively learn and model time-dependent patterns in sequential data. Traditional RNNs are limited by the vanishing gradient problem, which hinders their ability to capture long-term dependencies. LSTM addresses this challenge with its unique memory cell architecture, enabling it to retain both short-term and long-term dependencies efficiently.

An LSTM unit is composed of three key gates: the input gate, forget gate, and output gate. These gates control the flow of information within the unit, determining when to update, retain, or output information. This gating mechanism ensures that relevant data is preserved while irrelevant information is discarded, enhancing the model's performance on sequential tasks. The fundamental equations governing an LSTM cell are [12] as follows:

Forget Gate:

$$f_t = \sigma(W_f.\left[h_{t-1}, x_t\right] + b_f \quad (10.2)$$

where f_t is the forget gate's activation, h_{t-1} is the previous hidden state, x_t is the current input, and W_f and b_f are the weight matrix and bias for the forget gate. The forget gate decides which parts of the cell state should be kept or discarded.

Input Gate:

$$i_t = \sigma(W_i.[h_{t-1}, x_t] + b_i \tag{10.3}$$

$$\tilde{C}_t = \tanh(W_C.[h_{t-1}, x_t] + b_C \tag{10.4}$$

Here, i_t controls how much new information from $\tilde{C}_t$ (the candidate memory) should be added to the cell state. The new candidate $\tilde{C}_t$ is created using a non-linearity (tanh) to scale the incoming data.

Cell State Update:

$$C_t = f_t * C_{t-1} + i_t * \tilde{C}_t \tag{10.5}$$

The new cell state C_t is a combination of the previous state, controlled by the forget gate, and the updated candidate information controlled by the input gate.

Output Gate:

$$o_t = \sigma(W_O.[h_{t-1}, x_t + b_O] \tag{10.6}$$

$$h_t = o_t * \tanh(C_t) \tag{10.7}$$

The output gate determines the final hidden state h_t, which is derived by applying a tanh activation function to the updated cell state. This mechanism ensures that the hidden state reflects the most relevant aspects of the processed information.

These equations form the core of LSTM's capability to retain and utilize relevant information across long sequences, making it highly suitable for handling complex time series tasks and capturing intricate dependencies in sequential data.

10.2.3 Gated Recurrent Units (GRU)

The Gated Recurrent Units (GRU) model [13] is a variant of recurrent neural networks (RNNs) designed to overcome the limitations of traditional RNNs in learning long-term dependencies. Introduced by Cho et al. in 2014, GRUs provide a simpler yet effective alternative to Long Short-Term Memory (LSTM) networks, utilizing gating mechanisms to regulate the flow of information. The GRU architecture features two primary gates: the update gate, which determines how much of the past information is retained and new information is added, and the reset gate, which controls the influence of past data on the current computation. These mechanisms help address the vanishing gradient problem, enabling GRUs to efficiently model sequential data. The mathematical formulation of GRUs is as follows [14]:

Update Gate:

$$z_t = \sigma(W_z.\left[h_{t-1}, x_t\right] + b_z \tag{10.8}$$

where σ denotes the sigmoid activation function, W_z is the weight matrix for the update gate, h_{t-1} is the previous hidden state, and x_t is the current input. b_z is the bias term associated with the update gate.

Reset Gate:

$$r_t = \sigma(W_r.\left[h_{t-1}, x_t\right] + b_r \tag{10.9}$$

where W_r is the weight matrix for the reset gate, and b_r is the bias.

Candidate Hidden State:

$$\tilde{h}_t = \tanh(W_h.\left[r_t\odot, h_{t-1}, x_t\right] + b_h \tag{10.10}$$

where $\odot$ represents element-wise multiplication, W_h is the weight matrix for the candidate hidden state, and b_h is the bias.

Final Hidden State:

$$h_t = (1 - z_t) \odot h_{t-1} + z_t \odot \tilde{h}_t \tag{10.11}$$

These equations illustrate how GRUs control the flow of information through the network, allowing them to capture dependencies across varying time steps effectively [15].

10.2.4 Convolutional Neural Networks (CNN)

Convolutional Neural Networks (CNNs) [13] are a type of deep neural network specifically tailored for processing structured grid-like data, such as images and time series data. Inspired by the human visual system, CNNs have become fundamental in the field of computer vision and are increasingly applied to other domains like natural language processing and signal analysis. The cornerstone of CNNs is the convolutional layer, which performs convolution operations to extract meaningful features from the input data. The mathematical operation of a convolutional layer can be expressed as follows:

Convolution Operation:

$$K(m,n).(n+m,j+I(i\sum_{n}\sum_{m}=K)(i,j)*I) \tag{10.12}$$

where I is the input image, K is the convolutional kernel (or filter), and $*$ denotes the convolution operation. The indices (i, j) refer to the position in the output feature map, and mmm and n are the positions within the kernel.

Activation Function:

After the convolution operation, an activation function is applied to introduce non-linearity into the model. The Rectified Linear Unit (ReLU) is commonly used:

$$\max(0,x)=f(x) \tag{10.13}$$

where $f(x)$ is the activation function output, and x is the input to the function.

Pooling Operation:

Pooling layers are used to reduce the spatial dimensions of the feature maps, which helps in reducing the computational load and the risk of overfitting. A common pooling operation is *max* pooling:

$$(n+m,j+F(i\max_{m,n}=P(i,j) \tag{10.14}$$

where P is the output of the pooling layer, and F is the input feature map. The *max* operation is performed over a pooling window of size defined by m and n.

Fully Connected Layer:

After several convolutional and pooling layers, the output is typically flattened and passed through fully connected layers. This process can be expressed as follows:

$$b+x.W=y \tag{10.15}$$

where y is the output vector, W is the weight matrix, x is the flattened input vector, and b is the bias.

These components work together to automatically learn spatial hierarchies of features from input images, making CNNs highly effective for various image classification and detection tasks.

10.2.5 Transformer Model (TRM)

The primary function of the transformer encoder is to create contextualized representations of input tokens by capturing their relationships within the entire input sequence. The transformer, developed by a team of eight researchers at Google

[16], is a deep learning model designed to handle sequential data such as text, speech, or time series data. The model is based on the attention mechanism and represents the first fully attention-based sequence transduction model. Unlike traditional encoder–decoder architectures, which rely on recurrent layers, the transformer replaces them with multi-headed self-attention layers. This design enables significantly faster learning in tasks like translation compared to models based on recurrent or convolutional layers.

The key components of the transformer model include:

- **Multi-head attention**: Allows the model to focus on different parts of the input sequence simultaneously, enabling it to capture diverse contextual relationships.
- **Positional encoding**: Provides information about the order of the sequence, ensuring the model understands the positional relationships between tokens.
- **Feedforward network**: A fully connected neural network applied after the attention mechanism to enhance feature transformation.
- **Layer normalization and residual connections**: Help stabilize the training process and improve convergence by normalizing intermediate outputs and maintaining gradient flow.

These components collectively make the transformer a powerful and efficient architecture for handling sequential data.

10.2.6 Proposed Combination Scheme

In this chapter, we introduce a hybrid approach for photovoltaic (PV) power forecasting that leverages the strengths of different deep learning models—Long Short-Term Memory (LSTM), Convolutional Neural Networks (CNN), Gated Recurrent Units (GRU), and Transformer models—combined with the Variational Mode Decomposition (VMD) algorithm (Fig. 10.1). This integrated framework aims to reduce the inherent randomness in PV power time series and enhance forecasting accuracy by combining the advantages of various architectures.

10.2.6.1 Data Collection and Pre-Processing

The initial step involves collecting PV power data from real-world sources. The raw data undergoes a thorough pre-processing stage that includes [17]:

- **Data Cleaning**: Removing outliers and dealing with missing values (NaNs) using interpolation techniques.
- **Normalization**: Applying the Min–Max scaling method to ensure data is scaled within a specific range, enhancing the model's learning efficiency.

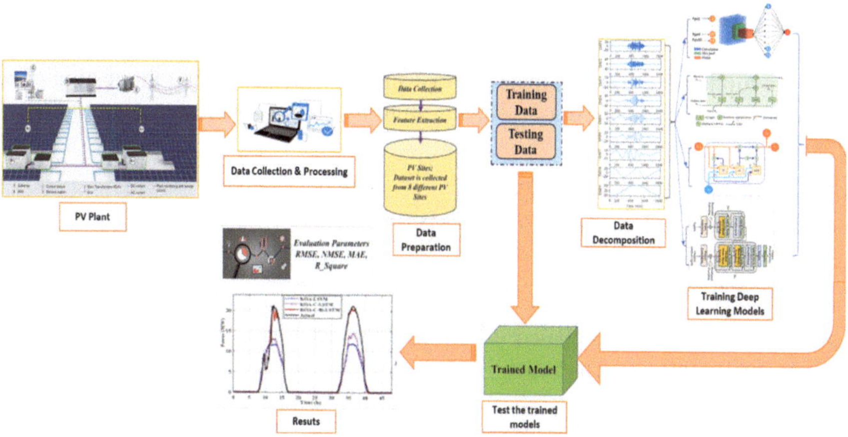

Fig. 10.1 Proposed hybrid model

The cleaned and normalized data is then split into two subsets: the training set and the testing set, ensuring an appropriate balance for effective model training and evaluation [9, 18].

10.2.6.2 Application of VMD

To address the random nature of PV power time series, the Variational Mode Decomposition (VMD) algorithm is applied to both the training and testing datasets. VMD decomposes the data into a finite number of Intrinsic Mode Functions (IMFs), capturing different frequency components. This step reduces noise and improves the quality of the input data for forecasting.

10.2.6.3 Deep Learning Model Training

The decomposed data is fed into four distinct deep learning models—LSTM, CNN, GRU, and Transformer. Each model is trained using the same input data prepared through VMD. The key highlights of the models are as follow:

- **LSTM**: Captures long-term temporal dependencies in the data, making it suitable for time series forecasting.
- **CNN**: Extracts spatial features and patterns in the data, complementing the temporal features.
- **GRU**: A simplified version of LSTM, GRU focuses on capturing short-term dependencies with lower computational requirements.
- **Transformer**: Efficiently handles long-range dependencies and non-linear relationships in the time series data using self-attention mechanisms.

10.2.6.4 Model Evaluation

Once the models are trained, their performance is evaluated using the testing data. To ensure a comprehensive assessment, multiple statistical metrics are employed, including [13–20]:

- Root Mean Squared Error (RMSE)
- Normalized Mean Squared Error (NMSE)
- Mean Absolute Error (MAE)
- Coefficient of Determination (R^2).

These metrics provide a detailed insight into the accuracy, reliability, and generalization capabilities of each model.

10.2.7 Combination Strategy

The hybrid approach leverages the unique strengths of the individual models by combining their predictions. This combination is performed through ensemble techniques, such as weighted averaging or optimization algorithms, to achieve a more robust and accurate forecast (Fig. 10.1).

10.3 Site Description and Data Exploration

The collected historical data were taken from three grid-connected solar photovoltaic plants lying within Algeria geographical location (26° 52' N−29° 14' N) and (−0° 10' O−0° 54' E). The three PV stations have a total capacity of 34 MW Fig. 10.2, where station 1 Aoulef has a capacity of 5 MW, station 2 Timimoune has a capacity of 9 MW, and station 3 Adrar has a capacity of 20 MW. Table 10.1 gives a short description of the PV systems. Data was recorded between January 1, 2018 and December 31, 2019, with a sample point every 15 min from station 1, every 25 min from station 2, and every 1 h from station 3. Training dataset for the study covers 74%, 1% reserved for validation, and 25% for testing.

10.4 Experimental Results and Discussions

In this section, we provide a detailed discussion of the achieved results.

The performance of standalone forecasting models (See Table 10.2)—CNN, LSTM, GRU, and Transformer—was analyzed across three photovoltaic (PV) plants (Station 1, Station 2, and Station 3) using R^2, RMSE, and NRMSE metrics. Among

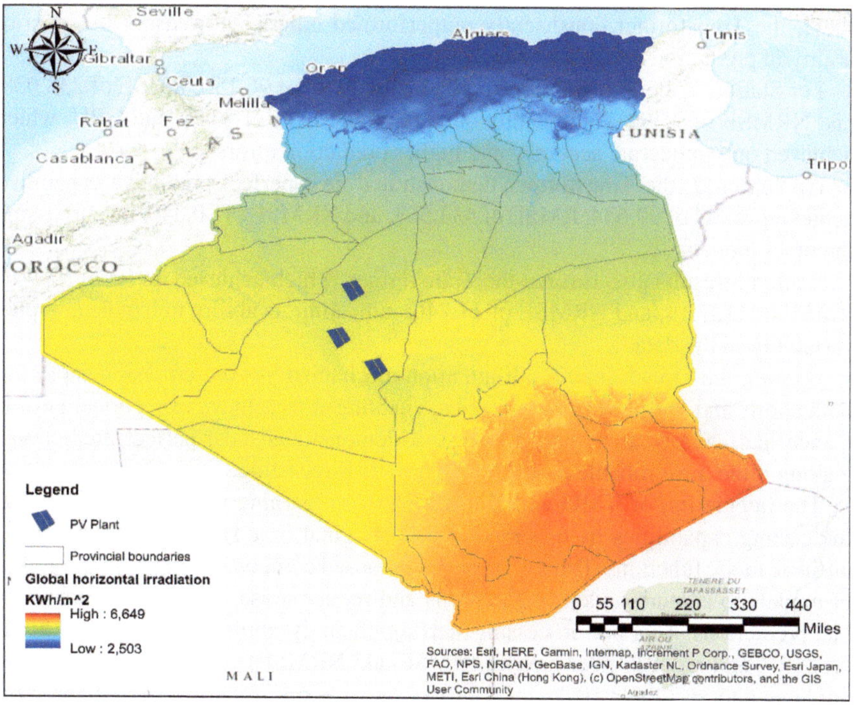

Fig. 10.2 Locations of the studied PV plants

Table 10.1 Data characteristics

Location	Aoulef	Timimoune	Adrar
Latitude	26° 52' N	29° 14' N	27° 48' N
Longitude	0° 54' E	0° 14' E	−0° 10' O
Elevation above sea	288 m	288 m	279 m
Number of panels	20 460	36,828	80,840
Panel tilt	26°	29°	27°
Panel orientation	South	South	South
Technology of panels	Polycrystalline	Polycrystalline	Polycrystalline
Total capacity	5 MW	9 MW	20 MW
Duration	2 Year's (January 2018–December 2019)		
Horizon	15 min	25 min	1 h
Time points	70,616	41,610	17,707

these, the Transformer consistently outperformed others, showcasing its superior ability in prediction modeling.

For Station 1, the Transformer excelled with an R^2 of 96.150, RMSE of 238.038, and NRMSE of 9.399, significantly outperforming CNN, LSTM, and GRU, which achieved only moderate accuracy and higher prediction errors.

At Station 2, the Transformer demonstrated near-perfect predictive capability, achieving an R^2 of 99.634, RMSE of 433.508, and NRMSE of 10.340, far surpassing the other models.

Station 3 results also favored the Transformer, which achieved an R^2 of 93.901, RMSE of 1137.19, and NRMSE of 11.246, indicating its ability to handle complex variability in the data.

Overall, the Transformer's self-attention mechanism proved crucial in capturing both short- and long-term dependencies, enabling it to achieve exceptional performance, particularly at Station 2, where it demonstrated near-perfect predictions, making it the most reliable model for PV power forecasting.

The integration of VMD with standalone deep learning models enhances their forecasting capabilities by decomposing the original time series data into simpler intrinsic mode functions (IMFs). These decomposed components improve the ability of models to capture underlying patterns and reduce noise, leading to more accurate predictions. This pre-processing step significantly improves the models' accuracy (R^2) and reduces error metrics (RMSE and NRMSE) compared to standalone implementations (Table 10.3).

The results clearly highlight the consistent superiority of the VMD-Transformer over other hybrid models across all three stations. At **Station 1**, the VMD-Transformer achieved an R^2 of 99.657, demonstrating exceptional accuracy, though it slightly trailed the VMD-GRU (99.684). However, with an RMSE of 69.494, the VMD-Transformer outperformed all other models except VMD-GRU (66.781). Its NRMSE of 2.744 was notably better than that of VMD-CNN and VMD-LSTM and was comparable to VMD-GRU (2.850), solidifying its robust performance.

For **Station 2**, the VMD-Transformer achieved remarkable results with an R^2 of 99.926, surpassing all other models and demonstrating its superior precision. It recorded the lowest RMSE of 84.11, significantly outperforming VMD-LSTM (317.385) and VMD-CNN (195.203). Its NRMSE of 2.006 was unparalleled, reflecting its exceptional ability to handle the complexities of this station's data.

At **Station 3**, the VMD-Transformer continued to dominate, achieving an R^2 of 99.733, slightly ahead of VMD-GRU and other models. It also recorded the lowest RMSE of 246.33, outperforming VMD-CNN (263.294) and VMD-GRU (300.198). With an NRMSE of 2.436, the VMD-Transformer proved its capacity to efficiently model intricate patterns, maintaining its superiority across all metrics.

Overall, the hybridization of deep learning models with the VMD algorithm significantly enhanced forecasting performance by reducing prediction errors and improving accuracy. Among these hybrid models, the VMD-Transformer consistently delivered superior precision, as demonstrated by its outstanding R^2, RMSE, and NRMSE metrics across all stations. Its ability to effectively manage both simple

Table 10.2 Forecasting results of standalone models for the three studied stations

Station 1					Station 2				Station 3			
Models	CNN	LSTM	GRU	Trans	CNN	LSTM	GRU	Trans	CNN	LSTM	GRU	Trans
R^2	91.17	91.166	91.904	**96.150**	85.989	91.079	90.845	**99.634**	91.353	91.432	91.489	**93.901**
RMSE	340.81	341.648	325.566	**238.038**	816.892	647.131	657.288	**433.508**	1380.32	1383.28	1398.6	**1137.19**
NRMSE	14.548	14.583	13.896	**9.399**	17.508	15.444	15.6865	**10.340**	14.753	14.785	14.949	**11.246**

Table 10.3 Forecasting results of hybrid models for the three studied stations

Station 1					Station 2				Station 3			
Models	VMD-CNN	VMD-LSTM	VMD-GRU	**VMD-Trans**	VMD-CNN	VMD-LSTM	VMD-GRU	**VMD-Trans**	VMD-CNN	VMD-LSTM	VMD-GRU	**VMD-Trans**
R^2	99.418	99.617	99.684	**99.657**	99.197	99.611	99.307	**99.926**	99.692	99.611	99.617	**99.733**
RMSE	87.291	77.584	66.781	**69.494**	195.203	317.385	179.405	**84.11**	263.294	317.385	300.198	**246.33**
NRMSE	3.726	3.311	2.850	**2.744**	4.183	3.392	4.281	**2.006**	2.654	3.392	3.208	**2.4362**

and complex patterns makes it the most reliable and efficient model for photovoltaic power forecasting.

A graphical representation of the achieved results (Figs. 10.3 and 10.4), highlighting the RMSE and NRMSE values for all employed models across the investigated stations, is presented in Figs. 10.3 and 10.4. These visualizations provide a clear and comparative understanding of the models' performances, illustrating the significant improvements achieved by the VMD-Transformer compared to other hybrid models. The figures also emphasize the enhanced accuracy and reduced error metrics, reinforcing the robust capability of the VMD-Transformer in photovoltaic power forecasting [21–25].

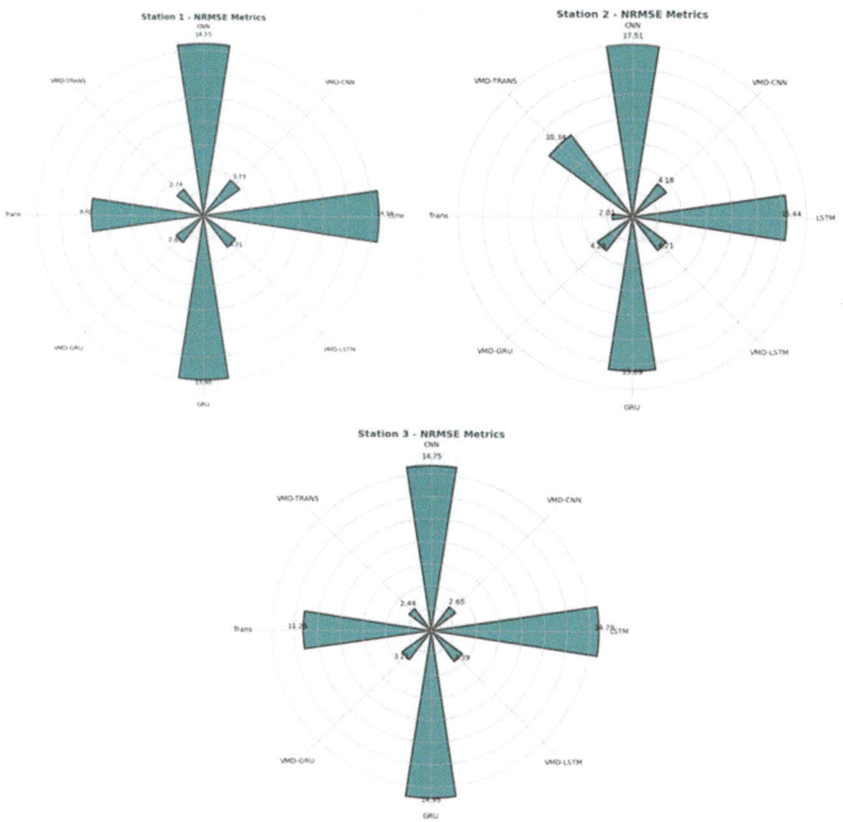

Fig. 10.3 Radial bar chart of NRMSE for all models across all stations

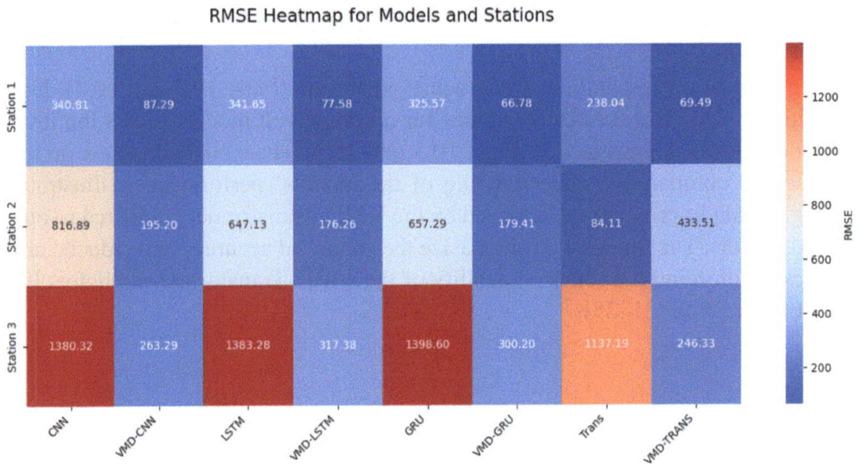

Fig. 10.4 Heatmap of RMSE for all models across all stations

10.5 Conclusion

The comprehensive analysis of forecasting models, both standalone and hybridized with the Variational Mode Decomposition (VMD) algorithm, underscores the critical advancements in photovoltaic power prediction. The results demonstrate that while standalone models such as CNN, LSTM, and GRU show moderate predictive capabilities, the Transformer model consistently outperforms them across all stations due to its superior ability to model complex dependencies. The hybridization of these models with VMD further enhances their performance by effectively decomposing time series data, improving prediction accuracy, and reducing error metrics. Among these, the VMD-Transformer emerges as the most reliable model, delivering exceptional results with significantly lower RMSE and NRMSE values and achieving near-perfect R^2 values.

These advancements hold profound implications for sustainable energy. Accurate and reliable forecasting models are vital for optimizing photovoltaic power generation, enhancing grid stability, and facilitating the integration of renewable energy sources into the power system. The ability of models like VMD-Transformer to precisely predict power output allows energy providers to better match supply with demand, reduce reliance on fossil fuels, and minimize energy wastage.

Moreover, such innovations contribute to the scalability and efficiency of renewable energy systems, paving the way for a more sustainable energy future. By addressing challenges in variability and prediction accuracy, these models support the broader adoption of photovoltaic systems, aligning with global efforts to combat climate change and transition toward cleaner, greener energy.

References

1. Guermoui M, Melgani F, Gairaa K, Mekhalfi ML (2020) A comprehensive review of hybrid models for solar radiation forecasting. J Clean Prod 258:120357. https://doi.org/10.1016/j.jclepro.2020.120357
2. Guermoui M, Bouchouicha K, Bailek N, Boland JW (2021) Forecasting intra-hour variance of photovoltaic power using a new integrated model. Energy Convers Manage 245:114569. https://doi.org/10.1016/j.enconman.2021.114569
3. Rabehi A, Guermoui M, Lalmi D (2020) Hybrid models for global solar radiation prediction: a case study. Int J Ambient Energy 41(1):31–40. https://doi.org/10.1080/01430750.2018.1443498
4. Guermoui M, Fezzani A, Mohamed Z, Rabehi A, Ferkous K, Bailek N, Ghoneim SS (2024) An analysis of case studies for advancing photovoltaic power forecasting through multi-scale fusion techniques. Sci Rep 14(1):6653. https://doi.org/10.1038/s41598-024-57398-z
5. Khelifi R, Guermoui M, Rabehi A, Taallah A, Zoukel A, Ghoneim SS, Zaitsev I (2023) Short-term PV power forecasting using a hybrid TVF-EMD-ELM strategy. Int Trans Electr Energy Syst 2023(1):6413716. https://doi.org/10.1155/2023/6413716
6. Guermoui M, Arrif T, Belaid A, Hassani S, Bailek N (2024) Enhancing direct normal solar irradiation forecasting for heliostat field applications through a novel hybrid model. Energy Convers Manag 304:118189. https://doi.org/10.1016/j.enconman.2024.118189. Get rights and content
7. Ferkous K, Guermoui M, Menakh S, Bellaour A, Boulmaiz T (2024) A novel learning approach for short-term photovoltaic power forecasting-a review and case studies. Eng Appl Artif Intell 133:108502. https://doi.org/10.1016/j.engappai.2024.108502
8. Guermoui M, Rabehi A, Gairaa K, Benkaciali S (2018) Support vector regression methodology for estimating global solar radiation in Algeria. Eur Phys J Plus 133:1–9. https://doi.org/10.1140/epjp/i2018-11845-y
9. Dragomiretskiy K, Zosso D (2013) Variational mode decomposition. IEEE Trans Signal Process 62(3):531–544. https://doi.org/10.1109/TSP.2013.2288675
10. Popel M, Bojar O (2018) Training tips for the transformer model. arXiv:1804.00247
11. Gers FA, Schmidhuber J, Cummins F (2000) Learning to forget: continual prediction with LSTM. Neural Comput 12(10):2451–2471. https://doi.org/10.1162/089976600300015015
12. Boulmaiz T, Guermoui M, Boutaghane H (2020) Impact of training data size on the LSTM performances for rainfall–runoff modeling. Model Earth Syst Environ 6:2153–2164. https://doi.org/10.1007/s40808-020-00830-w
13. Dey R, Salem FM (2017) Gate-variants of gated recurrent unit (GRU) neural networks. In: 2017 IEEE 60th international midwest symposium on circuits and systems (MWSCAS). IEEE, pp. 1597–1600. https://doi.org/10.1109/MWSCAS.2017.8053243
14. Guermoui M, Benkaciali S, Gairaa K, Bouchouicha K, Boulmaiz T, Boland JW (2022) A novel ensemble learning approach for hourly global solar radiation forecasting. Neural Comput Appl 1–23. https://doi.org/10.1007/s00521-021-06421-9
15. Ferkous K, Farouk C, Abdalah K, Belgacem B, Nacer H (2022) Wavelet packet-Gaussian process regression multivariate and univariate model for forecasting daily solar radiation. Int J Energy Clean Environ 23(2). https://doi.org/10.1615/InterJEnerCleanEnv.2021037413
16. Giuliari F, Hasan I, Cristani M, Galasso F (2021) Transformer networks for trajectory forecasting. In: 2020 25th international conference on pattern recognition (ICPR). IEEE, pp 10335–10342 https://doi.org/10.1109/ICPR48806.2021.9412190
17. Guermoui M, Gairaa K, Ferkous K, Santos DSDO Jr, Arrif T, Belaid A (2023) Potential assessment of the TVF-EMD algorithm in forecasting hourly global solar radiation: review and case studies. J Clean Prod 385:135680. https://doi.org/10.1016/j.jclepro.2022.135680
18. Benkaciali S, Haddadi M, Khellaf A, Gairaa K, Guermoui M (2016) Evaluation of the global solar irradiation from the artificial neural network technique. J Renew Energ 19(4):617–631. https://doi.org/10.54966/jreen.v19i4.599

19. Chauhan R, Ghanshala KK, Joshi RC (2018) Convolutional neural network (CNN) for image detection and recognition. In: 2018 first international conference on secure cyber computing and communication (ICSCCC). IEEE, pp 278–282. https://doi.org/10.1109/ICSCCC.2018.8703316
20. Guermoui M, Boland J, Rabehi A (2020) On the use of BRL model for daily and hourly solar radiation components assessment in a semiarid climate. Eur Phys J Plus 135(2):1–16. https://doi.org/10.1140/epjp/s13360-019-00085-0
21. ur Rehman N, Aftab H (2019) Multivariate variational mode decomposition. IEEE Trans Signal Process 67(23):6039–6052. https://doi.org/10.1109/TSP.2019.2951223
22. Guermoui M, Melgani F, Danilo C (2018) Multi-step ahead forecasting of daily global and direct solar radiation: a review and case study of Ghardaia region. J Clean Prod 201:716–734. https://doi.org/10.1016/j.jclepro.2018.08.006
23. Guermoui M, Gairaa K, Boland J, Arrif T (2021) A novel hybrid model for solar radiation forecasting using support vector machine and bee colony optimization algorithm: review and case study. J SolEnergy Eng 143(2):020801. https://doi.org/10.1115/1.4047852
24. Arrif T, Sánchez-González A, Bezza B, Belaid A (2022) Shadowing and blocking factors in heliostats: comparison between parallel and oblique projections. In: AIP conference proceedings, vol 2445, no 1. AIP Publishing. https://doi.org/10.1063/5.0085716
25. Guermoui M, Gairaa K, Rabehi A, Djafer D, Benkaciali S (2018) Estimation of the daily global solar radiation based on the Gaussian process regression methodology in the Saharan climate. Eur Phys J Plus 133:1–17. https://doi.org/10.1140/epjp/i2018-12029-7

Chapter 11
Role of Artificial Intelligence (AI) in Sustainable Biofuel Production from Lignocellulosic Biomass

Anjuman Ayub, Reena Singh, Atif Khurshid Wani, Rongrong Xie, Farhan Ahmad, Babbiker Mohammed Taher Gorish, Jianzhong Sun, Niknik Nurhayati, Radityo Pangestu, Ali Hashemizadeh, and Mudasir A. Dar

Abstract In view of the climate change concerns, there is a growing shift toward the advancement of green and renewable energy solutions to ensure a sustainable future. Renewable energy sources hold the potential to substantially supplant non-renewable fossil fuels, thereby promoting sustainable growth, advancing energy innovation, and mitigating reliance on conventional energy systems. However, the conventional techniques utilized in biorefineries to analyze and optimize biofuel production systems are frequently complex, time-intensive, and laborious. This highlights the pressing need for the development of more efficient and streamlined approaches to enhance the feasibility and scalability of biofuel technologies. The modern artificial intelligence (AI) holds immense potential for addressing high-dimensional, and complex scientific challenges while enhancing decision-making processes within renewable energy sectors. AI can revolutionize the biofuel industry by addressing critical challenges in the production of biofuels from lignocellulosic biomass, including feedstock optimization, supply chain management, and process optimization. By AI-driven approaches, such as crop yield prediction, waste biomass

A. Ayub · R. Singh · A. K. Wani (✉)
School of Bioengineering and Biosciences, Lovely Professional University, Jalandhar 144411, Punjab, India
e-mail: atifkhurshid61200216@gmail.com

R. Xie · F. Ahmad · B. M. T. Gorish · J. Sun · M. A. Dar (✉)
School of the Environment and Safety Engineering, Biofuels Institute, Jiangsu University, Zhenjiang, China
e-mail: muddar7@ujs.edu.cn

N. Nurhayati · R. Pangestu
Research Centre for Genetic Engineering, National Research and Innovation Agency of Indonesia, Bogor, West Java, Indonesia

A. Hashemizadeh
Key Laboratory for Resilient Infrastructures of Coastal Cities, Ministry of Education, Shenzhen University, Shenzhen, China

M. Nasr et al. (eds.), *Artificial Intelligence Applications for a Sustainable Environment*, Green Chemistry and Sustainable Technology,
https://doi.org/10.1007/978-3-031-91199-6_11

utilization, and genetic engineering, the chapter demonstrates how AI can maximize the potential of lignocellulosic feedstocks. Furthermore, AI's transformative role is explored in biofuel production processes, including fermentation, hydrolysis, and transesterification, showcasing significant improvements in efficiency, cost reduction, and product quality. Through predictive modeling and in silico testing, AI accelerates the discovery of novel methods for producing biofuels from lignocellulosic biomass, underscoring its pivotal role in advancing scalable and sustainable energy solutions. This chapter establishes the critical linkage between biofuel production from lignocellulosic biomass and artificial intelligence, illustrating the transformative capabilities in overcoming technical and operational barriers.

Keywords Artificial intelligence · Sustainable biofuel production · Feedstock optimization · CRISPR technology · Enzymatic efficiency · AI models · Biorefinery

11.1 Introduction

The global demand for energy has surged dramatically with industrialization and urbanization, leading to a growing reliance on fossil fuels. However, this dependency has increased environmental issues such as greenhouse gas emissions, climate change, and resource depletion [1]. In response, the pursuit of sustainable energy solutions has gained significant momentum, with biofuels emerging as a viable alternative. Derived from renewable biological resources, biofuels promise to reduce carbon emissions while meeting energy needs [2]. Despite their enormous potential, traditional biofuel production faces several challenges, including high production costs, inconsistent feedstock supply, and suboptimal process efficiencies (Fig. 11.1). These hurdles limit the scalability and sustainability of biofuels as a mainstream energy solution [3]. As industries strive to overcome these obstacles, integrating advanced technologies has become essential to revolutionize biofuel production processes. On recently developed technology, AI has emerged as a transformative tool in addressing the challenges of sustainable biofuel production. AI-driven algorithms, data analytics, and predictive modeling provide innovative solutions to optimize feedstock selection, enhance production efficiency, and reduce environmental impacts [4]. By machine learning (ML) and deep learning, AI enables precise monitoring, automation, and decision-making, driving the biofuel industry toward a more sustainable future [5]. In this chapter, we examined recent advancements in the integration of artificial intelligence (AI) within biofuel production processes, focusing on trends from 2010 to 2024. By utilizing databases such as Scopus and Web of Science (WoS), this study analyzed the most frequently referenced research articles and developments in AI-driven innovations to identify key breakthroughs and challenges. The chapter illustrates the role of AI in optimizing feedstock selection, process efficiencies, and environmental sustainability, providing a comprehensive outlook on the application of AI in biofuel systems. This chapter aims to explore the transformative

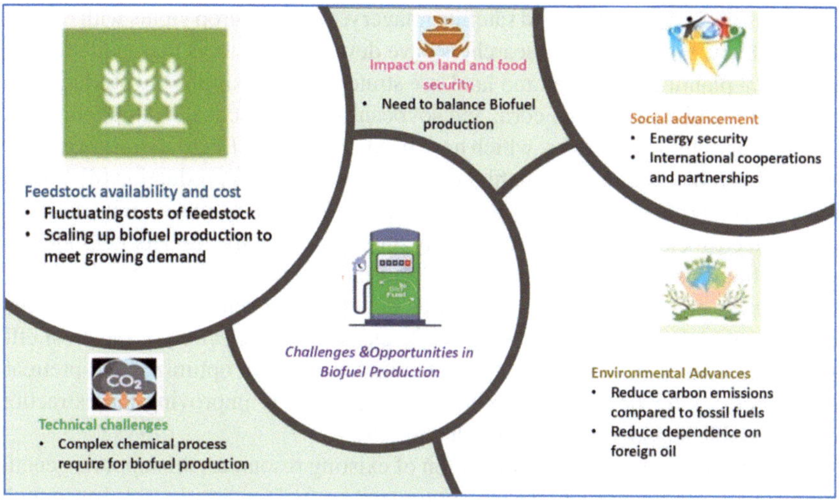

Fig. 11.1 Key challenges and opportunities in biofuel development: Exploring feedstock availability and costs, balancing biofuel production with land and food security, advancing social and environmental sustainability, and addressing technical barriers for a greener future. (*Source* Author created)

potential of AI in advancing sustainable biofuel production systems. Specifically, it focuses on addressing the major challenges of traditional biofuel production, such as high costs, inefficient processes, and environmental concerns, with an emphasis on AI-driven innovations to optimize these systems. The objectives include examining AI applications in feedstock selection, process optimization, and environmental impact reduction, particularly in the context of lignocellulosic biofuels. By investigating these AI-based advancements, the chapter seeks to demonstrate how they can improve economic viability, scalability, and sustainability, thereby contributing significantly to the global transition toward renewable energy.

11.2 AI-Driven Innovations in Biofuel Feedstock Optimization

AI is reshaping the biofuel industry, particularly in feedstock optimization, which forms the foundation of sustainable biofuel production (Fig. 11.2). By utilizing ML algorithms and big data analytics, AI is enabling precise predictions and strategic decision-making to optimize crop yields, select waste biomass, and engineer feedstocks for maximum efficiency [6]. These advancements not only reduce costs but also minimize environmental impacts, addressing two critical challenges in the sector. One significant application of AI in feedstock optimization is crop yield prediction and land optimization [7]. Advanced ML models analyze historical data, weather patterns, soil characteristics, and satellite imagery, to predict crop yields with remarkable accuracy. For example, researchers have developed AI tools that guide farmers on optimal planting schedules and land-use strategies to maximize the production of biofuel crops like sugarcane, corn, and soybeans [6]. A practical example is IBM's AI-based agricultural systems, which have been deployed in Brazil to improve sugarcane yield predictions, leading to higher efficiency in ethanol production [8]. AI also plays a crucial role in the selection and processing of waste biomass, an increasingly favored feedstock due to its sustainability and low cost. ML models help identify suitable biomass types from agricultural and industrial waste, such as rice husks, wheat straw, and sawdust [7]. AI systems analyze the chemical composition and energy potential of various waste materials, ensuring the selection of the most efficient feedstocks. For example, AI-driven simulations have optimized the pretreatment processes for lignocellulosic biomass, significantly improving the extraction of fermentable sugars for biofuel production [9].

In addition to improving the utilization of existing resources, AI supports genetic and metabolic engineering to develop optimized biofuel feedstocks. AI-driven tools in genomics and proteomics enable the identification of genetic traits associated with higher biomass yield and resilience to environmental stress [7]. By integrating deep learning algorithms, researchers have successfully engineered bioenergy crops such as switchgrass and *Miscanthus* to increase their photosynthetic efficiency and biomass output. A notable case is the application of AI in identifying gene-editing targets in algae, enhancing their lipid content for biodiesel production [10]. AI's

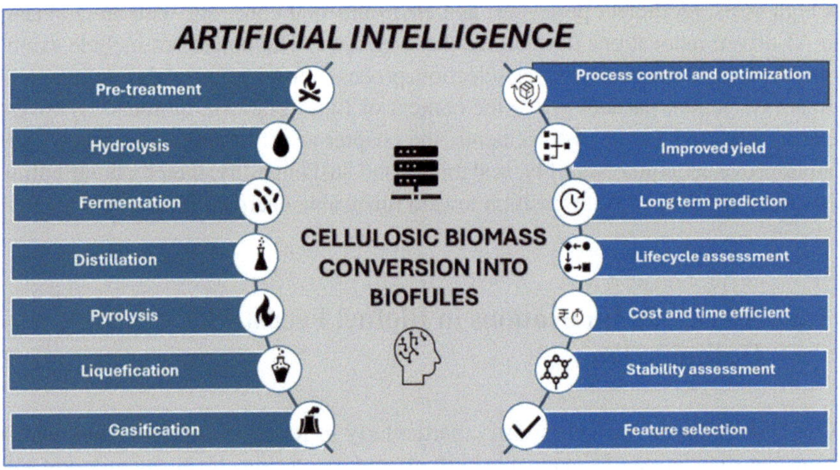

Fig. 11.2 A visual representation of AI-driven optimization in the biofuel production process, encompassing key stages such as pretreatment, hydrolysis, fermentation, distillation, and pyrolysis. The image highlights AI's role in process control, optimization, long-term prediction, cost reduction, time efficiency, and sustainability assessment, showcasing its transformative potential for advancing sustainable biofuel production from cellulosic biomass. (*Source* Author created)

impact extends to reducing resource wastage and improving environmental sustainability. For example, smart AI-driven irrigation systems optimize water use based on crop needs, while predictive models guide fertilizer application to prevent overuse and pollution. These systems are particularly relevant in regions like India and China, where AI technologies are being tested to optimize the production of jatropha and other biofuel crops under dry conditions [6]. In the context of waste-to-biofuel initiatives, AI supports the circular economy by transforming waste streams into valuable energy resources [9] leading to zero waste concepts. Companies like Bio Transformers leverage AI to automate the sorting, processing, and conversion of food waste into biofuels, demonstrating the scalability of such technologies for urban and industrial waste management systems. The integration of AI in feedstock optimization has also opened doors to personalized farming solutions, wherein small-scale farmers can access AI-driven APPs for guidance on crop selection, harvesting schedules, and market predictions (Table 11.1) [6, 7]. This democratization of technology ensures that even remote farming communities can contribute to and benefit from the biofuel economy. Moreover, AI has facilitated the development of hybrid feedstock systems, where multiple biomass sources are combined to improve biofuel yield and quality. By analyzing feedstock compatibility and synergistic effects, AI tools optimize blend ratios, resulting in higher efficiency in biofuel conversion processes. AI-driven innovations in feedstock optimization are revolutionizing the biofuel industry [6, 11, 12]. From crop yield prediction to waste biomass processing and genetic engineering, AI technologies are making biofuel production more efficient, cost-effective, and sustainable. These advancements not only strengthen the economic viability of biofuels but also support global efforts to transition toward renewable energy, offering a promising path to mitigate the energy and environmental crises of current time [3].

11.3 AI in Enhancing Biofuel Production Processes

AI is driving significant advancements in biofuel production by optimizing key processes, ensuring precision in monitoring, and enhancing operational efficiency. From fermentation to hydrolysis and transesterification, AI-powered systems are transforming traditional methods into cutting-edge, efficient workflows [20]. These improvements are making biofuel production more sustainable and economically viable, enabling the industry to meet increasing global energy demands. Process optimization is a core area where AI excels in biofuel production [4]. Likewise, ML algorithms are used to refine critical processes such as fermentation, hydrolysis, and transesterification. For example, in fermentation, AI analyzes microbial activity, nutrient availability, and environmental factors to maximize ethanol or biogas yield [16]. Similarly, hydrolysis processes, which involve breaking down complex biomass into fermentable sugars, are optimized using AI-driven predictive models that adjust enzyme concentrations and reaction conditions enabling real-time process monitoring [21]. Transesterification, crucial for biodiesel production, benefits from AI

Table 11.1 AI-driven innovations in biofuel feedstock optimization and application. The table illustrates diverse AI-driven methodologies, including machine learning, neural networks, reinforcement learning, and predictive analytics, demonstrating their transformative potential in optimizing feedstock utilization, enhancing biofuel production processes, and promoting environmental sustainability

AI technique	Study aim	Feedstock source	Practical applications	Implications	References
ML models	Predicting crop yield and land utilization	Agricultural feedstocks	GIS-based crop rotation optimization in Brazil	Enhanced agricultural efficiency: achieved a 15% increase in yield and a 10% reduction in operational costs, promoting sustainable farming practices	[13]
Neural networks	Biomass yield estimation	Lignocellulosic biomass	Predictive modeling for miscanthus and switchgrass cultivation	Accurate yield prediction: achieved 92% precision, enabling optimized harvest schedules, reducing waste, and increasing overall production efficiency	[14]
Deep learning algorithms	Biomass waste optimization	Agro-industrial residues	Waste sorting in bioethanol production	Improved waste utilization: enhanced biomass utilization efficiency by 25% and reduced processing time by 30%, leading to lower production costs	[9]
Genetic algorithm framework	Genetic and metabolic engineering of feedstocks	Energy crops	CRISPR-enhanced sorghum for bioethanol	Advanced crop engineering: developed drought-tolerant, high-yield energy crops, resulting in a 20% increase in bioethanol production efficiency	[15]

(continued)

Table 11.1 (continued)

AI technique	Study aim	Feedstock source	Practical applications	Implications	References
Predictive analytics models	Biomass quality assessment	Non-food energy crops	Quality control in jatropha-based biodiesel	Enhanced biodiesel quality: improved fuel consistency and efficiency by 12%, addressing key challenges in scaling-up biodiesel production	[13]
Reinforcement learning systems	Cultivation optimization for algal feedstocks	Microalgae lipid	Maximization in photobioreactors	Boosted biodiesel potential: achieved a 40% increase in lipid content for biodiesel applications, showcasing high potential for algal biofuels	[16]
Support vector machines (SVM)	Geospatial optimization for crop cultivation	Perennial energy crops	Oil palm location analysis in Sub-Saharan Africa	Reduced environmental impact: identified optimal regions for cultivation, minimizing deforestation and maximizing crop yield sustainably	[17]
Decision tree-based algorithms	Regional feedstock suitability assessment	Organic waste streams	Waste sorting for biogas production in Germany	Improved waste-to-energy conversion: boosted efficiency by 18%, demonstrating the feasibility of biogas as a sustainable energy source	[18]

(continued)

Table 11.1 (continued)

AI technique	Study aim	Feedstock source	Practical applications	Implications	References
Simulation-driven AI models	Environmental impact assessment of feedstocks	Leguminous energy crops	Lifecycle analysis of soybean-derived biofuel	Reduced carbon footprint: achieved a 25% reduction in lifecycle greenhouse gas (GHG) emissions, aligning biofuels with global sustainability goals	[6]
AI-enhanced image recognition	Pathogen and pest detection in energy crops	Root crops	sugar beet pest monitoring in European farms	Enhanced crop protection: reduced crop losses by 30%, minimized pesticide use, and increased overall productivity while lowering environmental risks	[19]

by automating the identification of optimal catalyst types and reaction parameters, ensuring consistent and high-quality fuel output. Companies like Novozymes are leveraging AI-based tools to enhance enzyme performance in biomass breakdown, reducing costs and improving efficiency [22].

AI enables real-time parameter control and monitoring across production systems, ensuring precision and minimizing energy and material wastage. Smart sensors, integrated with AI platforms, continuously collect and analyze data on temperature, pH levels, pressure, and substrate concentration [7]. This real-time feedback allows immediate adjustments, preventing process deviations that could compromise biofuel quality [23]. For example, AI-enabled fermentation tanks dynamically control oxygen and nutrient levels to maintain the ideal environment for microbial growth, leading to optimal ethanol production [24]. Systems like Siemens' AI-powered automation solutions are already being used in biorefineries to monitor these variables and maintain consistent production quality. Another crucial contribution of AI is in predictive maintenance and automation in biorefineries. Predictive maintenance, driven by AI algorithms, analyzes equipment performance data to forecast potential failures and recommend proactive measures [25]. This reduces downtime, extends equipment life, and ensures uninterrupted biofuel production. For example, biorefineries employing AI for predictive maintenance report significant cost savings by addressing issues before they escalate into costly repairs. Additionally, automation systems powered by AI streamline operations, from biomass input to fuel output, reducing the need for manual intervention and enhancing overall efficiency [24].

AI also improves energy efficiency in biorefineries by optimizing resource use. For example, AI systems monitor energy consumption and suggest adjustments to reduce power usage during high-energy processes like distillation and drying [26]. In bioethanol plants, AI tools analyze distillation data to minimize heat energy requirements without compromising production rates, significantly reducing operational costs [27]. Integration of AI in biorefineries has demonstrated success in advancing operational scalability [28]. Large-scale biofuel producers use AI to model complex production scenarios, test variables virtually, and implement the most efficient setups. For example, in biohydrogen production, AI has been employed to optimize anaerobic digestion processes, improving hydrogen yields and reducing waste output [29]. Furthermore, AI fosters collaboration between biofuel producers and researchers by simulating production models and identifying areas for innovation. For example, research teams use AI-driven simulations to study the effects of alternative catalysts or substrates in transesterification, offering valuable insights that drive experimental breakthroughs. The environmental benefits of AI in biofuel production are equally noteworthy. AI systems minimize greenhouse gas emissions by optimizing processes that reduce waste generation and energy consumption [16, 30]. Real-time monitoring ensures compliance with environmental regulations, further enhancing the sustainability of biofuel production. AI's integration into biofuel production processes is transforming the industry. By optimizing fermentation, hydrolysis, and transesterification, enabling real-time monitoring, and facilitating predictive maintenance, AI significantly enhances operational efficiency and sustainability. These advancements ensure that biofuel production remains a viable, eco-friendly alternative to fossil fuels, contributing to the global shift toward renewable energy solutions [31].

11.4 AI for Sustainable Biofuel Supply Chain Management

AI is transforming the biofuel supply chain by addressing challenges in logistics, emissions reduction, and waste utilization. In supply chain logistics, AI enhances the efficiency of feedstock collection and transportation, a critical step in biofuel production [16]. ML algorithms analyze variables like weather conditions, biomass availability, and road traffic to create optimized routes for feedstock transportation. For example, studies have shown that AI-integrated route optimization in Brazil's sugarcane industry reduced transportation costs by 15% while improving delivery times for bioethanol production facilities. AI systems also ensure better resource allocation by predicting feedstock availability based on real-time data and historical trends [32]. Predictive analytics tools help biofuel producers plan harvesting schedules and ensure continuous feedstock supply. A practical example is the use of AI in India's jatropha biodiesel projects, where algorithms monitor crop growth and determine the ideal harvest time, minimizing wastage and ensuring higher yields [4]. In emissions reduction and energy optimization, AI-driven systems monitor and adjust transportation and processing stages to lower greenhouse gas emissions. For example, AI-powered fleet management solutions, such as those implemented by companies like DHL, analyze vehicle performance and fuel consumption to suggest eco-friendly driving practices and route changes. Similar technology has been applied in biorefineries to optimize the energy use in storage and distribution processes, reducing carbon footprints by up to 20%.

AI also plays a significant role in enabling circular economy practices within the biofuel supply chain. Waste-to-energy technologies, powered by AI, identify and process usable by-products from agricultural and industrial sources. In one case, AI was employed to assess the energy potential of municipal solid waste in urban areas, leading to the successful conversion of food waste into bioethanol in pilot projects across Europe [30, 33]. Another innovative application is AI's ability to integrate blockchain technology into supply chains, ensuring transparency and traceability. By recording every stage of the feedstock lifecycle, from collection to fuel production, AI and blockchain systems provide detailed data for quality assurance and compliance with sustainability regulations. For example, this approach has been adopted in Argentina's soybean-based biodiesel production to track sustainability metrics and improve market credibility [34]. The economic benefits of AI in the supply chain are equally impactful. Automation of scheduling, inventory management, and distribution reduces labor costs and enhances operational efficiency. Companies such as Neste are using AI to streamline their waste-based biofuel supply chains, significantly lowering production costs and making biofuels more competitive with fossil fuels [35]. AI's integration into biofuel supply chains not only optimizes logistics and reduces emissions but also fosters sustainable practices like waste recycling and energy efficiency. These advancements are helping the biofuel industry align with global sustainability goals, while also driving cost-effective and scalable energy solutions [16].

11.5 AI-Driven Modeling and Simulation in Biofuel Research

AI-driven modeling and simulation are redefining research in biofuel production by enabling accurate predictions, optimizing processes, and accelerating innovation. Predictive modeling, a key AI application, utilizes data from past and ongoing experiments to forecast outcomes and identify optimal conditions for biofuel production [36]. For example, AI models have been used to predict enzymatic efficiency in hydrolysis, allowing researchers to maximize sugar extraction from lignocellulosic biomass [7]. This has significantly enhanced ethanol production in pilot-scale operations across the United States. In silico testing is another transformative application of AI, enabling the virtual exploration of biofuel production methods. These simulations reduce the time and cost associated with traditional laboratory experiments [5]. For example, researchers at the University of Cambridge employed AI-driven molecular simulations to study the effects of different catalysts on biodiesel transesterification. The findings identified low-cost catalysts with high conversion efficiency, paving the way for more sustainable production methods. AI models also aid in optimizing complex multi-step processes, such as anaerobic digestion for biogas production [6]. ML algorithms predict the interaction of various parameters, such as temperature, pH, and substrate composition, to enhance methane yields. In a notable study in Germany, AI-guided anaerobic digestion improved biogas production by 25%, demonstrating the potential of data-driven optimization [14].

Another exciting development is the use of AI for metabolic pathway analysis in microbial biofuel production. Deep learning algorithms map metabolic networks in microorganisms, identifying bottlenecks and suggesting modifications for enhanced biofuel yield [37]. A prominent example is the use of AI in engineering yeast strains to produce higher amounts of ethanol, a breakthrough achieved by researchers at MIT. AI also supports collaborative research by enabling global data sharing and model integration [16]. AI models require extensive and high-quality datasets for accurate predictions, and obtaining such data from diverse feedstocks and processes is a significant hurdle [16]. However, efforts are underway to standardize data collection in biofuel research, creating comprehensive repositories for future AI applications. AI-driven modeling and simulation are powerful tools for advancing biofuel research. By predicting outcomes, optimizing processes, and enabling virtual experimentation, these technologies are accelerating the development of efficient and sustainable biofuel production methods [28].

11.6 AI Applications in Biofuel Production

AI has proven its transformative potential in biofuel production, with numerous real-world implementations delivering measurable outcomes. One significant application is in enzymatic efficiency optimization (Fig. 11.3). AI-driven platforms like DeepMind's AlphaFold have advanced enzyme engineering by predicting protein structures with high accuracy [28]. This has allowed researchers to design enzymes tailored for efficient biomass hydrolysis, boosting bioethanol yields [16]. For example, Novozymes employed AI to develop enzymes that significantly enhance the breakdown of agricultural residues into fermentable sugars, reducing production costs. AI has also improved process efficiency in biodiesel production through advanced transesterification modeling [24]. BASF leveraged ML algorithms to determine the optimal conditions for catalyst selection, reaction temperature, and time, resulting in higher biodiesel purity and reduced energy consumption [38]. This approach has been adopted in large-scale facilities, such as those operated by Green Plains Inc., to enhance operational efficiency and fuel quality [39]. The integration of AI in real-time monitoring systems has streamlined biofuel production processes. IoT sensors combined with AI algorithms provide continuous data on temperature, pressure, and substrate levels, enabling instant adjustments to maintain optimal production conditions. A successful example is found in ethanol plants in Brazil, where AI-powered systems monitor fermentation tanks and automatically correct deviations, preventing production losses and improving yield consistency [32].

AI's role extends to microbial biofuel production, where it aids in identifying high-performing strains. AI-guided genome editing, such as CRISPR-based techniques, has been used to engineer microorganisms capable of producing advanced biofuels

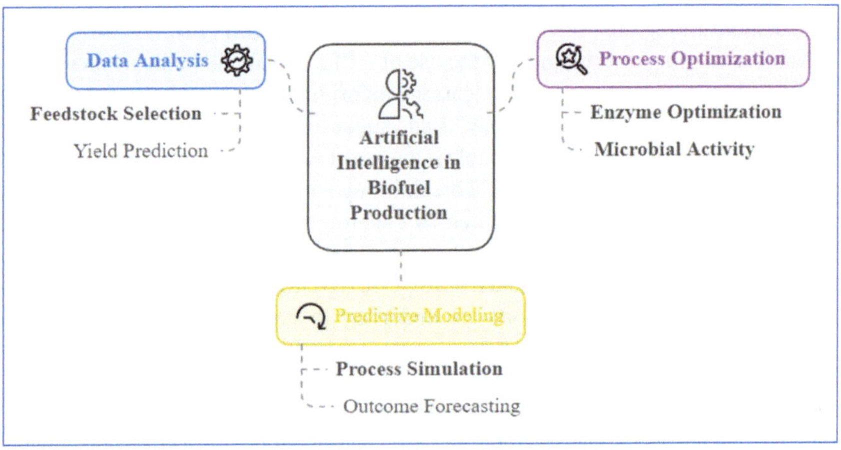

Fig. 11.3 Applications of artificial intelligence (AI) in biofuel production, showcasing the integration of AI technologies in optimizing feedstock selection, biomass conversion processes, and enhancing overall efficiency in biofuel production. (*Source* Author created)

like biobutanol and bio-propanol [7]. Companies like Ginkgo Bioworks utilize AI to analyze microbial genomes and design synthetic pathways, accelerating the development of industrial-scale biofuel production. Another critical application is in supply chain traceability [40]. AI, integrated with blockchain technology, has enabled transparent tracking of feedstock from farm to fuel [41]. In Argentina, soybean biodiesel producers use AI-driven blockchain platforms to ensure sustainable sourcing and compliance with environmental regulations, strengthening consumer confidence in biofuels [42]. Key lessons from these implementations highlight the importance of tailored AI solutions. Each application requires adaptation to specific feedstocks, processes, and market conditions [43]. For example, while AI excels in enzymatic optimization for lignocellulosic biomass, its application in algal biofuels requires further refinement due to the complexity of algal systems [30]. AI applications in biofuel production have demonstrated substantial benefits, from improving process efficiency to enabling advanced microbial engineering and supply chain transparency. These successes provide a blueprint for scaling AI technologies across diverse biofuel systems, fostering a more sustainable energy future [30].

11.7 Limitations of AI in Sustainable Biofuel Production

While AI offers significant advantages for biofuel production, it also faces several challenges and limitations. One major technical barrier is the availability of high-quality and comprehensive datasets [10]. AI models require diverse data inputs to produce accurate predictions, yet biofuel production data often vary widely depending on feedstock types, geographic regions, and production scales [44]. For example, small-scale biofuel producers in rural areas may lack the infrastructure for consistent data collection, limiting the effectiveness of AI tools [45]. The complexity of biofuel production processes presents another significant limitation. Biofuel production involves multiple interdependent steps, such as fermentation, enzymatic hydrolysis, and distillation, each influenced by numerous biological and chemical variables [21]. While AI can optimize parameters in controlled environments, it often struggles to model the unpredictable variations encountered during industrial-scale operations. For example, feedstock quality and environmental conditions can vary significantly, leading to inefficiencies that undermine the benefits of AI-driven optimization. This highlights the gap between lab-scale successes and real-world applications, where unexpected interactions between variables can complicate the scaling-up of biofuel production [46]. Economic and technical barriers further hinder the adoption of AI-driven innovations, particularly for small- and medium-sized enterprises. The high costs associated with implementing AI technologies including software development, hardware installation, and staff training can be prohibitive [1]. Large-scale producers, such as Neste, have successfully integrated AI due to their substantial resources, but smaller companies often struggle to justify the initial investment. Additionally, technical expertise is often lacking, with many biofuel producers unable to recruit or train personnel skilled in AI systems. For example, small biofuel industries in developing regions frequently face challenges in accessing the required sensors and IoT devices for real-time data acquisition, further limiting the applicability of AI-driven solutions. Regulatory challenges also complicate AI adoption in biofuel production [47]. Biofuel production is governed by stringent environmental and safety regulations that vary by region. AI systems must be designed to comply with these regulations, adding complexity to their implementation. For example, ensuring that AI-driven processes meet carbon reduction standards in Europe requires thorough validation and certification, delaying deployment [30]. Addressing these regulatory hurdles demands multi-stakeholder collaboration and proactive government involvement to create streamlined, AI-compliant policies.

Ethical concerns and environmental considerations are another avenue that demands further exploration and possible addressal. AI-driven land-use planning for biofuel crops can inadvertently impact food security by prioritizing energy crops over food crops [48]. This issue has been observed in regions like Sub-Saharan Africa, where biofuel expansion has raised concerns about food availability [49]. Ensuring ethical AI use requires balancing energy and food needs, emphasizing the importance of multi-stakeholder collaboration. Another challenge is the risk of over-automation [3]. While AI enhances efficiency, excessive reliance on automated systems can reduce the need for skilled human labor, potentially displacing workers in rural areas where biofuel feedstocks are sourced (Fig. 11.4). This calls for strategies to integrate AI in a way that supports job creation and community development [45]. Addressing these challenges requires a multifaceted approach, including investing in data standardization, providing financial incentives for AI adoption, and fostering collaboration between stakeholders. Governments can play a key role by funding AI research in biofuels and creating policies that promote ethical and sustainable practices [5]. While AI holds immense potential for advancing biofuel sustainability, overcoming its challenges is crucial to unlocking its full benefits. Proactive strategies and collaborative efforts are essential to address technical, economic, and ethical barriers, ensuring that AI contributes to a sustainable and inclusive biofuel industry [10].

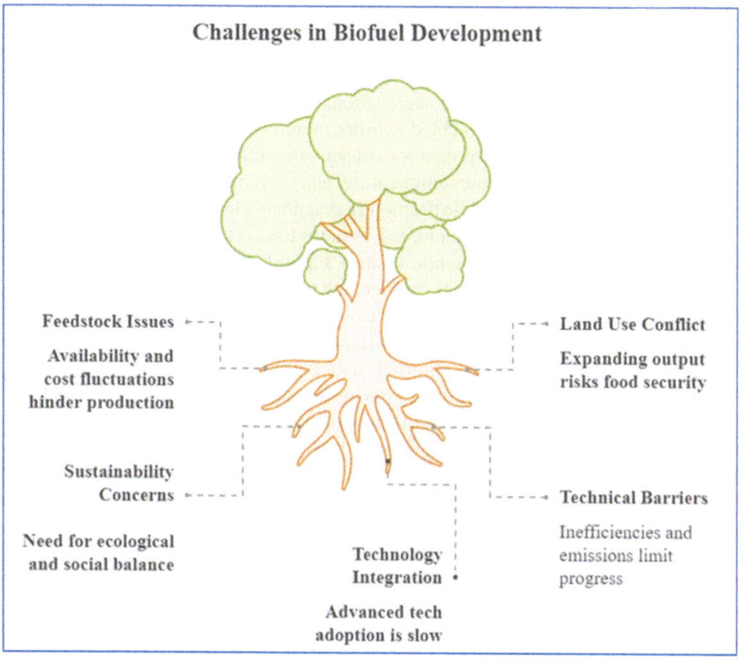

Fig. 11.4 A conceptual illustration of key challenges in biofuel development, including, feedstock availability, land-use competition, water resource constraints, technological inefficiencies, and environmental concerns. (*Source* Author created)

Acknowledgements This study was funded by the National Natural Science Foundation of China (grant number 32250410285), the National Key R&D Program of China (2023YFC3403600), and the Foreign Expert Scientist Program under the Ministry of Science and Technology of China (WGXZ2023020L).

Declarations

The Future of AI in Biofuel Production: Opportunities and Conclusions The future of AI in biofuel production is poised for rapid innovation, with emerging technologies offering transformative opportunities for sustainability and efficiency. Advanced neural networks (ANNs) and joined learning are anticipated to enhance predictive modeling by enabling AI systems to learn from decentralized datasets without compromising data privacy [1]. These advancements will improve AI's ability to optimize biofuel production processes across diverse regions and feedstocks. Integration with the Internet of Things (IoT) represents another promising avenue [20], encompassing smart devices like nanosensors that can collect real-time data on feedstock quality, environmental conditions, and production metrics. This data enables AI to make comprehensive, data-driven decisions. For example, IoT-enabled biorefineries in the Netherlands are already utilizing AI to optimize energy consumption and reduce waste, setting a benchmark for sustainable practices [50]. IoT technologies are increasingly used in environmental applications, providing significant opportunities for monitoring and addressing ecological challenges. In air quality monitoring, sensors are deployed in urban and industrial areas to measure pollutants such as particulate matter (PM2.5 and PM10), nitrogen dioxide (NO_2), sulfur dioxide (SO_2), and ozone (O_3), enabling authorities to take timely actions to reduce harmful emissions and improve public health [51]. In water quality monitoring, IoT devices equipped with sensors can assess parameters such as pH levels, turbidity, dissolved oxygen, and contaminants, enabling the management of water resources, particularly in critical sectors like agriculture and public health [52]. Waste management also benefits from IoT, with smart bins and waste tracking systems improving waste segregation, collection efficiency, and reducing landfill overflow. In wildlife tracking, GPS-enabled sensors monitor animal movements, behaviors, and migration patterns, providing essential data for conservation efforts [52]. Additionally, in agriculture, IoT sensors measure soil moisture, temperature, and nutrient levels, helping farmers optimize irrigation and reduce water wastage while promoting sustainable farming practices [53]. For climate change monitoring and carbon management, IoT systems track greenhouse gas emissions, temperature fluctuations, and other environmental data to support efforts aimed at mitigating climate change. Renewable energy optimization is another area where IoT plays a critical role, with devices monitoring solar panel performance, wind turbine efficiency, and energy storage systems, ensuring that renewable energy sources are used to their maximum potential. Through these applications, IoT technologies facilitate real-time monitoring, data collection, and automation, enabling more informed decision-making and contributing to environmental sustainability [54]. Blockchain technology, when combined with AI, offers significant potential for supply chain transparency. AI can analyze blockchain data to track feedstock origins, verify sustainability certifications, and predict market trends [10]. This integration is being explored in biofuel projects in India, where blockchain ensures traceability of jatropha and other feedstocks used in biodiesel production. Similarly, AI-powered CRISPR technology is set to revolutionize feedstock bioengineering by identifying genetic modifications that enhance crop yields or improve biomass quality, enabling more efficient biofuel production. A recent study in China demonstrated the use of AI-guided CRISPR editing to develop drought-resistant sorghum, a promising feedstock for bioethanol [42].

Emerging AI technologies also temporary collaborative global research. Platforms like OpenAI enable researchers to share AI models and datasets, fostering innovation across borders [10]. This collaborative approach is critical for addressing global challenges like climate change, where sustainable biofuels play a pivotal role. To fully realize these opportunities, investments in AI education, infrastructure, and regulatory frameworks are essential [55]. Training programs can equip the biofuel workforce with the necessary skills to leverage AI tools effectively, while supportive policies can encourage innovation and ethical AI use. AI is a cornerstone for advancing sustainable biofuel production. From optimizing feedstock supply chains to enhancing production processes and driving research innovations, AI has demonstrated its ability to address critical challenges in the biofuel industry [56]. Real-world implementations and emerging technologies highlight AI's potential to transform biofuel production into a more efficient, cost-effective, and environmentally sustainable endeavor. Despite challenges such as data limitations, economic barriers, and ethical concerns, collaborative efforts and proactive strategies are paving the way for broader AI adoption. The integration of AI with IoT, blockchain, and CRISPR amplifies its potential, creating opportunities for transparency, efficiency, and innovation in biofuel production [6]. Looking ahead, AI is not merely a tool but a fundamental pillar for achieving global energy sustainability. Its role in fostering a circular economy, reducing greenhouse gas emissions, and ensuring energy security underscores its importance in addressing climate change and promoting renewable energy. Although AI's journey in biofuel production is still in its nascent stage, continued investment in technology and collaborative research will enable AI to shape the future of sustainable energy, empowering the biofuel industry to meet the demands of a greener, more sustainable world [50].

The authors declare that they have no conflict of interest.

References

1. Aggrey J, Smith J, Smith M (2024) Solvent induced membrane stress in biofuel production: an experimental and computational study

2. Bhushan S, Jayakrishnan U, Shree B, Bhatt P, Eshkabilov S, Simsek H (2023) Biological pretreatment for algal biomass feedstock for biofuel production. J Environ Chem Eng 11:109870
3. Shelare SD, Belkhode PN, Nikam KC, Jathar LD, Shahapurkar K, Soudagar MEM, Veza I, Khan TY, Kalam MA, Nizami A-S (2023) Biofuels for a sustainable future: examining the role of nano-additives, economics, policy, internet of things, artificial intelligence and machine learning technology in biodiesel production. Energy 128874
4. Pahwa MS, Dadhich M, Saini JS, Saini DK (2022) Use of artificial intelligence (AI) in the optimization of production of biodiesel energy. In: Vyas AK, Balamurugan S, Hiran KK, Dhiman HS (eds) Artificial intelligence for renewable energy systems, 1st edn. Wiley, pp 229–238. https://doi.org/10.1002/9781119761686.ch11
5. Okolie JA (2024) Introduction of machine learning and artificial intelligence in biofuel technology. Curr Opinion Green Sustain Chem 100928
6. Shuai X, Zhan N (2024) AI-driven innovations in microalgae biotechnology: enhancing biofuel production and environmental health. In: The 3rd international scientific and practical conference "modern trends in the development of science and information technologies" (September 17–20, 2024). International Science Group, Sofia, Bulgaria, p 25. https://books.google.com/books?hl=en&lr=&id=PrAlEQAAQBAJ&oi=fnd&pg=PA25&dq=AI-Driven+Innovations+in+Biofuel+Feedstock+Optimization&ots=1TAbfcMkzg&sig=Pj4ISbiNIhEehbXw_HhNZgLyetM. Accessed 28 Nov 2024
7. Damian CS, Devarajan Y, Thandavamoorthy R, Jayabal R (2024) Harnessing artificial intelligence for enhanced bioethanol productions: a cutting-edge approach towards sustainable energy solution. Int J Chem Reactor Eng 22:719–727. https://doi.org/10.1515/ijcre-2024-0074
8. Pelissari MR, San Martín Cañas SJ, Ciotta MR, Tassinari CCG (2020) BECCUS: a digital approach for a bioethanol plant in Brazil. In: Rio oil & gas expo and conference. https://www.researchgate.net/profile/Mariana-Ciotta/publication/346572049_BECCUS_a_digital_approach_for_a_bioethanol_plant_in_Brazil/links/5fc7c4db299bf188d4e91a6e/BECCUS-a-digital-approach-for-a-bioethanol-plant-in-Brazil.pdf. Accessed 28 Nov 2024
9. Bhat MA, Bhat MA, Jan S, Shah AA, Jan AT (2024) Lignocellulosic biomass in circular economy: a techno-transition in carbon neutrality towards sustainable energy production. Biomass Bioenerg 189:107349
10. Khare S, Ahalya N, Prakash KJ, Devi PB, Bhattacharya S (2024) AI-IoT-enabled biomass-to-biofuel conversion: advancements in synthetic processes and bio-remediation. In: Intelligent solutions for sustainable power grids. IGI Global, pp 195–219. https://www.igi-global.com/chapter/ai-iot-enabled-biomass-to-biofuel-conversion/346140. Accessed 28 Nov 2024
11. Chui M, Evers M, Manyika J, Zheng A, Nisbet T (2023) The bio revolution: Innovations transforming economies, societies, and our lives. In: Augmented education in the global age. Routledge, pp 48–74. https://doi.org/10.4324/9781003230762-5. Accessed 9 Oct 2024
12. Ameen M, Ahmad M, Zafar M, Dawood S, Ahmad S, Kazmi SIM, Amin MF, Elshikh MS, Iqbal M (2024) Sustainable innovations in biodiesel plants: from production to environmental impact. https://www.intechopen.com/online-first/1195241. Accessed 28 Nov 2024
13. Shi Z, Ferrari G, Ai P, Marinello F, Pezzuolo A (2023) Artificial intelligence for biomass detection, production and energy usage in rural areas: a review of technologies and applications. Sustain Energy Technol Assess 60:103548
14. Soudagar MEM, Shelare S, Marghade D, Belkhode P, Nur-E-Alam M, Kiong TS, Ramesh S, Rajabi A, Venu H, Khan TY (2024) Optimizing IC engine efficiency: a comprehensive review on biodiesel, nanofluid, and the role of artificial intelligence and machine learning. Energy Convers Manage 307:118337
15. Mathieu A, Leclercq M, Sanabria M, Perin O, Droit A (2022) Machine learning and deep learning applications in metagenomic taxonomy and functional annotation. Front Microbiol 13:811495
16. Meena M, Shubham S, Paritosh K, Pareek N, Vivekanand V (2021) Production of biofuels from biomass: predicting the energy employing artificial intelligence modelling. Biores Technol 340:125642

17. Messinese E, Pitirollo O, Grimaldi M, Milanese D, Sciancalepore C, Cavazza A (2024) By-products as sustainable source of bioactive compounds for potential application in the field of food and new materials for packaging development. Food Bioprocess Technol 17:606–627
18. Saju L, Selvaraj D, Vairaperumal T (2025) Artificial intelligence and machine intelligence: modeling and optimization of bioenergy production. In: Computer vision and machine intelligence for renewable energy systems. Elsevier, pp 163–176. https://www.sciencedirect.com/science/article/pii/B9780443289477000094. Accessed 28 Nov 2024
19. Ahmad Kamal AH, Mohd Hamidi NF, Zakaria MF, Ahmad A, Harun MR, Chandra Segaran T, Jusoh M (2024) Genetically engineered microalgae for enhanced bioactive compounds. Discov Appl Sci 6:482. https://doi.org/10.1007/s42452-024-06116-5
20. Habib MS, Hwang S-J (2024) Developing sustainable, resilient, and responsive biofuel production and distribution management system: a neutrosophic fuzzy optimization approach based on artificial intelligence and geographic information systems. Appl Energy 372:123683
21. Sharma V, Tsai M-L, Chen C-W, Sun P-P, Nargotra P, Dong C-D (2023) Advances in machine learning technology for sustainable biofuel production systems in lignocellulosic biorefineries. Sci Total Environ 886:163972
22. Singh G, Sahu S, Bharti S, Arya SK (2024) Significance of enzymes for the recycling of wasted non-food biomass to value added products: a sustainable stewardship towards the cleaner environment. Proc Saf Environ Protect https://www.sciencedirect.com/science/article/pii/S095758202400898X. Accessed 28 Nov 2024
23. Wani AK, Chopra C, Ansari MA, Dar MA, Américo-Pinheiro JHP, Singh R (2024) Characterization of thermostable carboxypeptidase from high-altitude hot spring metagenome. Int J Biol Macromol 133974
24. Queiroz LP, Nogueira IBR, Ribeiro AM (2024) Flavor engineering: a comprehensive review of biological foundations, AI integration, industrial development, and socio-cultural dynamics. Food Res Int 115100
25. Izquierdo-Bueno I, Moraga J, Cantoral JM, Carbú M, Garrido C, González-Rodríguez VE (2024) Smart viniculture: applying AI for improved winemaking and risk management. https://www.preprints.org/manuscript/202409.0580. Accessed 28 Nov 2024
26. Miraei Ashtiani S-H, Martynenko A (2024) Toward intelligent food drying: integrating artificial intelligence into drying systems. Dry Technol 42:1240–1269. https://doi.org/10.1080/07373937.2024.2356177
27. Wani AK, Akhtar N, ul T, Mir G, Chopra C, Singh R, Hong JC, Kadam US (2024) CRISPR/Cas12a-based biosensors for environmental monitoring and diagnostics. Environ Technol Innov 34:103625. https://doi.org/10.1016/j.eti.2024.103625
28. Arshad I, Ahsan M, Zafar I, Sajid M, Sehgal SA, Yousaf W, Noor A, Rashid S, Garai S, Moovendhan M (2023) Bioinformatics approaches in upgrading microalgal oil for advanced biofuel production through hybrid ORF protein construction. Biomass Convers Biorefine 1–21
29. Zhang L, Zhao R, Jia D, Jiang W, Gu Y (2020) Engineering Clostridium ljungdahlii as the gas-fermenting cell factory for the production of biofuels and biochemicals. Curr Opin Chem Biol 59:54–61
30. Ananthi V, Raja R, Carvalho IS, Brindhadevi K, Pugazhendhi A, Arun A (2021) A realistic scenario on microalgae based biodiesel production: third generation biofuel. Fuel 284:118965
31. Ahirwar A, Rai A, Sirotiya V, Khandelwal P, Singh G, Jhadav D, Marchand HJ, Schoefs B, Varjani S (2024) Fermentation of algal biomass for its nutritional value: perspectives and revolutions in the food industry. Environ Technol Rev 13: 1–28
32. Setiawan AAR, Bardant TB, Ariesca R, Wiloso EI, Ahamed T, Noguchi R (2024) Development of IoT-based platform for biomass utilization toward low-carbon economic society: case of oil palm residue. In: Ahamed T (ed) IoT and AI in agriculture. Springer Nature Singapore, Singapore, pp 401–420. https://doi.org/10.1007/978-981-97-1263-2_25
33. Charoenying T, Patrojanasophon P, Ngawhirunpat T, Rojanarata T, Akkaramongkolporn P, Opanasopit P (2020) Three-dimensional (3D)-printed devices composed of hydrophilic cap and hydrophobic body for improving buoyancy and gastric retention of domperidone tablets. Eur J Pharm Sci 155:105555

34. Shahid A, Rehman AU, Usman M, Ashraf MUF, Javed MR, Khan AZ, Gill SS, Mehmood MA (2020) Engineering the metabolic pathways of lipid biosynthesis to develop robust microalgal strains for biodiesel production. Biotech App Biochem 67:41–51. https://doi.org/10.1002/bab.1812
35. Thompson D, Cognat V, Goodfellow M, Koechler S, Heintz D, Carapito C, Van Dorsselaer A, Mahmoud H, Sangal V, Ismail W (2020) Phylogenomic classification and biosynthetic potential of the fossil fuel-biodesulfurizing Rhodococcus strain IGTS8. Front Microbiol 11:1417
36. Zhao H, Hillson N, Kleese van Dam K, Tanjore D (2022) Artificial intelligence and machine learning for bioenergy research: opportunities and challenges. https://www.osti.gov/biblio/1968870. Accessed 28 Nov 2024
37. Kaur D, Gupta S (2024) Biosynthetic pathways involved in the synthesis of next-generation biofuels in microbial cells. Proc Biochem
38. Luo Y (2024) Techno-economic analysis, life-cycle assessment, and optimization of sustainable chemical production, PhD Thesis, University of Delaware. https://search.proquest.com/openview/e25a9d235e5b993a4df2f2f91e2c398c/1?pq-origsite=gscholar&cbl=18750&diss=y. Accessed 29 Nov 2024
39. Wani AK, Rahayu F, Fauziah L, Suhara C (2023) Advances in safe processing of sugarcane and bagasse for the generation of biofuels and bioactive compounds. J Agric Food Res 12:100549. https://doi.org/10.1016/j.jafr.2023.100549
40. Wani AK, Chopra C, Singh R, Ahmad S, Américo-Pinheiro JHP (2023) Mining microbial tapestry using high-throughput sequencing and In silico analysis of Trehalose synthase (TreS) derived from hot spring metagenome. Biocatal Agric Biotechnol 52:102829
41. Bendaoud A, Belkhiri A, Hmamou A, Tlemcani S, Eloutassi N, Lahkimi A (2024) Efficient bioethanol production from lignocellulosic biomass using diverse microbial strains. J Ecol Eng 25:353–362
42. Kaladhar DS, Kenar SK, Meesala S, Sahu S (2023) Biofuel production, phytochemical analysis and in silico analysis for active compound valorization from jatropha gossypifolia seeds. J Pharmaceut Negat Res 1555–1565
43. Wani AK, Rahayu F, Yustina I, Fatah GSA, Kariada IK, Antarlina SS, Jufri A, Pamungkas D (2023) Contribution of yeast and its biomass for the preparation of industrially essential materials: a boon to circular economy. Bioresource Technol Rep 23:101508
44. Wani AK, Rahayu F, Ben Amor I, Quadir M, Murianingrum M, Parnidi P, Ayub A, Supriyadi S, Sakiroh S, Saefudin S (2024) Environmental resilience through artificial intelligence: innovations in monitoring and management. Environ Sci Pollut Res 31:18379–18395
45. Arias DM, García-Valladares O, Besagni G, Markides CN (2023) A vision of renewable thermal technologies for drying, biofuels production and industrial waste, gas or water recovery. Appl Therm Eng 223:120022
46. Godbole V, Pal MK, Gautam P (2021) A critical perspective on the scope of interdisciplinary approaches used in fourth-generation biofuel production. Algal Res 58:102436
47. Bharadwaj SV, Ram S, Pancha I, Mishra S (2020) Recent trends in strain improvement for production of biofuels from microalgae. In: Microalgae cultivation for biofuels production. Elsevier, pp 211–225
48. Santana HS, de Sousa MR, Júnior JLS (2022) Intensification of biodiesel production through computational fluid dynamics. Biofuels and biorefining. Elsevier, pp 231–271
49. Jha P, Schmidt S (2021) State of biofuel development in sub-Saharan Africa: how far sustainable? Renew Sustain Energy Rev 150:111432
50. Durlik I, Miller T, Kostecka E, Łobodzińska A, Kostecki T (2024) Harnessing AI for sustainable shipping and green ports: challenges and opportunities. Appl Sci 14:5994
51. Panduman YYF, Funabiki N, Fajrianti ED, Fang S, Sukaridhoto S (2024) A survey of AI techniques in IoT applications with use case investigations in the smart environmental monitoring and analytics in real-time IoT platform. Information 15:153
52. Nandanwar H, Katarya R (2024) Deep learning enabled intrusion detection system for Industrial IOT environment. Expert Syst Appl 249:123808

53. Al-Ambusaidi M, Yinjun Z, Muhammad Y, Yahya A (2024) ML-IDS: an efficient ML-enabled intrusion detection system for securing IoT networks and applications. Soft Comput 28:1765–1784. https://doi.org/10.1007/s00500-023-09452-7
54. Kaur G, Balyan V, Gupta SH (2024) Experimental analysis of low-duty cycle campus deployed IoT network using LoRa technology. Res Eng 23:102844
55. Wani AK, Akhtar N, Sher F, Navarrete AA, Américo-Pinheiro JHP (2022) Microbial adaptation to different environmental conditions: molecular perspective of evolved genetic and cellular systems. Arch Microbiol 204:144
56. Bediaga H, Moreno-Benítez I, Arrasate S, Vilas-Vilela JL, Orbe L, Gómez-Martín JP, Unzueta E, González-Díaz H (2023) AI-driven cheminformatics models of chemical mixtures for sustainable design of drop-in biofuel blends. https://www.researchsquare.com/article/rs-2897107/latest. Accessed 29 Nov 2024

Chapter 12
Leveraging Artificial Intelligence for Enhanced Efficiency and Sustainability in Geothermal Energy Systems

Yurany Villada Villada, Alejandro Moncayo-Riascos, Lady Giraldo, Iván Moncayo-Riascos, Farid B. Corteś, Camilo A. Franco, and Oscar E. Medina

Abstract This chapter provides a comprehensive investigation into geothermal energy, with a focus on its potential for sustainable energy recovery and the transformative role of Artificial Intelligence (AI) and the Internet of Things (IoT) in optimizing geothermal facilities. The chapter systematically explores the various stages of geothermal energy production, from resource exploration and reservoir characterization to drilling, plant operation, and maintenance. Emphasis is placed on how AI, through machine learning, deep learning, and predictive analytics, enhances efficiency across these stages by improving subsurface temperature prediction, optimizing fluid flow, and managing equipment health in real-time. Additionally, the integration of IoT technologies is highlighted as a key enabler for continuous monitoring, data collection, and dynamic control of geothermal systems, fostering more responsive and adaptive facility management. Specific case studies demonstrate how AI and IoT solutions synergize to address geological uncertainties, predict reservoir performance, and reduce operational risks, contributing to sustainable and cost-effective energy recovery. The chapter also delves into advancements in enhanced geothermal systems (EGS), innovative drilling technologies, and the application of nanomaterials to boost thermal efficiency. A final focus is placed on AI-driven multi-objective optimization, which balances energy output, operational costs, and environmental

Y. Villada Villada · A. Moncayo-Riascos · L. Giraldo · I. Moncayo-Riascos · F. B. Corteś · C. A. Franco · O. E. Medina (✉)
Grupo de Investigación en Fenómenos de Superficie Michael Polanyi, Facultad de Minas, Universidad Nacional de Colombia Sede Medellín, Medellín, Colombia
e-mail: oemedinae@unal.edu.co

O. E. Medina
Instituto de Química, Química de Recursos Energéticos y Medio Ambiente, Universidad de Antioquia UdeA, Medellín, Colombia

Grupo de Ciencia y Tecnología de Gases y Uso Racional de La Energía, Facultad de Ingeniería, Universidad de Antioquia, Medellín, Colombia

M. Nasr et al. (eds.), *Artificial Intelligence Applications for a Sustainable Environment*, Green Chemistry and Sustainable Technology,
https://doi.org/10.1007/978-3-031-91199-6_12

sustainability. This synthesis underscores the role of AI and IoT in revolutionizing geothermal energy systems, driving them toward greater resilience and long-term viability within the renewable energy landscape.

Keywords Artificial intelligence (AI) applications · Enhanced geothermal systems (EGS) · Geothermal energy optimization · Reservoir management · Sustainable energy systems

12.1 Introduction

Geothermal energy is increasingly recognized as a key component in the global transition toward low-carbon energy systems, owing to its capacity to provide continuous, baseload power with minimal environmental impact [1]. As a renewable energy source, geothermal offers significant advantages over intermittent alternatives like solar and wind, primarily due to its stability and scalability. However, the technical complexities associated with subsurface exploration, reservoir management, and optimizing geothermal plant operations present considerable challenges [2]. These challenges, coupled with the high upfront costs and the inherent uncertainty of geothermal resource availability, have historically constrained the broader adoption of geothermal energy technologies. Advanced technological interventions are required to overcome these obstacles and maximize the potential of geothermal energy [3]. Among these, artificial intelligence (AI) stands out as a powerful tool with the potential to significantly enhance the efficiency, reliability, and sustainability of geothermal energy systems [4].

AI mainly through the application of machine learning (ML) and deep learning (DL) algorithms, is rapidly gaining traction across a wide range of industries [5]. In geothermal energy, AI can facilitate more accurate subsurface characterization, improve well placement strategies, optimize plant performance, and predict maintenance needs, all contributing to lowering operational costs and enhancing system efficiency [6]. By leveraging large datasets derived from geophysical surveys, well logs, and operational parameters, AI systems can analyze complex, multidimensional relationships that are often difficult for traditional modeling approaches to capture [7]. This capacity for handling vast amounts of data and generating predictive insights is crucial for successfully exploring and exploiting geothermal resources, often characterized by high geological variability and uncertainty [8].

Geothermal resource exploration is one of the most critical areas in which AI can substantially impact [9]. Geothermal reservoirs are located deep within the Earth's crust, and accurately predicting the location, size, and properties of these reservoirs is essential for project success [4]. While valuable, traditional exploration methods, such as seismic imaging and magnetotelluric surveys, are limited in their ability to provide a comprehensive understanding of subsurface conditions [10]. These methods often rely on assumptions and models that do not fully account for the complex, heterogeneous nature of geothermal reservoirs. AI, by contrast, can process data from multiple sources, including geological, geophysical, and geochemical datasets, to build more accurate models of subsurface conditions [11]. Machine learning algorithms can be trained to recognize patterns in historical data, enabling the identification of regions with high geothermal potential and reducing the risks associated with exploration [12].

Once a geothermal resource has been identified and a well has been drilled, the focus shifts to optimizing the operation of the geothermal plant [13]. Here, AI can provide significant value by enabling real-time monitoring and control of plant operations. Geothermal plants are dynamic systems where temperature, pressure, and flow rates fluctuate. These variations can affect the efficiency of energy conversion processes and equipment longevity [14]. AI-based systems can continuously monitor these operational parameters, using real-time data to optimize key processes such as fluid flow regulation, heat extraction, and power generation [15]. By adapting to changes in reservoir conditions, AI can ensure that the plant operates at maximum efficiency, thereby reducing energy losses and prolonging the life of critical infrastructure [16].

In addition to operational optimization, AI can play a crucial role in predictive maintenance. Like all energy infrastructure, geothermal plants are subject to wear and tear over time, with equipment such as pumps, turbines, and heat exchangers requiring regular maintenance to avoid failures. Unplanned equipment failures can lead to significant downtime and financial losses [17]. Traditional maintenance approaches, often based on predetermined schedules or reactive repairs, are only sometimes effective in preventing unexpected failures [18]. AI-based predictive maintenance systems, however, can analyze historical performance data, along with real-time sensor inputs, to detect early signs of equipment degradation [19]. By identifying potential issues before they result in catastrophic failures, AI can help reduce downtime, lower maintenance costs, and extend the lifespan of geothermal plants. This capability is particularly important for geothermal systems, which often operate in remote or difficult-to-access locations, where maintenance interventions can be costly and logistically challenging [20].

Another important application of AI in geothermal energy systems is in the field of reservoir management [21]. Geothermal reservoirs are complex and dynamic, with their performance influenced by various geological, chemical, and physical [22, 23]. Maintaining the long-term sustainability of a geothermal resource requires careful reservoir management, including regulating fluid injection and extraction rates to avoid resource depletion or the induction of seismic events [24, 25]. AI models can be used to simulate reservoir behavior under different operational scenarios, allowing operators to optimize fluid management strategies and ensure the sustainable use of the resource [26–28]. By integrating AI with reservoir simulation models, operators can predict how changes in operational parameters will affect reservoir pressure, temperature, and overall productivity, thereby enhancing the long-term viability of geothermal projects [29, 30]. Therefore, the primary objective of this chapter is to explore how AI can enhance the efficiency, reliability, and sustainability of geothermal energy systems across various stages of production. This includes AI applications in geothermal resource exploration, subsurface modeling, reservoir management, and predictive maintenance. By presenting state-of-the-art AI methodologies applied to various stages of geothermal energy production, from resource exploration to plant operation and maintenance, this chapter aims to provide a comprehensive understanding of how AI can enhance the efficiency, sustainability, and economic viability of geothermal projects. The chapter is structured to provide a logical and progressive understanding of the integration of AI into geothermal energy as follows. First, the fundamentals of geothermal energy are discussed, offering a detailed overview of geothermal systems, subsurface reservoirs, and the technologies currently employed in geothermal exploration and operation. In the next section, the advanced topics and emerging technologies in geothermal energy are explored, focusing on innovations in drilling, well placement, and enhanced geothermal systems. Additionally, the application of nanomaterials in geothermal operations is discussed, reflecting the ongoing advancements in materials science to improve the efficiency and sustainability of geothermal systems. Then, it introduces the fundamentals of AI, providing an overview of the key AI techniques, including machine learning, neural networks, and data-driven approaches. In the following section, AI applications in geothermal energy systems are shown. It outlines how AI can be used in various stages of geothermal energy production, including resource exploration, drilling, petrophysics, reservoir characterization, reservoir engineering, and production/injection engineering. Finally, future remarks and challenges were discussed, where opportunities for further AI innovation in geothermal energy are explored, along with the technical and operational challenges that need to be addressed to realize the potential of AI-enhanced geothermal systems.

12.2 Fundamentals of Geothermal Energy

12.2.1 Importance of Sustainable Energy

For decades, energy generation has been preserving human and economic development worldwide. Although fossil fuel-based energy systems are critical for supplying energy security, governmental policies, global warming, and decarbonization have promoted energy transition, including the exploitation/use of renewable energy sources [31]. Sustainable energy is the key to transitioning to a new energy model that addresses three global challenges: environmental conservation, energy security, and socio-economic development. This concept implies that it should be consumed efficiently and affordable [31]. In recent years, studies on renewable energy sources have increased in absolute and relative terms. In this sense, renewable energy can play an important role in addressing the issues of fossil fuel depletion and global warming. Fossil fuels, nuclear, and renewable resources are the three main energy sources. In particular, using fossil fuels promotes the production of greenhouse gases, which significantly impact the environment and contribute to global warming [32]. Thus, carbon dioxide (CO_2) emissions are expected to reach 40 thousand million

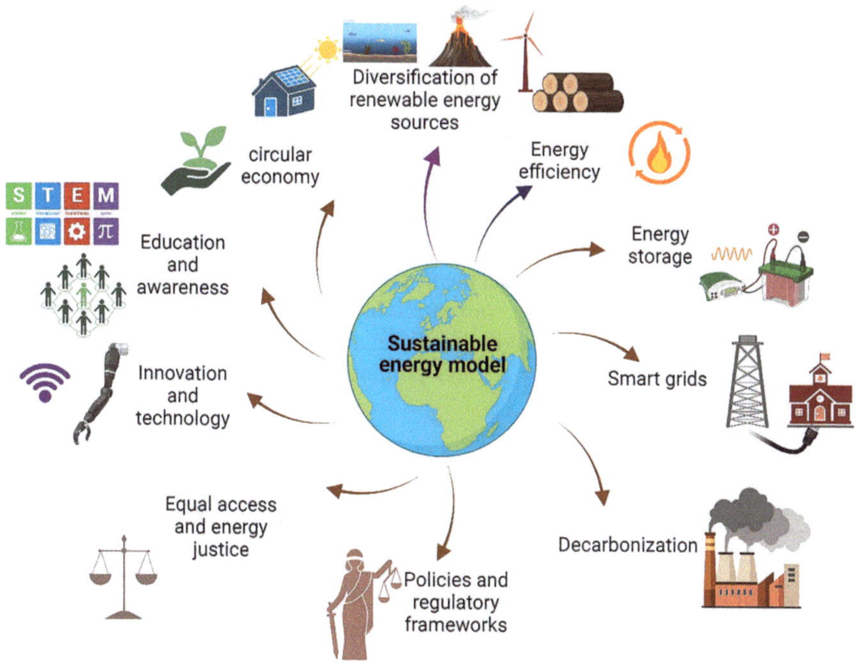

Fig. 12.1 Schema for the application of sustainable energy model. Own source

kg around 2030 [33]. The assisted state and the government are pursuing energy efficiency measures and substitution by low-carbon emitting renewables. However, decisions for transforming energy systems to reduce carbon emissions significantly should consider scalability, deployment timing, reliability, and capital and operating costs. Furthermore, several source options should be analyzed using an energy system framework that provides the best selection strategies [33].

Figure 12.1 presents the main parameters that should be considered in a sustainable energy model to ensure that energy production and use are environmentally friendly, economically viable, and socially responsible over the long term. It can be observed that the model incorporates different aspects, from techniques to social. Specifically, the model includes (*i*) the diversification of the energy model pretending the use of other sources of energy, such as solar, geothermal, wind, hydroelectric, biomass, and biogas, which diversify the energy matrix of each country worldwide; (*ii*) the energy efficiency evaluation of the reduction in energy demand through technologies that optimize consumption, such as efficient appliances, electric vehicles, and low-consumption industrial processes. Likewise, improving the thermal isolation of buildings to minimize energy losses; (*iii*) the integration of storage technologies, such as lithium-ion batteries, thermal storage to improve the intermittency of renewable sources such as solar and wind to energy storage; (*iv*) the implementation of networks that allow management of energy demand and supply, integrating distributed sources and offering real-time responses to development efficient smart grids; (*v*) implementation of new technologies for a progressive reduction in the use of fossil fuels and their replacement by renewable sources; (*vi*) the application of carbon capture and storage (CCS) technologies to reduce greenhouse gas emissions from industries based on fossil fuels; (*vii*) establish the development of public policies that encourage the adoption of renewable energy and energy efficiency; (*viii*) ensuring access to clean and affordable energy is available to all people, including rural and marginalized communities; (*ix*) promoting research and development of new technologies to improve the efficiency and reduce the costs of renewable energy; (*x*) using AI to optimize energy management; (*xi*) promoting of educational programs related to the responsible energy consumption, culture of sustainability in sectors such as transport, construction, and agriculture; and (*xii*) promoting the efficient use of resources, minimizing the waste and promoting recycling in industrial energy processes.

12.2.2 Role of Geothermal Energy in the Energy Transition

Geothermal energy is the oldest energy source for humans worldwide. This energy occupies a unique role in the energy matrix owing to advantages such as the independence of weather and fossil fuels. Furthermore, geothermal plants have operated in continuous mode for years, promoting efficient resources. Due to the necessity of replacing electric plants based on fossil fuels, country policies alternatives to stabilize electric rates recharged with energy renewables such as wind and solar, which are intermittent by nature. Considering the reports by the main organizations of energy regulations, the projection regarding the use of energy from geothermal sources has increased during this time [34]. Figure 12.2 shows the energy consumption of renewable and non-renewable sources worldwide. It can be observed that despite fossil fuels supplying the world's population as a primary energy source, renewable energies have an important role in projections until 2025. Figure 12.3 shows the increase in the use of geothermal energy projected for 2050. The projected trend suggests using geothermal energy as a renewable source and incorporating it into the energy for each country, which allows energy diversification.

Based on this information, the installed geothermal capacity worldwide was 16,318 MW until December 2022, spread across 32 countries and about 198 geothermal fields. This result represents a small fraction (0.16%) of the world's installed electrical capacity. Table 12.1 shows the evolution of geothermal capacity and electricity generation over the last 43 years, from 1980 to 2023. As expected, the use of geothermal energy slowly increased over time. However, increased applications based on geothermal sources have become more significant in recent years.

Figure 12.4 shows the evolution of the installed geothermal capacity and generation over 43 years. The installed capacity increased with time. In this sense, the information reported corresponds with the projections for 2050 performed by the Energy Information Administration (EIA) and IRENA, diversifying the energy matrix and

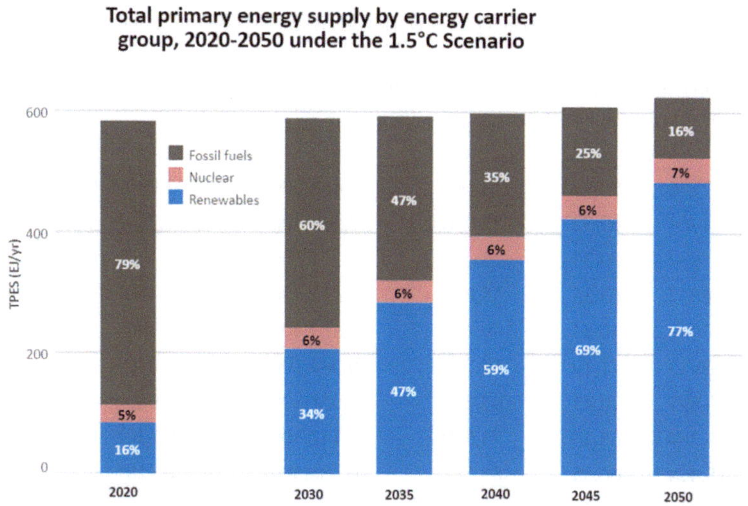

Fig. 12.2 Projection of energy resources to the 2050 year. Adapted from IRENA, Copyright, 2023 [35]

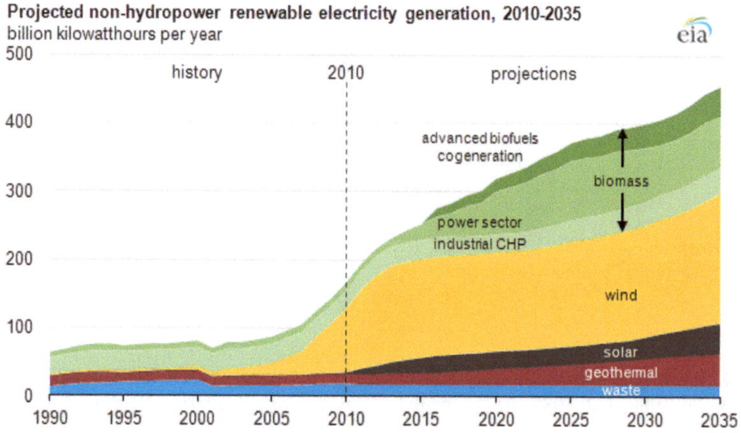

Fig. 12.3 Projection of geothermal energy from 1990 to 2035. Adapted from Energy Information Administration (EIA), Copyright 2023 [36]

promoting energy transition. On the other hand, Table 12.2 presents the reported list of countries with geothermal power plants in operation and their current status from 2018 to 2022. It can be observed that the country with the highest capacity and generation of energy is the United States, with 3919.4 and 18,702.2, respectively [37].

As reported by several researchers, geothermal energy plays a critical role in the global energy transition because it provides a reliable and renewable power source with minimal greenhouse emissions. Its importance highlights geothermal sources'

Table 12.1 Evolution of geothermal power worldwide in the last 43 years. Adapted from Gutiérrez-Negrín [37], Copyright Springer, 2024

Year	Capacity (MW)	Capacity factor (%)	Generation (GWh)			Share of (%)	
			Geothermal	Clean energy	Total	Clean energy	Total
1980	2110	70.87	13,100	2,438,000	5,633,000	0.54	0.23
1985	4764	69.32	28,930	3,547,900	9,831,380	0.82	0.29
1990	5834	69.33	35,430	4,351,600	11,853,110	0.81	0.30
1995	6832	63.55	38,035	5,022,400	13,382,480	0.76	0.28
2000	7972	70.54	49,261	5,485,400	14,971,790	0.90	0.33
2005	8933	71.19	55,709	6,044,100	17,763,940	0.92	0.31
2010	10,897	70.45	67,246	6,958,800	20,938,060	0.97	0.32
2015	12,284	68.35	73,550	8,142,800	23,660,330	0.90	0.31
2020	15,414	66.35	89,587	10,186,500	26,276,910	0.88	0.34
2023	16,318	67.54	96,552	11,143,100	28,254,250	0.87	0.34

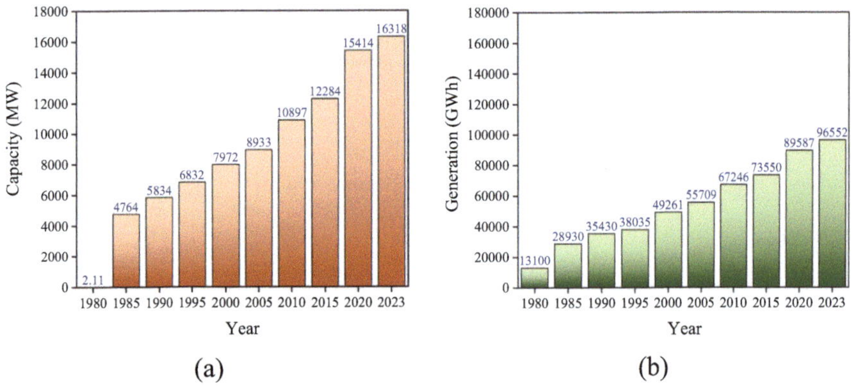

Fig. 12.4 World geothermal capacity and electric generation between 1980 and 2023. Electric generation in 1980 and 1985 are estimated. Adapted from Gutiérrez-Negrín [37], Copyright Springer, 2024

ability to provide electricity and heat, which can replace fossil fuel-based systems. Geothermal energy is especially valuable because it is always available, unlike other renewable sources that depend on weather conditions. Therefore, Germany and the United States have implemented geothermal energy, showing promising results in reducing carbon emissions. Germany has seen significant growth in geothermal district heating systems, which supply heat to cities through deep geothermal resources. These systems are supported by government programs that promote renewable energy development. Likewise, the United States has incorporated renewable

Table 12.2 Changes in the geothermal installed capacity and the generation in the last 43 years. Adapted from Gutiérrez-Negrín [37], Copyright Springer, 2024

Country	2018–2019		2021–2022		Net changes in	
	Capacity (MW)	Generation (GWh)	Capacity (MW)	Generation (GWh)	Capacity (MW)	Generation (GWh)
Australia	0.3	1.7	0.3	0	0	−1.7
Austria	1.2	2.2	1.3	0.5	0	−1.7
Belgium	0.8	2.0	0	0	−0.8	−2.0
Canada	0	0	6.3	0	6.3	0
Chile	48.0	400.0	81.0	600.0	33.0	200.0
China	29.1	174.6	45.1	131.2	16.0	−43.4
Colombia	0	0	0.1	0	0.1	0
Costa Rica	207.0	969.0	252.5	1599.0	45.5	630.0
Croatia	16.5	76.0	16.5	74.7	0	−1.4
El Salvador	204.4	1442.0	204.4	1575.0	0	133.0
Ethiopia	7.3	0	7.3	0	0	0
France	17.0	136.0	17.2	127.0	0.2	−9.0
Germany	46.1	165.6	47.0	207.7	1.0	42.1
Guatemala	52.0	237.0	46.5	310.0	−.5	73.0
Honduras	35.0	270.0	35.0	297.0	0	27.0
Hungary	3.0	15.6	3.0	18.1	0.0	2.5
Iceland	755.0	6010.0	755.3	5788.4	0.3	−221.6
Indonesia	2138.5	15,315.0	2384.4	16,588.2	245.9	1273.2
Italy	915.5	6015.0	915.8	5917.0	0.3	−98.0
Japan	545.2	2409.0	545.7	2660.8	0.6	251.8
Kenya	865.4	5240.0	972.5	5590.0	107.1	350.0
Mexico	1005.8	5375.0	1001.9	4511.5	−3.9	−863.5
New Zealand	1054.8	7564.0	1054.8	7820.0	0.0	256.0
Nicaragua	155.0	492.0	165.2	780.0	10.4	288.0
New Guinea	50.0	118.3	50.0	118.3	0	0
Philippines	1918.0	9893.0	1951.0	11,670.0	33.8	1777.0
Portugal	33.0	216.0	31.8	158.9	−1.2	−57.1
Romania	0.05	0	0.1	0.8	0.05	0.8
Russia	81.9	440.7	81.9	440.7	0	0
Taiwan	0.3	2.6	6.6	25.0	6.3	22.4
Thailand	0.3	1.6	0.3	0.2	0	−1.4
Türkiye	1549.0	8168.0	1717.3	10,840.0	168.3	2672.0
United States	3678.2	18,435.0	3919.4	18,702.2	241.2	267.2
Total	15,413.5	89,586.9	16,318.4	96,552.1	904.9	6965.2

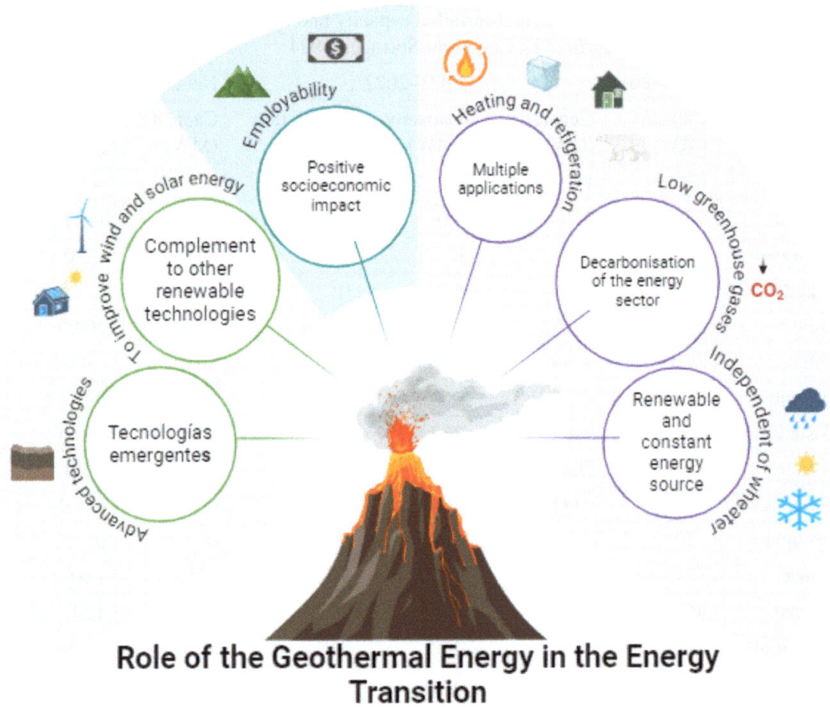

Fig. 12.5 Representative scheme for the role of geothermal energy in the energy transition. Own source

sources into decarbonization efforts as a broader strategy to reduce reliance on fossil fuels.

Geothermal energy is important in energy transition, promoting a sustainable and carbon-free system. Figure 12.5a shows the scheme, including the main aspect of geothermal energy's role in energy transitions. Geothermal energy is a fundamental strategy for achieving a global energy transition, providing a clean, constant, and versatile source of energy that can help reduce emissions and stabilize electricity grids.

12.2.3 Challenges in Optimizing Geothermal Energy Systems

Figure 12.6 depicts the main challenges in the optimization of geothermal projects. The issues are related to the activities implemented from exploration to production. Sharmin et al. [38] compared the production of geothermal energy with other renewable energies and found that the obtention of geothermal energy generates greater capital investment expenditures, longer repayment times, and difficulties accessing the source, as well as issues regarding modularization. Sircar et al. [39], Walmsley

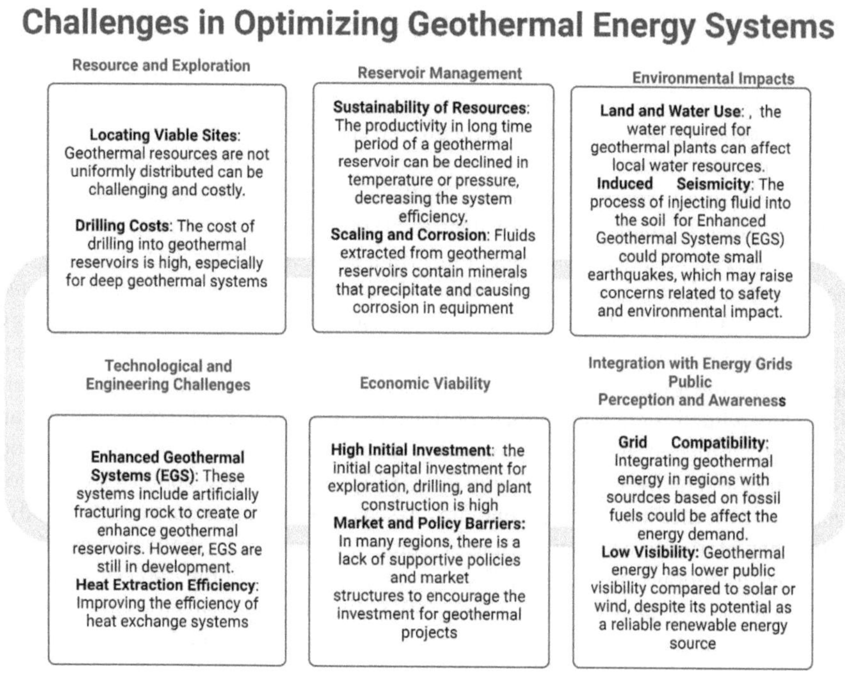

Fig. 12.6 Representative scheme for the challenges in the obtention of geothermal energy. Own source

et al. [40], Kabeyi et al. [41], and Song et al. [24] mentioned the challenges in the drilling of geothermal wells associated mainly with the high temperatures of reservoirs. Drilling geothermal wells is difficult because of the loss of circulating fluid, fracturing of abrasive rocks, and elevated temperatures that cause mechanical failures in drilling rigs. Furthermore, the degradability of certain materials is involved in the formulation of drilling fluids under harsh conditions. On the other hand, silica scaling is a common problem at geothermal power facilities. The traditional method to avoid the presence of silica is to keep the head pressure high enough to condense the debris out of the fluid [42]. Furthermore, enhanced geothermal system (EGS) techniques that operate by injecting a circulating fluid into cracked rock can cause seismicity or seismic disruptions. This seismic activity produces harmful structural changes and hazardous situations [43]. The challenges associated with environmental impacts include the air effect, property, and freshwater. These effects can be considered during the development and operation stages. Operating geothermal fields may cause local dangers, generating natural occurrences such as micro-earthquakes, hydrothermal eruptions, and land subsidence [33].

12.2.4 Overview of Geothermal Systems

According to the EIA, *geothermal energy* is the energy obtained from heat produced in Earth's core, and it is considered a renewable energy source because heat is generated continuously. Based on previous studies, geothermal energy spans approximately 4–5 billion years. In addition, geothermal energy storage in 10 km of the Earth's surface contains heat equivalent to 50,000 times more energy than global oil and gas reserves [44]. Compared with wind, solar, and hydroelectric energy, the energy obtained from geothermal sources does not depend on climatic variations, generating additional advantages for exploitation. Furthermore, this type of energy is considered environmentally friendly because of its lower environmental impact, highlighting the lower emission of greenhouse gases and suspended solids in the air [45]. Depending on the temperature of the extracted fluids, geothermal energy can be used for efficient electricity production, heating and cooling, balneology, food processing, and textile drying, among others [46]. Geothermal resources are categorized into low-, medium-, or high-enthalpy based on their temperature and thermodynamic characteristics. The United States Geological Survey (USGS) has defined these categories as follows: low-enthalpy resources have temperatures <90 °C, medium-enthalpy resources range between 90 and 150 °C, and high-enthalpy resources exceed 150 °C. A typical geothermal system comprises four fundamental elements: reservoir rock, caprock, heat source, and permeable structures such as fractures [47]. High-enthalpy geothermal systems may derive their heat from plutonic intrusions. In contrast, low-enthalpy systems rely on the heat flux from Earth's interior, generated by the radioactive decay of crustal minerals. This process creates a geothermal gradient in sedimentary basins. In geothermal systems, permeable structures, such as faults or fractures, lead to surface manifestations like geysers, fumaroles, or hot springs [48].

A geothermal system includes three main elements: a heat source, a reservoir, and fluid, the carrier that transfers heat, as shown in Fig. 12.7. Geothermal systems are classified based on how underground heat is harnessed, the characteristics of the reservoir, and the technology used. The main classification of geothermal systems is shown in Fig. 12.8 [49].

Figure 12.9 presents a scheme for using geothermal resources based on the Lindal diagram [50]. Their use was divided as follows: (*i*) Direct heat Use: The oldest geothermal energy utilization is used directly. The common uses worldwide are district heating, space heating, agricultural and greenhouse applications, and industrial applications [51]. Recently, heat pumps have been employed worldwide, particularly in the USA, Germany, and Switzerland; and (*ii*) Electricity Generation: Deep geothermal resources with temperatures exceeding 90 °C have demonstrated potential for power generation. The US Energy Information Administration states that geothermal hydrothermal resources suitable for electricity production should have temperatures ranging from 300 to 700 °F, existing as water, steam, or a combination of both. Power plants utilize these resources by either directly channeling steam from geothermal fields through a turbine or employing hot, pressurized geothermal

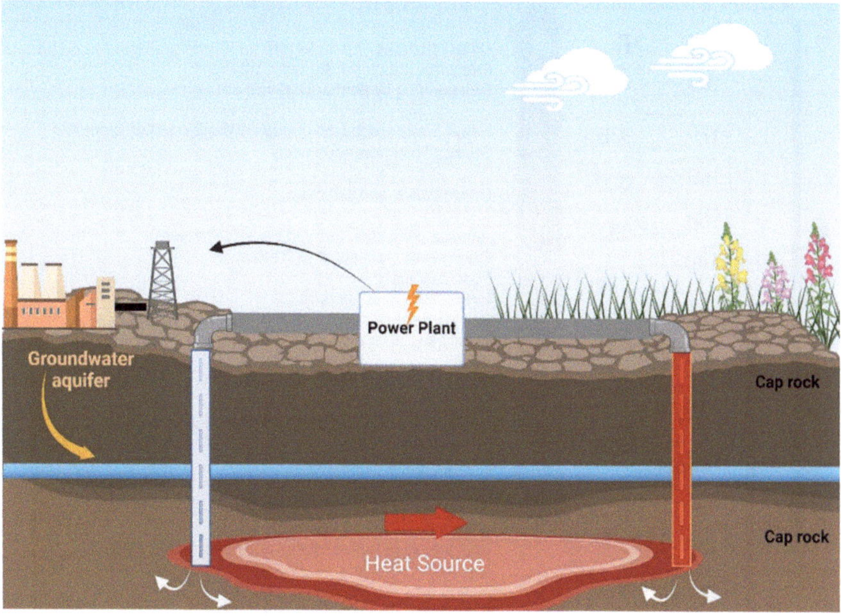

Fig. 12.7 Representative scheme for a conventional geothermal system. Own source

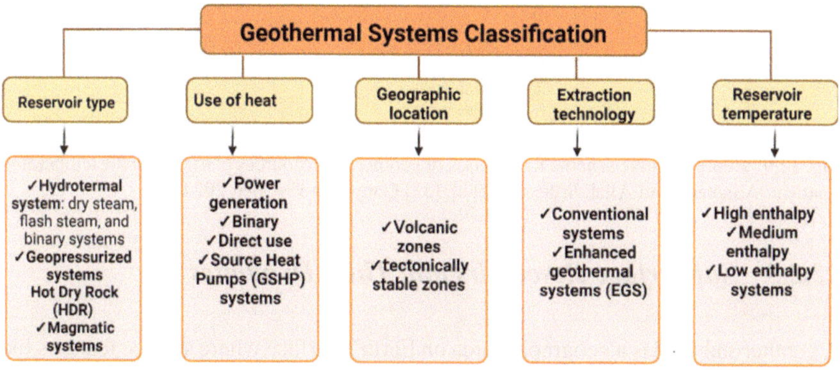

Fig. 12.8 Concept map for geothermal system classification based on the reservoir type, use of heat, geographic location, extraction technology, and reservoir temperature. Own source

fluid to vaporize a low boiling point organic fluid for use in a binary turbine cycle. High-temperature resources typically employ steam turbines, while intermediate-temperature geothermal systems often use binary plants. In these electricity generation systems, pressurized vapor expands through turbine stages, causing the turbine to rotate. This rotation drives a generator coupled to the turbine shaft, resulting in electricity production. After passing through the turbine, the vapor is condensed in a condense [52].

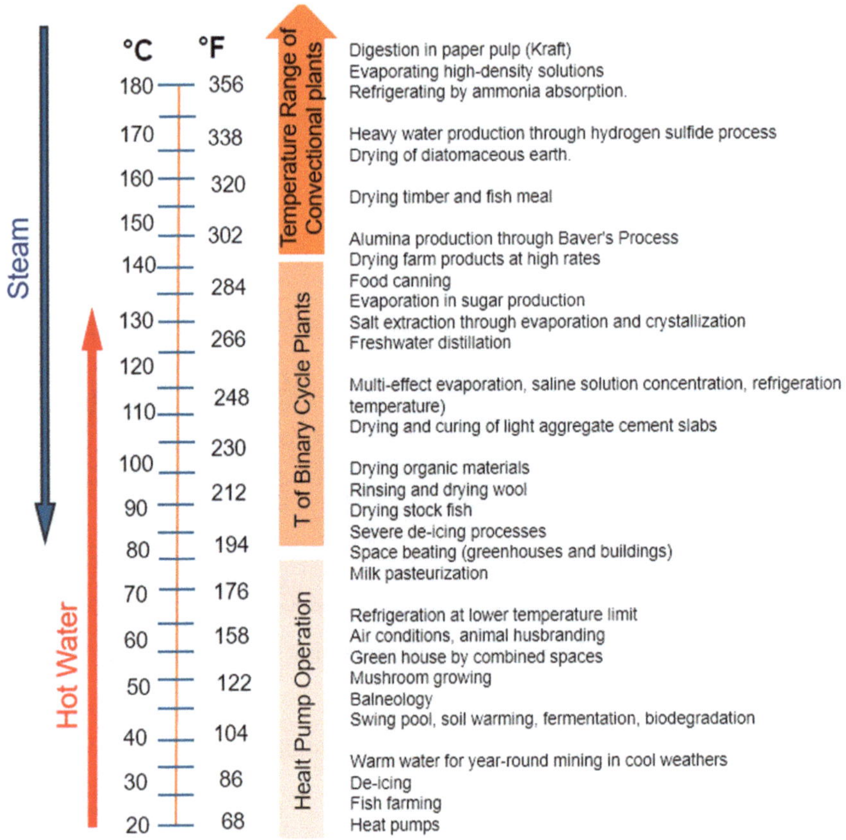

Fig. 12.9 Representative scheme for the uses of geothermal resources considering the temperature gradient. Adapted from Allahvirdizadeh et al. [53] Copyright Elsevier, 2024

12.2.5 Subsurface Reservoirs and Their Dynamics

A geothermal field is a geographic area on Earth's surface where surface features like geysers, fumaroles, or boiling mud ponds evidence underground geothermal activity [54]. These indicators are typically found in regions with active volcanism. Within a geothermal field, the geothermal reservoir is a specific zone from which thermal energy can be extracted via reservoir fluids, which may be steam, hot water, or a combination of both. The geothermal reservoir is a natural or man-made sub-system explored to produce energy. The geothermal reservoir includes hydrothermal, hot, dry rock (HRD), and artificial reservoirs (Enhanced Geothermal Reservoir, EGR) (See Fig. 12.10).

In general, the use of these reservoirs depends on different variables, such as fluid circulation, heat transfer, and reservoir recharge. Water is conducted through fractures and pores in rocks, and the flow is controlled by pressure and temperature

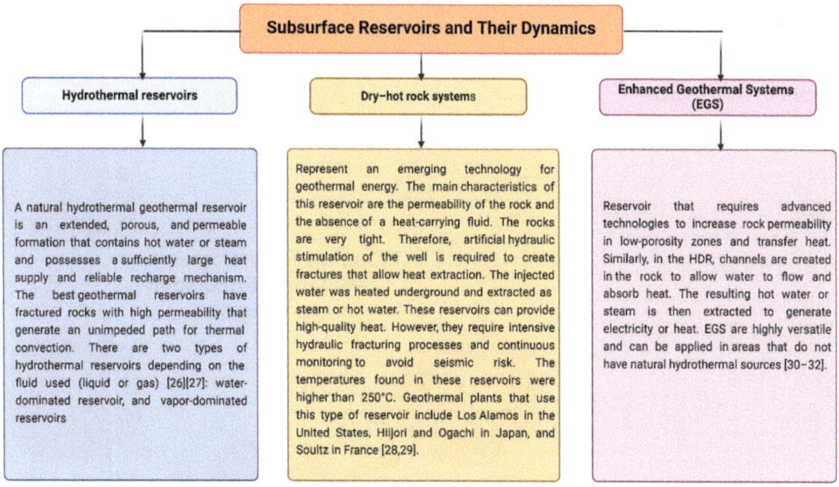

Fig. 12.10 Concept map for subsurface reservoirs and their dynamics. Own source

gradients. Therefore, the flow allows water to heat and transport heat to the surface. Circulation occurs naturally in hydrothermal systems. In contrast, flow is induced in the EGS and HDR systems. Heat transfer depends on the rocks' conductivity and the reservoir's temperature. In high-temperature reservoirs, heat flows efficiently into the fluids. In contrast, more energy is required to extract accumulated heat in low-conductivity rocks. Thus, the cooling rate and sustainability of the resource depend on the efficiency of the transfer. However, the cooled water after heat extraction is reinjected into the sustainable reservoirs to maintain the pressure and temperature in the system. This balance is essential for the stability of the reservoir, allowing for the continued availability of geothermal energy without depleting resources for a short time.

12.2.6 Current Technologies in Geothermal Exploration and Operation

Geothermal energy is considered an attractive renewable energy source owing to its multiple benefits, such as producing low emissions to the environment and independence of the weather, among others. Compared to other renewable options, the extraction efficiency can be affected. For this reason, new technologies have been developed to improve their efficiency. These technologies present advanced geophysical exploration methods, drilling, reservoir monitoring, and pumping and generation systems under harsh conditions. The actual techniques are shown in Fig. 12.11 [55].

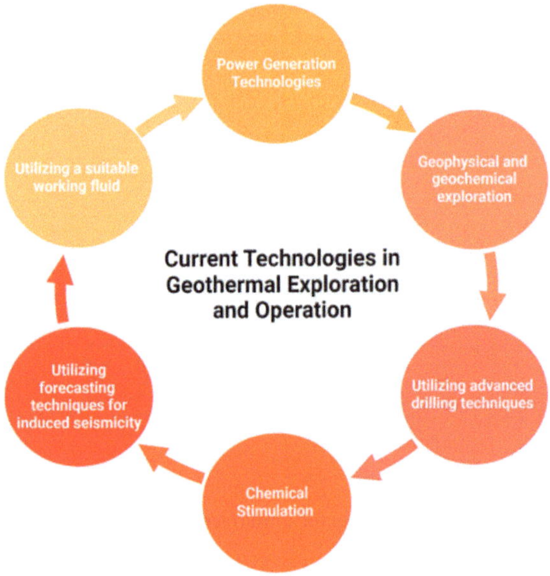

Fig. 12.11 Scheme for the current technologies in geothermal exploration and operation. Own source

12.3 Advanced Topics and Emerging Technologies in Geothermal Energy

12.3.1 Innovations and Challenges in Drilling Operations

The drilling operations of geothermal projects are very similar to those of Oil and Gas well drilling; however, there are important differences related to the characteristics of the target rock. The productive zones of geothermal projects are commonly volcanic rocks or hard sedimentary rocks, which are recognized by their excellent heat conductivity, hardness, and abrasiveness [56, 57]. Geothermal wells are often drilled in "live" production zones, meaning the holes produce steam or hot brine during drilling. Saleh et al. [58] reported that the drilling operation can spend between 30 and 70% of the total cost of geothermal projects linked to well depth [59, 60]. Song et al. [60] presented a deep cost analysis for geothermal wells considering the complexity of drilling operations generated by the inner properties of the reservoir and its fluids, which affect the rate of bit penetration and the bit life, the lost circulation, the corrosive problems, among others [61]. Thus, the circulation losses can promote blowouts that can result in stuck pipes and cause challenges in wellbore integrity [58, 62]. Allahvirdizadeh et al. [63] estimated that lost circulation treatments represent approximately 15% of well costs. Song et al. [60] described that the three main well cost drives in geothermal drilling are the in-situ rock properties, the diameter of the borehole to allow enough mass flow, and the uniqueness related to the particularity of each geothermal well.

On the other hand, some harsh reservoir conditions can lead to issues and challenges mainly associated with well integrity, good control, and lost circulation. High-temperature and hard reservoir (rock substrate) conditions lead to technical limitations for the selection of drilling fluids (mud), completion fluids, casing design, and drill bits (see Fig. 12.12). Thus, these issues encourage the need for innovative technological developments to overcome these challenges and diminish the possibility of drilling problems. Teodoriu and Bello [64] presented an outlook on drilling technologies, highlighting the need for technologies that enhance efficiency, safety, and costs. They mentioned novelty and existing technologies, such as multilateral drilling processes, horizontal drilling, extended reach drilling, automated drilling, casing drilling technologies, complex path drilling, positive displacement motor technology, and laser technology [64]. In addition, the author highlighted the opportunity for AI, robotics, and data analysis to address current drilling problems. In addition, the author highlighted the opportunity for AI, robotics, and data analysis to address current drilling problems [65].

The design and development of drilling fluids play a key role in the success of geothermal drilling projects. Erge et al. [66] described an excellent drilling fluid based on its capacity to provide the minimum non-productive time and prevent formation damage to achieve the maximum production injection rates. The authors mentioned the importance of drilling fluids, considering the experience of geothermal projects in Turkey. The authors remarked some possible risks for well drilling operations, including wellbore instability in tight holes due to reactivity with shales, shallow water flow (non-usual, but could be experienced at a depth of approximately 30

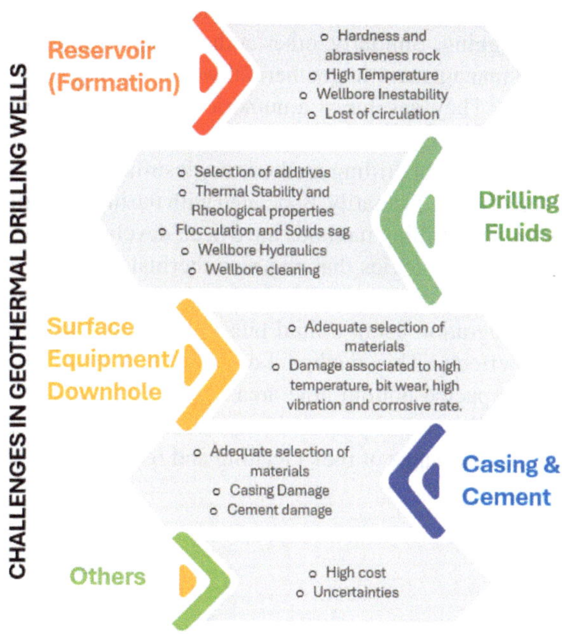

Fig. 12.12 Challenges in geothermal well drilling considering the main stages in the operation. Own source

to 150 m), differential sticking (with a high impact on costs and efficiency), lost circulation, poor mud removal during cement jobs, and high torque and drag [66]. According to the Turkish Association of Geothermal Investors (JESDER), an estimated cost of USD 3–4 million for a geothermal project at a depth of 4000 m with a duration of approximately 60–80 days. In this case, the budget for the drilling was less than 7% of the total cost of the project [67].

Special drilling fluid formulations with high thermal stability and good rheological properties are required for harsh reservoir conditions at high temperatures and pressures. Mohamed et al. [68] argued the importance of rheological properties and their impact on wellbore stability, wellbore hydraulics, hole cleaning, torque, and drag. They mentioned how rheological properties directly affect the efficiency and total cost of drilling.

Otherwise, the water temperature can exceed the critical temperature, leading to challenges in the drilling and completion operations for geothermal projects. Some studies have shown that conventional geothermal well productivity may increase by up to 10% when extracted from a supercritical source [69]. Kruszewski and Witting [69] described the advances and challenges for high-enthalpy well drilling and supercritical conditions of fluids, including the analysis of different cases worldwide. The authors remarked that when wells are drilled at an extremely high temperature, the drilling fluid, casing and its connections, cement sheaths, and wellhead assembly can suffer critical failure owing to the induced fatigue caused by temperature variations, pushing the material to its technical limits. The authors argued the need to seek safe operations for fluid production and problem-free drilling based on developing new technologies for drilling operations, special drilling fluids, enhanced cementing operations, adequate selection of wellhead and casing materials, and adequate strategies for logging. Similarly, other authors like Khaled et al. [70] investigated downhole heat management in geothermal wells with high enthalpy (high-temperature conditions). They developed a numerical model to simulate the flow and heat behavior while drilling geothermal wells.

Traditional drilling methods are insufficient due to the complexity of geothermal well drilling, primarily associated with the high cost of drilling deep (3–8 km). Therefore, new drilling methods have been developed based on thermophysical processes and methodologies that combine thermal and mechanical impacts on drilled rocks [71, 72]. Dreus and Kozhevnykov [72] proposed a thermal factor method for rock disintegration using a fluid pulsation method that combined thermal and mechanical effects. This method favors the control of the heat exchange processes of the impacted bottom-hole area. The authors showed that it is possible to increase the drilling velocity up to the top 2.6 times by using a heat factor with advantages in the performance of rock cleaning and reduction of thermal stress in the formation.

12.3.2 Enhanced Geothermal Systems (EGS)

Geothermal projects have higher risks and challenges than the Oil and Gas industry. Geothermal wells are hotter and require stronger drill bits and deeper drilling technology. Figure 12.13 compares the current and new technologies for overcoming the challenges of geothermal projects. Conventional geothermal systems require geophysical conditions, such as underground heat close to the surface with the formation of high permeability (for better access) and a sufficient water source (liquid or steam form) as carrier fluids to allow heat transfer. The particularities of the location and properties of reservoirs and fluids bring challenges and complexities for exploiting geothermal potential, promoting the development of novel alternatives.

Enhanced Geothermal System (EGS) technologies have important precedents in the Oil and Gas industry and directional drilling and stimulation processes, such as fracking, to improve hydrocarbon recovery. The first EGS project was conducted in New Mexico in 1974. Different studies have reported the EGS as the future of geothermal energy based on its performance in hard rock reservoirs, which must be stimulated after good completion [73]. According to a CRS report in 2024, recent developments in EGS have increased geothermal potential to supply more electricity in the United States. The EGS process includes multiple injection and production wells, using geothermal fluid injected and heated by contact with rock, which is posteriorly extracted in production wells [73]. Geothermal fluid was injected as a stimulation fluid into the reservoir to create or widen existing fractures, allowing deeper access and contact with the reservoir. This operation generates a better heat transfer process.

Enhanced geothermal systems can allow access to new locations due to their combination with hydraulic fracturing (widely known in the oil and gas industry). Hydraulic fracturing techniques promote new channels and fluid paths inside rocks at greater depths than conventional geothermal wells. EGS has other advantages related to the possibility of being applied close to population centers, reducing the additional distance of electricity transmission capacity needed, and decreasing the quantities

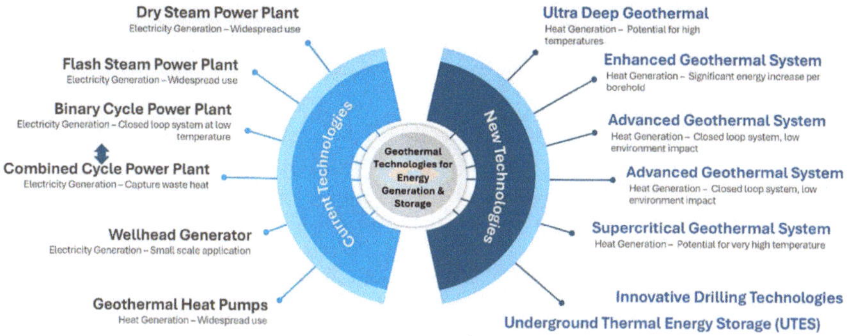

Fig. 12.13 Geothermal technologies for energy generation and storage. Own source

of natural gas pipelines or natural gas consumption [73]. An EGS project can be designed in different configurations depending on the position and orientation of the injection and production wells, the number of wells, and the spatial disposition of fractures (Fig. 12.14) [68].

Another process, such as EGS, is the CLG Advanced geothermal/closed-loop geothermal system, which does not use fluids for injections or extraction from the earth without a fracturing process. CLG wells overcome the conventional issues of low permeability by circulating fluid through a downhole heat exchanger or sealed pipelines. Heat is obtained from the fluid by conduction and transported to the surface for use in the direct generation of heat or electricity [75, 76]. However, CLG is considered more expensive than EGS because it requires two wells and long laterals. Different configurations are used in CLG: concentric pipe-in-pipe, multilateral, sealed, and U-loop configurations to optimize their performance and costs. Therefore, hybrid systems can be implemented using EGS and CLG technologies. In the first stage, the well was drilled, and the fractures propagated downward, increasing the depth and temperature. The fractures are filled with a highly convective slurry and then downhole a pipe-in-pipe closed-loop heat exchanger system using a supercritical fluid in circulation to promote better heat transfer [75].

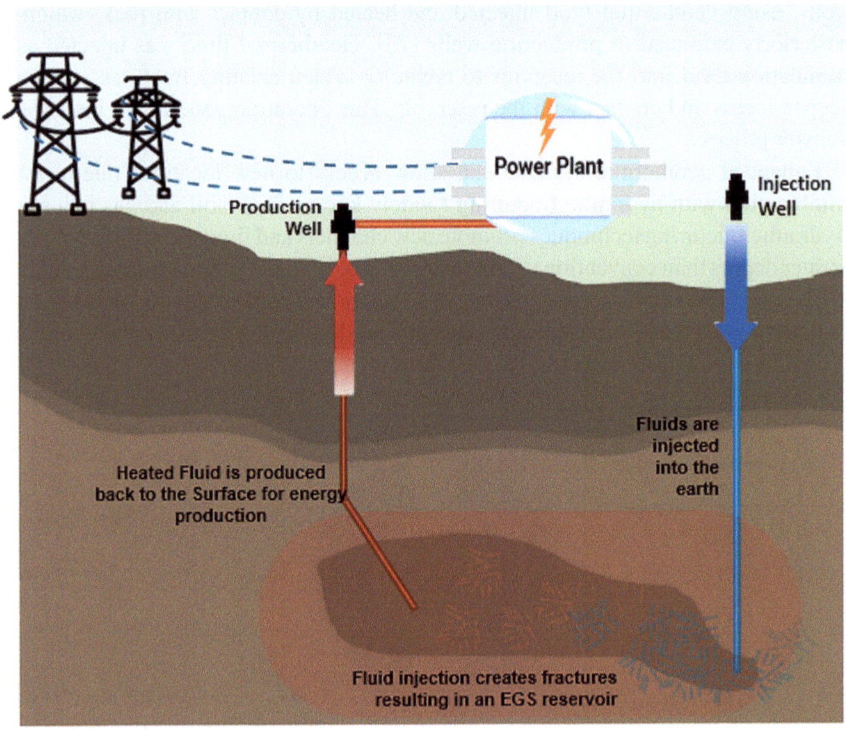

Fig. 12.14 Representation of an enhanced geothermal system. Adapted from Gelfgat and Geraskin [74], Copyright, SPE, 2021

12.3.3 Sustainable Management of Geothermal Resources

Geothermal heat is considered an energy source with a carbon-free impact and is of great interest in the global energy market in the energy transition route. The decline in the produced fluids mainly determines the geothermal potential. Therefore, sustainable management is crucial for adequate maintenance of geothermal projects for a long time. Sustainable development is defined as "development that meets the needs of the present without compromising the ability of future generations to meet their own needs" [76]. Geothermal resource management involves adequate control of energy extraction from geothermal systems to maximize its benefits without overexploiting resources. Overexploitation of geothermal sources often occurs because of inadequate monitoring and data analysis, leading to a poor understanding of systems and avoiding reliable modeling for their analysis.

Adequate reservoir management of geothermal projects must monitor flow rates, well pressures, and enthalpy distributions. Additionally, the chemistry of the produced fluids must be monitored for possible microseismic events. According to Rohit et al. [77] efficient and modern geothermal management must include energy-efficient utilization, careful monitoring, modeling, and reinjection. Modeling allows for calculating and predicting geothermal production potential, considering different scenarios for the outcome of geothermal management. Therefore, reinjection plays an important role in waste-water disposal (for environmental impact) and water recharge strategy to increase water source in-situ to favor the heat transfer process. However, reinjection must be carefully applied because there is the possibility of cooling production wells [78].

Authors like Zayed et al. [2] also argued the importance of numerical modeling for the thermodynamic behavior of underground heat exchange to optimize geothermal projects. Similarly, Khaled et al. [70] investigated downhole heat management while drilling shallow and ultradeep enthalpy. The authors highlighted the importance of accurately modeling the heat transfer between drilling fluids and a formation (rock) as vital information for developing and accessing geothermal energy utilization. Thus, the operation can be performed at temperatures within the limits of the downhole tool and drill bits, thereby preventing failure while drilling under harsh conditions.

12.3.4 Nanotechnology Applied to Geothermal System

As detailed in the above sections, geothermal projects face several challenges at each value chain stage. In this sense, nanotechnology, like the emergent technology in the last decades, offers important outcomes for its implementation with improved performance. Nanoparticle/nanofluid properties such as thermal conductivity, heat capacity, nanometric size, and surface energy, among others [3, 5, 6, 79], can improve the performance of drilling fluids, completion fluids, and heat transfer processes [4]. Furthermore, some authors have pointed out the functionality of nanofluids as cooling fluids to cool specific parts (sensors or electronics) of drilling equipment that are often exposed to high-temperature and high-friction environments [80]. In particular, some authors have studied the different benefits of nanoparticle inclusion in drilling fluids, such as control of filtrate loss, inhibition of formation damage, prevention of differential pipe sticking, reduction of clay swelling, mechanical, rheological, and thermal stability under (high-pressure and high-temperature) conditions [81–83]. Some of these nanoparticles include silica oxide, ceria, alumina carbon nanotubes, graphene, zinc oxide, copper oxide, and clays [83, 84]. Blaz et al. [81] investigated the increase of thermal resistance of water-based mud for drilling geothermal wells, adding some additives like carbon nanotubes and graphite. The results showed that using nanomaterials increased thermal conductivity by 27%. Likewise, Martin et al. [85] demonstrated an improvement in rheological properties of drilling fluid with the increase in temperature (149–232 °C) and reduction of filtrate loss using silica nanoparticles modified with CTAB (hexadecyltrimethylammonium bromide). In the same way that drilling fluids are affected by the harsh environmental conditions of geothermal reservoirs, cementing fluids can also be impacted. Conventional Portland cement degrades its performance under high-temperature and high-pressure conditions, negatively impacting its strength, rheological properties, porosity, and permeability. Thus, additives such as nanomaterials can overcome these drawbacks by improving the compressive strength, heat and electrical conductivity, rheological properties, porosity, permeability, and sulfate resistance at high temperatures [86, 87].

Thermal conductivity is an outstanding property of nanomaterials that can considerably enhance the heat transfer process in heat exchangers and for energy extraction. Nanofluids can act as superconductors and working fluids to extract energy from the earth [88, 89]. Soltani et al. [22] reviewed nanotechnology applications in closed geothermal heat pumps and heat exchangers, highlighting the thermophysical properties of nanofluids that directly affect heat transfer. The authors remarked on how using low-volume fractions of nanofluids can enhance thermal conductivity, which affects the heat transfer rate, heat transfer coefficient, and Nusselt number. Several authors have experimentally and analytically investigated the effectiveness of using nanofluids as heat-carrier fluids to enhance the conductive and convective heat transfer of geothermal heat exchangers [4, 90, 91]. More recently, the Colombian Workshop of Geothermal Energy held in Bogota-Colombia highlighted recent advances in nanomaterials for drilling and development of geothermal systems in co-production with hydrocarbons, highlighting novel nanomaterials as carbon quantum dots with striking properties of thermal conductivity to improve heat transfer. In particular, the differences in the physicochemical properties of the chemical nature of nanomaterials impact the thermal conductivity of nanofluids. Likewise, the results showed that nanofluids based on CQDs improved the efficiency of the Organic Rankine Cycle (ORC) by 15% [8].

12.4 Artificial Intelligence (AI) in the Geothermal Sector

AI has transformed multiple industries by automating complex processes, optimizing resources, and improving decision-making based on massive data. In the energy sector, and particularly in geothermal energy. AI plays a critical role in managing large volumes of geophysical and geochemical data, facilitating real-time pattern analysis and prediction of underground conditions that would otherwise be unattainable by traditional methods. This field has rapidly evolved in recent years, evidenced by the increase in academic production and interest in conferences and research projects oriented toward geothermal energy and AI [92, 93]. AI in geothermal energy includes machine learning and deep learning models, which enable optimizing processes such as resource exploration, reservoir characterization, and real-time system monitoring. AI improves the accuracy of thermal gradient predictions and soil quality assessments, which is essential for minimizing financial risks and enhancing efficiency in operational decision-making [94, 95].

In geothermal energy, machine learning and deep learning techniques have revolutionized how geothermal resources are explored, managed, and optimized. Below is an analysis of the main techniques used in the field, organized into five fundamental categories: dimensionality reduction algorithms, clustering techniques, regression and classification methods, deep neural networks, and multi-objective optimization methods. These techniques allow for a comprehensive approach to addressing the unique challenges of geothermal energy, from predicting temperatures and pressures in deep wells to optimizing energy production systems.

12.4.1 Dimensionality Reduction

When working with high-dimensional geothermal data, it is essential to simplify the analysis without losing critical information. Principal Component Analysis (PCA) is one of the most commonly used methods in geothermal energy, as it projects data into a new space with fewer dimensions while retaining as much variability as possible from the original data. Mathematically, PCA seeks to find a linear transformation that maximizes the variance of the projected data. A covariance matrix of the data is constructed to calculate PCA, and then an eigenvalue decomposition is performed to identify the directions of maximum variance (principal components). These directions (eigenvectors) allow for a compact representation of the data. In geothermal energy, Mejía-Fragoso et al. [95] applied PCA to analyze thermal gradients in multidimensional data, helping to identify the most influential geophysical variables, thereby reducing "noise" in the model and facilitating decision-making in well area selection. Moreover, Mejía-Fragoso compares the performance of different machine learning algorithms, finding that the Gradient-Boosted Regression Tree (GBRT) algorithm offers more accurate predictions of the geothermal gradient, with an error margin within 12% relative to independent validation data [95]. This supports the use of

PCA in the preprocessing phase as a technique to enhance the efficiency of advanced algorithms in geothermal studies and suggests that methods such as GBRT and PCA together can be particularly useful in data-limited environments, as is the case in some regions of Colombia [95].

12.4.2 Clustering

In geothermal applications, clustering, an unsupervised machine learning technique, allows for the grouping of data into categories based on inherent similarities, facilitating the analysis and organization of complex datasets. Algorithms such as k-means are particularly effective in segmenting geological and geophysical data, enabling the identification of homogeneous areas concerning subsurface composition and structure. This method optimizes the selection of potential sites for geothermal exploitation by grouping geographic regions with shared properties, such as seismic characteristics and geological conditions, thus improving accuracy and efficiency in terrain characterization Coro and Trumpy [96]. The k-means algorithm assigns each data point to a specific "cluster," optimizing the grouping so that points are as close as possible to a central "centroid.“ This process minimizes the total distances within clusters, ensuring that points within the same group exhibit similar characteristics, a crucial step for identifying patterns and achieving homogeneity in complex datasets. In geothermal applications, such clustering methods are instrumental in distinguishing geological and geochemical variations across different regions. Ahmmed and Vesselinov [97] utilized this approach to examine the chemical composition of groundwater in the Great Basin, USA, by applying the k-means algorithm. They effectively classified regions with similar geochemical characteristics, directly correlating these with areas of geothermal potential. Their findings showed that specific groundwater chemistry components, such as trace element concentrations, align closely with geothermal activity hotspots. This approach allows for a focused exploration strategy, targeting regions with high geothermal promise based on chemical signatures. As a result, the study contributes a more cost-effective and precise framework for geothermal exploration, reducing the need for extensive field surveys by using data-driven classification to identify high-potential geothermal sites.

12.4.3 Supervised Regression and Classification

These techniques are crucial for predicting continuous and discrete variables within geothermal systems, facilitating the precise modeling of essential operational parameters such as reservoir temperature and pressure. These techniques are also employed in classifying geothermal conditions that directly influence the efficiency and stability of energy generation systems. Two methods that stand out for their effectiveness in these contexts are Gradient Boosting and Decision Trees, which allow robust modeling tailored to the specific complexities of geothermal systems. This approach aligns with the findings of Frison et al. [98], who applied artificial neural networks to optimize district heating network (DHN) operations by forecasting heat demand across different seasonal periods, reducing prediction errors by 27% in summer and 17% in winter. These results underscore the significance of advanced predictive models in optimizing geothermal energy systems, where techniques like Gradient Boosting and Decision Trees can be crucial in enhancing operational efficiency and stability Frison et al. [98]. Gradient Boosting is an iterative method that builds sequential models, optimizing error reduction at each process step. In each iteration, the model adjusts a new estimator that minimizes residual errors from the previous model, using a differentiable loss function to guide this optimization. Combining multiple base models produces a final model that reduces bias and variance, enhancing prediction accuracy. In geothermal energy, Dashtgoli et al. [92], rigorously applied Gradient Boosting, specifically XGBoost, to predict temperatures within low-enthalpy geothermal reservoirs with exceptional accuracy. The model achieved an R^2 value of 0.993 on the test dataset and an RMSE of 0.788, underscoring its effectiveness in managing the nuanced and multidimensional data characteristic of geothermal environments. These results affirm the robustness of Gradient Boosting in capturing the complex relationships between temperature and multiple hydrochemical parameters essential to geothermal systems. Notably, Dashtgoli et al. [92] illustrated through their study in the Friulian Plain that features such as bicarbonate and magnesium levels are among the most significant predictors of subsurface temperatures. The correlation heatmap from their study (Fig. 12.15) vividly displays the strength of these associations, with temperature (T) showing strong positive correlations with both bicarbonate (HCO_3^-) and magnesium (Mg^{2+}), among other variables. This visual representation highlights the intricate interdependence within the data, demonstrating how the gradient-boosting model leverages these relationships to provide highly accurate predictions. By incorporating such detailed hydrochemical interactions, the approach improves temperature prediction in geothermal contexts and underscores the potential of data-driven models in advancing sustainable geothermal resource management.

Decision Trees are hierarchical machine learning models that systematically segment data into homogeneous subgroups by making a series of attribute-based decisions. In this structure, each node represents a test on a particular variable, while branches denote outcomes, optimizing information gain at each split. This approach is particularly advantageous for handling interactions among multiple geophysical and operational variables, which is crucial for complex geothermal applications. The structured decision paths in Decision Trees make them well-suited for assessing conditions within geothermal reservoirs, providing a robust framework for predicting and modeling operational scenarios and enhancing the accuracy and reliability of geothermal system analyses.

Bourhis et al. [99] applied Decision Trees to enhance the accuracy of thermal response tests (TRTs) in evaluating geothermal potential. This method allows for segmenting and analyzing geothermal data to uncover significant relationships between subsurface thermal conductivity and thermal resistance. Such insights were

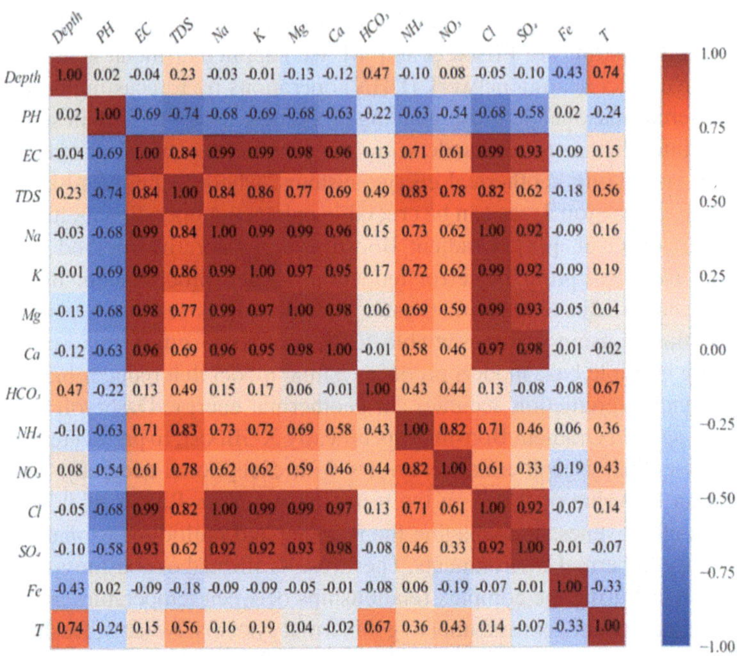

Fig. 12.15 Correlation heatmap showing the relationships between key hydrochemical parameters and temperature. Adapted from Dashtgoli et al. [92]. Copyright, Elsevier, 2024

instrumental in refining heat exchange system designs, tailoring them to the specific thermal characteristics of the soil. As a result, the predictive error in thermal estimations was reduced by approximately 20%, improving the effectiveness of heat transfer systems when adjustments were based on model-driven recommendations. Moreover, Decision Trees proved highly effective in identifying zones with similar thermal properties, which is critical for optimizing well placement and minimizing operational costs by avoiding areas with low thermal efficiency. The model provided a strong framework for characterizing geothermal parameters, thereby enhancing the accuracy of predictions in diverse geothermal contexts and promoting more efficient resource extraction strategies Bourhis et al. [99].

12.4.4 Deep Neural Networks

Long Short-Term Memory (LSTM) models play a pivotal role in processing time series and sequential data, which are critical in geothermal studies. These networks are a specialized form of Recurrent Neural Networks (RNNs) and excel at capturing long-term dependencies and intricate temporal patterns in data that change over time. By using memory cells with input, forget, and output gates, LSTMs selectively retain or discard information, allowing them to utilize past data efficiently for future predictions. This makes LSTM networks particularly well-suited to monitoring geothermal wells, where changes in temperature, pressure, and other conditions over time are essential metrics.

Park et al. [100] used LSTM models for real-time monitoring and operational control within geothermal wells. By leveraging LSTM's memory capabilities, the system could make precise short-term predictions on variables such as flow rate, thermal output, and reservoir pressure. The results demonstrated a significant reduction in system downtime by 15% due to the early detection of anomalies and automatic adjustments based on model predictions. Notably, the use of LSTMs allowed operators to optimize energy production while proactively managing risks associated with unexpected temperature or pressure fluctuations. This application highlights the LSTM model's advantage in improving system reliability and efficiency through real-time operational insights and adjustments.

In parallel, Physics-Informed Neural Networks (PINNs) offer a more advanced approach by integrating physical laws directly into the neural network architecture, making them well-suited for modeling subsurface systems where data is sparse or hard to obtain. Ishitsuka et al. [101] developed a PINN that modeled geothermal conditions using governing fluid dynamics and thermodynamics equations. By embedding principles such as energy and mass conservation, the model achieved high accuracy in predicting subsurface temperature and pressure distributions, with an R^2 of 0.97 compared to conventional neural networks. This accuracy is particularly advantageous in geothermal exploration, as it allows for better prediction of thermal gradients and resource potential with minimal data. The results indicated that PINNs could reduce exploration costs by up to 20% by minimizing the need for extensive in-field measurements while providing a robust model for resource assessment. This study underscores PINNs' potential to enhance predictive precision in geothermal applications, making them ideal for resource estimation in uncharted regions.

Similarly, Lindsey et al. [102] applied the Play Fairway Analysis (PFA) framework enhanced with AI to identify geothermal resource-rich zones in Argentina. This study integrated AI algorithms within a GIS-based system to analyze geological features and groundwater chemistry, pinpointing high-potential geothermal zones without extensive exploratory drilling. The model's predictions were validated against historical data, showing an 85% accuracy in identifying viable zones for geothermal exploration. This approach significantly reduced exploration costs and environmental impact by focusing on regions with favorable geological and geochemical conditions. Specifically, the PFA-AI hybrid approach helped prioritize zones with high geothermal gradients and reduced the need for exploratory wells by approximately 30%, highlighting the method's efficiency and cost-effectiveness in large-scale geothermal exploration.

These studies collectively demonstrate the effectiveness of DNN-based methods in geothermal energy applications. LSTM models contribute to operational efficiency by enabling precise real-time control, PINNs bring predictive accuracy in complex subsurface modeling, and AI-enhanced PFA optimizes resource allocation in exploration. These applications underscore how tailored neural network architectures can transform geothermal resource management, from cost reduction in exploration to improved system reliability and sustainability in operations.

12.4.5 Multi-objective Optimization

Conversely, this is crucial in geothermal systems due to balancing multiple objectives such as operational efficiency, costs, and sustainability. This approach identifies "Pareto optimal" solutions, where improving one objective can only be achieved by sacrificing another. This approach designs and operates hybrid systems, integrating various renewable energy sources and achieving an optimal balance between performance criteria. Multi-objective optimization identifies "Pareto optimal solutions," meaning that improving one objective can only be achieved at the expense of another. Gasmi et al. [93], used this method to simultaneously optimize heat and electricity production while minimizing the environmental impact of hybrid geothermal systems. This approach is essential for maximizing the system's profitability and sustainability, offering a practical and effective solution for managing geothermal resources comprehensively.

12.5 Methodology

12.5.1 Data Collection

To evaluate the adoption of machine learning techniques in the geothermal industry, an exhaustive search was conducted in academic databases and specialized publications in geothermal energy and technology. Scientific journal databases such as ScienceDirect were consulted, applying a set of specific keywords related to AI and geothermal energy, such as "AI," "machine learning," and "neural networks," among others. This resulted in an initial set of 71 peer-reviewed articles exclusively focused on the intersection between machine learning and geothermal applications. To ensure relevance, each article was manually reviewed, excluding those that only tangentially mentioned AI in non-relevant contexts.

12.5.2 Categorization of Research Areas

To structure a precise and technical methodology for analyzing and classifying the geothermal articles collected, a structured process was followed in several stages using advanced natural language processing (NLP) techniques and clustering methods. This methodological process was adapted considering the approaches detailed by Farea et al. [103], who utilize models like LDA (Latent Dirichlet Allocation) and BERTopic to identify and optimize the number of topics within complex textual data, and Sheikhkhoshkar et al. [104], who employ techniques such as SBERT and hierarchical clustering to classify functional concepts.

Stage 1: Text Preprocessing

Text preprocessing is essential for converting unstructured content into a format suitable for mathematical analysis. First, the full text of each article was extracted, focusing on titles, abstracts, and key sections. Subsequently, data-cleaning techniques were applied to remove special characters, unnecessary punctuation, and common words without analytical value (stop words), retaining only meaningful terms. Lemmatization and stemming were employed to reduce dimensionality without losing context, using libraries such as NLTK and spaCy. Sheikhkhoshkar et al. [104] demonstrate how using SBERT allows precise vectorization and the capture of key semantic similarities between concepts.

Stage 2: Numerical Representation

With the text in a homogeneous format, each fragment was transformed into feature vectors. Two main techniques were used for this task: TF-IDF (Term Frequency-Inverse Document Frequency) and word embeddings with pre-trained models like Word2Vec and BERT. Embeddings, in particular, proved helpful in detecting complex concepts and their context, a fundamental technique in technical areas like geothermal energy. Farea et al. [103] highlighted how implementing BERTopic can better capture contextual meaning within documents by applying a class-specific TF-IDF for document groupings.

Stage 3: Document Clustering

To categorize the articles into research areas, clustering algorithms were implemented to identify common patterns and themes. The k-means and DBSCAN algorithms were selected for their effectiveness in identifying consistent patterns and emerging themes. Farea et al. [103] recommend using methods like the elbow method to determine the optimal number of clusters for k-means, ensuring adequate categorization without redundancy. The methodology reported by Sheikhkhoshkar et al. [104] also suggests that hierarchical clustering can provide a detailed view of the distribution of functions and concepts, highlighting the underlying relationships between related topics in research articles.

12.5.3 Validation and Visualization of Results

The structured approach was validated through semantic cohesion evaluation methods and topic concordance analysis. To increase interpretability, word clouds, and inter-topic distance maps were generated using tools such as pyLDAvis, which allow visualization of the proximity between identified topics. This process was crucial to ensuring that the groups reflected coherent research areas, as suggested by Farea et al. [103] and Sheikhkhoshkar et al. [104], who emphasized the importance of integrating visual validation into the analysis of complex textual data.

12.6 Results

12.6.1 Research Areas

To assess the progress in adopting machine learning in the geothermal industry, an exhaustive search was conducted in academic databases such as ScienceDirect, utilizing keywords encompassing AI techniques and specific geothermal applications. This search resulted in the compilation of 71 peer-reviewed articles that directly address the application of AI in geothermal challenges. Each article was classified through natural language processing and clustering algorithms, enabling the identification of key research areas and current trends within this field. The findings were organized into research clusters that encompass optimizing and modeling geothermal systems, the integration of renewable energies into hybrid systems, monitoring, and risk management, exploring geothermal resources, and socio-technical analyses concerning policies and sustainability.

Figure 12.16 presents a schematic that summarizes these clusters and their primary focus areas, providing a structured overview of how AI has been utilized in various facets of geothermal energy. This visual representation aids in understanding the organization of the results and highlights the predominant research areas within the field.

12.6.1.1 Optimization and Modeling of Geothermal Systems Using Artificial Intelligence

This cluster encompasses studies that apply AI techniques, such as artificial neural networks (ANNs), genetic algorithms, and deep learning, to optimize and model geothermal and multi-generation systems. Researchers utilize AI algorithms to improve energy efficiency and reduce the operational costs of geothermal facilities [92, 95, 99, 101, 105]. For instance, ANNs have been applied for accurate geothermal temperature prediction in the lower Friulian Plain, enhancing the precision of subsurface temperature models [92]. Similarly, machine learning approaches have been used to predict the geothermal gradient in Colombia, aiding in the identification of potential geothermal resources [95].

In addition, machine learning techniques have been employed to enhance thermal response tests for geothermal potential evaluations at both site and regional scales, contributing to more accurate assessments of geothermal energy potential [99]. Physics-informed neural networks have also been utilized for inverse modeling of

Fig. 12.16 Research clusters in AI applications in geothermal energy. Own source

natural-state geothermal systems, providing novel approaches to model complex geothermal reservoirs [101].

These intelligent systems enable real-time prediction and control of geothermal systems, facilitating dynamic and adaptive management of available resources. The integration of AI in geothermal systems not only optimizes their performance but also contributes to more sustainable and cost-effective energy solutions [105].

In this cluster, a growing trend is observed in applying AI techniques to enhance energy efficiency and reduce costs in geothermal systems. Advanced predictive models allow for more effective real-time management, optimizing performance, and adapting to changing operational conditions. Notably, other authors [106] demonstrated the potential of AI-driven optimization by examining multiple development scenarios in geothermal reservoirs, where the selection of an optimal scenario can significantly impact the Net Present Value (NPV). The study's results underscore how strategic data acquisition decisions can reduce expected losses and increase profitability, showcasing that careful scenario-based planning leads to higher NPVs and

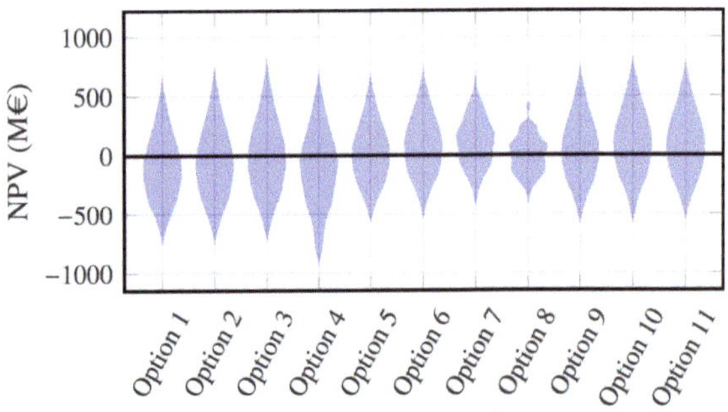

Fig. 12.17 Distribution of NPVs for each development option. Adapted from Corso et al. [106]. Copyright, Elsevier, 2024

more economically viable projects. Integrating machine learning algorithms facilitates the simultaneous optimization of multiple objectives, efficiently balancing energy efficiency, operational cost, and environmental sustainability. Corso et al. [106] illustrate this multi-objective approach through distribution of NPVs across 11 development options, highlighting how optimal development choices can enhance economic feasibility, as shown in Fig. 12.17. In particular, options with higher NPVs align with efficient geothermal exploitation, reducing financial risks while addressing sustainability goals.

From a practical standpoint, these techniques significantly improve decision-making in designing and operating geothermal systems, increasing their economic viability and energy performance. Theoretically, they contribute to developing novel approaches in applying AI to complex energy systems, expanding knowledge in multi-objective optimization, and advanced modeling of geothermal systems. This opens opportunities for future research that further explores the integration of AI in the energy sector.

12.6.1.2 Integration of Renewable Energies and Hybrid Systems

Recent advancements in geothermal exploration and drilling techniques have emphasized the integration of AI and machine learning (ML) to optimize resource assessment and decision-making under geological uncertainties. Machine learning models, particularly deep learning approaches, have proven highly accurate in predicting subsurface characteristics critical for geothermal drilling. For instance, ANNs and support vector machines (SVMs) have been used to estimate geothermal potential accurately, with some models achieving over 97% accuracy in subsurface temperature predictions when calibrated with magnetotelluric and borehole data [107]. Moreover, a study using partially observable Markov decision processes (POMDPs) improved data acquisition strategies, reducing time and cost while boosting decision-making accuracy by nearly 30% compared to traditional methods [106].

In hybrid systems, integrating geothermal with other renewables like solar and hydrogen storage has been shown to enhance system resilience and efficiency. Enhanced Geothermal Systems (EGS) utilizing CO_2 as a working fluid are promising innovations, serving dual roles of energy generation and carbon sequestration, which address energy demand and environmental objectives [108]. Additionally, models developed for multi-well geothermal fields, often deployed in complex well patterns like five-spot arrays, help optimize well placement and flow management, crucially increasing net present value (NPV) by maximizing thermal recovery while minimizing Capital Expenditures (CAPEX) and Operational Expenditures (OPEX) [108].

Overall, these studies illustrate the transformative impact of ML and AI on geothermal exploration and drilling. By addressing geological uncertainties and enabling optimized drilling configurations, hybrid systems combining geothermal with other renewable sources enhance energy production efficiency and contribute to sustainable practices through innovative resource management and carbon–neutral approaches. These findings underscore the necessity for high-quality data and robust predictive models in advancing geothermal and hybrid renewable energy systems, making them viable in diverse settings.

12.6.1.3 Monitoring, Safety, and Risk Management in Geothermal Systems

This group includes studies on using the Internet of Things (IoT), AI, and digital technologies for real-time monitoring, fault detection, and risk management in geothermal and energy networks. Recent advancements in digital twin (DT) technology have proven particularly valuable, allowing the creation of virtual replicas of physical geothermal assets that enhance maintenance and operational oversight. As highlighted by Onile et al. [109], digital twins provide continuous system updates, allowing for precise, real-time fault detection, improving predictive maintenance practices, and significantly reducing the risk of system failures. Moreover, digital twins empower operators to simulate potential disruptions, enabling scenario-based planning to mitigate risks.

Recent studies further underscore the transformative role of IoT, AI, and machine learning (ML) in enhancing geothermal systems' monitoring, safety, and risk management. IoT-enabled sensors play a critical role across different stages of geothermal energy production, including resource exploration, drilling, and reservoir management. For instance, IoT-based geophysical sensors measure temperature gradients, fluid pressures, and flow dynamics in real time, aiding in identifying optimal reservoir zones. Additionally, during drilling, IoT-enabled downhole tools equipped with pressure and temperature sensors provide actionable insights that help predict and address operational risks such as equipment wear or fluid instabilities [110]. Digital twins, for example, enable the creation of virtual replicas of geothermal systems, allowing for continuous monitoring and advanced simulations that anticipate potential failures, thus supporting proactive maintenance strategies and reducing downtime. This integration optimizes system performance and mitigates risks associated with seismic activities in high-risk zones.

Machine learning algorithms complement these efforts by analyzing the vast datasets generated by IoT sensors, enabling accurate and timely risk assessments. ML models, such as neural networks and supervised learning algorithms, are widely applied for anomaly detection and predictive maintenance. For example, these models forecast equipment degradation, identify thermal shifts in geothermal reservoirs, and predict potential hazards before they escalate into severe mechanical or operational failures. Such capabilities are particularly valuable in high-pressure, high-temperature environments, where mechanical risks and operational inefficiencies can have significant consequences.

In the context of seismic risk mitigation, IoT devices and AI-driven analytics provide an essential early warning mechanism. By integrating real-time seismic monitoring data into predictive models, operators can proactively adjust operational parameters to minimize risks associated with induced seismicity, which is a critical concern in geothermal energy production [110]. Furthermore, Industry 4.0 technologies, such as cloud-based analytics platforms, facilitate seamless data integration, enabling global visibility of geothermal operations while supporting more robust decision-making.

Together, these advancements optimize system performance, reduce downtime, and ensure operational safety in geothermal systems, reinforcing their long-term sustainability and economic viability.

12.6.1.4 Exploration and Assessment of Geothermal Resources Using Artificial Intelligence

This cluster focuses on applying machine learning algorithms to geothermal exploration. Moraga et al. [111] highlight the use of AI in geothermal exploration through remote sensing (RS) data, specifically by applying machine learning models to analyze surface temperature, fault density, and geothermal minerals as indicators of resource potential. At the Brady geothermal site, this model achieved predictive accuracy levels of 92–95% in detecting geothermal indicators, while a similar analysis at the Desert Peak site, which lacks visible surface signs, achieved 72–76% accuracy. This result underscores the ability of AI to improve resource targeting in manifest and blind geothermal systems.

Additionally, Zhang et al. [112] employed Gaussian Process Regression (GPR) within a Bayesian correction framework to probabilistically assess geothermal resources, focusing on reservoir characteristics like temperature, porosity, and thickness. Their findings in the Qingshankou and Quantou formations revealed substantial geothermal potential, with estimated resources of 5.80×10^{10} kJ/km^2 and 6.22×10^{10} kJ/km^2, and respective occurrence probabilities of 4.65 and 3.69%. This probabilistic approach enhances the precision of resource estimation and allows for robust decision-making under uncertainty.

In geothermal resource exploration, AI and machine learning (ML) frameworks offer powerful tools for optimizing system performance and sustainability. A recent study developed a highly accurate predictive model using artificial neural networks

(ANN) to forecast key metrics, such as energy and exergy efficiencies, environmental impact, and electricity costs, achieving an impressive R^2 of 0.9999. By simulating different operational scenarios, the model demonstrated optimal configurations: energy efficiency at 33.22%, exergy efficiency at 40.08%, and an economical setup that reduced electricity costs to $34 USD per hour. Additionally, sensitivity analysis identified critical factors—such as turbine inlet and geothermal inlet temperatures—as essential for multi-criteria optimization, indicating AI's value in guiding sustainable and cost-effective geothermal exploration and management [113].

Finally, Li et al. [114] utilized a 3D geomechanical model to provide valuable insights for geothermal resource exploration in the Guangdong-Hong Kong-Macao Greater Bay Area. This analysis encompassed fault kinematics, heat flow, and structural characteristics, revealing that NE-trending faults, such as the Wuchuan-Sihui fault with slip rates up to 0.12 mm/year, are favorable zones for geothermal fluid migration. Additionally, intersections of NW-trending and NE/ENE-trending faults were identified as high permeability areas, making them prime targets for geothermal exploration. This approach highlights how detailed fault analysis can effectively support geothermal resource assessments by identifying structurally favorable sites for fluid migration and heat extraction.

12.6.1.5 Socio-technical Analyses, Energy Policies, and Sustainability

This group gathers articles that address socio-technical, economic, and policy barriers to adopting geothermal technologies and other renewable sources, as well as their impact on sustainability and energy transition. In addressing socio-technical challenges, Entezari et al. [115] underscore the transformative role of AI and machine learning (ML) in energy systems. Their study highlights how AI-driven models can overcome barriers to sustainable energy by optimizing resource allocation, forecasting demand, and dynamically adapting energy usage. These capabilities enhance operational efficiency and support the integration of renewable sources, making AI an asset in policymaking. For instance, AI enables more resilient energy networks and improved energy management, aligning with policies to reduce emissions and promote environmental sustainability. By facilitating data-driven decisions, AI supports a smoother transition to sustainable energy systems, showing that overcoming socio-technical and policy barriers is essential for achieving widespread adoption of renewable.

Chu et al. [116] offer critical insights into the socio-technical, economic, and policy barriers affecting the adoption of renewable energy technologies, focusing on AI as a driver of corporate energy transition efforts. Their research reveals that AI facilitates corporate energy transitions through increased information transparency, thereby mitigating information asymmetry between companies and stakeholders and alleviating financial constraints by improving investor confidence. They highlight the role of AI in shaping corporate sustainable development strategies, encouraging companies to align with environmental, social, and governance (ESG) criteria. Companies achieve operational efficiency and a more resilient energy transition framework by implementing AI-powered predictive analytics and automation. This approach underscores AI's ability to accelerate sustainable energy practices, positioning it as a critical tool to overcome socio-technical and policy barriers within the corporate sector, thereby supporting broader energy transition goals.

12.6.2 Analysis of Extracted Data

The analysis of the extracted data is illustrated in Fig. 12.18, which shows the distribution of the number of articles across various categories of AI techniques applied to geothermal research. The number of articles per thematic area is presented, highlighting the areas of "Energy" and "Engineering" as the most represented, indicating a predominant focus on technical and scientific aspects within AI and geothermal research. Figure 12.18 shows the distribution of articles according to the categories of AI techniques used, allowing the identification of areas with the highest scientific production and techniques gaining relevance in the geothermal research field.

The Machine Learning and Neural Networks (ML/RN) category is the most represented, with 20 articles (Fig. 12.19). This suggests a growing adoption of these advanced techniques in the field due to their ability to model complex patterns and improve decision-making accuracy. The predominance of ML/RN indicates that machine learning and neural network techniques are highly valued for their versatility and applicability in various aspects of energy research.

The numerical modeling and simulation category, with 15 articles, is an essential technique, especially in geothermal. This technique allows researchers to create precise representations of geothermal systems, facilitating the understanding and prediction of complex phenomena. Its high presence in literature underscores its relevance for advancing knowledge in this field. Optimization algorithms, with 8 articles, are a key tool in research to improve energy systems' efficiency and effectiveness. These techniques enable the identification of optimal solutions in the design and operation of energy technologies, contributing to maximizing resources and minimizing costs. The socio-economic analysis category, with 8 articles, highlights the

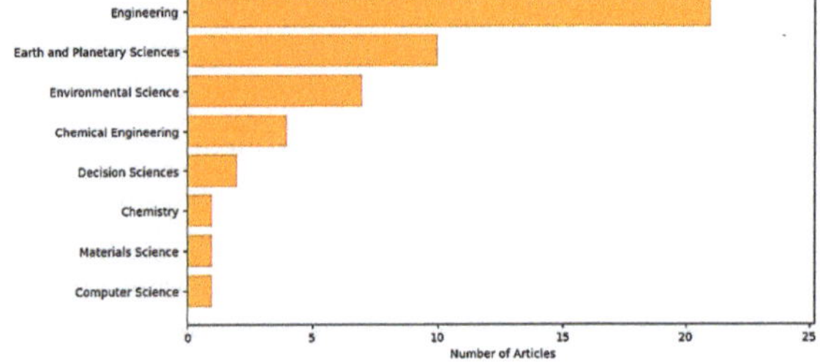

Fig. 12.18 Number of articles per thematic area. Own source. *Information subject to information search

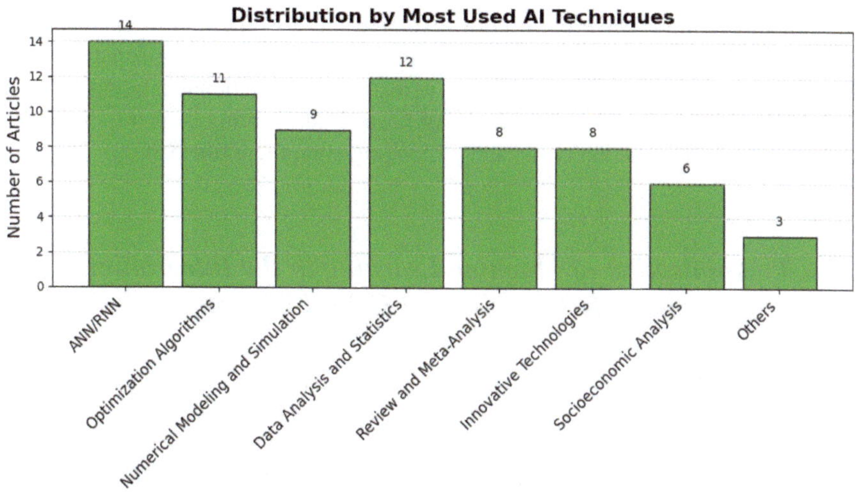

Fig. 12.19 Distribution of articles according to AI technique categories. *Own source* *Information subject to information search

importance of considering social and economic aspects in implementing new energy technologies. Research in this area addresses the impact of technologies on society and the economy, ensuring that proposed solutions are viable and sustainable in the long term. Although less represented, the remaining categories highlight the wide range of approaches and techniques that researchers are exploring. This reflects a diverse field where multiple methodologies are applied to address energy challenges from various perspectives, enriching the body of knowledge and fostering innovation in research.

12.6.3 Application of Machine Learning in Optimization and Modeling of Geothermal Systems Using Artificial Intelligence

In this category, AI is employed to enhance the efficiency and performance of geothermal systems through advanced modeling and optimization techniques. These tools allow researchers to simulate and adjust systems to maximize energy efficiency and reduce operational costs. In the case of the Combined Heat, Cold, and Power (CCHP) cogeneration system adapted for geothermal energy, Ling et al. [117] applied Artificial Neural Networks (ANN) and Genetic Algorithms to optimize its design. Using an ANN model, they simulated the system's behavior under different operational conditions, while Genetic Algorithms identified optimal parameters to maximize efficiency and minimize costs. This optimization resulted in a significant reduction in investment recovery time and an increase in energy production.

In another example, Noye et al. [118] explored the use of AI in the autonomous control of Geothermal Heat Pump Systems (GSHP). They implemented machine learning algorithms that allow the pumps to learn from operational data and adjust in real time. This facilitates the prediction of heating and cooling demand, improving energy efficiency, reducing costs, and prolonging equipment lifespan.

12.6.4 Application of Machine Learning in the Integration of Renewable Energies and Hybrid Systems

AI also plays a crucial role in integrating geothermal energy with other renewable sources, optimizing the performance of hybrid systems to achieve greater efficiency and resilience. This enables more effective utilization of available resources and contributes to energy sustainability. Yang et al. [114] applied an Artificial Neural Network (ANN) to enhance a geothermal-solar hybrid heating and hot water supply system. The ANN model, trained with operational data, predicted performance under varying conditions, optimizing energy efficiency, costs, and emissions. This approach led to a 22% increase in exergetic efficiency and a significant reduction in operational costs, underscoring the potential of AI to improve hybrid renewable energy systems. Gasmi et al. [93] also utilized machine learning techniques to optimize a geothermal hybrid system for maximizing heat and electricity production. The researchers evaluated the system's performance under various configurations through predictive models, enabling AI to identify optimal operational settings for simultaneous electricity and heat generation. This method effectively enhanced system efficiency and reduced costs, highlighting AI's value in hybrid geothermal applications. Together, these studies underscore the transformative potential of AI and machine learning in hybrid renewable systems. The application of ANN and predictive models in these geothermal hybrid setups increases energy efficiency and improves cost-effectiveness, making AI a crucial asset in advancing renewable energy integration and sustainability.

12.6.5 Machine Learning Applications for Monitoring, Safety, and Risk Management in Geothermal Systems

AI is also applied to enhance real-time monitoring and risk management in geothermal systems, contributing to their safety and efficiency. These applications enable early problem detection and rapid operational adjustments, ensuring safe and efficient operation. Ahmed et al. [48] developed a deep learning-based system to detect and analyze refrigerant bubbles in geothermal boiling systems. Using a Convolutional Neural Network (CNN) with a Mask R-CNN architecture, they could segment and detect bubbles in images. This facilitated real-time monitoring of changes in the boiling process, which helped identify potential failures or efficiency losses. Figure 12.20 illustrates how the Mask R-CNN model detects and segments bubbles in geothermal boiling systems. The visual illustrates the model's precision in bubble detection, enabling real-time tracking of changes in the boiling process. This capability significantly enhances the system's safety and operational efficiency by alerting them to potential faults or efficiency losses.

On the other hand, Pelzer and Schulte [120] applied a CNN to model how heat plumes generated by geothermal heat pumps interact in shallow aquifers. This system predicts thermal distribution in the aquifer, helping determine the optimal location for new geothermal installations. This ensures long-term efficiency and minimizes environmental impacts.

Applying machine learning techniques, particularly convolutional neural networks (CNNs), has shown significant potential in enhancing geothermal systems' monitoring, safety, and risk management. The ability to detect refrigerant bubbles in geothermal boiling systems, demonstrated by Ahmed et al. [119], provides a valuable tool for real-time monitoring and early detection of efficiency losses or failures, thereby supporting safer and more efficient operations. Similarly, the two-stage CNN model developed by Pelzer and Schulte [120] to predict heat plume interactions in aquifers offers a practical approach to optimize geothermal installation planning.

This model relies on a two-stage neural network approach to estimate temperature interactions in aquifers efficiently. The first stage uses a convolutional neural network (CNN) trained on simulated datasets generated with realistic subsurface parameters (e.g., permeability, pressure gradients, and temperature differences) to predict the localized thermal plume of a single heat pump. The CNN architecture incorporates spatially resolved input fields derived from borehole measurements and simulates the effects of key variables such as groundwater flow and thermal conductivity.

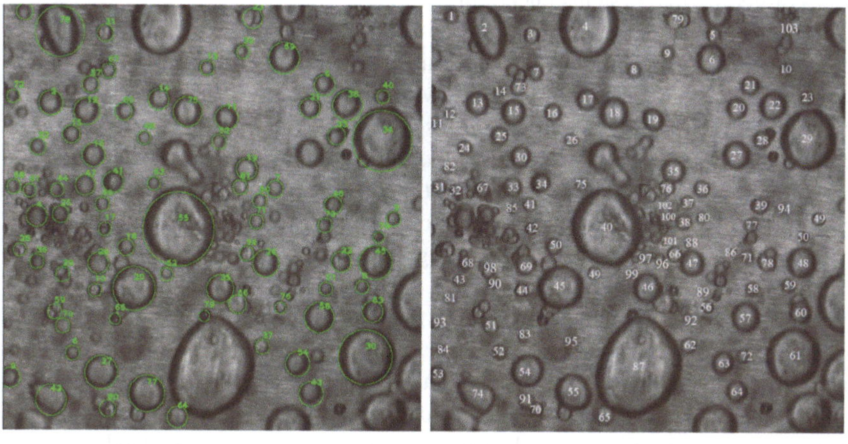

Fig. 12.20 Bubble detection and segmentation with mask R-CNN. Adapted from Ahmed et al. [119]. Copyright, Elsevier, 2023

In the second stage, a separate CNN refines the predictions by integrating overlapping heat plumes from multiple pumps, modeling their interactions dynamically. This stage employs combined input fields from the first-stage predictions, allowing the network to adjust the thermal distribution based on the spatial overlap of neighboring plumes. The methodology enables the accurate simulation of complex scenarios where individual heat pumps influence one another, achieving a Root Mean Square Error (RMSE) of approximately 0.1 °C across validation datasets. Validation was performed using simulated data generated with PFLOTRAN, ensuring consistency with established physical models while significantly reducing computational cost. By accurately forecasting thermal distribution, this model helps minimize environmental impacts and ensures the long-term efficiency of geothermal installations. Together, these applications highlight the transformative role of AI in advancing geothermal technology, supporting sustainable energy practices, and enhancing operational resilience. Figure 12.21 illustrates the two-stage modeling approach: the first stage predicts a single heat plume around one heat pump, while the second stage adjusts this prediction by incorporating the effects of other heat pumps. This result highlights how the integration of both stages allows for the efficient prediction of thermal interactions in aquifers, supporting effective planning by identifying optimal installation locations and avoiding thermal interference.

The approach proposed by Pelzer and Schulte [120] introduces an innovative two-stage modeling method using convolutional neural networks (CNNs) to predict heat plume interactions generated by geothermal heat pumps in shallow aquifers. While this method offers significant computational advantages over traditional simulations, it has notable limitations. One major constraint is its reliance on data derived from a specific geological region—the Munich area—which may restrict its applicability in regions with significantly different subsurface properties. Furthermore, the model is designed for open-loop geothermal heat pumps, making its application to other types of geothermal systems uncertain.

Another limitation of this approach is its simplification of certain physical processes. The model assumes a steady-state scenario, ignoring temporal variations, vertical dimensions, and heterogeneities within the subsurface. These simplifications may hinder their accuracy in real-world scenarios where seasonal changes or more complex interactions occur. Additionally, the model's ability to predict long-term and large-scale interactions between multiple heat pumps remains untested, raising concerns about its performance in dense or dynamic urban environments [120].

Finally, while the two-stage CNN design reduces computational demands, it still depends on high-fidelity simulation data, such as those generated by PFLOTRAN, for training. This requirement could pose challenges in regions where such detailed subsurface data are unavailable. Moreover, some inaccuracies persist in critical areas, such as near injection points and at simulation box boundaries, due to the smoothing effects inherent in CNNs and zero-padding techniques. These issues highlight the need for further refinements, such as incorporating transient data, expanding parameter variability, and exploring advanced architectures like physics-informed neural networks to improve the model's generalizability and robustness [120].

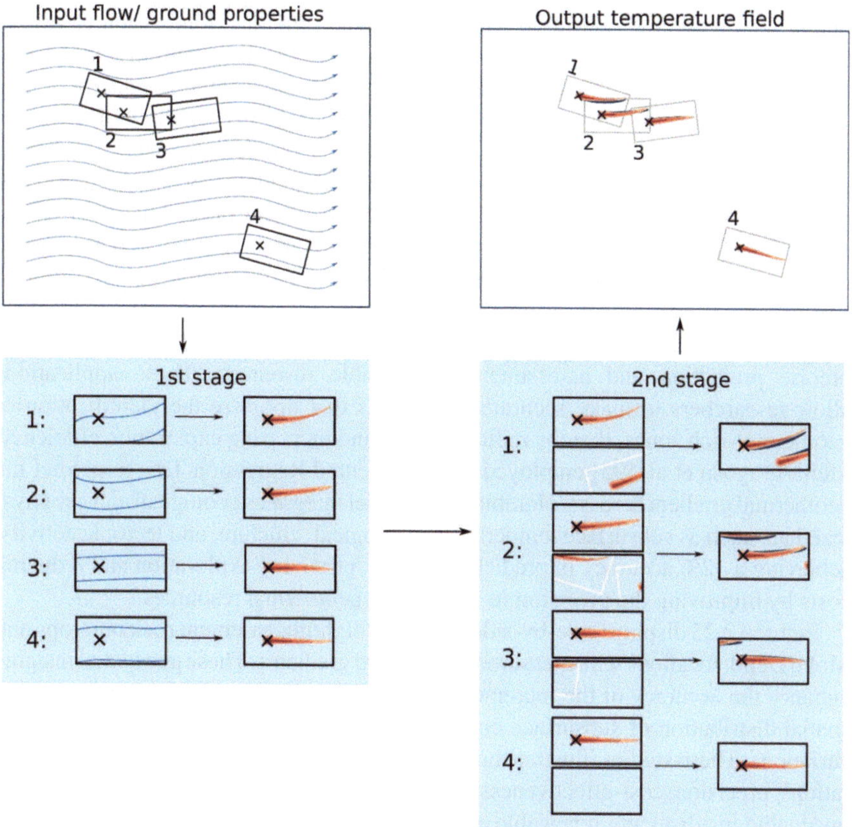

Fig. 12.21 Two-stage prediction of heat plume interactions. Adapted from Pelzer and Schulte et al. [120]. Copyright, Elsevier, 2024

This method allows for a dynamic assessment of temperature interactions in the aquifer, supporting effective planning and environmental sustainability by identifying optimal installation locations and avoiding thermal interference.

12.6.6 Application of Machine Learning in Exploration and Assessment of Geothermal Resources Using Artificial Intelligence

AI enhances geothermal exploration through advanced tools that enable more precise prediction and assessment of available resources. These applications allow researchers to make accurate predictions that optimize the identification of geothermal-rich zones, thereby reducing costs and increasing exploration efficiency.

Sheini Dashtgoli et al. [92] used the XGBoost algorithm and Regression Networks to predict temperatures in geothermal aquifers in the Friulian Plain in northeastern Italy. Incorporating geological variables such as depth, thermal conductivity, and geothermal gradients, they achieved an advanced predictive model with a coefficient of determination (R^2) of 0.993, facilitating the identification of areas with high geothermal potential. Figure 12.22 underscores the model's robustness and reliability, showing minimal deviation between predicted and observed values in the training and testing phases. This important level of accuracy supports the use of machine learning for identifying geothermal resource potential, thus contributing to cost-effective and efficient exploration efforts.

AI enhances geothermal exploration through advanced tools that enable more precise prediction and assessment of available resources. These applications allow researchers to make accurate predictions that optimize the identification of geothermal-rich zones, thereby reducing costs and increasing exploration efficiency. Mejía-Fragoso et al. [95] employed an Incremented Regression Tree to predict the geothermal gradient across Colombia. This model integrates geological and geophysical data, such as subsurface temperature, geological structure, and tectonic activity, achieving a 12% accuracy in predictions, thus optimizing exploration and reducing costs by improving the precision in identifying geothermal resources.

Figure 12.23 displays side-by-side maps highlighting basement rock outcrops and Moho Depth locations with measured geothermal gradients. These geospatial insights enhance the accuracy of the Incremented Regression Tree model by illustrating the spatial distribution of subsurface conditions that contribute to geothermal gradient variations. These studies illustrate how AI-driven models enhance geothermal exploration's precision, cost-effectiveness, and environmental sustainability, making them invaluable in advancing renewable energy resources.

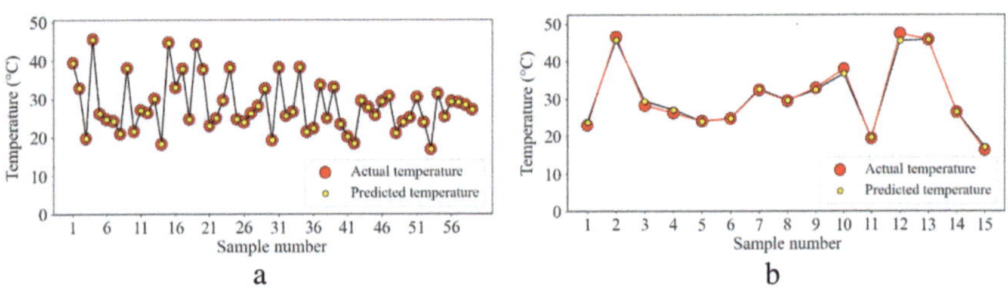

Fig. 12.22 Comparison of predicted and actual temperature measurements using XGBoost mode. Adapted from Dashtgoli et al. [92]. Copyright, Elsevier, 2024

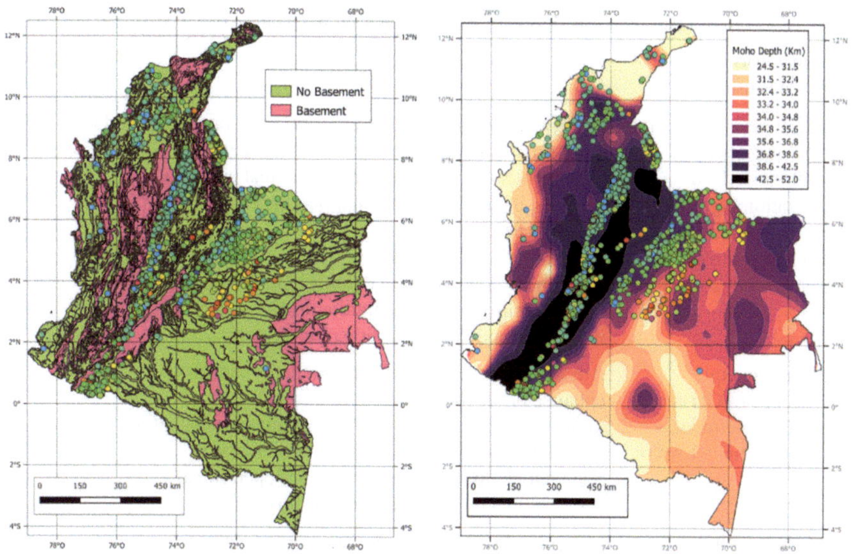

Fig. 12.23 Baseline data for geothermal gradient prediction. Adapted from Mejia et al. [95]. Copyright, Elsevier, 2024

12.6.7 Summary of Machine Learning Techniques Used in Each Research Area

Tables 12.3, 12.4, 12.5, 12.6 and 12.7 synthesize the machine learning techniques applied in different geothermal research clusters, relating them to the fundamental problems they address. Artificial Neural Networks (ANN) and Genetic Algorithms are used to improve system efficiency in optimization and modeling. Multi-objective optimization and neural networks facilitate the integration of geothermal energy into hybrid systems, achieving efficient adaptation to various demands. In safety and risk management, predictive analytics, IoT, and digital twins enable fault detection and risk mitigation in real time. Clustering algorithms and physics-informed neural networks optimize the exploration and assessment of geothermal resources, while socio-technical modeling and policy analysis address adoption barriers and strengthen regulatory frameworks.

12.7 Challenges and Opportunities

The application of AI in the geothermal sector presents challenges and opportunities crucial for its development and optimization. One of the main challenges is the availability and quality of data required to train AI models. Obtaining accurate and representative subsurface data is costly and complex due to the heterogeneity and

Table 12.3 Optimization and modeling of geothermal systems using AI

Machine learning techniques used	Problem(s) being addressed	References
Artificial neural networks (ANN) and genetic algorithms; hybrid optimization algorithms; data-driven models; intelligent control; exergetic analysis	Energy and geothermal systems optimization: focus on optimal sizing, modeling, and efficiency improvements for hybrid and geothermal energy systems, including geothermal rankine cycles for power generation Geothermal drilling and flow enhancements: address multiphase flow and fluid rheology challenges in high-pressure, high-temperature drilling with improved modeling and mitigation strategies	[121–125]
Artificial general intelligence (AGI) and large language models (LLMs); computer vision	Advanced applications in the geoenergy industry: enhance production optimization, safety improvements, and exploration techniques using AGI and LLMs in the upstream geoenergy industry. Segment anything model (SAM), for operational decision-making in geothermal energy	[126]
Machine learning and computational fluid dynamics (CFD)	Improvement in geothermal drilling and multiphase flow: address challenges of multiphase flow in drilling, completions, and injection in geothermal systems; Improve modeling and mitigation strategies in geothermal operations using ML and CFD. ML models applied with CFD for real-time drilling optimizations; neural networks for predicting flow behavior under extreme conditions	[127, 128]
Bayesian networks and supervised learning algorithms; neural networks, BRBNNs, LMBNNs	Risk management and uncertainty handling: handle uncertainties and complex problems in renewable energy systems using probabilistic models; Applications in forecasting, fault diagnosis, maintenance, operation, planning, and risk management in renewable energy, including geothermal energy. Energy forecasting and Prediction: improve accuracy in energy demand forecasting and production in wind, solar, and geothermal systems. Geothermal drilling and flow enhancements: address multiphase flow and fluid rheology challenges in high-pressure, high-temperature drilling with improved modeling and mitigation strategies. BRBNNs and LMBNNs for enhanced forecasting accuracy in geothermal energy systems; genetic algorithms for optimizing geothermal drilling strategies. BN for predictive fault diagnosis in renewable energy systems, including geothermal fields. Fuzzy-logic-based site assessment models for geothermal power plants; hybrid models combining neural networks and fuzzy systems for control strategies. Decision tree models for diagnosing faults in geothermal turbines; SVMs for reservoir characterization	[129, 130]

(continued)

Table 12.3 (continued)

Machine learning techniques used	Problem(s) being addressed	References
AI and optimization methods	Process and material optimization: improve efficiency and reduce costs in solar thermal desalination using AI; optimize processes and materials; use AI for design concepts in renewable energy systems. Optimization of energy storage systems: optimize sand-based thermal energy storage systems using AI/ML to enhance thermal performance and design	[131, 132]
Not specified (focus on drilling fluid modeling)	Improvement in geothermal drilling and multiphase flow: address challenges related to drilling fluid rheology in geothermal drilling; review advancements in drilling mud systems, rheology improvement, and modeling of fluid properties under high-pressure and high-temperature conditions	[133]

complexity of geological systems [92, 95]. Additionally, integrating AI solutions with existing geothermal infrastructures requires overcoming technical and operational barriers, including the need for advanced computational resources and trained personnel [117, 120].

Table 12.4 Integration of renewable energies and hybrid systems

Machine learning techniques used	Problem(s) being addressed	References
Machine learning methods and intelligent control techniques; demand prediction and energy dispatch; multi-objective optimization; control algorithms	Integration and optimization of renewable energies: energy management and control strategies for district heating and cooling systems with renewable integrations; integration of renewable sources (solar, geothermal, biomass) into energy systems and district heating; development of hybrid systems and carbon–neutral communities through integrated renewable energy systems (IRES)	[60, 134, 135]
Applications of AI in renewable energy; AI in renewable energy research; machine learning in hydrogen production; nanotechnology (nanoparticles, nanofluids)	Improvement in hydrogen production and storage: use ML to optimize hydrogen production processes and catalyst development; address technical and ethical challenges in hydrogen production and storage. Energy efficiency and sustainability: improve energy efficiency in data centers and airports using AI and data-driven systems; application of nanofluids to enhance heat transfer and efficiency in energy production; application of AI to improve renewable energy technologies, including geothermal and solar energies	[136–139]

(continued)

Table 12.4 (continued)

Machine learning techniques used	Problem(s) being addressed	References
Applications of AI in energy efficiency and sustainability; AI and data-driven systems for energy management	Energy efficiency and sustainability: improve energy efficiency in data centers and green airports using AI and data-driven systems; monitoring and controlling energy consumption; implementation of renewable energy sources and energy management techniques. Promotion of sustainable practices: advance resilient urban infrastructure using AI; reinforce the interconnection between social and ecological systems	[140, 141]
Not specified (hydrogen in energy transition)	Role of hydrogen in energy transition: role of hydrogen in the energy transition of the oil and gas industry; leverage oil and gas exploration experiences for hydrogen production and storage; address challenges in scaling up hydrogen production and infrastructure	[1]

Table 12.5 Monitoring, safety, and risk management in geothermal systems

Machine learning techniques used	Problem(s) being addressed	References
Predictive analytics, AI, and digital twins; machine learning (ML); internet of things (IoT)	Risk management and operational safety: improve safety protocols, predictive analytics, and risk management in geothermal systems; Real-time fault detection and mitigation strategies in geothermal operations; future perspectives on risk management approaches, including advanced technologies like IoT and digital twins	[142]
Life cycle assessment (LCA) with AI; AI in LCA	Environmental optimization and sustainability: minimize environmental impact of renewable energy-driven membrane distillation (MD) systems; application of LCA to evaluate MD systems; discuss the role of AI in optimizing LCA processes for sustainability in geothermal applications; enhance sustainability in geothermal operations through process optimization and environmental impact reduction	[2]
Not specified (focus on CO_2 leakage risk management)	Risk management and operational safety: risk management of CO_2 leakage in wellbores in CO_2 geological utilization and storage; detection and assessment of CO_2 leakage risks induced by cement failure; challenges in automated monitoring systems for risk mitigation in geothermal operations	[143]

Another challenge is the acceptance and trust in AI-based solutions by industry professionals. A lack of understanding or experience with these technologies can generate resistance to their adoption [8]. Furthermore, there are regulatory and policy challenges that may not be updated with technological advancements, potentially hindering the implementation of AI in geothermal energy [142]. However,

Table 12.6 Exploration and assessment of geothermal resources using AI

Machine learning techniques used	Problem(s) being addressed	References
Neural networks and deep learning algorithms; ANN, CNN, LSTM, GAN; support vector machines (SVM); fuzzy inference systems (FIS), ANFIS	Improvement in geothermal exploration: predict complex nonlinear relationships in geotechnical engineering, including applications in rock mechanics and slope stability; enhance accuracy in predicting geological features and geothermal resources using AI techniques; Identification of subsurface geological features from seismic data, such as faults and structures, relevant to geothermal resource assessment	[3, 144]
AI technologies and data mining; machine learning for seismic interpretation; AI for resource evaluation and management	Optimization of resource evaluation: use AI and geoscientific data to improve efficiency in utilizing shallow geothermal energy; Facilitate access and processing of large geoscientific datasets for societal needs and sustainable development goals; intelligent decision-making in sectors related to geothermal energy; applications in monitoring, control, and optimization	[4, 145]
AI Algorithms for fracture aperture prediction	Improvement in geothermal exploration: address challenges in measurement, inference, and modeling of fracture aperture in geological contexts; use AI algorithms to predict fracture aperture based on other known surface or subsurface data; enhance fracture network modeling for geothermal reservoirs	[7]

the opportunities provided by AI in this field are significant. AI can improve efficiency and reduce operational costs of geothermal systems, optimizing performance and adapting to changing conditions [93, 117]. In resource exploration and assessment, AI techniques allow for more accurate predictions of thermal gradients and subsurface characteristics, reducing the need for extensive and costly explorations and minimizing risks [92, 95].

Table 12.7 Socio-technical analysis, energy policy, and sustainability

Machine learning techniques used	Problem(s) being addressed	References
AI strategies and data mining; AI in the agri-food and aquaculture industries	Promotion of sustainable practices and efficient resource management: understand how AI can enhance key industries like agri-food and aquaculture, positively impacting sustainability; intelligent decision-making in aquaculture and fisheries from complex datasets, contributing to sustainable practices and resource management; propose research to advance AI application in agriculture and related sectors	[5, 146]
Computational and bibliometric analysis; computational social network analysis;	Identification of trends and future opportunities: trace evolution and trends in geothermal energy research and development, identifying future opportunities and challenges; map trends in research and development of geothermal energy and TES systems; employ bibliometric analysis to identify trends, evolution, and future research directions in sand-based thermal energy storage systems	[147]
Not specified (policy analysis and renewable energy status)	Policy analysis and regulatory frameworks: address barriers to the adoption of geothermal and renewable energy; offer strategies and discuss policy implications; assessment of renewable energy status, challenges, COVID-19 impacts, opportunities, and sustainable energy solutions in various regions. Current status and future prospects of renewable and sustainable energy in North America; discuss progress and challenges	[6, 148, 149]
Not specified (exploration of technological advances and sustainability)	Identification of trends and future opportunities: viability, challenges, and proposed solutions for establishing a sustainable human colony on mars; explore technological advances including AI; review advances in polyoxometalate-based materials for electrochemical water splitting; address challenges and present future perspectives; impact of contemporary technologies like AI on sustainable research and development	[7, 150]

Additionally, AI facilitates real-time monitoring and risk management, enhancing safety and extending the lifespan of installations [120]. The integration of geothermal energy with other renewable sources in hybrid systems is boosted by AI, enabling more efficient and sustainable energy resource management [114]. Finally, aligning AI-enhanced geothermal projects with sustainability and renewable energy policy goals can attract support and funding, driving the transition toward cleaner energy systems [151].

12.8 Study Limitations and Recommendations

While Artificial AI presents transformative potential for optimizing geothermal energy systems, its widespread adoption is constrained by several limitations that must be addressed to unlock its full capabilities. Geothermal energy, by nature, poses unique operational and geological challenges, including subsurface uncertainty, variable reservoir behavior, and the complex interplay between heat transfer and geological formations. Although AI offers innovative solutions to enhance efficiency, predict performance, and manage geothermal resources sustainably, the inherent complexities of geothermal environments introduce hurdles that require careful consideration. One of the foremost limitations lies in data availability and quality, which serve as the foundation for training and deploying AI models. Geothermal systems often operate in regions where subsurface exploration is limited by the cost and logistical difficulties of acquiring comprehensive geological data. The efficacy of machine learning (ML) and deep learning (DL) algorithms hinges on large, diverse, and high-resolution datasets encompassing geological surveys, well logs, seismic profiles, and operational parameters. However, many geothermal fields lack sufficient historical data, resulting in models that are prone to overfitting or failing to generalize across varying geothermal environments. This scarcity introduces uncertainty in AI-driven predictions of subsurface temperature, fluid dynamics, and reservoir potential. Additionally, geothermal reservoirs are highly heterogeneous, exhibiting significant variability in porosity, permeability, and thermal conductivity, which complicates the development of AI models that can accurately capture reservoir behavior under different geological conditions.

The geological complexity and variability of geothermal systems further exacerbate these challenges. Unlike oil and gas reservoirs, geothermal fields often contain fractured rock formations, high-pressure hydrothermal zones, and magmatic intrusions, all of which influence the heat extraction process. Traditional data acquisition methods, such as seismic imaging and magnetotelluric surveys, provide valuable insights but are limited in their ability to fully characterize the three-dimensional structure of subsurface reservoirs. AI, while capable of integrating data from multiple sources, struggles to compensate for the fundamental gaps in geological understanding. This results in suboptimal drilling strategies, increased exploration risks, and lower overall efficiency in geothermal energy extraction. Moreover, the integration of AI and IoT technologies into existing geothermal infrastructure poses significant technical and economic barriers. Many geothermal plants rely on legacy systems that lack the digital architecture required to support advanced monitoring, real-time data acquisition, and automated control processes. Retrofitting these facilities with IoT-enabled sensors, edge computing devices, and AI-based control algorithms necessitates substantial investment and may lead to operational disruptions during the transition. Additionally, the harsh operational environment of geothermal plants—characterized by extreme temperatures, high pressures, and corrosive fluids—can degrade sensor performance and reduce the reliability of IoT devices, thereby limiting the effectiveness of AI-driven operational management.

Another notable limitation is the computational intensity and energy demands associated with AI model development. Geothermal energy systems, particularly those involving enhanced geothermal systems (EGS) or deep drilling, require complex simulations to model fluid flow, heat transfer, and mechanical stress. Training AI models capable of handling these tasks often requires significant computational power, which may not align with the sustainability goals of geothermal operators. Furthermore, the environmental footprint of large-scale AI computations could offset some of the carbon-reduction benefits associated with geothermal energy production, raising concerns about the net environmental impact.

To overcome these limitations, several recommendations are proposed to facilitate the broader adoption of AI in geothermal energy systems. One key recommendation is the expansion of data acquisition efforts and the creation of open-access geothermal databases. The integration of satellite data, seismic networks, and IoT-enabled subsurface sensors can enhance data granularity, allowing AI models to develop more accurate predictions of reservoir behavior and thermal gradients. Hybrid modeling approaches represent another critical pathway to address the limitations of AI in geothermal applications. By combining AI techniques with physics-based models rooted in thermodynamics and fluid dynamics, it is possible to create hybrid models that leverage both data-driven insights and fundamental scientific principles. These models can compensate for sparse datasets by incorporating prior geological knowledge, thereby improving the accuracy and resilience of AI predictions. For instance, reservoir simulation models augmented with machine learning algorithms can provide more accurate forecasts of fluid flow patterns and reservoir depletion rates, enabling operators to optimize injection and extraction strategies in real time. The development of adaptive AI systems capable of continuous learning and self-correction is also essential for enhancing geothermal efficiency. Reinforcement learning algorithms, which iteratively refine their performance based on real-time feedback, can be deployed to dynamically adjust operational parameters in response to changing reservoir conditions. This adaptive capability is particularly valuable in geothermal environments where fluid flow rates, well pressures, and thermal output fluctuate over time. By embedding AI into digital twin models of geothermal plants, operators can simulate various operational scenarios, test optimization strategies, and proactively address potential failures before they occur.

To support these technological advancements, it is imperative to invest in digital infrastructure and workforce development. Governments and industry leaders should prioritize funding for the deployment of IoT devices, fiber-optic sensors, and advanced control systems in geothermal plants. Simultaneously, educational programs focusing on AI, data science, and geothermal engineering can equip the next generation of professionals with the skills necessary to implement and manage AI technologies effectively. Specialized training in AI-driven reservoir management, predictive maintenance, and energy optimization will foster a workforce capable of driving the digital transformation of the geothermal sector.

Finally, pilot projects and demonstration plants serve as critical testbeds for validating AI applications in real-world geothermal environments. By establishing pilot projects that integrate AI into geothermal exploration, drilling, and operational workflows, industry stakeholders can demonstrate the tangible benefits of AI adoption, build confidence among investors, and refine technological approaches based on practical insights. Successful pilot projects can pave the way for large-scale deployment, reducing risks and accelerating the transition toward AI-enhanced geothermal systems.

12.9 Conclusions and Future Work

The integration of AI into geothermal energy systems marks a transformative shift toward enhancing efficiency, sustainability, and operational resilience. This chapter has explored how AI technologies, particularly machine learning (ML), deep learning (DL), and clustering algorithms, are being applied across various stages of geothermal energy production–from resource exploration to reservoir management and plant optimization. AI's capacity to handle complex datasets, recognize patterns, and predict subsurface conditions has proven invaluable in reducing exploration risks, optimizing drilling strategies, and improving reservoir characterization. These innovations address long-standing challenges associated with geothermal variability, geological uncertainty, and operational inefficiencies. AI-driven models facilitate accurate subsurface temperature predictions, optimize well placement, and monitor equipment performance in real time. This enables predictive maintenance, minimizing unplanned downtime, reducing operational costs, and extending the lifespan of geothermal infrastructure. Furthermore, enhanced geothermal systems (EGS) and advanced drilling technologies benefit from AI's ability to simulate various operational scenarios, guiding decisions that balance energy output, reservoir sustainability, and environmental impact. Case studies highlighted in this chapter demonstrate how AI enhances fluid flow regulation, thermal extraction efficiency, and overall plant performance by adapting to changing reservoir conditions dynamically. The integration of AI with Internet of Things (IoT) technologies further amplifies these benefits by providing continuous monitoring and data collection, fostering smarter and more adaptive geothermal facilities.

However, several challenges must be addressed to fully realize the potential of AI in geothermal energy. One of the most pressing issues is the availability and quality of data, which underpins the accuracy and reliability of AI models. Sparse or incomplete data can lead to biased predictions and operational inefficiencies. Geological complexity, coupled with the heterogeneous nature of geothermal reservoirs, further complicates AI implementation, limiting the generalizability of models across different sites. Additionally, the retrofitting of legacy geothermal plants with IoT sensors and AI platforms requires significant capital investment and may disrupt ongoing operations. Overcoming these barriers will require collaborative efforts to enhance data acquisition, expand open-access geothermal databases, and develop hybrid AI-physics models that integrate domain knowledge with data-driven insights.

While the technical aspects of AI integration have been the primary focus, future work must also consider the socio-economic dimensions of geothermal energy expansion, particularly in the context of geothermal district heating (GDH). AI has the potential to play a pivotal role in promoting GDH adoption by addressing key barriers such as public awareness, economic feasibility, and infrastructure planning. By leveraging AI to analyze community energy needs, demographic data, and regional heat demand, stakeholders can identify high-potential areas for GDH deployment. Machine learning algorithms can simulate cost-saving scenarios, model long-term returns on investment, and demonstrate the environmental benefits of transitioning from fossil-fuel-based heating to geothermal solutions.

AI can further contribute to the economic viability of GDH by optimizing energy distribution networks, forecasting peak demand, and enhancing load management. This can help municipalities and energy providers design efficient district heating systems that minimize energy waste and reduce operational costs.

Acknowledgements The authors extend their gratitude to the Universidad Nacional de Colombia and the Universidad de Antioquia (UdeA) for their invaluable logistical support, which significantly contributed to the development.

Declaration of Competing Interest The authors declare that they have no known competing financial interests or personal relationships that could have appeared to influence the work reported in this paper.

References

1. Soltani M, Moradi Kashkooli F, Alian Fini M, Gharapetian D, Nathwani J, Dusseault MB (2022) A review of nanotechnology fluid applications in geothermal energy systems. Renew Sustain Energy Rev 167 (2022). https://doi.org/10.1016/j.rser.2022.112729
2. Sheini Dashtgoli D, Giustiniani M, Busetti M, Cherubini C (2024) Artificial intelligence applications for accurate geothermal temperature prediction in the lower Friulian plain (north-eastern Italy). J Clean Prod 460. https://doi.org/10.1016/j.jclepro.2024.142452
3. Mordensky SP, Lipor JJ, DeAngelo J, Burns ER, Lindsey CR (2023) When less is more: how increasing the complexity of machine learning strategies for geothermal energy assessments may not lead toward better estimates. Geothermics 110. https://doi.org/10.1016/j.geothermics.2023.102662
4. Ahmmed B, Vesselinov VV (2022) Machine learning and shallow groundwater chemistry to identify geothermal prospects in the great Basin, USA. Renew Energy 197:1034–1048. https://doi.org/10.1016/j.renene.2022.08.024
5. Coro G, Trumpy E (2020) Predicting geographical suitability of geothermal power plants. J Clean Prod 267. https://doi.org/10.1016/j.jclepro.2020.121874
6. Gasmi H, Abed AM, Dutta AK, Alhomayani FM, Mahariq I, Alturise F, Alkhalaf S, Alkhalifah T, Elmasry Y, Khan B (2024) Heat recovery integration in a hybrid geothermal-based system producing power and heating using machine learning approach to maximize outputs. Case Stud Therm Eng 63. https://doi.org/10.1016/j.csite.2024.105210
7. Okoroafor ER, Smith CM, Ochie KI, Nwosu CJ, Gudmundsdottir H, (Jabs) Aljubran M (2022) Machine learning in subsurface geothermal energy: two decades in review. Geothermics 102:102401. https://doi.org/10.1016/J.GEOTHERMICS.2022.102401

8. Moraga J, Duzgun HS, Cavur M, Soydan H (2022) The geothermal artificial intelligence for geothermal exploration. Renew Energy 192:134–149. https://doi.org/10.1016/J.RENENE.2022.04.113
9. Wardoyo GK, Pratama HB, Ashat SA, Yudhistira Y (2021) Application of artificial intelligence in forecasting geothermal production. In: IOP conference series: earth and environmental science. IOP Publishing Ltd. https://doi.org/10.1088/1755-1315/732/1/012022
10. AlGaiar MM (2024) Unlocking geothermal potential: advancing exploration with artificial intelligence for sustainable energy solutions. In: Abu Dhabi international petroleum exhibition and conference. SPE, p D041S133R002
11. Khosravi A, Syri S, Zhao X, Assad MEH (2019) An artificial intelligence approach for thermodynamic modeling of geothermal based-organic Rankine cycle equipped with solar system. Geothermics 80:138–154
12. Guan Y, Post S, Zhao D, Zhang S, Becker S (2024) Overview of the application status and development trends of hydropower and geothermal power in New Zealand. Energy Built Environ
13. Olabi AG, Elsaid K, Obaideen K, Abdelkareem MA, Rezk H, Wilberforce T, Maghrabie HM, Sayed ET (2023) Renewable energy systems: comparisons, challenges and barriers, sustainability indicators, and the contribution to UN sustainable development goals. Int J Thermofluids 20:100498
14. Wang J, Azam W (2024) Natural resource scarcity, fossil fuel energy consumption, and total greenhouse gas emissions in top emitting countries. Geosci Front 15:101757. https://doi.org/10.1016/J.GSF.2023.101757
15. Tester JW, Beckers KF, Hawkins AJ, Lukawski MZ (2021) The evolving role of geothermal energy for decarbonizing the United States. Energy Environ Sci 14:6211–6241
16. Nath F, Mahmood MN, Ofosu E, Khanal A (2024) Enhanced geothermal systems: a critical review of recent advancements and future potential for clean energy production. Geoenergy Sci Eng (2024):213370
17. World energy transitions outlook 2023 (2023) 1.5°C pathway. Int Renew Energy Agency IRENA
18. DeCarolis J, LaRose A (2023) Annual energy outlook 2023. US Energy Information Administration
19. Gutiérrez-Negrín LCA (2024) Evolution of worldwide geothermal power 2020–2023. Geothermal Energy 12:14
20. Sharmin T, Khan NR, Akram MS, Ehsan MM (2023) A state-of-the-art review on geothermal energy extraction, utilization, and improvement strategies: conventional, hybridized, and enhanced geothermal systems. Int J Thermofluids 18:100323. https://doi.org/10.1016/J.IJFT.2023.100323
21. Sircar A, Solanki K, Bist N, Yadav K (2022) Enhanced geothermal systems–promises and challenges. Int J Renew Energy Dev 11:333–346
22. Walmsley TG, Philipp M, Picón-Núñez M, Meschede H, Taylor MT, Schlosser F, Atkins MJ (2023) Hybrid renewable energy utility systems for industrial sites: a review. Renew Sustain Energy Rev 188:113802. https://doi.org/10.1016/J.RSER.2023.113802
23. Kabeyi MJB, Olanrewaju OA (2022) Sustainable energy transition for renewable and low carbon grid electricity generation and supply. Front Energy Res 9. https://doi.org/10.3389/fenrg.2021.743114
24. Song X, Li G, Huang Z, Shi Y, Wang G, Song G, Xu F (2023) Review of high-temperature geothermal drilling and exploitation technologies. Gondwana Res 122:315–330. https://doi.org/10.1016/J.GR.2022.10.013
25. Anderson A, Rezaie B (2019) Geothermal technology: trends and potential role in a sustainable future. Appl Energy. https://doi.org/10.1016/j.apenergy.2019.04.102
26. Luo W, Kottsova A, Vardon PJ, Dieudonné AC, Brehme M (2023) Mechanisms causing injectivity decline and enhancement in geothermal projects. Renew Sustain Energy Rev 185. https://doi.org/10.1016/j.rser.2023.113623

27. Kurnia JC, Shatri MS, Putra ZA, Zaini J, Caesarendra W, Sasmito AP (2022) Geothermal energy extraction using abandoned oil and gas wells: techno-economic and policy review. Int J Energy Res. https://doi.org/10.1002/er.6386
28. Shah M, Prajapati M, Yadav K, Sircar A (2024) A comprehensive review of geothermal energy storage: methods and applications. J Energy Storage 98:113019. https://doi.org/10.1016/J.EST.2024.113019
29. van der Zwaan B, Dalla Longa F (2019) Integrated assessment projections for global geothermal energy use. Geothermics 82:203–211
30. Céspedes S, Cano NA, Foo G, Jaramillo D, Martinez D, Gutiérrez M, Pataquiba J, Rojas J, Cortés FB, Franco CA (2022) Technical and environmental feasibility study of the co-production of crude oil and electrical energy from geothermal resources: first field trial in Colombia. Processes 10:568
31. Cano NA, Céspedes S, Redondo J, Foo G, Jaramillo D, Martinez D, Gutiérrez M, Pataquiba J, Rojas J, Cortés FB, Franco CA (2022) Power from geothermal resources as a co-product of the oil and gas industry: a review. ACS Omega 7:40603–40624. https://doi.org/10.1021/acsomega.2c04374
32. Davoodi S, Al-Shargabi M, Wood DA, Slivkin S, Shishaev G, Rukavishnikov VS (2024) Geothermal energy recovery from abandoned petroleum wells: a review of the challenges and opportunities. Sustain Energy Technol Assess 68:103870. https://doi.org/10.1016/J.SETA.2024.103870
33. Allahvirdizadeh P (2020) A review on geothermal wells: well integrity issues. J Clean Prod 275:124009. https://doi.org/10.1016/J.JCLEPRO.2020.124009
34. Lund JW, Toth AN (2021) Direct utilization of geothermal energy 2020 worldwide review. Geothermics 90:101915. https://doi.org/10.1016/J.GEOTHERMICS.2020.101915
35. Watban Khalid Fahmi AT, Reza Kashyzadeh K, Ghorbani S (2022) A comprehensive review on mechanical failures cause vibration in the gas turbine of combined cycle power plants. Eng Fail Anal 134:106094. https://doi.org/10.1016/J.ENGFAILANAL.2022.106094
36. Wang Y, Pang Z (2023) Heat flux in volcanic and geothermal areas: methods, principles, applications and future directions. Gondwana Res 122:260–278. https://doi.org/10.1016/J.GR.2022.09.010
37. Archer R (2020) Geothermal energy. In: Future energy. Elsevier, pp 431–445
38. Isania F, Elahifar B (2025) Advances in geothermal drilling: a comparative study with oil and gas techniques. Geoenergy Sci Eng 246:213638. https://doi.org/10.1016/J.GEOEN.2024.213638
39. Kumar L, Hossain MS, Assad MEH, Manoo MU (2022) Technological advancements and challenges of geothermal energy systems: a comprehensive review. Energies (Basel) 15. https://doi.org/10.3390/en15239058
40. Saleh F, Teodoriu C, Salehi S, Ezeakacha C (2020) Geothermal drilling: a review of drilling challenges with mud design and lost circulation problem. In: Proceedings of 45th annual Stanford geothermal workshop. Stanford University, Stanford, CA
41. Shamoushaki M, Fiaschi D, Manfrida G, Niknam PH, Talluri L (2021) Feasibility study and economic analysis of geothermal well drilling. Int J Environ Stud 78:1022–1036
42. Arbad N, Emadi H, Watson M (2022) A comprehensive review of geothermal cementing from well integrity perspective. J Pet Sci Eng 217:110869. https://doi.org/10.1016/J.PETROL.2022.110869
43. Jr Capuano LE (2025) Geothermal well drilling. Geotherm Power Gener 125–158 (2025). https://doi.org/10.1016/B978-0-443-24750-7.00030-0
44. Teodoriu C, Bello O (2021) An outlook of drilling technologies and innovations: present status and future trends. Energies (Basel) 14:4499
45. Jeffery C, Creegan A (2020) Adaptive drilling application uses AI to enhance on-bottom drilling performance. J Petrol Technol 72:45–47
46. Erge O, Sakaoglu K, Sonmez A, Bagatir G, Dogan HA, Ay A, Gucuyener IH (2020) Overview and design principles of drilling fluids systems for geothermal Wells in Turkey. Geothermics. https://doi.org/10.1016/j.geothermics.2020.101897

47. Think-Geonergy, Turkey targets 2,000 MW geothermal power generation capacity by 2020 (2020). https://www.thinkgeoenergy.com/turkey-targets-2000-mw-geothermal-power-generation-capacity-by-2020/
48. Mohamed A, Salehi S, Ahmed R (2021) Significance and complications of drilling fluid rheology in geothermal drilling: a review. Geothermics 93:102066. https://doi.org/10.1016/j.geothermics.2021.102066
49. Kruszewski M, Wittig V (2018) Review of failure modes in supercritical geothermal drilling projects. Geothermal Energy 6:28
50. Khaled MS, Wang N, Ashok P, van Oort E (2023) Downhole heat management for drilling shallow and ultra-deep high enthalpy geothermal wells. Geothermics 107:102604
51. Ndeda R, Sebusang SEM, Marumo R, Ogur EO (2022) Review of thermal surface drilling technologies. In: Proceedings of the sustainable research and innovation conference. pp 61–69
52. Dreus A, Kozhevnykov A (2019) Innovative approach to drilling of geothermal deep wells using the heat factor. In: 2019 IEEE 6th international conference on energy smart systems (ESS). IEEE, pp 192–195
53. CSR (2024) Enhanced geothermal systems (EGS): frequently asked questions. https://crsreports.congress.gov/product/pdf/R/R48090. Accessed 12 Nov 2024
54. Zhang Y, Zhang Y, Zhou L, Lei Z, Guo L, Zhou J (2022) Reservoir stimulation design and evaluation of heat exploitation of a two-horizontal-well enhanced geothermal system (EGS) in the Zhacang geothermal field, Northwest China. Renew Energy 183:330–350. https://doi.org/10.1016/J.RENENE.2021.10.101
55. Onay ME (2020) Analytical solutions for predicting fracture outlet temperature of produced fluid from enhanced geothermal systems with different well-completion configurations. In: SPE annual technical conference and exhibition. SPE, p D023S100R016
56. Gelfgat MY, Geraskin AS (2021) Deep geothermal well construction problems and possible solutions. In: SPE Russian petroleum technology conference? SPE, p D021S012R002
57. Feder J (2021) Geothermal well construction: a step change in oil and gas technologies. J Petrol Technol 73:32–35
58. Gola G, Di Sipio E, Facci M, Galgaro A, Manzella A (2022) Geothermal deep closed-loop heat exchangers: a novel technical potential evaluation to answer the power and heat demands. Renew Energy 198:1193–1209
59. Emina KA (2021) Sustainable development and the future generations. Soc Sci, Hum Educ J (SHE J) 2:57. https://doi.org/10.25273/she.v2i1.8611
60. Rohit RV, Vipin Raj R, Kiplangat DC, Veena R, Jose R, Pradeepkumar AP, Kumar KS (2023) Tracing the evolution and charting the future of geothermal energy research and development. Renew Sustain Energy Rev 184:113531. https://doi.org/10.1016/J.RSER.2023.113531
61. Kamila Z, Kaya E, Zarrouk SJ (2021) Reinjection in geothermal fields: an updated world-wide review 2020. Geothermics 89:101970. https://doi.org/10.1016/J.GEOTHERMICS.2020.101970
62. Zayed ME, Shboul B, Yin H, Zhao J, Zayed AAA (2023) Recent advances in geothermal energy reservoirs modeling: challenges and potential of thermo-fluid integrated models for reservoir heat extraction and geothermal energy piles design. J Energy Storage 62:106835
63. Xu M, Yang X, Dai Z, Wang R, Wang J, Xie J, Liu M, Shi Y, Chen S, Xue M, Cai J, Jiang G (2024) Experimental research of convective heat transfer between nanofluids and high-temperature dense granite in deep geothermal reservoirs. Renew Energy 235:121245. https://doi.org/10.1016/J.RENENE.2024.121245
64. Chappidi S, Kumar A, Singh J (2023) Geothermal energy extraction from abandoned oil and gas wells using mono and hybrid nanofluids. Geothermics 114:102794. https://doi.org/10.1016/J.GEOTHERMICS.2023.102794
65. Soltani M, Moradi Kashkooli F, Alian Fini M, Gharapetian D, Nathwani J, Dusseault MB (2022) A review of nanotechnology fluid applications in geothermal energy systems. Renew Sustain Energy Rev 167:112729. https://doi.org/10.1016/J.RSER.2022.112729.
66. Li X, Yuan F, Tian W, Dai C, Yang X, Wang D, Du J, Yu W, Yuan H (2022) Heat transfer enhancement of nanofluids with non-spherical nanoparticles: a review. Appl Sci (Switzerland) 12. https://doi.org/10.3390/app12094767

67. Cui Y, Tian S, Zoras S, Zhu J (2024) Recent advances in various nanomaterials utilized in geothermal heat exchangers. Nano Energy 109309
68. Alkalbani AM, Chala GT (2024) A comprehensive review of nanotechnology applications in oil and gas well drilling operations. Energies (Basel) 17. https://doi.org/10.3390/en17040798
69. Błaż S, Zima G, Jasiński B, Kremieniewski M (2024) Increasing the thermal resistance of water-based mud for drilling geothermal wells. Energ (Basel) 17:4537
70. Martin C, Babaie M, Nourian A, Nasr GG (2023) Designing smart drilling fluids using modified nano silica to improve drilling operations in geothermal wells. Geothermics 107:102600. https://doi.org/10.1016/J.GEOTHERMICS.2022.102600
71. Gokapai V, Pothana P, Ling K (2024) Nanoparticles in drilling fluids: a review of types. Mech, Appl, Futur Prospect, Eng 5:2462–2495. https://doi.org/10.3390/eng5040129
72. Mikhienkova EI, Lysakov SV, Neverov AL, Zhigarev VA, Minakov AV, Rudyak VY (2022) Experimental study on the influence of nanoparticles on oil-based drilling fluid properties. J Pet Sci Eng 208:109452. https://doi.org/10.1016/J.PETROL.2021.109452
73. Makwana D, Bellani J, Verma HK, Khatri D, Shah M (2021) Emergence of nano silica for oil and gas well cementing: application, challenges, and future scope. Environ Sci Pollut Res 28:37110–37119
74. Kaggwa A, Carson JK (2019) Developments and future insights of using nanofluids for heat transfer enhancements in thermal systems: a review of recent literature. Int Nano Lett 9:277–288
75. Ahmed W, Zhan Y, Zhang H, Zhou X, Shahid M, Mudasar F, Khan QU (2023) Preparation, applications, stability and improved thermal characteristics of sonochemically synthesized nanosuspension using varying heat exchangers, a review. J Mol Liq 387:122665. https://doi.org/10.1016/J.MOLLIQ.2023.122665
76. Du R, Jiang D, Wang Y, Shah KW (2020) An experimental investigation of CuO/water nanofluid heat transfer in geothermal heat exchanger. Energy Build 227:110402
77. Jahanbin A, Semprini G, Pulvirenti B (2022) Performance evaluation of U-tube borehole heat exchangers employing nanofluids as the heat carrier fluid. Appl Therm Eng 212:118625. https://doi.org/10.1016/J.APPLTHERMALENG.2022.118625
78. Villada-Villada Y (2024) Nanotecnología Aplicada a la perforación y desarrollo de Sistemas Geotérmicos en co-producción con hidrocarburos, Workshop Geotermia-Recurso Del Subsuelo Para Todos
79. Mejía-Fragoso JC, Flórez MA, Bernal-Olaya R (2024) Predicting the geothermal gradient in Colombia: a machine learning approach. Geothermics 122. https://doi.org/10.1016/j.geothermics.2024.103074
80. Frison L, Gölzhäuser S, Bitterling M, Kramer W (2024) Evaluating different artificial neural network forecasting approaches for optimizing district heating network operation. Energy 307. https://doi.org/10.1016/j.energy.2024.132745
81. Bourhis P, Cousin B, Rotta Loria AF, Laloui L (2021) Machine learning enhancement of thermal response tests for geothermal potential evaluations at site and regional scales. Geothermics 95. https://doi.org/10.1016/j.geothermics.2021.102132
82. Park CH, Shim BO, Park JW (2022) Open-source IoT monitoring system of a shallow geothermal system for heating and cooling year-round in Korea. Energy 250. https://doi.org/10.1016/j.energy.2022.123782
83. Ishitsuka K, Lin W (2023) Physics-informed neural network for inverse modeling of natural-state geothermal systems, Appl Energy 337. https://doi.org/10.1016/j.apenergy.2023.120855
84. Lindsey CR, Ayling BF, Asato G, Seggiaro R, Carrizo N, Larcher N, Marquetti C, Naón V, Serra AC, Faulds JE, Coolbaugh MF (2021) Play fairway analysis for geothermal exploration in north-western Argentina. Geothermics 95. https://doi.org/10.1016/j.geothermics.2021.102128
85. Farea A, Tripathi S, Glazko G, Emmert-Streib F (2024) Investigating the optimal number of topics by advanced text-mining techniques: sustainable energy research. Eng Appl Artif Intell 136:108877. https://doi.org/10.1016/j.engappai.2024.108877

86. Sheikhkhoshkar M, Bril H, Haouzi E, Aubry A, Hamzeh F (2024) From NLP to taxonomy: identifying and classifying key functionality from NLP to taxonomy: identifying and classifying key functionality concepts of multi-level project planning and control systems to cite this version: HAL Id: hal-04701660
87. Ghafariasl P, Mahmoudan A, Mohammadi M, Nazarparvar A, Hoseinzadeh S, Fathali M, Chang S, Zeinalnezhad M, Garcia DA (2024) Neural network-based surrogate modeling and optimization of a multigeneration system. Appl Energy 364. https://doi.org/10.1016/j.apenergy.2024.123130
88. Corso A, Chiotoroiu M, Clemens T, Zechner M, Kochenderfer MJ (2024) Sequentially optimized data acquisition for a geothermal reservoir. Geothermics 120. https://doi.org/10.1016/j.geothermics.2024.102983
89. Bennagi A, AlHousrya O, Cotfas DT, Cotfas PA (2024) Comprehensive study of the artificial intelligence applied in renewable energy. Energ Strat Rev 54:101446. https://doi.org/10.1016/j.esr.2024.101446
90. Onile AE, Machlev R, Petlenkov E, Levron Y, Belikov J (2021) Uses of the digital twins concept for energy services, intelligent recommendation systems, and demand side management: a review. Energy Rep 7:997–1015. https://doi.org/10.1016/j.egyr.2021.01.090
91. Naeem G, Asif M, Khalid M (2024) Industry 4.0 digital technologies for the advancement of renewable energy: functions, applications, potential and challenges. Energy Convers Manag X 24:100779. https://doi.org/10.1016/j.ecmx.2024.100779
92. Zhang J, Xiao C, Yang W, Liang X, Zhang L, Wang X (2023) Probabilistic Geothermal resources assessment using machine learning: Bayesian correction framework based on Gaussian process regression. Geothermics 114:102787. https://doi.org/10.1016/j.geothermics.2023.102787
93. Khanmohammadi S, Musharavati F, Tariq R (2022) A framework of data modeling and artificial intelligence for environmental-friendly energy system: application of Kalina cycle improved with fuel cell and thermoelectric module. Process Saf Environ Prot 164:499–516. https://doi.org/10.1016/j.psep.2022.06.029
94. Li X, Huang S, Hergert T, Henk A, Du Q (2024) Contemporary crustal kinematics in the Guangdong-Hong Kong-Macao greater Bay area, SE China: implications for the geothermal resource exploration. J Asian Earth Sci 263:106041. https://doi.org/10.1016/j.jseaes.2024.106041
95. Entezari A, Aslani A, Zahedi R, Noorollahi Y (2023) Artificial intelligence and machine learning in energy systems: a bibliographic perspective. Energ Strat Rev 45:101017. https://doi.org/10.1016/j.esr.2022.101017
96. Chu Z, Zhang Z, Tan W, Chen P (2024) Revolutionizing energy practices: Unleashing the power of artificial intelligence in corporate energy transition. J Environ Manage 357:120806. https://doi.org/10.1016/j.jenvman.2024.120806
97. Ling W, Abed AM, Farouk N, Fouad Y, Abduvalieva D, Islam S, AL Garalleh H, Alrawashdeh AI (2024) Modeling of geothermal tailored CCHP system with heat recovery centered thermal design/analysis; ANN-based optimization and economic study. Case Stud Therm Eng 5. https://doi.org/10.1016/j.csite.2024.104511
98. Noye S, Mulero Martinez R, Carnieletto L, De Carli M, Castelruiz Aguirre A (2022) A review of advanced ground source heat pump control: artificial intelligence for autonomous and adaptive control. Renew Sustain Energy Rev 153. https://doi.org/10.1016/j.rser.2021.111685
99. Yang Y, Wang S, Zhou C, Tian C, Yu J (2024) Enhancing economic and performance effectiveness of a geo-solar system using multi aspect artificial neural network. Case Stud Therm Eng 59. https://doi.org/10.1016/j.csite.2024.104588
100. Ahmed M, Habib A, Nawal MM, Saikot MMH, Chowdhury MAH, Hoque MA, Asaduzzaman AKM, Pálsson H, Björnsson P (2023) Deep learning-based approach to R-134a bubble detection and analysis for geothermal applications. Case Stud Therm Eng 49. https://doi.org/10.1016/j.csite.2023.103377

101. Pelzer J, Schulte M (2024) Efficient two-stage modeling of heat plume interactions of geothermal heat pumps in shallow aquifers using convolutional neural networkss. https://doi.org/10.18419/darus-3
102. Zahraee SM, Khalaji Assadi M, Saidur R (2016) Application of artificial intelligence methods for hybrid energy system optimization. Renew Sustain Energy Rev 66:617–630. https://doi.org/10.1016/j.rser.2016.08.028
103. Thirunavukkarasu M, Sawle Y, Lala H (2023) A comprehensive review on optimization of hybrid renewable energy systems using various optimization techniques. Renew Sustain Energy Rev 176:113192. https://doi.org/10.1016/j.rser.2023.113192
104. Adebayo P, Beragama Jathunge C, Darbandi A, Fry N, Shor R, Mohamad A, Wemhöner C, Mwesigye A (2024) Development, modeling, and optimization of ground source heat pump systems for cold climates: a comprehensive review. Energy Build 320. https://doi.org/10.1016/j.enbuild.2024.114646
105. Ahmadi A, El Haj Assad M, Jamali DH, Kumar R, Li ZX, Salameh T, Al-Shabi M, Ehyaei MA (2020) Applications of geothermal organic Rankine cycle for electricity production. J Clean Prod 274. https://doi.org/10.1016/j.jclepro.2020.122950
106. Loni R, Mahian O, Najafi G, Sahin AZ, Rajaee F, Kasaeian A, Mehrpooya M, Bellos E, le Roux WG (2021) A critical review of power generation using geothermal-driven organic Rankine cycle. Therm Sci Eng Prog 25:101028. https://doi.org/10.1016/j.tsep.2021.101028
107. Li JX, Zhang T, Zhu Y, Chen Z (2024) Artificial general intelligence for the upstream geoenergy industry: a review. Gas Sci Eng 131:205469. https://doi.org/10.1016/j.jgsce.2024.205469
108. Obi CE, Hasan AR, Rahman MA, Banerjee D (2024) Multiphase flow challenges in drilling, completions, and injection: part 1. Petroleum. https://doi.org/10.1016/j.petlm.2024.05.001
109. Obi CE, Hasan AR, Badejo A, Rahman MA, Banerjee D (2024) Multiphase flow challenges in drilling, completions, and injection: part 2. Petroleum. https://doi.org/10.1016/j.petlm.2024.05.002
110. Borunda M, Jaramillo OA, Reyes A, Ibargüengoytia PH (2016) Bayesian networks in renewable energy systems: a bibliographical survey. Renew Sustain Energy Rev 62:32–45. https://doi.org/10.1016/j.rser.2016.04.030
111. Ahmad T, Chen H (2020) A review on machine learning forecasting growth trends and their real-time applications in different energy systems. Sustain Cities Soc 54:102010. https://doi.org/10.1016/j.scs.2019.102010
112. Huang J, Zheng H, Kong H (2024) Key pathways for efficient solar thermal desalination. Energy Convers Manag 299:117806. https://doi.org/10.1016/j.enconman.2023.117806
113. Odoi-Yorke F, Opoku R, Davis F, Obeng GY (2024) Employing bibliometric analysis to identify the trends, evolution, and future research directions of sand-based thermal energy storage systems. J Energy Storage 94:112343. https://doi.org/10.1016/j.est.2024.112343
114. Bin Abu Sofian ADA, Lim HR, Chew KW, Khoo KS, Tan IS, Ma Z, Show PL (2024) Hydrogen production and pollution mitigation: Enhanced gasification of plastic waste and biomass with machine learning & storage for a sustainable future. Environ Pollut 342:123024. https://doi.org/10.1016/j.envpol.2023.123024
115. Zhou Y, Zheng S, Hensen JLM (2024) Machine learning-based digital district heating/cooling with renewable integrations and advanced low-carbon transition. Renew Sustain Energy Rev 199:114466. https://doi.org/10.1016/j.rser.2024.114466
116. Das HS, Salem M, Zainuri MAAM, Dobi AM, Li S, Ullah MH (2022) A comprehensive review on power conditioning units and control techniques in fuel cell hybrid systems. Energy Rep 8:14236–14258. https://doi.org/10.1016/j.egyr.2022.10.407
117. Sarbu I, Dorca A, Mirza M (2024) Exhaustive overview of advances in integrating renewable energy sources into district heating systems. Sustain Cities Soc 116:105897. https://doi.org/10.1016/j.scs.2024.105897
118. Jha SK, Bilalovic J, Jha A, Patel N, Zhang H (2017) Renewable energy: present research and future scope of artificial intelligence. Renew Sustain Energy Rev 77:297–317. https://doi.org/10.1016/j.rser.2017.04.018

119. Romanov D, Leiss B (2022) Geothermal energy at different depths for district heating and cooling of existing and future building stock. Renew Sustain Energy Rev 167. https://doi.org/10.1016/j.rser.2022.112727
120. Zhu Y, Wu S, Li J, Jia Q, Zhang T, Zhang X, Han D, Tan Y (2024) Towards a carbon-neutral community: integrated renewable energy systems (IRES)–sources, storage, optimization, challenges, strategies and opportunities. J Energy Storage 83. https://doi.org/10.1016/j.est.2024.110663
121. Khosravi A, Sandoval OR, Taslimi MS, Sahrakorpi T, Amorim G, Garcia Pabon JJ (2024) Review of energy efficiency and technological advancements in data center power systems. Energy Build 323:114834. https://doi.org/10.1016/j.enbuild.2024.114834
122. Zhou S, Diao H, Wang M, Jia W, Wang Y, Liu Z, Gan W, Zhou M, Wu Z, Zhao Z (2024) Knowledge mapping and emerging trends of urban resilient infrastructure research in urban studies: precedent work, current progress and future perspectives. J Clean Prod 452:142087. https://doi.org/10.1016/j.jclepro.2024.142087
123. Zhang L, Dieudonné AC, Daniilidis A, Dong L, Cao W, Thibaut R, Tas L, Hermans T (2025) Thermo-hydro-mechanical modeling of geothermal energy systems in deep mines: uncertainty quantification and design optimization. Appl Energy 377. https://doi.org/10.1016/j.apenergy.2024.124531
124. Abedsoltan H, Abedsoltan A, Zoghi Z (2024) Future of process safety: Insights, approaches, and potential developments. Process Saf Environ Prot 185:684–707. https://doi.org/10.1016/j.psep.2024.03.034
125. Kaczmarczyk M, Mukti M, Ghaffour N, Soukane S, Bundschuh J, Tomaszewska B (2024) Renewable energy-driven membrane distillation in the context of life cycle assessment. Renew Sustain Energy Rev 192. https://doi.org/10.1016/j.rser.2023.114249
126. Su X, Liu S, Zhang L, Liu G, Wang Y, Gan M, Li X (2023) Wellbore leakage risk management in CO2 geological utilization and storage: a review. Energy Rev 2:100049. https://doi.org/10.1016/j.enrev.2023.100049
127. Baghbani A, Choudhury T, Costa S, Reiner J (2022) Application of artificial intelligence in geotechnical engineering: a state-of-the-art review. Earth Sci Rev 228:103991. https://doi.org/10.1016/j.earscirev.2022.103991
128. Lin L, Zhong Z, Li C, Gorman A, Wei H, Kuang Y, Wen S, Cai Z, Hao F (2024) Machine learning for subsurface geological feature identification from seismic data: methods, datasets, challenges, and opportunities. Earth Sci Rev 257:104887. https://doi.org/10.1016/j.earscirev.2024.104887
129. Xu YS, Wang XW, Shen SL, Zhou A (2020) Distribution characteristics and utilization of shallow geothermal energy in China. Energy Build 229:110479. https://doi.org/10.1016/j.enbuild.2020.110479
130. Hinsby K, Négrel P, de Oliveira D, Barros R, Venvik G, Ladenberger A, Griffioen J, Piessens K, Calcagno P, Götzl G, Broers HP, Gourcy L, van Heteren S, Hollis J, Poyiadji E, Čápová D, Tulstrup J (2024) Mapping and understanding earth: open access to digital geoscience data and knowledge supports societal needs and UN sustainable development goals. Int J Appl Earth Obs Geoinformation 130. https://doi.org/10.1016/j.jag.2024.103835
131. Pouraskarparast Z, Aghaei H, Colombera L, Masoero E, Ghaedi M (2024) Fracture aperture: a review on fundamental concepts, estimation methods, applications, and research gaps. Mar Pet Geol 164:106818. https://doi.org/10.1016/j.marpetgeo.2024.106818
132. Rejeb A, Rejeb K, Zailani S, Keogh JG, Appolloni A (2022) Examining the interplay between artificial intelligence and the agri-food industry. Artif Intell Agric 6:111–128. https://doi.org/10.1016/j.aiia.2022.08.002
133. Gladju J, Kamalam BS, Kanagaraj A (2022) Applications of data mining and machine learning framework in aquaculture and fisheries: a review. Smart Agric Technol 2:100061. https://doi.org/10.1016/j.atech.2022.100061
134. Amir M, Khan SZ (2022) Assessment of renewable energy: Status, challenges, COVID-19 impacts, opportunities, and sustainable energy solutions in Africa. Energy Built Environ 3:348–362. https://doi.org/10.1016/j.enbenv.2021.03.002

135. Azarpour A, Mohammadzadeh O, Rezaei N, Zendehboudi S (2022) Current status and future prospects of renewable and sustainable energy in North America: progress and challenges. Energy Convers Manag 269:115945. https://doi.org/10.1016/j.enconman.2022.115945
136. Soltani M, Moradi Kashkooli F, Souri M, Rafiei B, Jabarifar M, Gharali K, Nathwani JS (2021) Environmental, economic, and social impacts of geothermal energy systems. Renew Sustain Energy Rev 140:110750. https://doi.org/10.1016/j.rser.2021.110750
137. Neukart F (2024) Towards sustainable horizons: a comprehensive blueprint for Mars colonization. Heliyon 10:e26180. https://doi.org/10.1016/j.heliyon.2024.e26180
138. Dashtian K, Shahsavarifar S, Usman M, Joseph Y, Ganjali MR, Yin Z, Rahimi-Nasrabadi M (2024) A comprehensive review on advances in polyoxometalate based materials for electrochemical water splitting. Coord Chem Rev 504:215644. https://doi.org/10.1016/j.ccr.2023.215644
139. Gordon JA, Balta-Ozkan N, Nabavi SA (2023) Socio-technical barriers to domestic hydrogen futures: Repurposing pipelines, policies, and public perceptions. Appl Energy 336. https://doi.org/10.1016/j.apenergy.2023.120850
140. Bashir MA, Dengfeng Z, Filipiak BZ, Bilan Y, Vasa L (2023) Role of economic complexity and technological innovation for ecological footprint in newly industrialized countries: does geothermal energy consumption matter? Renew Energy 217. https://doi.org/10.1016/j.renene.2023.119059
141. Quang DV, Milani D, Abu Zahra M (2023) A review of potential routes to zero and negative emission technologies via the integration of renewable energies with CO2 capture processes. Int J Greenh Gas Control 124:103862. https://doi.org/10.1016/J.IJGGC.2023.103862
142. Hoang AT, Pham VV, Nguyen XP (2021) Integrating renewable sources into energy system for smart city as a sagacious strategy towards clean and sustainable process. J Clean Prod 305:127161. https://doi.org/10.1016/J.JCLEPRO.2021.127161
143. Elahi M, Afolaranmi SO, Martinez Lastra JL, Perez Garcia JA (2023) A comprehensive literature review of the applications of AI techniques through the lifecycle of industrial equipment. Discov Artif Intell 3:43. https://doi.org/10.1007/s44163-023-00089-x
144. Cao X, Liu Z, Hu C, Song X, Quaye JA, Lu N (2024) Three-dimensional geological modelling in earth science research: an in-depth review and perspective analysis. Minerals 14. https://doi.org/10.3390/min14070686
145. Muther T, Syed FI, Lancaster AT, Salsabila FD, Dahaghi AK, Negahban S (2022) Geothermal 4.0: AI-enabled geothermal reservoir development-current status, potentials, limitations, and ways forward. Geothermics 100:102348. https://doi.org/10.1016/J.GEOTHERMICS.2022.102348
146. Bourdeau JE, Zhang SE, Nwaila GT, Ghorbani Y (2024) Data generation for exploration geochemistry: past, present and future. Appl Geochem 172:106124. https://doi.org/10.1016/J.APGEOCHEM.2024.106124
147. Jolie E, Scott S, Faulds J, Chambefort I, Axelsson G, Gutiérrez-Negrín LC, Regenspurg S, Ziegler M, Ayling B, Richter A, Zemedkun MT (2021) Geological controls on geothermal resources for power generation. Nat Rev Earth Environ 2:324–339. https://doi.org/10.1038/s43017-021-00154-y
148. Verma J, Sandys L, Matthews A, Goel S (2024) Readiness of artificial intelligence technology for managing energy demands from renewable sources. Eng Appl Artif Intell 135:108831. https://doi.org/10.1016/J.ENGAPPAI.2024.108831
149. Ahmed N, Assadi M, Ahmed AA, Banihabib R (2023) Optimal design, operational controls, and data-driven machine learning in sustainable borehole heat exchanger coupled heat pumps: key implementation challenges and advancement opportunities. Energy Sustain Dev 74:231–257. https://doi.org/10.1016/J.ESD.2023.04.004
150. Yazdi M (2024) Maintenance strategies and optimization techniques. In: Yazdi M (ed) Advances in computational mathematics for industrial system reliability and maintainability. Springer Nature Switzerland, Cham, pp 43–58. https://doi.org/10.1007/978-3-031-53514-7_3

151. Ukoba K, Olatunji KO, Adeoye E, Jen T-C, Madyira DM (2024) Optimizing renewable energy systems through artificial intelligence: review and future prospects. Energy & Environ 35:3833–3879. https://doi.org/10.1177/0958305X241256293

Part IV
AI in Environmental Monitoring and Prediction

Chapter 13
Artificial Intelligence-Based Flood Mapping: A Case Study of El Tarf Governorate, Algeria

Tayeb Boulmaiz, Khaled Ferkous, Mawloud Guermoui, Radia Hafsi, Kada Riche, Bilel Zerouali, and Hamouda Boutaghane

Abstract Floods are among the most destructive natural disasters, causing significant socioeconomic and environmental impacts worldwide. This chapter examines how Artificial Intelligence (AI), encompassing Machine Learning (ML), Deep Learning (DL), and hybrid models, can be leveraged for flood susceptibility mapping in the governorate of El Tarf, northeastern Algeria. The study area is particularly prone to flooding due to its diverse climatic and hydrological characteristics, including heavy rainfall and river overflows. The study utilized twelve flood-conditioning factors, including land cover, rainfall, SPI, TWI, and elevation, to train and evaluate various AI models, including LightGBM, Random Forest, CNN, and CNN-LSTM. Among these, LightGBM achieved the highest classification performance, with an accuracy of 90% and an AUC of 0.96, demonstrating its effectiveness in identifying flood-prone areas with minimal false positives and negatives. Despite

T. Boulmaiz · K. Ferkous
Materials, Energy Systems Technology and Environment Laboratory, Ghardaia University, Ghardaia, Algeria

M. Guermoui (✉)
Centre de Développement Des Energies Renouvelables, CDER, Unité de Recherche Appliquée en Energies Renouvelables, URAER, Ghardaïa, Algeria
e-mail: mawloud.guermoui@gmail.com

Telecommunications and Smart Systems Laboratory, University of Ziane Achour, Djelfa, Algeria

R. Hafsi
Hydraulic and Civil Engineering Department, Ghardaia University, Ghardaia, Algeria

K. Riche
University of Sciences and Technology Houari Boumediene, Alger, Algeria

B. Zerouali
Laboratory of Architecture, Cities and Environment, Department of Hydraulic, Faculty of Civil Engineering and Architecture, Hassiba Benbouali University of Chlef, Chlef, Algeria

H. Boutaghane
Laboratory of Soil and Hydraulic, Badji Mokhtar Annaba University, Annaba, Algeria

M. Nasr et al. (eds.), *Artificial Intelligence Applications for a Sustainable Environment*, Green Chemistry and Sustainable Technology,
https://doi.org/10.1007/978-3-031-91199-6_13

the promising results, the study highlights the challenges posed by limited data availability and the need for larger, diverse datasets to improve model generalization and applicability to other regions. This research underscores the potential of AI-driven approaches in flood risk management, offering an innovative and scalable solution for mitigating flood impacts in vulnerable areas like El Tarf.

Keywords Flood mapping · Classification · Deep learning · Machine learning · Hybrid model

13.1 Introduction

As climate change continues to intensify, flooding has emerged as one of the most destructive and life-threatening natural disasters, posing persistent and substantial risks to communities worldwide. This pervasive hazard disrupts lives and livelihoods, with far-reaching consequences that extend beyond the immediate impacts. Floods are responsible for tragic losses of human life, extensive damage to infrastructure and property, and the disruption of essential services [1–3]. They also frequently lead to population displacement, compounding the social, economic, and emotional challenges faced by affected communities. The cascading effects of flooding underscore the urgent need for comprehensive strategies to mitigate risks and enhance resilience in the face of this growing global threat. Central to addressing this challenge is the development and deployment of effective flood-detection systems, which are crucial for minimizing the impact of such events. As extreme weather events, including storms and intense rainfall, become more frequent and severe due to climate change, the likelihood of flooding, particularly in vulnerable areas, continues to rise. This escalating risk highlights the critical importance of proactive, reliable, and innovative solutions to manage and mitigate the dangers posed by floods.

Flooding requires diverse approaches for effective management, particularly for river and flash floods. Traditional techniques, such as physics-based models like HEC-HMS/RAS and SWAT, simulate flood scenarios and evaluate susceptibility based on physical processes. These models provide reliable predictions but require extensive input data, significant computational resources, and struggle with scalability for large watersheds exceeding 1,000 km^2 [4–6]. Multi-Criteria Decision Analysis (MCDA) methods, such as the Analytic Hierarchy Process (AHP), have proven effective across varied regions, systematically evaluating flood risk using multiple factors [7, 8]. However, MCDA techniques depend heavily on expert knowledge, introducing potential bias and limitations in addressing nonlinear flood dynamics [9, 10]. Statistical methods, including bivariate and multivariate analyses, are valuable for identifying influential flood factors but share similar constraints due to their linear assumptions [11–13]. Remote sensing techniques, leveraging Synthetic Aperture Radar (SAR) and optical imagery, are instrumental in real-time flood detection. Advances, such as SAR-optical fusion enhance accuracy, yet challenges like cloud cover and limited temporal imagery persist [14–17].

Recent advancements in Flood Susceptibility Mapping (FSM) have increasingly harnessed the potential of ML techniques, offering a robust alternative to traditional approaches such as physical flood models, statistical methods, and remote sensing. Unlike conventional methods, ML techniques do not require predefined assumptions about physical processes. Instead, they extract complex relationships directly from input data, allowing for greater adaptability and scalability across varying contexts. This adaptability has been validated through numerous studies showcasing the effectiveness of diverse ML algorithms in different regions worldwide. For example, [18] reported a significant breakthrough in China using the Naive Bayes Tree algorithm, while [19] demonstrated remarkable accuracy in Austria with an integrated approach combining Ensemble AHP-ANP_RF-SVM methods. Similarly, [20] highlighted the efficacy of hybrid Dagging models in addressing flood susceptibility challenges in Bangladesh. Further evidence of ML reliability comes from studies conducted in India, where [21, 22] emphasized the robustness of XGBoost and Stochastic Gradient Boosting (SGB) models, respectively. Complementing these findings, [23] illustrated how Adaptive Boosting-Random Forest (AB-RF) approaches enhanced flood prediction accuracy in the same region. Investigations in other areas have also validated the versatility of ML methods; for instance, [3] demonstrated high accuracy using RF in China, while [24] achieved similar results with Cascade Forest models in Iran. Moreover, studies conducted in Egypt by [25, 26] revealed commendable success with RF models, further emphasizing the capability of ML techniques to model complex flood dynamics effectively.

These collective findings underscore the transformative role of ML in FSM, consistently outperforming traditional methods in diverse geographical and environmental contexts. By leveraging ML's ability to adapt to varying datasets and regional characteristics, researchers have paved the way for more accurate and scalable flood susceptibility assessments globally.

DL approaches extend these capabilities, using neural networks to analyze intricate relationships within flood data. Notable examples include Deep Belief Networks (DBN), Convolutional Neural Networks (CNN), and autoencoder-MLP hybrids, which have achieved high precision in various countries, including Vietnam, India, and Kenya [26–28]. These methods efficiently process complex datasets, minimize reliance on domain-specific expertise, and enhance flood prediction accuracy across diverse environments. As computational technologies advance, ML and DL driven FSM continue to revolutionize flood risk management, offering scalable, real-time solutions for proactive disaster mitigation.

The paper is structured as follows: Sect. 13.2 introduces the study area, focusing on the governorate of El Tarf in northeastern Algeria. Section 13.3 describes the database used in the study. Section 13.4 outlines the methodology, explaining the use of ML, DL, and hybrid models for flood classification, as well as the evaluation metrics employed to assess model performance. Section 13.5 presents the results and discussion, including the performance analysis of the models and the generated flood maps. Section 13.6 highlights the study's limitations, Finally, Sect. 13.7 concludes the paper by summarizing the key findings and emphasizing the potential of AI-driven approaches in flood mapping and risk management.

This chapter aims to address the pressing challenge of flood susceptibility in the governorate of El Tarf by leveraging AI-based methodologies. The primary objectives are to develop and evaluate ML, DL, and hybrid models for accurately mapping flood-prone areas using twelve flood-conditioning factors. Furthermore, the study seeks to identify the most effective AI model in terms of predictive accuracy and applicability, while also highlighting limitations such as data availability and generalization challenges. Ultimately, this work aims to contribute to the growing body of research on AI-driven solutions for flood risk management, offering scalable approaches to mitigate flood impacts in vulnerable regions.

13.2 Study Area

The governorate of El Tarf is located in the northeastern region of Algeria, near the Tunisian border. With a population of 507,772 and an area of 2,904 km^2, it has a population density of 174.9 inhabitants per km^2. This region is characterized by a variety of geographical, climatic, hydrological, and geological features that contribute to its vulnerability to flooding, which is the central concern of this study.

This region is strategically situated at the extreme northeastern edge of Algeria. It is bordered to the north by the Mediterranean Sea, to the east by Tunisia (specifically the Jendouba Governorate), to the south by the Wilaya of Souk Ahras, to the southwest by the Wilaya of Guelma, and to the west by the Wilaya of Annaba. This positioning places El Tarf in a crucial climatic and hydrological zone, making it prone to flood risks, particularly from heavy rainfall and the overflow of riverbeds.

The topographical relief of El Tarf is diverse, with varying altitudes that influence the flow of water during heavy rainfall. Its topography varies from low plains prone to flooding due to river overflow and sediment accumulation, to steeper slopes that cause rapid runoff, increasing flood risks. Higher elevations, such as hills and mountains, influence water flow dynamics and act as catchments, funneling rainfall into lower areas, contributing to downstream flooding.

The governorate of El Tarf experiences a Mediterranean climate, characterized by hot, dry summers and mild, wet winters. The region's precipitation varies significantly, with annual rainfall totals often exceeding 500 mm. In extreme cases, the area has experienced intense rainfall events exceeding 100 mm within a 24-h period, contributing to sudden flooding. The climate's unpredictability and irregular rainfall make the area particularly susceptible to flash floods, which have caused considerable damage in past events, such as the catastrophic flooding in 2012 (see Fig. 13.1).

The region has a long history of flooding, which has caused significant social and economic impacts. Notable flood events include:

- February 25, 2012: A catastrophic flood resulted in the loss of three lives and widespread property damage. Heavy rainfall, totaling 120 mm over 48 h, led to rivers bursting their banks and flooding agricultural land. The damage to infrastructure, including bridges and roads, isolated entire villages and required

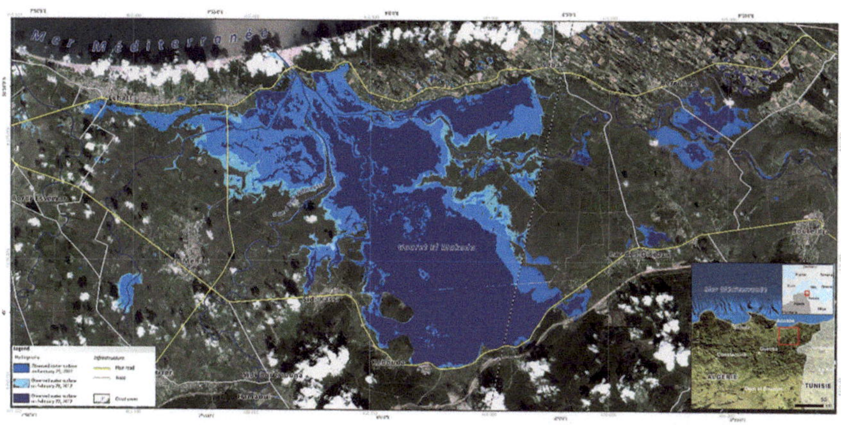

Fig. 13.1 Flood event of February 28, 2012 in El Tarf governorate

emergency response operations, including the use of military helicopters for aid distribution.

- February 10, 2017: A flood caused by 70 mm of rainfall within 48 h led to the inundation of roads, homes, and several administrative buildings. The flood also affected the region's infrastructure, causing significant disruption.
- August 24, 2018: Severe rainfall in the El Kala area led to widespread flooding in urban areas, including the main roundabout, where water submerged roads, hampering traffic.
- Ongoing Floods: Throughout the years, El Tarf has seen multiple floods with varying degrees of severity. In some cases, such as the floods of 2019, water levels in dams reached dangerous levels, causing authorities to release water in order to prevent dam overflows.

The combination of heavy precipitation, steep topography, and urbanization makes El Tarf particularly vulnerable to flooding. Low-lying areas, especially in the plains and river valleys, are at the highest risk during intense rainfall. Furthermore, the presence of large dams, including Chaffia, Bougous, and Meksa, while providing flood control, also pose a risk if dam levels are not carefully monitored and managed during heavy rainfall periods.

13.3 Database

Flood events are shaped by a diverse range of factors encompassing physical, environmental, meteorological, and hydrological dimensions. This study examined 12 such factors (Fig. 13.2), building on insights from prior research [29–33] and the availability of data for our study region. These factors were systematically grouped into three primary categories:

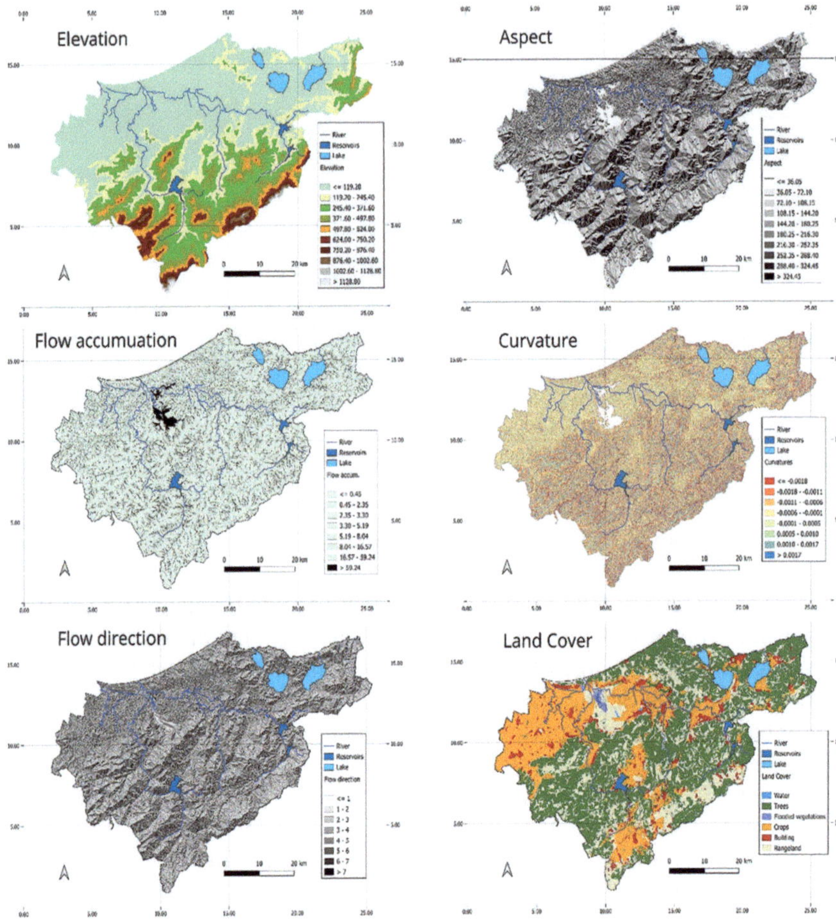

Fig. 13.2 The twelve flood-conditioning factors used in this study

13.3.1 Topographical Factors

This group includes key variables such as the Digital Elevation Model (DEM), slope, flow accumulation, flow direction, Topographic Wetness Index (TWI), and Topographic Roughness Index (TRI). All topographical data were derived from the Shuttle Radar Topography Mission (SRTM) DEM at a 30-m resolution.

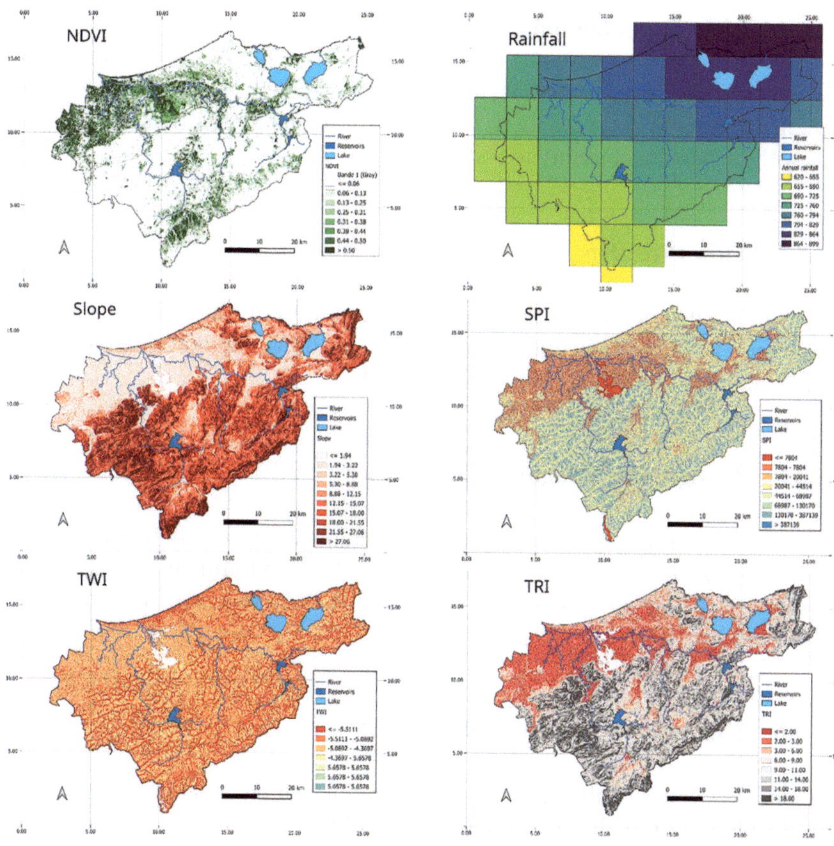

Fig. 13.2 (continued)

13.3.1.1 Digital Elevation Model (DEM)

The DEM provides critical elevation data that underpins terrain analysis. Elevation directly affects water movement, with low-lying areas more susceptible to water accumulation and flooding, while higher terrains influence downstream water flow.

13.3.1.2 Slope

Derived from the SRTM DEM, slope measures the steepness of the terrain, which impacts the velocity and volume of runoff. Steeper gradients promote rapid runoff, reducing soil infiltration and increasing flood risks downstream.

13.3.1.3 Flow Accumulation

This metric identifies where water converges, highlighting potential flood-prone zones like valleys or river channels. Calculated from the SRTM DEM, it pinpoints areas with higher susceptibility to water pooling [17].

13.3.1.4 Flow Direction

Using SRTM DEM, flow direction models the pathways of water movement across the terrain, enabling predictions of water flow patterns and the identification of downstream areas at risk.

13.3.1.5 Topographic Roughness Index (TRI)

TRI evaluates terrain ruggedness through elevation variations. It is derived from SRTM DEM data, and it is crucial for understanding landform characteristics that influence water flow and accumulation. Its role in flood risk assessment has been well-documented [34].

13.3.1.6 Topographic Wetness Index (TWI)

TWI identifies zones likely to experience saturation and potential flooding, making it a valuable tool in hydrological studies, including flood susceptibility modeling (FSM) [3, 26, 32]. Defined by the formula:

$$\mathrm{TWI} = \ln(\alpha/\tan\beta) \tag{13.1}$$

where a is the upslope area draining through a point per unit contour length and β is the slope angle.

13.3.1.7 Curvature

Curvature refers to the rate of change of slope and is derived from the DEM. It helps identify convex and concave landforms, which are crucial for understanding water flow patterns. Convex areas typically direct water flow away, while concave areas collect and retain water, making them more prone to flooding.

13.3.1.8 Aspect

Aspect represents the orientation or direction of the slope. It is critical in determining the exposure of a slope to sunlight and wind, which indirectly affect soil moisture and evaporation. Additionally, aspect influences the flow direction and can help identify areas where water is likely to accumulate.

13.3.2 Environmental Factors

Environmental parameters considered in this study include Land Use and Land Cover (LULC), Normalized Difference Vegetation Index (NDVI). These factors were derived from Sentinel-2 imagery, which provides high-resolution multispectral data ideal for monitoring vegetation cover and land surface conditions.

13.3.2.1 Land Cover

Land cover classifications, generated using the method of classification such as SVM method, delineate land types and their hydrological impacts. For instance, forests and wetlands facilitate infiltration and water storage, reducing runoff, whereas urbanized areas with impervious surfaces intensify runoff, exacerbating flood risk [35].

13.3.2.2 Normalized Difference Vegetation Index (NDVI)

Introduced in 1974 [36], NDVI measures vegetation density using the formula:

$$\mathrm{NDVI} = \left(\mathrm{NIR} - \frac{\mathrm{RED}}{\mathrm{NIR}} + \mathrm{RED}\right), \tag{13.2}$$

where SWIR is shortwave infrared reflectance. NDBI highlights impervious urban surfaces that restrict water infiltration and heighten flood risks by increasing surface runoff.

13.3.3 Hydrological Factors

These factors include precipitation, Stream Power Index (SPI).

13.3.3.1 Stream Power Index (SPI)

SPI quantifies the erosive energy of flowing water using the formula:

$$\mathrm{SPI} = \mathrm{A} * \tan(\beta), \tag{13.3}$$

where A is upslope contributing area (flow accumulation) and β is slope. Higher SPI values indicate stronger water flow, increasing erosion and flood potential [37].

13.3.3.2 Rainfall

Rainfall is a primary driver of flooding, as it directly adds water to the surface, saturating the ground and potentially overwhelming natural or artificial drainage systems. The amount, intensity, and duration of rainfall significantly influence the likelihood and severity of floods. Heavy or prolonged rainfall can exceed the soil's infiltration capacity, leading to increased surface runoff.

13.4 Methodology

13.4.1 Feature Selection

The selection of relevant features is a critical step in ML modeling, particularly for FSM. An improper choice of input factors or the inclusion of redundant variables can significantly affect the model's performance and predictive accuracy [38]. Consequently, the feature selection process in this study employed techniques such as Spearman's rank correlation, Information Gain Ratio (IGR), and multicollinearity analysis to identify and exclude irrelevant or highly correlated features, as demonstrated in similar approaches carried out by [26, 39].

Reference [40] identified that a correlation coefficient exceeding 0.7 signifies a strong relationship between variables. This threshold was utilized in this study to identify significant correlations among flood-influencing factors.

13.4.1.1 Spearman's Correlation Coefficient

The nonparametric Spearman's rank correlation coefficient is a statistical measure used to evaluate the strength and direction of a monotonic relationship between two variables, X and Y. Its values range from -1 to 1, where values closer to -1 or 1 indicate stronger correlations, and values near 0 suggest weaker relationships. A coefficient above 0.7 indicates significant collinearity between variables.

13.4.1.2 Multicollinearity Test

While Spearman's correlation evaluated pairwise associations, the Variance Inflation Factor (VIF) quantified the interrelatedness of variables in a multivariate setting. The VIF quantified the degree to which the variance of a regression coefficient is inflated due to multicollinearity among predictor variables. The formula for VIF is as follows:

$$\text{VIF} = \frac{1}{1 - R^2}, \tag{13.4}$$

where R^2 is the coefficient of determination obtained by regressing one predictor variable against the others. Consistent with prior studies in flood susceptibility modeling [41, 42], a VIF threshold was used to identify collinearity. Although some research applies a threshold of 5, other studies (e.g., [43, 44]) adopt a more lenient threshold of 10, which was adopted in this study to exclude collinear variables and ensure model stability.

13.4.1.3 Information Gain Ratio (IGR)

The IGR test was used to evaluate the relative significance of conditioning factors in flood susceptibility modeling. As established by [45] and expanded in recent works [46, 47], IGR is a widely used feature selection method in environmental modeling.

IGR measures the reduction in entropy (uncertainty) when a dataset is split based on a particular feature. The formula for IGR is as follows:

$$\text{IGR}(X, Y) = \frac{IG(X, Y)}{H(Y)}, \tag{13.5}$$

where

- $\text{IG}(X, Y)$ is the information gain, calculated as the difference between the entropy of the target variable Y before and after splitting by feature X.
- $H(Y)$ is the entropy of the target variable Y, calculated as:

$$H(Y) = -\sum_{i=1}^{n} p(y_i)\log_2 p(y_i) \tag{13.6}$$

Here, $p(y_i)$ represents the probability of class y_i. A higher IGR value indicates that the feature provides more information for classifying flood-prone areas, while a value close to zero suggests that the feature is less relevant.

Inputs with an IGR value of zero indicate no correlation with the output, suggesting their inclusion in the model would add noise rather than predictive value. Thus, such variables are excluded to improve model performance and reliability.

13.4.2 ML Models

13.4.2.1 Random Forest

Random Forest (RF) models have gained prominence in classification tasks due to their capacity to manage high-dimensional datasets and address overfitting challenges [48]. As an ensemble learning technique, RF consolidates the outputs of multiple decision trees, thereby enhancing the robustness and accuracy of predictions. This methodological strength has positioned RF as a key tool in environmental studies, particularly in flood prediction and mapping, where it excels in capturing complex relationships between predictor variables and target outcomes [49, 50]. In this research, RF was employed to distinguish between flood and nonflood points by leveraging a suite of conditioning variables encompassing topographic, hydrologic, and climatic factors.

The RF algorithm achieves diversity among its constituent decision trees through two principal mechanisms: bootstrapping and random feature selection. Bootstrapping generates multiple training subsets by sampling with replacement, introducing variability across the datasets used for tree construction. Simultaneously, random feature selection limits each tree to a subset of predictor variables, mitigating the risk of overfitting while fostering model generalizability. The final classification is determined via majority voting, wherein the mode of predictions across all trees forms the aggregated output. This process ensures the model's resilience to data noise, enhancing its reliability for flood mapping applications.

Despite its advantages, RF is not devoid of limitations. While its interpretability surpasses that of certain black-box algorithms, such as Artificial Neural Networks (ANNs), the complexity of its ensemble structure can pose challenges [51]. Furthermore, the computational demands of RF increase proportionally with the number of trees and input features, which can hinder its scalability for large datasets [52]. Nonetheless, RF remains one of the most extensively utilized ML algorithms for environmental classification tasks, frequently outperforming traditional methods like logistic regression in diverse applications [53, 54].

13.4.2.2 LightGBM

LightGBM, a gradient boosting decision tree (GBDT) variant, was introduced by Microsoft to enhance the efficiency and scalability of GBDT models [55]. Like traditional GBDT methods, LightGBM builds a robust model by combining weak learners; however, it incorporates innovative algorithms, including histogram-based computation, leaf-wise tree growth, gradient-based one-side sampling (GOSS), and exclusive feature bundling (EFB). These advancements enable LightGBM to process large datasets with high-dimensional features more efficiently than conventional GBDT methods.

One of the key innovations in LightGBM is its adoption of a histogram-based algorithm for split finding. Traditional GBDT models use a presorted algorithm, which evaluates all potential split points based on information gain—a computationally expensive process. By contrast, LightGBM groups continuous features into discrete bins, constructing histograms to approximate the data distribution. The split point is then selected from the histogram, significantly reducing computational time and complexity without compromising accuracy.

Another critical improvement in LightGBM is its leaf-wise tree growth strategy, which departs from the level-wise approach employed by traditional GBDT methods. In level-wise growth, the decision tree expands symmetrically by splitting all nodes at the same level, often resulting in trees that are less optimized for minimizing error. LightGBM instead adopts a leaf-wise approach, where only the leaf node with the highest error reduction is split, leading to asymmetric trees optimized for accuracy.

LightGBM further optimizes computation and memory usage through GOSS and EFB [26]. GOSS enhances efficiency by selecting a subset of samples based on their gradient magnitude, prioritizing those with larger gradients that contribute more to the model's learning process. EFB reduces dimensionality by combining mutually exclusive features into a single feature, further accelerating computation.

These advancements make LightGBM particularly effective for handling large-scale datasets and complex prediction tasks, outperforming traditional GBDT implementations in terms of both speed and accuracy.

13.4.2.3 CatBoost

CatBoost, developed by [56], is an advanced gradient boosting decision tree model designed to improve regression and classification tasks. It refines traditional gradient boosting techniques to enhance estimation accuracy while minimizing the risk of overfitting. CatBoost has demonstrated high performance in a wide range of fields, including environmental parameter modeling, geospatial analysis, and meteorological forecasting. A key strength of CatBoost lies in its ability to handle categorical features effectively, a common challenge in ML models. By leveraging gradient boosting and a binary tree classification structure, CatBoost ensures robust and accurate predictions.

CatBoost distinguishes itself from other gradient boosting methods through several unique features and innovations:

- Efficient Handling of Categorical Variables: CatBoost employs a sophisticated approach to encode categorical features into numerical values. Reference [57] highlighted the effectiveness of target statistics in reducing information errors, enabling the model to process categorical attributes with high precision.
- Exploitation of Feature Relationships: By combining categorical variables, CatBoost capitalizes on the relationships between features, allowing the model to extract additional predictive value from complex datasets.

- Symmetrical Tree Structure: To address overfitting and enhance classification performance, CatBoost uses a symmetrical tree-building strategy. This approach ensures balanced splits and contributes to the model's stability and efficiency.

These features make CatBoost a powerful tool for diverse applications, particularly in datasets with significant categorical attributes. Its ability to mitigate overfitting and extract meaningful patterns from complex data further reinforces its utility in ML tasks.

13.4.2.4 Support Vector Machine (SVM)

Support Vector Machines (SVMs) are among the most effective and theoretically grounded classification methods in ML. The key idea behind SVM is to find an optimal hyperplane that not only separates data points of different classes but does so with the maximum margin, thereby reducing the risk of overfitting. For linearly separable data, the hyperplane is a linear decision boundary. However, for more complex, nonlinear datasets, SVM employs the "kernel trick" to project the data into a higher-dimensional space where linear separation becomes feasible. Common kernel functions include polynomial, radial basis function (RBF), and sigmoid, enabling SVM to adapt to a wide range of classification problems.

The method's strength lies in its reliance on support vectors, the data points closest to the decision boundary, making the algorithm computationally efficient and robust against outliers. The concept was first formalized in [58]. This paper introduced the fundamental principles of SVMs, including the idea of maximizing the margin and using Lagrange multipliers for optimization. The research demonstrated SVM's superior performance on classification tasks [59], particularly with small or noisy datasets [60], and laid the groundwork for its widespread adoption in fields such as bioinformatics, image recognition, and natural language processing.

13.4.3 DL Models

13.4.3.1 Convolutional Neural Networks (CNNs) and 1D CNNs

Convolutional Neural Networks (CNNs) are a class of DL models specifically designed for processing structured data, such as images, sequences, and signals. They leverage the concept of convolution to extract hierarchical spatial or temporal features from the input data. A CNN architecture typically consists of convolutional layers, pooling layers, and fully connected layers. The convolutional layers apply a set of filters to extract local patterns, while pooling layers downsample the data, reducing its dimensionality and computational complexity. Fully connected layers then map the extracted features to the output space, often used for classification or regression tasks.

CNNs originated from the work of Yann LeCun [61]. Later, the paper [62], by Yann LeCun et al., solidified CNNs as a foundational DL technique, particularly in tasks like image recognition.

While 2D CNNs are widely used for image data, 1D CNNs are tailored for processing sequential data such as time-series signals, audio, and text. In 1D CNNs, the convolution operation is performed along one spatial dimension, making them ideal for feature extraction from temporal sequences. Instead of 2D filters, 1D CNNs use kernels that slide along the sequence to capture patterns such as trends, periodicities, and local dependencies. For example, in time-series analysis, 1D CNNs can identify patterns related to fluctuations, peaks, or anomalies [63].

The strength of 1D CNNs lies in their efficiency and simplicity compared to recurrent architectures like LSTMs, especially when long-term dependencies are not critical. They also benefit from their ability to stack multiple layers, allowing for the extraction of increasingly complex features as data propagates through the network. The adoption of 1D CNNs has been particularly successful in domains such as speech recognition, bioinformatics (e.g., protein sequence analysis), and ECG signal classification, showcasing their versatility in real-world applications.

13.4.3.2 Long Short-Term Memory (LSTM) Networks

Long Short-Term Memory (LSTM) networks are a type of recurrent neural network (RNN) designed to effectively capture long-term dependencies in sequential data. Standard RNNs struggle with the vanishing gradient problem, which makes it difficult for them to learn and retain information over long sequences. LSTMs address this limitation through a carefully designed memory cell and gating mechanism, allowing them to maintain and update information selectively over extended time steps.

LSTMs were introduced in 1997 by Sepp Hochreiter and Jürgen Schmidhuber in their seminal research [64]. The architecture introduced the concept of a memory cell that stores information across time steps and gates—input, forget, and output gates—that regulate the flow of information into, out of, and within the cell. The input gate determines how much of the new information to store, the forget gate decides how much of the existing memory to discard, and the output gate controls what part of the memory to expose to the next layer or time step.

LSTMs are particularly well-suited for tasks where context or sequential patterns are crucial, such as natural language processing, time-series forecasting, and speech recognition. For instance, in language modeling, LSTMs can understand dependencies across words in a sentence, enabling more coherent predictions or text generation [65].

13.4.3.3 GRU Architecture

Unlike LSTMs, GRUs do not have a separate memory cell; instead, they integrate the memory mechanism directly into the hidden state. This simplification reduces the number of parameters in GRUs, making them computationally efficient while still capturing dependencies over long sequences. With fewer gates and parameters, GRUs are computationally lighter than LSTMs, leading to faster training and inference. Despite their simpler architecture, GRUs often perform comparably to LSTMs on many tasks, including language modeling, speech recognition, and time-series forecasting. GRUs are particularly effective when the dataset is not excessively large, or when the task requires relatively short- to medium-term dependencies [66–68].

GRUs have found extensive use in natural language processing (NLP) tasks such as text generation, machine translation, and sentiment analysis. They are also applied in speech signal modeling, bioinformatics, and real-time systems where computational efficiency is a priority.

While GRUs are not as flexible as LSTMs in handling extremely long sequences, their simplicity and effectiveness make them a popular choice in many scenarios, especially when computational resources are constrained.

In the hybridization of CNN with other models like SVM, LSTM, and GRU, the main idea is that CNN acts as a feature extractor, identifying and extracting new pertinent features from the original input data. These extracted features, which represent high-level patterns or relevant characteristics of the input, are then fed into SVM, LSTM, or GRU for further processing and classification. This approach is particularly effective for tasks like flood classification, where CNN captures spatial or temporal features from raw data, and the subsequent models refine these features for accurate classification.

13.5 Results and Discussion

13.5.1 Selection of Flood-Influencing Factors

Twelve conditioning factors (Land cover, Annual Rainfall, SPI, TWI, TRI, NDVI, Flow accumulation, Flow direction, Slope, Curvature, Aspect, and Elevation) were analyzed for correlation and multicollinearity (Table 13.1). Spearman's correlation analysis revealed strong relationships among several factors, particularly between TRI and Slope, and SPI and Flow accumulation, indicating potential redundancy. Furthermore, the VIF for TRI (= 41.60), Slope (= 41.39), SPI (= 11.97), and Flow accumulation (= 12.36) exceeded the acceptable threshold (>5), confirming multicollinearity issues (Fig. 13.3a).

To assess the relative importance of these factors in flood generation, the IGR was computed (Fig. 13.3b). Most factors had an IGR value above 0.05, except for TWI, Curvature, and Flow direction, which showed negligible contributions.

The selection of conditioning factors was performed as follows:

- Based on multicollinearity analysis, TRI and SPI were excluded due to their high correlation with Slope and Flow accumulation, respectively, in addition to their high VIF values.

Table 13.1 Spearman's coefficient analysis for the initial conditioning factors

	Land cover	Annual R	SPI	TWI	TRI	NDVI	Flow acc.	Flow dir.	Slope	Curv.	Aspect	Elev.
Land cover	1	0.10	−0.15	0.00	−0.23	0.27	−0.10	0.12	−0.23	−0.01	0.06	−0.27
Annual R	0.10	1	−0.22	0.05	−0.47	0.25	−0.17	−0.02	−0.48	−0.05	0.03	−0.39
SPI	−0.15	−0.22	1	0.36	0.32	−0.23	0.86	0.03	0.28	−0.35	0.00	0.25
TWI	0.00	0.05	0.36	1	−0.29	0.11	0.68	0.02	−0.27	−0.31	−0.01	−0.28
TRI	−0.23	−0.47	0.32	−0.29	1	−0.47	0.15	−0.07	0.94	0.00	0.01	0.71
NDVI	0.27	0.25	−0.23	0.11	−0.47	1	−0.13	−0.02	−0.46	−0.02	−0.01	−0.52
Flow acc.	−0.10	−0.17	0.86	0.68	0.15	−0.13	1	0.01	0.14	−0.46	−0.02	0.08
Flow dir.	0.12	−0.02	0.03	0.02	−0.07	−0.02	0.01	1	−0.06	0.03	0.43	−0.01
Slope	−0.23	−0.48	0.28	−0.27	0.94	−0.46	0.14	−0.06	1	0.00	0.03	0.69
Curv.	−0.01	−0.05	−0.35	−0.31	0.00	−0.02	−0.46	0.03	0.00	1	0.03	0.10
Aspect	0.06	0.03	0.00	−0.01	0.01	−0.01	−0.02	0.43	0.03	0.03	1	0.01
Elev.	−0.27	−0.39	0.25	−0.28	0.71	−0.52	0.08	−0.01	0.69	0.10	0.01	1

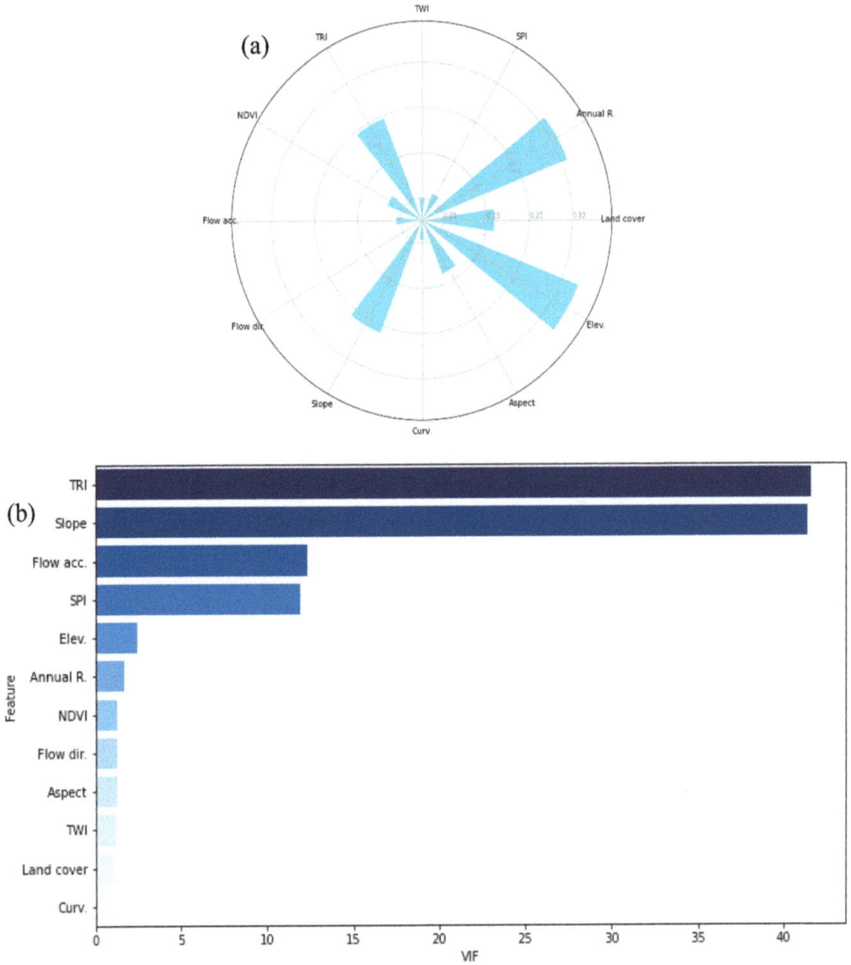

Fig. 13.3 **a** VIF and **b** IGR analysis for conditioning factors

- Using IGR as a selection criterion, Curvature, TWI, and Flow direction were excluded due to their low IGR values.
- Despite their high VIF values, Slope and Flow accumulation were retained due to their critical role in flood generation processes.

After this selection process, the following factors were retained for flood prediction: Land cover, Annual Rainfall, NDVI, Flow accumulation, Slope, Aspect, and Elevation.

13.5.2 *Performances of Models*

The performance of ML and DL models for flood classification was evaluated using metrics such as Accuracy, Sensitivity, Precision, F1-Score, and Area Under the Curve (AUC) for both training and testing datasets (Table 13.2).

13.5.2.1 ML Models

Among the ML models, the RF classifier achieved the highest accuracy on the training set (0.99), along with near-perfect metrics for sensitivity, precision, and F1-Score (all at 0.99). However, its performance on the test set was slightly lower, with an accuracy of 0.87, sensitivity of 0.87, precision of 0.87, and an AUC of 0.95, demonstrating reliable but slightly less robust generalization. The LightGBM model delivered a balanced performance, achieving 0.90 accuracy, 0.95 sensitivity, and an AUC of 0.96 on the test set, reflecting strong predictive power with a slight trade-off in precision (0.86). Similarly, CatBoost demonstrated perfect metrics during training (1.00 across all metrics) but exhibited a decline in the test set, with an accuracy of 0.89, sensitivity of 0.95, and an AUC of 0.89, suggesting possible overfitting during training.

The SVM achieved moderate performance, with a test set accuracy of 0.84 and an AUC of 0.91. Its sensitivity (0.92) and F1-Score (0.85) were competitive, but its precision (0.78) was lower compared to ensemble models like RF, LightGBM, and CatBoost. Despite this, SVM demonstrated reasonable generalizability and performed well considering its simplicity compared to ensemble techniques.

13.5.2.2 DL Models

DL models demonstrated variable results, with some outperforming others significantly. CNN and CNN-SVM showed relatively higher performance, achieving test set accuracies of 0.85 and 0.82, respectively, with AUC values around 0.95. The CNN-LSTM hybrid model achieved promising results as well, with a test accuracy of 0.83 and an AUC of 0.92.

In contrast, LSTM and GRU performed less effectively, with test set accuracies of 0.79 and 0.71, respectively. These models struggled with precision, likely due to limited data for capturing complex temporal patterns effectively. Similarly, the CNN-GRU model exhibited slightly better results than standalone GRU but still lagged behind CNN-LSTM in overall performance.

Table 13.2 Performance of ML and DL models for flood classification during training and testing datasets

Models		Train					Test				
		Accuracy	Sensitivity	Precision	F1-Score	AUC	Accuracy	Sensitivity	Precision	F1-Score	AUC
ML	RF	0.99	0.99	0.99	0.99	0.99	0.87	0.87	0.87	0.87	0.95
	LightGBM	0.92	0.95	0.9	0.92	0.98	0.9	0.95	0.86	0.9	0.96
	CatBoost	1	1	1	1	1	0.89	0.95	0.83	0.89	0.89
	SVM	0.90	0.97	0.86	0.91	0.97	0.84	0.92	0.78	0.85	0.91
DL	CNN	0.90	0.95	0.86	0.90	0.96	0.85	0.94	0.79	0.86	0.95
	LSTM	0.79	0.92	0.73	0.82	0.85	0.79	0.92	0.72	0.81	0.88
	GRU	0.74s	0.88	0.69	0.77	0.81	0.71	0.88	0.65	0.75	0.80
	CNN-LSTM	0.86	0.95	0.81	0.88	0.95	0.83	0.93	0.76	0.84	0.92
	CNN-GRU	0.85	0.96	0.79	0.87	0.93	0.81	0.95	0.73	0.83	0.92
	CNN-SVM	0.86	0.94	0.82	0.88	0.95	0.82	0.95	0.75	0.84	0.92

13.5.2.3 Comparison of ML and DL Models

ML models, particularly RF, demonstrated high stability and required relatively smaller datasets and minimal fine-tuning to achieve reliable predictions. On the other hand, DL models generally required larger datasets to capture complex relationships among flood-conditioning factors effectively. While DL models like CNN and CNN-LSTM exhibited competitive performance, their reliance on extensive training data and fine-tuning processes highlights a significant challenge.

DL models are particularly advantageous in capturing spatial and temporal complexities, as seen in CNN's and CNN-LSTM's higher AUC values (0.95 and 0.92, respectively). However, their sensitivity to hyperparameter tuning and the need for substantial computational resources may hinder their applicability in scenarios with limited data availability. ML models, with their simplicity and robustness, remain more practical for applications requiring quick deployment and lower computational costs.

LightGBM demonstrated the highest overall test performance among the ML models, achieving an accuracy of 0.90, sensitivity of 0.95, precision of 0.86, F1-Score of 0.90, and an AUC of 0.96. These results highlight its exceptional capability in accurately classifying flood-prone areas while minimizing both false positives and false negatives. Moreover, the model exhibited a strong balance between training and testing performance, indicating minimal overfitting.

13.6 Flood Susceptibility Map

Using the trained LightGBM model, a flood susceptibility map (Fig. 13.4) was generated to spatially predict flood-prone areas based on the input variables. The final model was applied to the study area, integrating the spatial layers of the predictor variables into the ML workflow. The output of the model is a continuous probability map, where each pixel represents the likelihood of flood susceptibility. These probabilities were subsequently categorized into distinct susceptibility classes, enabling a clear visualization of areas at no flood, low, moderate, and high flood.

The flood susceptibility map reflects El Tarf's vulnerability to flooding, which can be attributed to its coastal geography and hydrological features. The northern part of the governorate, shown in red and orange, corresponds to low-lying areas near rivers and coastal plains, where water accumulation is more likely during heavy rainfall events. These areas are particularly susceptible to floods due to the combined effects of the presence of rivers, lakes, and reservoirs in the northern part, which increases the likelihood of flooding, especially during periods of excessive precipitation. Additionally, being a coastal zone, El Tarf is prone to the combined risks of riverine floods and coastal surges, which exacerbate flood susceptibility in flat terrains. Finally, the southern parts, depicted in gray, are less susceptible due to their higher elevations, which naturally reduce water retention and flood risk.

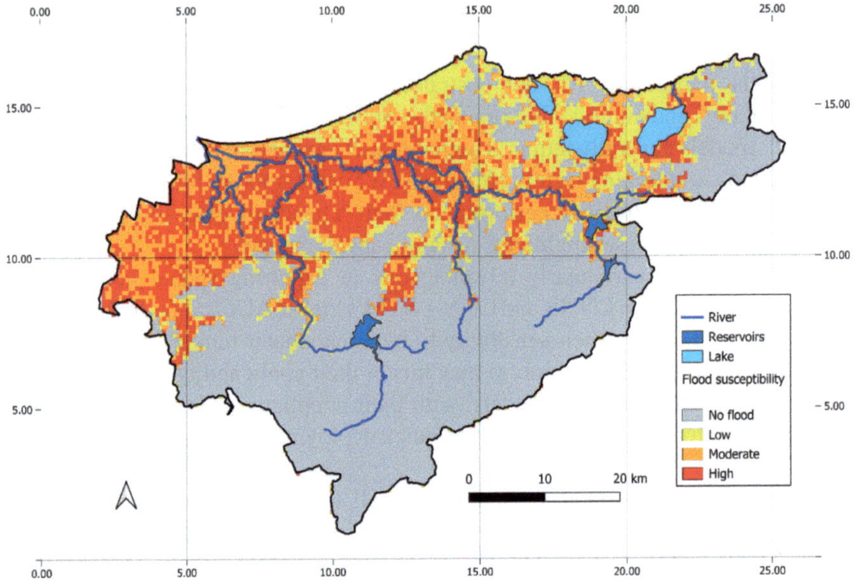

Fig. 13.4 Flood susceptibility map developed with LightGBM model

The flood susceptibility pattern observed is consistent with natural drainage networks, as the high-risk zones align closely with the rivers. This suggests that proper river management and monitoring are critical to mitigating floods. Additionally, the presence of lakes and reservoirs in the northeast highlights the need to regulate water flow and prevent overflow during extreme weather events.

13.7 Study Limitations and Recommendations

While this study demonstrated the potential of ML, DL, and hybrid models for flood classification, there are a few key limitations worth noting.

13.7.1 Limited Dataset Size

The models were trained and tested on a relatively small dataset, which might limit their robustness. As a result, the models may struggle to generalize when applied to larger or more diverse datasets. Expanding the dataset in future research could help improve their performance and reliability.

13.7.2 Data Availability

The study relied on specific flood-related conditioning factors such as SPI, TWI, and flow accumulation. However, the most significant limitation was the partial availability of a comprehensive and accurate flood inventory map, which is critical for validating model predictions and improving their reliability. This issue is particularly common in remote or underdeveloped regions, where data collection infrastructure is often lacking. The absence of such key datasets underscores the need for standardized and consistent flood inventory mapping to support robust and scalable flood modeling efforts.

13.7.3 Challenges with Generalization

Since the models were developed using data from a specific region, their performance in other areas with different environmental and hydrological conditions is uncertain. Future studies should focus on testing these models across diverse regions to ensure they can generalize effectively.

13.7.4 Dynamic Environmental Factors

Flood patterns are influenced by changes in land use, urbanization, and climate variability, but these factors were not considered in the study. Including dynamic and evolving data in future research could enhance the models' ability to predict floods in real-world scenarios.

13.7.5 Requirements for DL Models

DL and hybrid models, while showing potential, require large amounts of data and computational resources for proper training and fine-tuning. In this study, the limited dataset and resource constraints may have restricted their full potential, making it necessary to address these challenges in future work.

13.8 Conclusion

This study evaluated the performance of various ML, DL, and hybrid models for flood classification using multiple flood-conditioning factors. Among the models tested, LightGBM emerged as the most effective, achieving the highest overall performance with a test accuracy of 90%, sensitivity of 95%, precision of 86%, F1-Score of 90%, and an AUC of 0.96. This result highlights its ability to accurately classify flood-prone areas while maintaining a low rate of false positives and negatives. Other ML models, such as RF and CatBoost, also demonstrated strong performance, with RF providing a robust alternative for scenarios requiring simpler implementation and interpretability.

DL models, including CNN and hybrid architectures like CNN-LSTM, showed promise in capturing complex spatial and temporal relationships. However, their performance was constrained by the limited size of the dataset and their reliance on extensive computational resources and fine-tuning. Despite this, CNN and CNN-LSTM achieved competitive results with AUC values exceeding 0.91, making them viable options when sufficient data and computational resources are available.

The study also highlighted several limitations, including the dependence on a relatively small dataset, the challenges of data availability in remote regions, and the difficulty in generalizing results to other regions with different environmental and hydrological conditions. Additionally, the exclusion of dynamic factors such as climate variability and land use changes points to opportunities for future research.

In conclusion, while LightGBM proved to be the best model for flood classification in this study, the results demonstrate the potential of both ML and DL approaches for flood prediction. Future studies should aim to address current limitations by incorporating larger, more diverse datasets, dynamic environmental factors, and advanced optimization techniques to further improve the accuracy, scalability, and generalization of these models for real-world applications.

While ML models like LightGBM and CNN-LSTM have demonstrated high accuracy, their "black-box" nature can be a barrier to adoption in practical flood risk management. Future research could focus on improving the explainability and interpretability of these models, making their predictions more transparent and actionable for decision-makers.

References

1. Boutaghane H et al (2022) Flood analysis and mitigation strategies in Algeria. In: Sumi T, Kantoush SA, Saber M (eds) Wadi flash floods. Natural Disaster Science and Mitigation Engineering: DPRI reports. Springer, Singapore. https://doi.org/10.1007/978-981-16-2904-4_3
2. Chen W, Li Y, Xue W, Shahabi H, Li S, Hong H, Wang X, Bian H, Zhang S, Pradhan B et al (2020) Modeling flood susceptibility using data-driven approaches of naïve Bayes tree, alternating decision tree, and random forest methods. Sci Total Environ 701:134979. https://doi.org/10.1016/j.scitotenv.2019.134979

3. Fang L, Huang J, Cai J, Nitivattananon V (2022) Hybrid approach for flood susceptibility assessment in a flood-prone mountainous catchment in China. J Hydrol 612:128091. https://doi.org/10.1016/j.jhydrol.2022.128091
4. Pham BT, Luu C, Van Phong T et al (2021) Can deep learning algorithms outperform benchmark machine learning algorithms in flood susceptibility modeling? J Hydrol 592:125615. https://doi.org/10.1016/j.jhydrol.2020.125615
5. Rajib A, Liu Z, Merwade V, Tavakoly AA, Follum ML (2020) Towards a large-scale locally relevant flood inundation modeling framework using SWAT and LISFLOOD-FP. J Hydrol 581:124406. https://doi.org/10.1016/j.jhydrol.2019.124406
6. Seenu PZ, Venkata Rathnam E, Jayakumar KV (2020) Visualisation of urban flood inundation using SWMM and 4D GIS. Spat Inf Res 28:459–467. https://doi.org/10.1007/s41324-019-00306-9
7. Hammami S, Zouhri L, Souissi D, Souei A, Zghibi A, Marzougui A, Dlala M (2019) Application of the GIS based multi-criteria decision analysis and analytical hierarchy process (AHP) in the flood susceptibility mapping (Tunisia). Arab J Geosci 12:653. https://doi.org/10.1007/s12517-019-4754-9
8. Shadmehri Toosi A, Calbimonte GH, Nouri H, Alaghmand S (2019) River basin-scale flood hazard assessment using a modified multi-criteria decision analysis approach: a case study. J Hydrol 574:660–671. https://doi.org/10.1016/j.jhydrol.2019.04.072
9. Msabi MM, Makonyo M (2021) Flood susceptibility mapping using GIS and multi-criteria decision analysis: a case of Dodoma region, central Tanzania. Remote Sens Appl Soc Environ 21:100445. https://doi.org/10.1016/j.rsase.2020.100445
10. Shahiri Tabarestani E, Afzalimehr H (2022) A comparative assessment of multi-criteria decision analysis for flood susceptibility modelling. Geocarto Int 37:5851–5874. https://doi.org/10.1080/10106049.2021.1923834
11. Dey H, Shao W, Moradkhani H et al (2024) Urban flood susceptibility mapping using frequency ratio and multiple decision tree-based machine learning models. Nat Hazards 1–29. https://doi.org/10.1007/s11069-024-06609-x
12. Youssef AM, Pradhan B, Sefry SA (2016) Flash flood susceptibility assessment in Jeddah city (Kingdom of Saudi Arabia) using bivariate and multivariate statistical models. Environ Earth Sci 75:12. https://doi.org/10.1007/s12665-015-4830-8
13. Costache R (2019) Flood susceptibility assessment by using bivariate statistics and machine learning models—a useful tool for flood risk management. Water Resour Manag 33:3239–3256. https://doi.org/10.1007/s11269-019-02301-z
14. Cao H, Zhang H, Wang C, Zhang B (2019) Operational flood detection using sentinel-1 SAR data over large areas. Water 11:786. https://doi.org/10.3390/w11040786
15. Hamidi E, Peter BG, Munoz DF, Moftakhari H, Moradkhani H (2023) Fast flood extent monitoring with SAR change detection using google earth engine. IEEE Trans Geosci Remote Sens 61:4201419. https://doi.org/10.1109/TGRS.2023.3240097
16. Tavus B, Kocaman S, Nefeslioglu HA, Gokceoglu C (2020) A fusion approach for flood mapping using sentinel-1 and sentinel-2 datasets. Int Arch Photogramm Remote Sens Spat Inf Sci XLIII-B3-2020:641–648. https://doi.org/10.5194/isprs-archives-XLIII-B3-2020-641-2020
17. Riche A, Drias A, Guermoui M, Gherib T, Boulmaiz T, Souissi B, Melgani F (2024) A novel hybrid deep-learning approach for flood susceptibility mapping. Remote Sens 16:3673. https://doi.org/10.3390/rs16193673
18. Khosravi K, Shahabi H, Pham BT, Adamowski J, Shirzadi A, Pradhan B, Dou J, Ly H-B, Gróf G, Ho HL et al (2019) A comparative assessment of flood susceptibility modeling using multi-criteria decision-making analysis and machine learning methods. J Hydrol 573:311–323. https://doi.org/10.1016/j.jhydrol.2019.03.073
19. Gudiyangada Nachappa T, Tavakkoli Piralilou S, Gholamnia K, Ghorbanzadeh O, Rahmati O, Blaschke T (2020) Flood susceptibility mapping with machine learning, multi-criteria decision analysis and ensemble using Dempster Shafer theory. J Hydrol 590:125275. https://doi.org/10.1016/j.jhydrol.2020.125275

20. Towfiqul Islam ARM, Talukdar S, Mahato S, Kundu S, Eibek KU, Pham QB, Kuriqi A, Linh NTT (2021) Flood susceptibility modelling using advanced ensemble machine learning models. Geosci Front 12:101075
21. Madhuri R, Sistla S, Srinivasa Raju K (2021) Application of machine learning algorithms for flood susceptibility assessment and risk management. J Water Clim Chang 12:2608–2623. https://doi.org/10.2166/wcc.2021.051
22. Saravanan S, Abijith D, Reddy NM, Kss P, Janardhanam N, Sathiyamurthi S, Sivakumar V (2023) Flood susceptibility mapping using machine learning boosting algorithms techniques in Idukki district of Kerala India. Urban Clim 49
23. Prasad P, Loveson VJ, Das B, Kotha M (2022) Novel ensemble machine learning models in flood susceptibility mapping. Geocarto Int 37:4571–4593. https://doi.org/10.1080/10106049.2021.1892209
24. Seydi ST, Kanani-Sadat Y, Hasanlou M, Sahraei R, Chanussot J, Amani M (2022) Comparison of machine learning algorithms for flood susceptibility mapping. Remote Sens 15:192. https://doi.org/10.3390/rs15010192
25. Youssef AM (2023) Optimal flood susceptibility model based on performance comparisons of LR, EGB, and RF algorithms. Nat Hazards 115:1071–1096
26. Saber M, Boulmaiz T, Guermoui M, Abdrabo KI, Kantoush SA, Sumi T, Boutaghane H, Nohara D, Mabrouk E (2021) Examining LightGBM and CatBoost models for Wadi flash flood susceptibility prediction. Geocarto Int 37:7462–7487. https://doi.org/10.1080/10106049.2021.1974959
27. Tien Bui D, Hoang N-D, Martínez-Álvarez F, Ngo P-TT, Hoa PV, Pham TD, Samui P, Costache R (2020) A novel deep learning neural network approach for predicting flash flood susceptibility: a case study at a high frequency tropical storm area. Sci Total Environ 701:134413. https://doi.org/10.1016/j.scitotenv.2019.134413
28. Saha S, Gayen A, Bayen B (2022) Deep learning algorithms to develop flood susceptibility map in data-scarce and ungauged river basin in India. Stoch Environ Res Risk Assess 36:3295–3310. https://doi.org/10.1007/s00477-022-02195-1
29. Askar S, Zeraat Peyma S, Yousef MM, Prodanova NA, Muda I, Elsahabi M, Hatamiafkoueieh J (2022) Flood susceptibility mapping using remote sensing and integration of decision table classifier and metaheuristic algorithms. Water 14:3062. https://doi.org/10.3390/w14193062
30. Samanta S, Pal DK, Palsamanta B (2018) Flood susceptibility analysis through remote sensing, GIS and frequency ratio model. Appl Water Sci 8:66. https://doi.org/10.1007/s13201-018-0710-1
31. Amiri A, Soltani K, Ebtehaj I, Bonakdari H (2024) A novel machine learning tool for current and future flood susceptibility mapping by integrating remote sensing and geographic information systems. J Hydrol 632:130936. https://doi.org/10.1016/j.jhydrol.2024.130936
32. Liu J, Liu K, Wang M (2023) A residual neural network integrated with a hydrological model for global flood susceptibility mapping based on remote sensing datasets. Remote Sens 15:2447. https://doi.org/10.3390/rs15092447
33. Burayu DG, Karuppannan S, Shuniye G (2023) Identifying flood vulnerable and risk areas using the integration of analytical hierarchy process (AHP), GIS, and remote sensing: a case study of southern Oromia region. Urban Clim 51:101640. https://doi.org/10.1016/j.uclim.2023.101640
34. Werner MGF, Hunter NM, Bates PD (2005) Identifiability of distributed floodplain roughness values in flood extent estimation. J Hydrol 314:139–157. https://doi.org/10.1016/j.jhydrol.2005.03.012
35. Riche A, Drias A, Ricci R, Souissi B, Melgani F (2023) Predicting LULC changes and assessing their impact on surface runoff with machine learning and remote sensing data. Res Sq https://doi.org/10.21203/rs.3.rs-3511051/v1
36. Rouse JW, Haas RH, Schell JA, Deering DW (1974) Monitoring vegetation systems in the great plains with ERTS. NASA Spec Publ 351:309
37. Moore ID, Grayson RB (1991) Terrain-based catchment partitioning and runoff prediction using vector elevation data. Water Resour Res 27:1177–1191

38. Öztürk F, Akdeniz F (2000) Ill-conditioning and multicollinearity. Linear Algebra Appl 321(1–3):295–305
39. Saber M, Boulmaiz T, Guermoui M, Abdrabo KI, Kantoush SA, Sumi T, Mabrouk E et al (2023) Enhancing flood risk assessment through integration of ensemble learning approaches and physical-based hydrological modeling. Geomat Nat Hazards Risk 14(1). https://doi.org/10.1080/19475705.2023.2203798
40. Chen T-HK, Qiu C, Schmitt M, Zhu XX, Sabel CE, Prishchepov AV (2020) Mapping horizontal and vertical urban densification in Denmark with Landsat time-series from 1985 to 2018: a semantic segmentation solution. Remote Sens Environ 251:112096
41. Khosravi K, Shahabi H, Pham BT, Adamowski J, Shirzadi A, Pradhan B, Dou J, Ly H-B, Grof G, Ho HL et al (2019) A comparative assessment of flood susceptibility modeling using multi-criteria decision-making analysis and machine learning methods. J Hydrol 573:311–323
42. Rahman M, Ningsheng C, Islam MM, Dewan A, Iqbal J, Washakh RMA, Shufeng T (2019) Flood susceptibility assessment in Bangladesh using machine learning and multi-criteria decision analysis. Earth Syst Environ 3(3):585–601
43. Dou J, Yunus AP, Bui DT, Merghadi A, Sahana M, Zhu Z, Chen C-W, Khosravi K, Yang Y, Pham BT (2019) Assessment of advanced random forest and decision tree algorithms for modeling rainfall-induced landslide susceptibility in the Izu-Oshima Volcanic Island, Japan. Sci Total Environ 662:332–346
44. Wang Y, Hong H, Chen W, Li S, Panahi M, Khosravi K, Shirzadi A, Shahabi H, Panahi S, Costache R (2019) Flood susceptibility mapping in Dingnan County (China) using adaptive neuro-fuzzy inference system with biogeography based optimization and imperialistic competitive algorithm. J Environ Manag 247:712–729
45. Quinlan JR (1986) Induction of decision trees. Mach Learn 1(1):81–106
46. Xu Y, Dai Y, Dong ZY, Zhang R, Meng K (2013) Extreme learning machine-based predictor for real-time frequency stability assessment of electric power systems. Neural Comput Applic 22(3–4):501–508
47. Shahabi H, Shirzadi A, Ronoud S, Asadi S, Pham BT, Mansouripour F, Geertsema M, Clague JJ, Bui DT (2021) Flash flood susceptibility mapping using a novel deep learning model based on deep belief network, back propagation and genetic algorithm. Geosci Front 12(3):101100
48. Cutler DR, Edwards TC, Beard KH, Cutler A, Hess KT, Gibson J, Lawler JJ (2007) Random forests for classification in ecology. Ecology 88(11):2783–2792. https://doi.org/10.1890/07-0539.1
49. Voges LF, Jarren LC, Seifert S (2023) Exploitation of surrogate variables in random forests for unbiased analysis of mutual impact and importance of features. Bioinformatics 39(8):btad471
50. Meghana P, Annepu V, Jweeg MJ, Bagadi K, Aljibori HSS, Mohammed MN, Alfiras M et al (Jan 2024) Analysis of neural network algorithm in comparison to multiple linear regression and random forest algorithm. In: 2024 ASU international conference in emerging technologies for sustainability and intelligent systems (ICETSIS). IEEE, pp 437–443
51. Rodriguez-Galiano VF, Ghimire B, Rogan J, Chica-Olmo M, Rigol-Sanchez JP (2012) An assessment of the effectiveness of a random forest classifier for land-cover classification. ISPRS J Photogramm Remote Sens 67:93–104. https://doi.org/10.1016/j.isprsjprs.2011.11.002
52. Belgiu M, Drăguţ L (2016) Random forest in remote sensing: a review of applications and future directions. ISPRS J Photogramm Remote Sens 114:24–31. https://doi.org/10.1016/j.isprsjprs.2016.01.011
53. Wahba M, Essam R, El-Rawy M, Al-Arifi N, Abdalla F, Elsadek WM (2024) Forecasting of flash flood susceptibility mapping using random forest regression model and geographic information systems. Heliyon 10(13)
54. Daviran M, Shamekhi M, Ghezelbash R, Maghsoudi A (2023) Landslide susceptibility prediction using artificial neural networks, SVMs and random forest: hyperparameters tuning by genetic optimization algorithm. Int J Environ Sci Technol 20(1):259–276
55. Ke G, Meng Q, Finley T, Wang T, Chen W, Ma W, Ye Q, Liu T-Y (2017) Lightgbm: a highly efficient gradient boosting decision tree. Adv Neural Inf Process Syst 30:3146–3154

56. Dorogush AV, Ershov V, Gulin A (2018) CatBoost: gradient boosting with categorical features support. ArXiv181011363. 1–7
57. Prokhorenkova L, Gusev G, Vorobev A, Dorogush AV, Gulin A (2018) CatBoost: unbiased boosting with categorical features. Adv Neural Inf Process Syst 31
58. Cortes C (1995) Support-vector networks. Mach Learn
59. Guermoui M, Mekhalfi ML, Ferroudji K (2013) Heart sounds analysis using wavelets responses and support vector machines. In: 2013 8th international workshop on systems, signal processing and their applications (WoSSPA). IEEE
60. Guermoui M, Abdelaziz R, Gairaa K, Djemoui L, Benkaciali S (2022) New temperature-based predicting model for global solar radiation using support vector regression. Int J Ambient Energy 43(1):1397–1407
61. LeCun Y et al (1989) Backpropagation applied to handwritten zip code recognition. Neural Comput 1(4):541–551
62. LeCun Y et al (1998) Gradient-based learning applied to document recognition. Proc IEEE 86(11):2278–2324
63. Guermoui M, Benkaciali S, Gairaa K, Bouchouicha K, Boulmaiz T, Boland JW (2022) A novel ensemble learning approach for hourly global solar radiation forecasting. Neural Comput Appl 1–23
64. Hochreiter S (1997) Long short-term memory. Neural Computation MIT-Press
65. Boulmaiz T, Guermoui M, Boutaghane H (2020) Impact of training data size on the LSTM performances for rainfall–runoff modeling. Model Earth Syst Environ 6:2153–2164
66. Ferkous K, Guermoui M, Menakh S, Bellaour A, Boulmaiz T (2024) A novel learning approach for short-term photovoltaic power forecasting-A review and case studies. Eng Appl Artif Intell 133:108502
67. Guermoui M, Arrif T, Belaid A, Hassani S, Bailek N (2024) Enhancing direct Normal solar Irradiation forecasting for heliostat field applications through a novel hybrid model. Energy Convers Manag 304:118189. https://doi.org/10.1016/j.enconman.2024.118189
68. Guermoui M, Fezzani A, Mohamed Z, Rabehi A, Ferkous K, Bailek N, Ghoneim SS et al (2024) An analysis of case studies for advancing photovoltaic power forecasting through multi-scale fusion techniques. Sci Rep 14(1):6653. https://doi.org/10.1038/s41598-024-57398-z

Chapter 14
Predicting Soil Hydraulic Conductivity: A Review of Artificial Neural Networks Applications

Hasan Mozaffari, Morteza Pakjoo, Mohammad Amin Nematollahi, Shekoufeh Forouzan, and Ali Akbar Moosavi

Abstract Soil is one of the most critical parts of the environment that supplies human food. Characterizing the soil parameters, especially hydraulic ones, is crucial for protecting soil and water resources, and environmental health. Hydraulic conductivity (K) is a water movement-related attribute in the soil, which is a vital factor in managing irrigation, drainage, flood protection, and erosion control. Direct measurement of K in the field and laboratory is hard and time-consuming and needs special equipment. Therefore, numerous studies have been performed to indirectly predict this important parameter using easily measurable soil properties. However, due to its high variability, the straightforward modeling procedures normally did not predict K, accurately. It seems that complicated relationships between K and some basic and easily measurable soil parameters as predictors exist. Therefore, advanced modeling procedures are needed to discover these relationships. Artificial intelligence (AI) can recognize the complex relationships between soil parameters. On the other hand, artificial neural networks (ANNs), as a well-known branch of AI, present robust modeling procedures, which can be used for acceptable prediction of K using easily measurable soil parameters. In the present chapter, we collected and summarized the results of important studies that applied the ANNs approaches to predict K. Most of the studies in the literature predicted saturated and unsaturated K with good, very good, and excellent accuracies ($0.66 \leq R^2 \leq 0.997$) using easily measurable and basic soil properties by applying both multilayer perceptron neural network (MLPNN) and radial basis function neural network (RBFNN) algorithms. While a few single studies predicted saturated and unsaturated K values with poor and fair (acceptable) accuracies ($0.33 \leq R^2 \leq 0.63$) using the mentioned methodologies. In addition, saturated K was predicted with good to excellent accuracies ($0.71 \leq R^2$

H. Mozaffari · M. Pakjoo · S. Forouzan · A. A. Moosavi (✉)
Department of Soil Science, College of Agriculture, Shiraz University, Shiraz, Iran
e-mail: aamousavi@gmail.com

M. A. Nematollahi
Department of Biosystems Engineering, College of Agriculture, Shiraz University, Shiraz, Iran

M. Nasr et al. (eds.), *Artificial Intelligence Applications for a Sustainable Environment*, Green Chemistry and Sustainable Technology,
https://doi.org/10.1007/978-3-031-91199-6_14

$\leq$ 0.974) by inputting easily measurable soil properties and applying some lesser-used ANN-based algorithms, including group method of data handling (GMDH), adaptive neuro-fuzzy inference system (ANFIS), combination of group method of data handling and harmony search (GMDH-HS), extreme learning machine (ELM), multiple model integration scheme driven by ANN (MM-ANN), deep learning (DL), generalized regression neural network (GRNN), genetic algorithm neural network (GANN), and particle swarm optimization neural network (PSONN) in some single studies. A study also fairly predicted saturated K ($R^2 = 0.524$) by easily measurable soil properties and using the cascade forward network (CFN) algorithm. Generally, the present chapter can lead scientists to investigate ANN-based algorithms and their capabilities to find scientific gaps in predicting saturated and unsaturated K.

Keywords Artificial intelligence · Environment health · Hydrology · Modeling · Multilayer perceptron neural network · Radial basis function neural network

14.1 Introduction

Soil is the basis of earth ecosystems [27], the core of life's complex network [50], and a vital element in the biogeochemical cycles of our planet [22]. Soil, as a limited renewability natural resource [65], supports nearly every aspect of human life and ecological operation [30]. The diverse functions of soil in supporting plants growth [103], controlling water movement [23], hosting a wide range of biodiversity [24], and cycling nutrients [89, 110], make it to be the foundation of our ecosystems. In addition, soil serves as a carbon reservoir [63], playing a crucial role in climate control and the reduction of greenhouse gas emissions [46, 82]. However, this vital natural resource is threatened to be lost by erosion phenomenon and degradation [77, 112]

Nowadays, due to concerns regarding food product safety, soil contamination and health issues related to heavy metals and other pollutants have become an environmental topic of great interest worldwide [47]. Among soil contaminants, heavy metal pollution holds significant importance due to its nonbiodegradable nature, long lifespan, persistence, toxicity, and high mobility, in which they enter easily to the food chain and pose risks to the living organisms health [37, 111, 113]. Human activities frequently lead to groundwater contamination [1]. Any activity that results in the release of chemicals (including heavy metals) or wastes into the environment can contribute to groundwater contamination [19]. Once the groundwater is contaminated, its treatment becomes costly and challenging [11]. Many processes can influence the fate and movement of groundwater contamination. Some of these crucial processes include transport, adsorption, and biological degradation. The transport process occurs when contaminants move through the porous soil medium toward groundwater [92], highlighting the importance of a well-known parameter named soil hydraulic conductivity (K), while the adsorption process happens when pollutants bind to soil colloidal particles [101]. This process can either impede or expedite the movement of contaminants in the soil. Biological degradation takes place when microorganisms utilize pollutants as their energy source [96].

Similarly, the movement of colloidal particles (clay) within the porous soil environment is a significant engineering challenge in groundwater contamination [26]. These colloidal particles originate from clay minerals, aluminum and iron oxides, silicates, and natural organic materials [13]. Numerous researchers have investigated the movement of contaminants, colloidal particles, and their interplay, with the majority of studies suggesting that colloidal particles aid in the transportation of pollutants in groundwater [14, 18, 51, 95]. Nevertheless, certain scientists have suggested that under specific circumstances, colloidal particles may impede the movement of pollutants by adsorbing pollutants onto the colloidal particles, which consequently reduces hydraulic conductivity [10, 26, 40, 80].

Hydraulic conductivity (K) is a crucial factor in water resources management that is used for describing different aspects of water movement in the soil, like runoff, infiltration, migration of contaminants, pesticides, and nutrients, aquifer recharge, irrigation, and water balance [67, 83]. In addition, this parameter plays a crucial role in addressing environmental issues related to leachate transport in designing landfills [100], water movement in the root zone for plant growth [31], and in the design of earth dams [76]. In fact, K is an indicator used to measure the soil's ability to transport water and solutes when subjected to a hydraulic gradient. Generally, K is defined in two statuses, including saturated and unsaturated. When all pores in soil are filled by water, the soil is under saturated conditions [34]. At these conditions, all pores cooperate in transporting water movement, and K at this status is the highest value. The saturated K is extensively used in drainage-related studies and management [28]. While, at unsaturated conditions, some pores with specific sizes are filled by water and cooperate in water movement [53]. At unsaturated conditions, the K value depends on soil–water tension or matric potential [70]. Increasing tension (decreasing matric potential) causes decreasing unsaturated K values [69]. Therefore, most soil–water processes, like irrigation and rainfall infiltration, initially happen at unsaturated conditions [48].

The definition of K is conceptually simple, but in practice, it has many complex aspects, especially when we attempt to measure it accurately. There are different laboratory- and field-based methods for measuring K. Yet, directly measuring K is time-consuming and difficult [61] and needs special equipment. In fact, the K is a hardly measurable soil parameter. Therefore, many attempts were made to indirectly predict saturated and unsaturated K values using some easily measurable properties of soil as inputs and applying various straightforward and conceptual modeling procedures. Various soil physical and chemical factors affect soil hydraulic parameters, specifically K, like particle size distribution and texture [62], pore size distribution [115], bulk density [42], structure [102], organic matter [7], calcium carbonate [67], sodium content [49], etc. These and their related properties can be considered as easily measurable soil indicators (inputs) for modeling and predicting K.

Although the straightforward modeling procedures present valuable insights for predicting K, normally these performances remain low, due to the high variability of K, even in soil with the same components. However, complicated machine learning

(ML) algorithms have strong potential to predict K [64]. Different ML algorithms like random forest [94], regression tree [79], fuzzy logic [109], support vector regression [66], and artificial neural networks, ANNs [55] are available in the literature for predicting a target parameter (e.g., K) using basic and easily measurable inputs. These procedures operate as black-box tools, and their interpretations remain unknown.

Recently, ANN-based algorithms have been extensively used for simulation, pattern recognition, classification, and optimization in various engineering fields [72]. The capabilities of neural network (NN) models have made them suitable tools for modeling soil hydraulic properties [55]. The ANNs have been applied in various studies, such as simulating soil cation exchange capacity [107], sorptivity coefficient [59], infiltration rate [2], and water content at different matric potential [43]. In the past decade, there has been a significant increase in the number of practical methods used for hydrological modeling and prediction, including conventional intelligent and data-driven methods. It seems that the application of artificial intelligence (AI) models in predicting hydrological processes is essential due to their efficiency in modeling complex physical processes based on available data and information. Unlike field and laboratory tests, these intelligent methods can provide acceptable estimates of hydrological parameters in a shorter time, with lower costs, and using limited available data.

The objective of the present chapter is to investigate how ANN-based modeling procedures can enhance the efficiency and accuracy of estimating saturated and unsaturated K in different soils around the world. This chapter presents a literature review on the application of various ANN models and shows the potential of these models to predict saturated and unsaturated K. The idea can help researchers and scientists to find study gaps and develop comprehensive plans for accurately predicting K using the best ANN algorithm.

14.2 Methodology

14.2.1 Machine Learning (ML) Models

The ML algorithms are usually categorized into four main types, including supervised, unsupervised, semi-supervised, and reinforcement algorithms [21]. The mentioned types can be implemented for some applications such as classification, regression, and clustering tasks [86].

Regarding the literature, many studies were conducted to predict K by different NNs such as multilayer perceptron (MLP), radial basis function (RBF), group method of data handling (GMDH), cascade forward network (CFN), and the other ML models such as random forest (RF), support vector machine (SVM), decision tree (DT), k-nearest neighbors (KNN), Gaussian process regression (GPR), and also traditional regression methods like multiple linear regression (MLR). Each of the mentioned techniques has its unique advantages and their implementations depend on the nature and complexity of the relation(s) between input(s) and output data. Here, a brief explanation of the structure of the well-known and common NN algorithms including MLP, RBF, and GMDH is given in the following subsections.

14.2.1.1 Multilayer Perceptron Neural Network (MLPNN)

Many NNs are established and categorized as feedforward NNs. The well-known MLPNN, as a type of supervised ML [6], is one of the most powerful and popular models among the feedforward NNs, which is widely used in many sciences and engineering fields for function approximation (regression) and classification [36] due to their adaptable structure, and high capabilities in complex problems, and large numbers of available training algorithms. This model is developed by several input, hidden, and output layers [12]. Each layer consists of some basic building elements called neurons. The neighboring layers are connected through weights. The inputs of neurons in various layers consist of the outputs from neurons in the preceding layer, as well as a single input called bias (Fig. 14.1).

Indeed, the objective of the NN training is to adjust the mentioned two parameters (weights and biases). For this purpose, an optimization problem is formulated, in which a defined error function (difference between NN outputs and actual ones, Eq. 14.1) should be minimized to find the proper weights and biases [64, 75].

$$E = \frac{1}{2}\sum_{j=1}^{N}\sum_{k=1}^{M}\left[(P_p)_j^k - (P_m)_j^k\right] \tag{14.1}$$

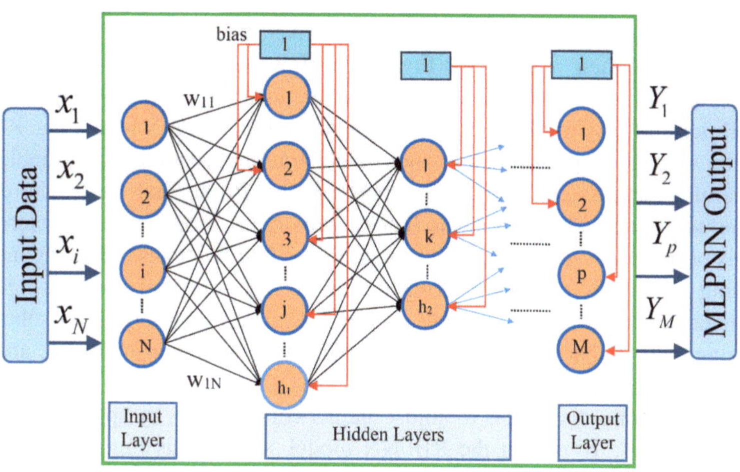

Fig. 14.1 A schematic structure of the multilayer perceptron neural network (MLPNN)

where P_p is NN output (predicted) values and is a function of weights and biases. P_m is the measured (actual) data that is presented as inputs to the NN model. N and M are the numbers of measured data and outputs, respectively. Two well-known techniques including traditional (e.g., Levenberg–Marquardt, steepest descent, conjugate gradient algorithms, etc.) and nature-inspired optimization (learning) (e.g., genetic, artificial bee colony, particle swarm optimization, cuckoo search, etc.) algorithms are employed to minimize the output error function.

For example, the $\Delta w_{ji}\ (q)$ related to the connection values between the hidden and input layers for an MLPNN with a single hidden layer based on the steepest descent method and backpropagation approach considering the momentum term is calculated [74]:

$$\Delta w_{ji}\,(q) = \gamma f_j'(.)\, x_i \sum_{k=1}^{M} \left\{ \left[(P_m)_k - (P_p)_k \right] f_k'(.)\, w_{kj}\,(q) \right\} + \alpha\, \Delta w_{ji}\,(q-1) \quad (14.2)$$

where γ, $f'(.)$, x_i, and α are learning rate, a derivative of the transfer function concerning its input, ith input (from input layer), and momentum value (a positive value between 0 and 1), respectively. The q is also the number of iterations. To achieve the best NN performance, the main hyperparameters such as the number of hidden layers and their corresponding neurons, transfer function types, learning algorithms and other fine-tuning parameters like learning rate, number of epochs, etc. also should be selected and regularized, appropriately [59].

14.2.1.2 Radial Basis Function Neural Network (RBFNN)

RBFNN represents another type of feedforward NNs with a structure of three layers, namely input, hidden, and output (Fig. 14.2). However, the RBFNN requires more neurons than the MLPNN, but it enjoys unique advantages such as rapid learning capabilities and easy implementation compared to the MLPNN [105]. The RBFNN also has very good performance when many training vectors are available [52]. Its' learning process consists of two main phases. In the first phase, the centers and the width (spread) of the basis (kernel-activation) functions were determined. For this purpose, several unsupervised techniques like K-means clustering methods including dynamic K-means clustering algorithm [29], improved K-means algorithm [17], fuzzy clustering [41], particle swarm optimization (PSO)-based subtractive clustering [56], K-harmonic clustering [114], self-organizing map [38], and self-additive clustering [114] have been implemented.

The second phase attempts to adjust the network weights by applying the least-squares methods. To design the RBFNN model, some basis functions, including the Gaussian function, generalized multiquadric function, generalized inverse multiquadric function, thin plate spline basis function, cubic function, and linear function, can be used. The well-known Gaussian function (φ) is commonly employed as the basis function in the RBFNN structure [57]. The network output (Y) is a function of

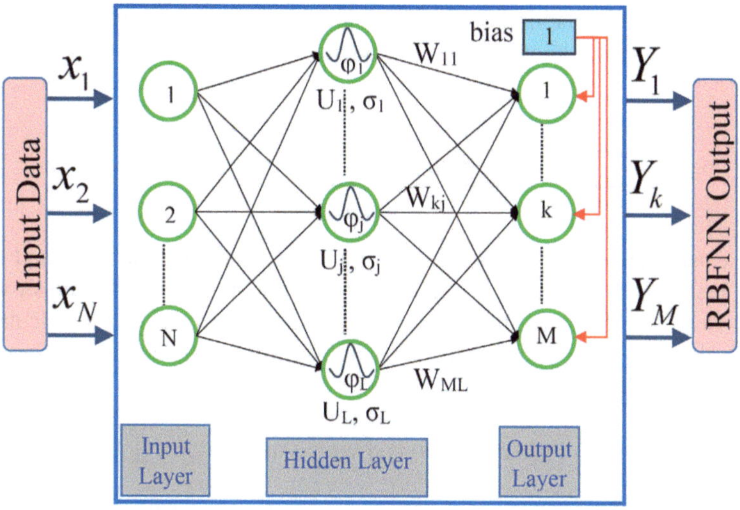

Fig. 14.2 A schematic structure of the radial basis function neural network (RBFNN)

the Euclidean difference between the input vector (X) and the center of the neurons (U) as well as the parameters weights (w) and biases (b) of each neuron that are calculated using the following equation:

$$Y_k(X) = \sum_{j=1}^{n} w_{kj}\,\varphi_j\,(\|X - U_j\|) + b_k \tag{14.3}$$

$$\varphi_j(X) = \exp\left(\frac{-\|X - U_j\|}{\sigma_j^2}\right) \tag{14.4}$$

where σ is the width (spread) of the Gaussian function that controls the smoothness of the kernel function [59, 75].

For more information about the structure and training process of MLPNN and RBFNN, one can refer to the references in Moosavi et al. [59] and Mozaffari et al. [64].

14.2.1.3 Group Method of Data Handling (GMDH)

GMDH, as an inductive algorithm, was first introduced by Ivakhnenko [33] and developed based on the heuristic self-organizing principle [45]. This algorithm can be organized as a type of feedforward NN and employed as a good alternative to the standard regression algorithms. Because of its self-organizing and inductive characteristics, GMDH automatically finds the best network configuration [73]. The mentioned algorithm considers all the pairwise combinations of input variables. The general form of the relation between input and output variables can be represented by the Volterra series, known as the Kolmogorov–Gabor polynomial as follows [3, 8]:

$$\hat{y} = a_0 + \sum_{i=1}^{m} a_i x_i + \sum_{i=1}^{m}\sum_{j=1}^{m} a_{ij} x_i x_j + \sum_{i=1}^{m}\sum_{j=1}^{m}\sum_{k=1}^{m} a_{ijk} x_i x_j x_k + \ldots \tag{14.5}$$

where x parameters with m samples and $\hat{y}$ are the input data and predicted output model, respectively. The a parameters denote the polynomial coefficients or weight values. However, the quadratic forms of Eq. (14.5) are mostly used to calculate the output value only for two variables in each layer, for example with two inputs (x_1 and x_2) as the following form:

$$\hat{y} = a_0 + a_1 x_1 + a_2 x_2 + a_3 x_1^2 + a_4 x_2^2 + a_5 x_1 x_2 \tag{14.6}$$

It is noted that the output of each neuron for all pairs of the input variables is computed using Eq. (14.6) and the differences between the actual outputs and the estimated ones (Eq. 14.6) can be obtained by some external criteria like root mean square error (RMSE). Indeed, the aim of the GMDH training is formulated as an optimization problem to find the appropriate a values by minimization of RMSE. The least-squares algorithms were employed to find the mentioned a values. The traditional GMDH architecture is shown in Fig. 14.3. Generally, the training of the GMDH algorithm is performed in four steps. In the first step, $\binom{m}{2}$ neurons are built up in the first hidden layer from the input data. Then by two inputs at the time for any $\binom{m}{2}$ combination, the a values are computed and the approximate output values are found. In the second step, based on the applied external criteria, the accuracy of each output neuron is computed and ranked. Therefore, the neurons with the most accurate estimations are selected and the rest of them are eliminated from NN and the first hidden layers with the most efficient neurons are formed, systematically. In the third step, by using the new inputs (the outputs from selected neurons) resulting from the second step, the a values are determined and as mentioned in the second step, the best neurons are selected based on their accuracy to form the second hidden layer. The second step is repeated to produce the last hidden layer [32, 71]. Therefore, as Fig. 14.3 shows, only one neuron is selected.

14.2.2 Performance Criteria

In the present review, we used two well-known and unitless performance criteria, including determination coefficient (R^2) and normalized root mean square error (NRMSE), to compare the results of available studies in the literature, which

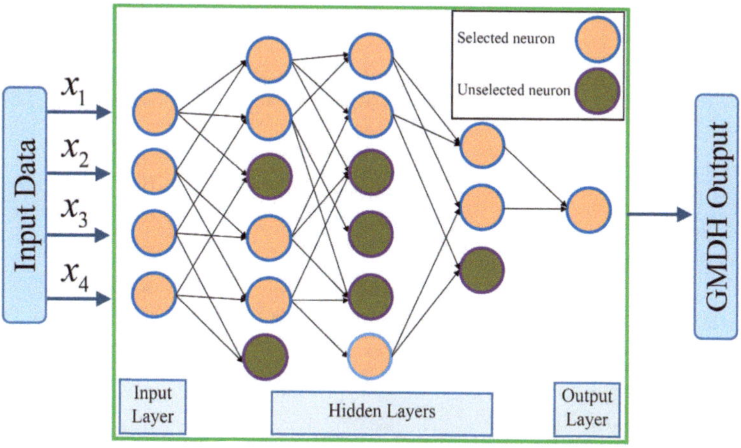

Fig. 14.3 A schematic structure of the group method of data handling (GMDH) neural network

predicted different forms of parameter K using various ANNs models. According to Mozaffari et al. [68], five performance classes can be considered for R^2 including (i) poor ($R^2 < 0.50$); (ii) fair ($0.50 \leq R^2 < 0.65$); (iii) good ($0.65 \leq R^2 < 0.75$); (iv) very good ($0.75 \leq R^2 < 0.90$); and (v) excellent ($0.90 \leq R^2 \leq 1$). In addition, the NRMSE values can fall into four performance classes including (i) poor (NRMSE > 0.30); (ii) fair (0.20 < NRMSE ≤ 0.30); (iii) good (0.10 < NRMSE ≤ 0.20); and (iv) excellent (0 < NRMSE ≤ 0.10) [9].

14.3 Results and Discussion

Figure 14.4 shows case studies that predicted saturated and unsaturated K values using easily measurable soil properties and various ANN algorithms in different world countries. As can be seen, more extensive studies were performed in Iranian and American soils to predict different forms of K using ANN-based methods. However, some studies predicted K values using the well-known global Unsaturated Soil Database, UNSODA [25, 87] or available data in different studies in the literature [16, 104], which were not presented in Fig. 14.4. We summarized the performed studies, which predicted the K values, in three sections and based on the applied different ANN models (see sections below).

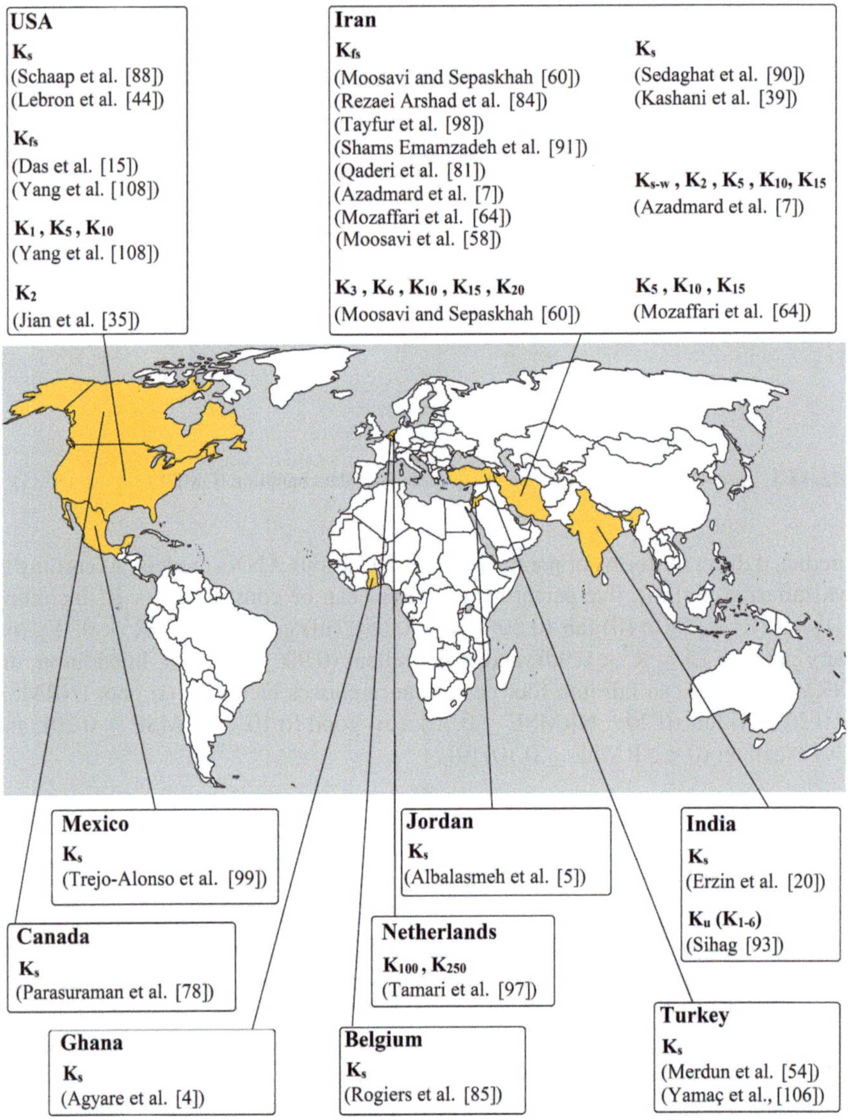

Fig. 14.4 Performed studies in soils of different world countries to predict soil hydraulic conductivity (K) using easily measurable soil properties and applying artificial neural network-based algorithms (Abbreviations: K_s, laboratory-measured saturated K; K_{fs}, field-measured saturated K; $K_{1\text{-}6}$, unsaturated K at a tension between 1 and 6 cm; K_{number}, unsaturated K at a specific tension number (cm); $K_{s\text{-}w}$, calculated saturated K from Wooding's equation)

14.3.1 Multilayer Perceptron Neural Network (MLPNN)

Among ANNs, the MLPNN was the most used algorithm for predicting K using easily measurable soil properties. As can be seen in Table 14.1, the major studies around the world accurately predicted different forms of the parameter K by the MLPNN model. According to the R^2 criterion, different forms of K were excellently predicted by Das and Basudhar [16], Lebron et al. [44], Erzin et al. [20], Ghanbarian-Alavijeh et al. [25], Trejo-Alonso et al. [99], Williams and Ojuri [104], and Yamaç et al. [106] for K_s (laboratory-measured of saturated K) ($0.903 \leq R^2 \leq 0.997$); Shams Emamzadeh et al. [91] for $K_{fs\text{-}h}$ and $K_{fs\text{-}v}$ (horizontal and vertical field-measured of saturated K, respectively) (R^2 of 0.94 and 0.97); and Yang et al. [108] for K_1 and K_5 (unsaturated K at tensions 1 and 5 cm, respectively) (R^2 of 0.965 and 0.984) using easily measurable soil properties and the MLPNN model.

While using MLPNN, Rogiers et al. [85] predicted K_s, Shams Emamzadeh et al. [91] predicted $K_{fs\text{-}hv}$ (horizontal-vertical field-measured saturated K); Azadmard et al. [7], Qaderi et al. [81], Yang et al. [108], Mozaffari et al. [64], and Moosavi et al. [58] predicted K_{fs} (field-measured of saturated K); Sihag [93] predicted $K_{1\text{-}6}$ (unsaturated K at a tension between 1 to 6 cm); Yang et al. [108] and Mozaffari et al. [64] predicted K_{10} (unsaturated K at tension 10 cm); and Mozaffari et al. [64] predicted K_5 with very good accuracy ($0.75 \leq R^2 \leq 0.89$) by easily measurable soil properties.

In addition, the mentioned NN algorithm with easily measurable soil indicators was used to predict K_s [88], K_{fs} [15, 84], and K_{15} (unsaturated K at tension 15 cm) [64] with good accuracy ($0.66 \leq R^2 \leq 0.72$) and K_{fs} [98], $K_{s\text{-}w}$ (calculated saturated K from wooding's equation) and K_{10} [7] with fair accuracy ($0.51 \leq R^2 \leq 0.63$).

Furthermore, Agyare et al. [4] predicted K_s, Azadmard et al. [7] predicted K_2 (unsaturated K at tension 2 cm), K_5, and K_{15}; and Jian et al. [35] predicted K_2 with poor accuracy ($0.33 \leq R^2 \leq 0.43$) by applying MLPNN and using easily measurable soil properties.

Parasuraman et al. [78], Schaap and Leij [87] and Sedaghat et al. [90] did not present the R^2 or NRMSE values for their prediction of K by MLPNN. However, according to the NRMSE classification, Moosavi and Sepaskhah [60] by using MLPNN and selected physical and chemical parameters predicted K_{10} excellently (NRMSR = 0.084) and K_{fs}, unsaturated K at tensions 3, 6, and 20 cm (K_3, K_6, K_{20}, respectively), and K_{15} poorly ($0.403 \leq \text{NRMSE} \leq 0.577$).

14.3.2 Radial Basis Function Neural Network (RBFNN)

Compared to MLPNN, fewer studies were performed to predict different forms of K using RBFNN (Table 14.2). Among the summarized studies in Table 14.2, Tamari et al. [97] and Sedaghat et al. [90] did not calculate unitless and common performance criteria (R^2 or NRMSE) for predicting their studied K forms using the RBFNN

Table 14.1 A literature review on predicting different forms of hydraulic conductivity (K) using easily measurable soil parameters and applying the multilayer perceptron neural network (MLPNN) method

Country of study	n	Parameter	R^2	NRMSE	References
UNSODA database	315	K_s	NA	NA	[87]
UNSODA database	245	K_u	NA	NA	[87]
USA	620	K_s	0.72	NA	[88]
USA	36	K_s	0.91	NA	[44]
Canada	126 and 78	K_s	NA	NA	[78]
Data collected from the literature	42	K_s	0.903	NA	[16]
Ghana	1126	K_s	0.41	NA	[4]
India	NA	K_s	0.997	NA	[20]
UNSODA database	311	K_s	0.98	NA	[25]
Iran	138	K_{fs}	NA	0.577	[60]
Iran	138	K_3	NA	0.477	[60]
Iran	138	K_6	NA	0.403	[60]
Iran	138	K_{10}	NA	0.084	[60]
Iran	138	K_{15}	NA	0.506	[60]
Iran	138	K_{20}	NA	0.483	[60]
Belgium	173	K_s	0.89	NA	[85]
USA	41	K_{fs}	0.717	NA	[15]
Iran	175	K_{fs}	0.66	NA	[84]
Iran	132	K_{fs}	0.63	NA	[98]
Iran	260	K_s	NA	NA	[90]
Iran	123	$K_{fs\text{-}h}$	0.94	NA	[91]
Iran	119	$K_{fs\text{-}v}$	0.97	NA	[91]
Iran	85	$K_{fs\text{-}hv}$	0.85	NA	[91]
Iran	151	K_{fs}	0.851	NA	[81]
India	46	$K_{1\text{-}6}$	0.75	NA	[93]
USA	44	K_{fs}	0.870	NA	[108]
USA	44	K_1	0.965	NA	[108]
USA	44	K_5	0.984	NA	[108]
USA	44	K_{10}	0.817	NA	[108]
Iran	212	K_{fs}	0.80	0.12	[7]
Iran	212	$K_{s\text{-}w}$	0.51	0.26	[7]
Iran	212	K_2	0.43	0.18	[7]
Iran	212	K_5	0.35	0.51	[7]
Iran	212	K_{10}	0.53	0.29	[7]
Iran	212	K_{15}	0.40	0.09	[7]

(continued)

Table 14.1 (continued)

Country of study	n	Parameter	R^2	NRMSE	References
Data collected from the literature	144	K_s	0.955	NA	[104]
USA	307	K_2	0.33	NA	[35]
Mexico	900	K_s	0.978	NA	[99]
Turkey	291	K_s	0.940	NA	[106]
Iran	102	K_{fs}	0.80	0.213	[64]
Iran	102	K_5	0.82	0.131	[64]
Iran	102	K_{10}	0.78	0.236	[64]
Iran	102	K_{15}	0.71	0.237	[64]
Iran	100	K_{fs}	0.870	NA	[58]

* Abbreviations: n, number of the used data; R^2, determination coefficient; NRMSE, normalized root mean square error; NA, not available; K_s, laboratory-measured saturated K; K_u, unsaturated K; K_{fs}, field-measured saturated K; K_{number}, unsaturated K at a specific tension number (cm); $K_{fs\text{-}h}$, horizontal field-measured saturated K; $K_{fs\text{-}v}$, vertical field-measured saturated K; $K_{fs\text{-}hv}$, horizontal-vertical field-measured saturated K; $K_{1\text{-}6}$, unsaturated K at a tension between 1 to 6 cm; $K_{s\text{-}w}$, calculated saturated K from Wooding's equation

method. While, Shams Emamzadeh et al. [91] predicted $K_{fs\text{-}v}$ excellently (R^2 = 0.95) and $K_{fs\text{-}h}$ and $K_{fs\text{-}hv}$ with very good accuracy (R^2 of 0.88 and 0.83, respectively) using selected soil attributes as inputs and applying the RBFNN method. Similarly, Mozaffari et al. [64] predicted K_{fs} and K_5 with very good accuracy (R^2 of 0.77 and 0.78, respectively); however, their prediction accuracy for K_{10} and K_{15} was classified as fair (R^2 of 0.62 and 0.58, respectively) when they used the RBFNN model with selecting some basic physical and chemical parameters as predictors. In addition, Rezaei Arshad et al. [84] and Moosavi et al. [58] respectively predicted K_{fs} with good ($R^2 = 0.68$) and very good ($R^2 = 0.857$) accuracies by the mentioned methodology.

Generally, comparing studies that used both MLPNN and RBFNN methods for predicting the same form of hydraulic conductivity [58, 64, 84, 91], the MLPNN method showed better potential for predicting K values in comparison with RBFNN. However, the RBFNN method is a faster method than MLPNN.

14.3.3 Less-Used ANN Models

Table 14.3 shows a summary of single studies, which predicted different forms of K by different ANN-based approaches and using easily measurable soil properties. As demonstrated, Das and Basudhar [16] used the GMDH algorithm to predict K_s without reporting the R^2 or NRMSE values. Qaderi et al. [81] excellently predicted K_{fs} using the GMDH algorithm and its combination with harmony search (HS) by R^2 values of 0.971 and 0.974, respectively. In addition, excellent prediction of K_s was

Table 14.2 A literature review on predicting different forms of hydraulic conductivity (K) using easily measurable soil parameters and applying the radial basis function neural network (RBFNN) method

Country of study	n	Parameter	R^2	NRMSE	References
Netherlands	178	K_{100}	NA	NA	[97]
Netherlands	178	K_{250}	NA	NA	[97]
Iran	175	K_{fs}	0.68	NA	[84]
Iran	260	K_s	NA	NA	[90]
Iran	123	$K_{fs\text{-}h}$	0.88	NA	[91]
Iran	119	$K_{fs\text{-}v}$	0.95	NA	[91]
Iran	85	$K_{fs\text{-}hv}$	0.83	NA	[91]
Iran	102	K_{fs}	0.77	0.225	[64]
Iran	102	K_5	0.78	0.171	[64]
Iran	102	K_{10}	0.62	0.243	[64]
Iran	102	K_{15}	0.58	0.247	[64]
Iran	100	K_{fs}	0.857	NA	[58]

* Abbreviations: n, number of the used data; R^2, determination coefficient; NRMSE, normalized root mean square error; NA, not available; K_s, laboratory-measured saturated K; K_{fs}, field-measured saturated K; K_{number}, unsaturated K at a specific tension number (cm); $K_{fs\text{-}h}$, horizontal field-measured saturated K; $K_{fs\text{-}v}$, vertical field-measured saturated K; $K_{fs\text{-}hv}$, horizontal-vertical field-measured saturated K

reported by Kashani et al. [39] and Yamaç et al. [106] using a multiple model integration scheme driven by ANN (MM-ANN) and deep learning (DL) algorithms by R^2 values of 0.912 and 0.919, respectively. Kashani et al. [39] and Albalasmeh et al. [5] also predicted K_s with very good accuracy using an extreme learning machine (ELM) and generalized regression neural network (GRNN) algorithms with R^2 values of 0.889 and 0.86, respectively. Moosavi et al. [58] reported excellent predictions of K_{fs} by applying a genetic algorithm neural network (GANN) and particle swarm optimization neural network (PSONN) with R^2 values of 0.901 and 0.918, respectively. Rezaei Arshad et al. [84] used an adaptive neuro-fuzzy inference system (ANFIS) algorithm to predict K_{fs} with good accuracy ($R^2 = 0.71$). Furthermore, K_s was predicted fairly ($R^2 = 0.524$) by Merdun et al. [54] using the CFN algorithm.

The mentioned algorithms were used in limited studies to predict different forms of K. It seems that the majority of them can predict the K values, accurately. However, it suggests these methods be tested in different soils worldwide to predict K and compared to more used ANN algorithms, like MLPNN, and RBFNN.

Table 14.3 A literature review on predicting different forms of hydraulic conductivity (K) using easily measurable soil parameters and applying different artificial neural network (ANN) algorithms

Country of study	n	Parameter	ANN model	R^2	NRMSE	References
Turkey	195	K_s	CFN	0.524	NA	[54]
Data collected from the literature	42	K_s	GMDH	NA	NA	[16]
Iran	175	K_{fs}	ANFIS	0.71	NA	[84]
Iran	151	K_{fs}	GMDH	0.971	NA	[81]
Iran	151	K_{fs}	GMDH-HS	0.974	NA	[81]
Iran	245	K_s	ELM	0.889	NA	[39]
Iran	245	K_s	MM-ANN	0.912	NA	[39]
Turkey	291	K_s	DL	0.919	NA	[106]
Jordan	165	K_s	GRNN	0.86	NA	[5]
Iran	100	K_{fs}	GANN	0.901	NA	[58]
Iran	100	K_{fs}	PSONN	0.918	NA	[58]

* Abbreviations: n, number of the used data; R^2, determination coefficient; NRMSE, normalized root mean square error; NA, not available; K_s, laboratory-measured saturated K; K_{fs}, field-measured saturated K; CFN, cascade forward network; GMDH, group method of data handling; ANFIS, adaptive neuro-fuzzy inference system; GMDH-HS, combination of group method of data handling and harmony search; ELM, extreme learning machine; MM-ANN, multiple model integration scheme driven by ANN; DL, deep learning; GRNN, generalized regression neural network, GANN: genetic algorithm neural network; PSONN: particle swarm optimization neural network

14.4 Limitations and Recommendations

Generally, the parameter K is a highly variable and hardly measurable soil hydraulic parameter. Therefore, its accurate prediction needs careful selection of the predictors (independent input variables) and modeling procedure and algorithm. Normally, those parameters, which affect the size and geometry of the pores, are very effective for predicting K. According to Tables 14.1, 14.2, and 14.3, the ANN-based models successfully predicted saturated and unsaturated forms of K. However, in some studies, poor results were obtained, which may be due to considering the unrelated and less related input soil parameters for predicting the K values.

The ANN-based models have a black-box nature. They are more difficult to understand, less applicable, and harder to use and need complicated computer software to process compared to straightforward modeling procedures. Despite these limitations, they are effectively used for modeling hardly measurable soil parameters, like K. In this regard, we observed significant increase of R^2 values for predicting K_{15}, K_{10}, K_5, and K_0 when the ANN-based models were used (R^2 of 0.71, 0.78, 0.82, and 0.80 for MLPNN and 0.58, 0.62, 0.78, and 0.77 for RBFNN, respectively) compared to using stepwise multiple linear regression method (R^2 of 0.55, 0.52, 0.56, and 0.63, respectively) [64].

It is recommended to apply the straightforward machine learning modeling approaches to predict soil parameters. However, when straightforward models cannot accurately predict hardly measurable soil parameters, it is suggested that the advanced machine learning models, like the ANN-based models, be used. This potential for predicting some highly variable parameters (like saturated and unsaturated K) is beneficial.

14.5 Conclusion

K is a highly variable soil parameter and its value depends on pores size and geometry. Therefore, acceptable prediction of K depends on two factors including independent input variables (easily measurable soil properties) and selected modeling procedure. The ANN algorithms have enormous potential for modeling hardly measurable soil parameters (e.g., K), however, they are conceptual modeling procedures and have a black-box nature. Most of the available studies in the literature acceptably to excellently ($0.62 \leq R^2 \leq 0.997$) predicted saturated and unsaturated K using frequently used ANNs (i.e., MLPNN and RBFNN) and some lesser-used ones (i.e., GMDH, ANFIS, GMDH-HS, ELM, MM-ANN, DL, GRNN, GANN, and PSONN). This chapter presents a beneficial overview for soil scientists to use the ANN-based algorithms for predicting hardly measurable soil parameters, specifically hydraulic ones. Future works can compare the potential of different machine learning algorithms (like RF, SVM, DT, KNN, GPR, MLR, etc.) with the ANN-based models for predicting K and discovering the most appropriate algorithm. Finally, it is suggested that soil scientists raise public awareness about the benefits of using machine learning methods, especially the ANN-based algorithms, to predict different environmental parameters with respect to socioeconomic impacts.

Acknowledgements The authors gratefully acknowledge Shiraz University.

References

1. Abanyie SK, Apea OB, Abagale SA, Amuah EEY, Sunkari ED (2023) Sources and factors influencing groundwater quality and associated health implications: a review. Emerg Contamin 9(2):100207
2. Abdalrahman G, Lai SH, Kumar P, Ahmed AN, Sherif M, Sefelnasr A, Chau KW, Elshafie A (2022) Modeling the infiltration rate of wastewater infiltration basins considering water quality parameters using different artificial neural network techniques. Eng Appl Comput Fluid Mecha 16(1):397–421
3. Aghelpour P, Bahrami-Pichaghchi H, Karimpour F (2022) Estimating daily rice crop evapotranspiration in limited climatic data and utilizing the soft computing algorithms MLP, RBF, GRNN, and GMDH. Complexity 2022(1):4534822
4. Agyare WA, Park SJ, Vlek PLG (2007) Artificial neural network estimation of saturated hydraulic conductivity. Vadose Zone J 6(2):423–431

5. Albalasmeh A, Mohawesh O, Gharaibeh M, Deb S, Slaughter L, El Hanandeh A (2022) Artificial neural network optimization to predict saturated hydraulic conductivity in arid and semi-arid regions. CATENA 217:106459
6. Alghamdi FA, Almanaseer H, Jaradat G, Jaradat A, Alsmadi MK, Jawarneh S, Alfagham H (2024) Multilayer perceptron neural network with arithmetic optimization algorithm-based feature selection for cardiovascular disease prediction. Mach Learn Knowl Extract 6(2):987–1008
7. Azadmard B, Mosaddeghi MR, Ayoubi S, Chavoshi E, Raoof M (2020) Estimation of near-saturated soil hydraulic properties using hybrid genetic algorithm-artificial neural network. Ecohydrol Hydrobiol 20(3):437–449
8. Azizi E, Yosefvand F, Yaghoubi B, Izadbakhsh MA, Shabanlou S (2024) Prediction of groundwater level using GMDH artificial neural network based on climate change scenarios. Appl Water Sci 14(4):77
9. Bannayan M, Hoogenboom G (2009) Using pattern recognition for estimating cultivar coefficients of a crop simulation model. Field Crop Res 111(3):290–302
10. Bekhit HM, El-Kordy MA, Hassan AE (2009) Contaminant transport in groundwater in the presence of colloids and bacteria: model development and verification. J Contam Hydrol 108(3–4):152–167
11. Burri NM, Weatherl R, Moeck C, Schirmer M (2019) A review of threats to groundwater quality in the anthropocene. Sci Total Environ 684:136–154
12. Chan KY, Abu-Salih B, Qaddoura R, Ala'M AZ, Palade V, Pham DS, Muhammad K (2023) Deep neural networks in the cloud: Review, applications, challenges and research directions. Neurocomputing 545:126327
13. Chen A, Arai Y (2020) Current uncertainties in assessing the colloidal phosphorus loss from soil. In: Sparks DL (ed) Advances in agronomy. Academic Press Inc., pp 117–151
14. Cheng D, Liao P, Yuan S (2016) Effects of ionic strength and cationic type on humic acid facilitated transport of tetracycline in porous media. Chem Eng J 284:389–394
15. Das SK, Samui P, Sabat AK (2012) Prediction of field hydraulic conductivity of clay liners using an artificial neural network and support vector machine. Int J Geomech 12(5):606–611. https:\\doi.org\10.1061/(ASCE)GM.1943-5622.0000129
16. Das SK, Basudhar PK (2007) Prediction of hydraulic conductivity of clay liners using artificial neural network. Lowland Technol Int 9(1):50–58
17. Dash CSK, Behera AK, Dehuri S, Cho SB (2016) Radial basis function neural networks: a topical state-of-the-art survey. Open Comput Sci 6(1):33–63
18. Emerson HP, Hickok KA, Powell BA (2016) Experimental evidence for ternary colloid-facilitated transport of Th (IV) with hematite (α-Fe2O3) colloids and Suwannee River fulvic acid. J Environ Radioact 165:168–181
19. Enjavinejad SM, Zahedifar M, Moosavi AA, Khosravani P (2024) Integrated application of multiple indicators and geographic information system-based approaches for comprehensive assessment of environmental impacts of toxic metals-contaminated agricultural soils and vegetables. Sci Total Environ 926:171747
20. Erzin Y, Gumaste SD, Gupta AK, Singh DN (2009) Artificial neural network (ANN) models for determining hydraulic conductivity of compacted fine-grained soils. Can Geotech J 46(8):955–968
21. Galal A, Talal M, Moustafa A (2022) Applications of machine learning in metabolomics: disease modeling and classification. Front Genet 13:1017340
22. da Gama JT (2023) The role of soils in sustainability, climate change, and ecosystem services: challenges and opportunities. Ecologies 4(3):552–567
23. Gavrilescu M (2021) Water, soil, and plants interactions in a threatened environment. Water 13(19):2746
24. Geisen S, Briones MJ, Gan H, Behan-Pelletier VM, Friman VP, de Groot GA, Wall DH (2019) A methodological framework to embrace soil biodiversity. Soil Biol Biochem 136:107536
25. Ghanbarian-Alavijeh B, Liaghat AM, Sohrabi S (2010) Estimating saturated hydraulic conductivity from soil physical properties using neural networks model. Int J Environ Ecol Eng 4(2):58–63

26. Ghiasi B, Niksokhan MH, Mahdavi Mazdeh A (2020) Effect of bentonite particles' presence on two-dimensional chromium transmission. Environ Sci Pollut Res 27:21692–21701
27. Guerra CA, Berdugo M, Eldridge DJ, Eisenhauer N, Singh BK, Cui H, Delgado-Baquerizo M (2022) Global hotspots for soil nature conservation. Nature 610(7933):693–698
28. Gyan GS, Verma CL, Denis DM (2023) A new model to measure in situ saturated hydraulic conductivity using auger hole with flat bottom: model to measure in situ saturated hydraulic conductivity. J Soil Salin Water Qual 15(1):52–60
29. He J, Liu H (2009). The application of dynamic K-means clustering algorithm in the center selection of RBF neural networks. In: 2009 Third International conference on genetic and evolutionary computing. IEEE, pp 488–491
30. Hosseini E, Zarei M, Moosavi AA, Ghasemi-Fasaei R, Baghernejad M, Mozaffari H (2024) Feasibility of Vis-NIR spectroscopy approach to predict soil biological attributes in arid land soils. PLoS ONE 19(9):e0311122
31. Hu T, Kang S, Li F, Zhang J (2011) Effects of partial root-zone irrigation on hydraulic conductivity in the soil–root system of maize plants. J Exp Bot 62(12):4163–4172
32. Iliyasu AM, Mayet AM, Hanus R, El-Latif AAA, Salama AS (2022) Employing GMDH-type neural network and signal frequency feature extraction approaches for detection of scale thickness inside oil pipelines. Energies 15(12):4500
33. Ivakhnenko AG (1968) The group method of data of handling; A rival of the method of stochastic approximation. Soviet Automatic Control 13:43–55
34. Jasoliya D, Untaroiu A, Untaroiu C (2024) A review of soil modeling for numerical simulations of soil-tire/agricultural tools interaction. J Terrramech 111:41–64
35. Jian J, Shiklomanov A, Shuster WD, Stewart RD (2021) Predicting near-saturated hydraulic conductivity in urban soils. J Hydrol 595:126051
36. Jin H, Kim YG, Jin Z, Rushchitc AA, Al-Shati AS (2022) Optimization and analysis of bioenergy production using machine learning modeling: Multi-layer perceptron, Gaussian processes regression, K-nearest neighbors, and Artificial neural network models. Energy Rep 8:13979–13996
37. Jing F, Chen X, Yang Z, Guo B (2018) Heavy metals status, transport mechanisms, sources, and factors affecting their mobility in Chinese agricultural soils. Environ Earth Sci 77:1–9
38. Kamalabady AS, Salahshoor K (2008) New siso and miso adaptive nonlinear predictive controllers based on self organizing rbf neural networks. In: 2008 3rd international symposium on communications, control and signal processing. IEEE, pp 703–708
39. Kashani MH, Ghorbani MA, Shahabi M, Naganna SR, Diop L (2020) Multiple AI model integration strategy—application to saturated hydraulic conductivity prediction from easily available soil properties. Soil Tillage Res 196:104449
40. Kheirabadi M, Niksokhan MH, Omidvar B (2017) Colloid-associated groundwater contaminant transport in homogeneous saturated porous media: mathematical and numerical modeling. Environ Model Assess 22:79–90
41. Kong F, Zhang Z, Liu Y (2008) Selection of logistics service provider based on fuzzy RBF neural networks. In: 2008 IEEE International conference on automation and logistics. IEEE, pp 1189–1192
42. Kotlar AM, Iversen BV, de Jong Van Lier Q (2019) Evaluation of parametric and nonparametric machine-learning techniques for prediction of saturated and near-saturated hydraulic conductivity. Vadose Zone J 18(1):1–13
43. Lamorski K, Pachepsky Y, Sławiński C, Walczak RT (2008) Using support vector machines to develop pedotransfer functions for water retention of soils in Poland. Soil Sci Soc Am J 72(5):1243–1247
44. Lebron I, Schaap MG, Suarez DL (1999) Saturated hydraulic conductivity prediction from microscopic pore geometry measurements and neural network analysis. Water Resour Res 35(10):3149–3158
45. Li D, Moghaddam MR, Monjezi M, Jahed Armaghani D, Mehrdanesh A (2020) Development of a group method of data handling technique to forecast iron ore price. Appl Sci 10(7):2364

46. Li H, Wu Y, Liu S, Xiao J, Zhao W, Chen J, Cao Y (2022) Decipher soil organic carbon dynamics and driving forces across China using machine learning. Glob Change Biol 28(10):3394–3410
47. Li C, Zhou K, Qin W, Tian C, Qi M, Yan X, Han W (2019) A review on heavy metals contamination in soil: effects, sources, and remediation techniques. Soil Sediment Contamin Int J 28(4):380–394
48. Liu G, Zha XY, Guan JK, Tong FG (2021) Field experiment of rainfall infiltration on a soil slope and simulations based on a water-air two-phase flow model. J Mt Sci 18(8):2159–2167
49. Liu X, Zhu Y, Bennett JM, Wu L, Li H (2022) Effects of sodium adsorption ratio and electrolyte concentration on soil saturated hydraulic conductivity. Geoderma 414:115772
50. Lybrand RA (2023) Connecting soils to life in conservation planning, nutrient cycling, and planetary science. Earth Sci Rev 237:104247
51. Ma J, Guo H, Lei M, Li Y, Weng L, Chen Y, Xiu W (2018) Enhanced transport of ferrihydrite colloid by chain-shaped humic acid colloid in saturated porous media. Sci Total Environ 621:1581–1590
52. Markopoulos AP, Georgiopoulos S, Manolakos DE (2016) On the use of back propagation and radial basis function neural networks in surface roughness prediction. J Indus Eng Int 12:389–400
53. Matula S, Miháliková M, Lufinková J, Bát'ková K (2015) The role of the initial soil water content in the determination of unsaturated soil hydraulic conductivity using a tension infiltrometer. Plant Soil Environ 61(11):515–521
54. Merdun H, Çınar Ö, Meral R, Apan M (2006) Comparison of artificial neural network and regression pedotransfer functions for prediction of soil water retention and saturated hydraulic conductivity. Soil Tillage Res 90(1–2):108–116
55. Mojid MA, Hossain ABMZ, Ashraf MA (2019) Artificial neural network model to predict transport parameters of reactive solutes from basic soil properties. Environ Pollut 255:113355
56. Montazer GA, Giveki D (2015) An improved radial basis function neural network for object image retrieval. Neurocomputing 168:221–233
57. Montazer GA, Giveki D, Karami M, Rastegar H (2018) Radial basis function neural networks: a review. Comput Rev J 1(1):52–74
58. Moosavi AA, Nematollahi MA, Omidifard M (2024) Comparing machine learning approaches for estimating soil saturated hydraulic conductivity. PLoS ONE 19(11):e0310622
59. Moosavi AA, Nematollahi MA, Rahimi M (2021) Predicting water sorptivity coefficient in calcareous soils using a wavelet–neural network hybrid modeling approach. Environ Earth Sci 80(6):226
60. Moosavi AA, Sepaskhah A (2012) Artificial neural networks for predicting unsaturated soil hydraulic characteristics at different applied tensions. Arch Agron Soil Sci 58(2):125–153
61. Morbidelli R (2020) On the determination of soil hydraulic properties. J Hydrol 580:124362
62. Mozaffari H, Moosavi AA, Baghernejad M, Cornelis W (2024) Revisiting soil texture analysis: introducing a rapid single-reading hydrometer approach. Measurement 228:114330
63. Mozaffari H, Moosavi AA, Cornelis W (2023) Vis-NIR-spectroscopy-and loss-on-ignition-based functions to estimate organic matter content of calcareous soils. Arch Agron Soil Sci 69(6):962–980
64. Mozaffari H, Moosavi AA, Nematollahi MA (2024) Predicting saturated and near-saturated hydraulic conductivity using artificial neural networks and multiple linear regression in calcareous soils. PLoS ONE 19(1):e0296933
65. Mozaffari H, Moosavi AA, Nematollahi MA, Ostovari Y (2024) Potential of spectroscopy-based approaches for predicting soil erosion-related parameters: a short review. In: Pourghasemi HR, Kariminejad N (eds) Advanced tools for studying soil erosion processes. Elsevier, pp 397–411
66. Mozaffari H, Moosavi AA, Ostovari Y (2024) Feasibility of proximal sensing for predicting soil loss tolerance. CATENA 247:108503
67. Mozaffari H, Moosavi AA, Ostovari Y, Cornelis W (2022) Comparing visible-near-infrared spectroscopy with classical regression pedotransfer functions for predicting near-saturated and saturated hydraulic conductivity of calcareous soils. J Hydrol 613:128412

68. Mozaffari H, Moosavi AA, Ostovari Y, Nematollahi MA, Rezaei M (2022) Developing spectrotransfer functions (STFs) to predict basic physical and chemical properties of calcareous soils. Geoderma 428:116174
69. Mozaffari H, Moosavi AA, Sepaskhah AR (2021) Land use-dependent variation of near-saturated and saturated hydraulic properties in calcareous soils. Environ Earth Sci 80:769
70. Mozaffari H, Moosavi AA, Sepaskhah A, Cornelis W (2022) Long-term effects of land use type and management on sorptivity, macroscopic capillary length and water-conducting porosity of calcareous soils. Arid Land Res Manag 36(4):371–397
71. Mulashani AK, Shen C, Nkurlu BM, Mkono CN, Kawamala M (2022) Enhanced group method of data handling (GMDH) for permeability prediction based on the modified Levenberg Marquardt technique from well log data. Energy 239:121915
72. Mumali F (2022) Artificial neural network-based decision support systems in manufacturing processes: a systematic literature review. Comput Ind Eng 165:107964
73. Nasir V, Nourian S, Avramidis S, Cool J (2019) Prediction of physical and mechanical properties of thermally modified wood based on color change evaluated by means of "group method of data handling" (GMDH) neural network. Holzforschung 73(4):381–392
74. Nematollahi MA, Jamali B, Hosseini M (2020) Fluid velocity and mass ratio identification of piezoelectric nanotube conveying fluid using inverse analysis. Acta Mech 231(2):683–700
75. Nematollahi MA, Mousavi Khaneghah A (2019) Neural network prediction of friction coefficients of rosemary leaves. J Food Process Eng 42(6):e13211
76. Nourani V, Ojaghi A, Zhang Y (2022) Saturated and unsaturated seepage analysis of earth-fill dams using fractal hydraulic conductivity function and its verification. J Hydrol 612:128302
77. Pakjoo M, Moosavi AA, Mozaffari H, Karami A, Ostovari Y (2024) Modeling spatial variability of soil loss tolerance (T-value) using geostatistical approaches (case study: Dorudzan Watershed, Fars Province, Iran). In: Pourghasemi HR, Kariminejad N (eds) Advanced tools for studying soil erosion processes. Elsevier, pp 369–393
78. Parasuraman K, Elshorbagy A, Si BC (2006) Estimating saturated hydraulic conductivity in spatially variable fields using neural network ensembles. Soil Sci Soc Am J 70(6):1851–1859
79. Pekel E (2020) Estimation of soil moisture using decision tree regression. Theoret Appl Climatol 139(3):1111–1119
80. Peng S, Wu D, Ge Z, Tong M, Kim H (2017) Influence of graphene oxide on the transport and deposition behaviors of colloids in saturated porous media. Environ Pollut 225:141–149
81. Qaderi K, Jalali V, Etminan S, Masoumi Shahr-babak M, Homaee M (2018) Estimating soil hydraulic conductivity using different data-driven models of ANN, GMDH and GMDH-HS. Paddy Water Environ 16(4):823–833
82. Rahman S, Jama Ali A, Raihan A (2022) Soil carbon sequestration in agroforestry systems as a mitigation strategy of climate change: a case study from Dinajpur, Bangladesh. Adv Environ Eng Res 3(4):15
83. Rahmati M, Weihermüller L, Vanderborght J, Pachepsky YA, Mao L, Sadeghi SH, Vereecken H (2018) Development and analysis of the soil water infiltration global database. Earth Syst Sci Data 10(3):1237–1263
84. Rezaei Arshad R, Sayyad G, Mosaddeghi M, Gharabaghi B (2013) Predicting saturated hydraulic conductivity by artificial intelligence and regression models. Int Scholar Res Notices 2013(1):308159
85. Rogiers B, Mallants D, Batelaan O, Gedeon M, Huysmans M, Dassargues A (2012) Estimation of hydraulic conductivity and its uncertainty from grain-size data using GLUE and artificial neural networks. Math Geosci 44:739–763
86. Sarker IH (2021) Machine learning: algorithms, real-world applications and research directions. SN Comput Sci 2(3):160
87. Schaap MG, Leij FJ (1998) Using neural networks to predict soil water retention and soil hydraulic conductivity. Soil Tillage Res 47(1–2):37–42
88. Schaap MG, Leij FJ, Van Genuchten MT (1998) Neural network analysis for hierarchical prediction of soil hydraulic properties. Soil Sci Soc Am J 62(4):847–855

89. Schröder JJ, Schulte RPO, Creamer RE, Delgado A, Van Leeuwen J, Lehtinen T, Wall DP (2016) The elusive role of soil quality in nutrient cycling: a review. Soil Use Manag 32(4):476–486
90. Sedaghat A, Bayat H, Safari Sinegani AA (2016) Estimation of soil saturated hydraulic conductivity by artificial neural networks ensemble in smectitic soils. Eurasian Soil Sci 49:347–357
91. Shams Emamzadeh ES, Soltani J, Mashal M, Kalanaki M, Asadolahzadeh T (2016) Performance evaluation of MLP and RBF neural networks to estimate the soil saturated hydraulic conductivity. Mod Appl Sci 11(3):1
92. Sharma PK, Mayank M, Ojha CSP, Shukla SK (2020) A review on groundwater contaminant transport and remediation. ISH J Hydraulic Eng 26(1):112–121
93. Sihag P (2018) Prediction of unsaturated hydraulic conductivity using fuzzy logic and artificial neural network. Model Earth Syst Environ 4:189–198
94. Siqueira RG, Moquedace CM, Fernandes-Filho EI, Schaefer CE, Francelino MR, Sacramento IF, Michel RF (2024) Modelling and prediction of major soil chemical properties with Random Forest: machine learning as tool to understand soil-environment relationships in Antarctica. CATENA 235:107677
95. Snousy MG, Rashad AM, Ebiad MAES, Helmy HM, Abd El Bassier MA (2018) Lead and associated micropollutant propagations in the North Suez Gulf, Egypt. Int J Environ Res 12:357–371
96. Speight JG (2019).Natural water remediation: chemistry and technology. Butterworth-Heinemann
97. Tamari S, Wösten JHM, Ruiz-Suarez JC (1996) Testing an artificial neural network for predicting soil hydraulic conductivity. Soil Sci Soc Am J 60(6):1732–1741
98. Tayfur G, Nadiri AA, Moghaddam AA (2014) Supervised intelligent committee machine method for hydraulic conductivity estimation. Water Resour Manage 28:1173–1184
99. Trejo-Alonso J, Fuentes C, Chávez C, Quevedo A, Gutierrez-Lopez A, González-Correa B (2021) Saturated hydraulic conductivity estimation using artificial neural networks. Water 13(5):705
100. Usoh GA, Ahaneku IE, Horsfall IT, Alaneme GU, Itam DH (2022) Numerical modeling and simulation of leachate transport in MSW contaminated soil: Impact on seasonal changes. Clean Mater 4:100089
101. Wang J, Han Z, Zhang C, Wang M, Li H, Gao D (2024) Effects of soil colloids on adsorption and migration of benzo (a) pyrene on contaminated sites under runoff infiltration processes. Environ Pollut 353:124150
102. Wang Y, Ma R, Zhu G (2023) Representation of the influence of soil structure on hydraulic conductivity prediction. J Hydrol 619:129330
103. Wijerathna-Yapa A, Pathirana R (2022) Sustainable agro-food systems for addressing climate change and food security. Agriculture 12(10):1554
104. Williams CG, Ojuri OO (2021) Predictive modelling of soils' hydraulic conductivity using artificial neural network and multiple linear regression. SN Appl Sci 3(2):152
105. Wu Y, Wang H, Zhang B, Du KL (2012) Using radial basis function networks for function approximation and classification. Int Scholar Res Notices 2012(1):324194
106. Yamaç SS, Negiş H, Şeker C, Memon AM, Kurtuluş B, Todorovic M, Alomair G (2022) Saturated hydraulic conductivity estimation using artificial intelligence techniques: a case study for calcareous alluvial soils in a semi-arid region. Water 14(23):3875
107. Yang J, Wang X, Wang R, Wang H (2020) Combination of convolutional neural networks and recurrent neural networks for predicting soil properties using Vis–NIR spectroscopy. Geoderma 380:114616
108. Yang Y, Wendroth O, Kreba S, Liu B (2019) Estimating near-saturated soil hydraulic conductivity based on its scale-dependent relationships with soil properties. Vadose Zone J 18(1):1–15
109. Yünkül K, Karaçor F, Gürbüz A, Budak TÖ (2024) Prediction of the undrained shear strength of remolded soil with non-linear regression, fuzzy logic, and artificial neural network. J Mt Sci 21(9):3108–3122

110. Zahedifar M (2017) Sequential extraction of zinc in the soils of different land use types as influenced by wheat straw derived biochar. J Geochem Explor 182:22–31
111. Zahedifar M (2020) Effect of biochar on cadmium fractions in some polluted saline and sodic soils. Environ Manage 66(6):1133–1141
112. Zahedifar M (2023) Assessing alteration of soil quality, degradation, and resistance indices under different land uses through network and factor analysis. CATENA 222:106807
113. Zahedifar M, Moosavi AA (2017) Modeling desorption kinetics of the native and applied zinc in biochar-amended calcareous soils of different land uses. Environ Earth Sci 76:1–11
114. Zainuddin Z, Lye WK (2010) Improving RBF networks classification performance by using K-harmonic means. Int J Comput Inform Eng 4(2):335–339
115. Zhai Q, Rahardjo H, Satyanaga A, Zhu Y, Dai G, Zhao X (2021) Estimation of wetting hydraulic conductivity function for unsaturated sandy soil. Eng Geol 285:106034

Chapter 15
Statistical AI Models for Environmental Sustainability: ARIMA, LSTM, and CNN-LSTM in Climate Prediction

Serifat A. Folorunso, Saheed A. Afolabi, Olawale Dele Osanyintupin, Morufu Aderemi Folorunso, Richard O. Kehinde, Aliu Adeniran Adebiyi, Abdulsamad Adeola Folorunso, Sukurat Salam, and Timothy Tomiwa Oyelakin

Abstract **Introduction**: This study analyzes long-term relative humidity (RH) data from the Des Moines Station (DSM), Iowa, USA, from 1936 to 2024, using traditional and artificial intelligence (AI) approaches to improve forecasting accuracy. Understanding RH trends using AI is crucial for climate science, agriculture, and public health.

S. A. Folorunso (✉)
School of Computing, Engineering and Digital Technologies, Teesside University, Middlesbrough, UK
e-mail: serifatf005@gmail.com

S. A. Afolabi
Department of Industrial and Systems Engineering, College of Computing and Mathematics, King Fahd University of Petroleum and Minerals, Dhahran, Saudi Arabia
e-mail: g202115770@kfupm.edu.sa

O. D. Osanyintupin · A. A. Adebiyi
Department of Statistics, University of Ibadan, Ibadan, Nigeria

M. A. Folorunso
Department of General Studies, Federal School of Statistics, Ibadan, Nigeria
e-mail: folorunsomoruf@fssibadan.edu.ng

R. O. Kehinde
Department of Statistics, Federal College of Animal Health and Production Technology, Ibadan, Nigeria

A. A. Folorunso
Department of Data and Information Science, University of Ibadan, Ibadan, Nigeria

S. Salam
Department for Works and Pension, London, UK

T. T. Oyelakin
Department of Epidemiology and Medical Statistics, University of Ibadan, Ibadan, Nigeria

M. Nasr et al. (eds.), *Artificial Intelligence Applications for a Sustainable Environment*, Green Chemistry and Sustainable Technology,
https://doi.org/10.1007/978-3-031-91199-6_15

Objectives: The research aims to predict long-term RH changes using both conventional statistical and advanced AI techniques. Key objectives include preprocessing daily RH data into a monthly time series format; applying ARIMA as a baseline model for linear trend forecasting; implementing deep learning models, LSTM and Bidirectional LSTM (Bi-LSTM), for nonlinear pattern analysis; using CNN to identify short-term variations though pattern recognition and developing hybrid models (CNN-LSTM and CNN-BiLSTM) for enhanced accuracy; and evaluating model performance using MAE, RMSE, MAPE, and R-squared metrics to identify the most effective model for predicting RH trends.

Methodology: The study converted daily RH data into a monthly series, using ARIMA as a baseline. LSTM, Bi-LSTM, CNN, and hybrid CNN-LSTM models were applied to capture both short-term and long-term patterns. Model accuracy was evaluated using MAE, RMSE, MAPE, and R-squared.

Findings: Results showed that hybrid AI techniques CNN-LSTM and CNN-Bi-LSTM models outperformed ARIMA, accurately capturing RH trends. This approach offers valuable insights for environmental monitoring and sectors sensitive to humidity changes.

Keywords Relative humidity · ARIMA · LSTM · CNN · Environmental monitoring

15.1 Introduction

Environmental science, as defined by [1], is an interdisciplinary field integrating natural and social sciences to study the Earth's environment and the dynamic interactions between natural systems and human activities. This discipline draws on concepts from biology, chemistry, geology, and social sciences to address pressing global challenges, including climate change, pollution, and biodiversity loss. Environmental scientists strive to optimize resource utilization while minimizing adverse effects on ecosystems and human health [2].

In recent years, deep learning has emerged as a transformative tool in environmental science, offering advanced capabilities for analyzing complex and voluminous datasets [1]. According to Raihan [3], applications of deep learning in this field span diverse areas, from remote sensing and climate modeling to conservation biology and pollution management. For instance, neural networks can process satellite imagery to detect environmental changes such as deforestation and land use shifts [4]. Additionally, these models enhance climate predictions by analyzing historical data to uncover inherent patterns, thereby improving the accuracy of forecasts for temperature and rainfall. In conservation biology, deep learning techniques facilitate species monitoring and biodiversity assessments by automating the analysis of ecological data, such as thermograms and camera trap images [5].

Deep learning plays a pivotal role in identifying pollution sources and monitoring environmental hazards. By integrating data from sensors and satellite imagery, neural networks provide real-time insights into pollution levels, facilitating timely interventions and informed policy adjustments [6]. In agriculture, deep learning algorithms enhance resource optimization by analyzing drone and satellite imagery to guide decisions on crop health, soil conditions, and water requirements [7]. These applications demonstrate how deep learning not only advances scientific understanding but also promotes sustainable development and efficient resource management. Additionally, authors such as [8] propose exploring neural networks within deep learning to evaluate facial expressions linked to health conditions, including traits associated with autism spectrum disorder (ASD).

Climate change remains a critical challenge in environmental science, with profound effects on human health, agriculture, and ecosystems. Factors such as rising greenhouse gas levels, extreme weather events, and shifting precipitation patterns intensify health risks, including heat stress, respiratory issues, and the spread of infectious diseases [9]. Relative humidity (RH), a key climatic variable, plays a significant role in these processes by influencing atmospheric moisture, cloud formation, and precipitation. These impacts extend to agriculture, public health, and disaster management. Furthermore, the environment's essential role in sustaining human existence highlights the need for careful and informed stewardship of natural resources, particularly in relation to human activities. Understanding the environment is crucial for both physical and emotional well-being [10].

Given its importance, accurate RH prediction is essential for climate resilience and environmental sustainability. This study explores the integration of traditional statistical methods and advanced AI techniques, including ARIMA, LSTM, and hybrid models like CNN-LSTM, to forecast long-term RH trends using data from Des Moines Station, Iowa (1936–2024). By evaluating model performance through metrics such as MAE, RMSE, and R-squared, this research identifies optimal approaches for environmental monitoring and policy formulation. The findings aim to enhance climate prediction accuracy, inform sustainable practices, and contribute to global efforts in combating climate change.

Therefore, the following are the specific objectives of this work include to transform daily RH data into a monthly time series data, to model RH data with the known ARIMA model for capturing linear changes and to use LSTM and Bi-LSTM for capturing the nonlinear trends in RH data. Furthermore, the research also intends to utilize CNN for detecting short term fluctuations and to create new models based on CNN with LSTM and Bi-LSTM methods to improve the forecasting precision. These objectives bring out the study's focus of promoting climate prediction approaches and improving the environmental conservation practices.

15.2 Literature Review

15.2.1 Insights into Relative Humidity Data

Atmospheric scientists use a variety of variables to characterize atmospheric moisture based on research goals, with relative humidity (RH) being a critical measure. However, its dependency on temperature and its diurnal and seasonal nature often limits its standalone application in environmental or epidemiological research [11]. RH quantifies the air's proximity to saturation and is linked to climatic impacts on mortality, morbidity, heat stress, and hydration levels [11].

Examining long-term trends in RH is crucial for understanding interannual-to-decadal climate variations. Long-term meteorological series provide insights into climate variability but face challenges such as spatial data gaps and measurement discontinuities [12]. Recent efforts to merge satellite records have improved global coverage of climate parameters, including RH, yet the mechanisms driving long-term variability remain poorly understood [13, 14].

For instance, Fernandez-Duque et al. [15] analyzed temporal patterns in RH and found distinct seasonal trends. While annual RH trends were largely nonsignificant, significant patterns emerged during specific months, suggesting the role of local climatic and topographical influences. These findings underscore the need to explore elevation-dependent climate trends and their implications for long-term variability.

Short-term climatic oscillations also play a pivotal role in climate dynamics. Studies such as that by [16] illustrate how oscillations influence environmental processes, including sediment transport and water cycles, revealing hydroclimatic controls on ecosystem dynamics.

New perspectives on climatic variances highlight opportunities for future research, such as investigating the interplay of elevation gradients and monthly climatic patterns to advance our understanding of RH variability [17].

15.2.2 Internet of Things (IoT) and Climate Change Environmental Applications

The Internet of Things (IoT) plays a transformative role in addressing climate change and promoting environmental sustainability by enabling real-time monitoring, data collection, and adaptive interventions. These applications leverage real-time data collection and advanced analytics to enhance environmental resilience and resource management.

IoT technologies have proven invaluable for real-time environmental monitoring, offering innovative solutions to mitigate climate change and environmental challenges. In Bangladesh, IoT towers monitor critical environmental parameters such as temperature, humidity, and soil moisture, providing timely alerts for extreme weather events [18]. Similarly, IoT-enabled sensor networks are employed for forest monitoring, where they collect data on temperature and air quality. This technology is instrumental in detecting forest fires and illegal logging activities, enabling prompt interventions to protect ecosystems [19].

In the realm of predictive analytics and resource management, the integration of IoT with machine learning has enhanced adaptive resource allocation and predictive modeling, leading to improved climate change management and response strategies [20]. IoT-powered frameworks for ecosystem-based adaptation continuously monitor and assess the effectiveness of strategies over time, ensuring long-term resilience to climate challenges [21, 22]. Additionally, water resource management has benefited from IoT applications such as ConnecSenS, which monitors water systems and integrates the data into cloud platforms, making it accessible for analysis by diverse stakeholders [23]. Despite these advancements, challenges remain in implementing IoT solutions for environmental applications. Issues such as data privacy, system integration, and the need for robust infrastructure must be addressed to fully realize the potential of these technologies in combating climate change and promoting sustainability.

15.2.3 Predictive Modeling of Long-Term Trends: ARIMA

Predictive models like ARIMA are foundational for analyzing and forecasting time series data. By integrating autoregressive and moving average approaches, ARIMA provides a robust framework for capturing linear trends and seasonal patterns [24]. Its simplicity and reliability make it a widely adopted model in environmental applications, including climate, pollution, and resource management [25].

Jaiswal et al. [26] demonstrated ARIMA's efficacy in analyzing air quality trends, identifying it as the optimal model for forecasting pollutant levels. Similarly, Kumar et al. [19] applied ARIMA to forecast ozone concentrations, showcasing its precision for short-term environmental predictions.

Despite its strengths, ARIMA has limitations in modeling non-linear trends and seasonal behaviors, which are integral to real-world climatic systems. Advanced derivatives, such as seasonal ARIMA, address some of these challenges, while hybrid models combining ARIMA with machine learning techniques enhance its capability for complex environmental forecasting [27].

15.2.4 Convolutional Neural Networks (CNNs) for Short-Term Variation

Artificial intelligence (AI) has revolutionized data analysis across diverse fields, with Convolutional Neural Networks (CNNs) emerging as a powerful tool for pattern recognition and time series forecasting [28]. Originally developed for image recognition, CNNs have demonstrated their utility in forecasting tasks, outperforming traditional models like ARIMA in capturing non-linear and complex data dynamics [29, 30].

Emerging studies have applied CNNs to environmental monitoring, such as predicting climate parameters like RH. Accurate RH forecasts are critical for addressing climate impacts on agriculture, public health, and ecosystem stability. By leveraging CNNs' ability to identify short-term oscillations, this study proposes an innovative approach to enhance climate forecasting accuracy.

15.3 Methodology

15.3.1 Data Source and Description

The data for the study were downloaded from the Iowa Environmental Mesonet of the Iowa State University through https://mesonet.agron.iastate.edu/request/daily.phtml. The original dataset was collected through Des Moines Station (DSM), Iowa State in the United States of America. The dataset is a multidimensional relative humidity (RH) dataset. The dataset has 32,462 rows and different RH attributes which involve features like minimum, average, and maximum RH. Following the discussions in the Literature Review, RH was chosen for its sensitivity to temperature and moisture changes, making it a reliable indicator for monitoring climate variability.

15.3.2 Analytical Framework

The methodology employs a hybrid modeling approach, combining autoregressive integrated moving average (ARIMA) models with advanced deep learning techniques, specifically convolutional neural networks (CNNs) and long short-term memory networks (LSTMs).

15.3.2.1 ARIMA for Long-Term Trends

Overview of ARIMA Model

The first model developed for time series analysis was the autoregression model by [15], which serve as building block for the ARIMA model proposed by statistician George Box and Gwilym Jenkins in 1970 where the name Box–Jenkins method was formed for the model. The simplest form of the ARIMA model is best explained in three main components.

Autoregressive (AR): In this model, the variable of interest is forecast based on the linear combination of the past variables, this approach is similar to linear regression which uses its past values as an input to the regression [31]. This approach captures the dependency between an observation and a number of previous time steps (called lagged values). This can be denoted with the following equation.

$$y_t = C + \#_1 y_{t-1} + \#_2 y_{t-2} + \ldots \#_q y_{t-q} + \epsilon_t \tag{15.1}$$

where the ϵ_t represents the white noise and the represent the predicted/output values of the q lagged/previous values $y_{t\text{-}q}$ predictors. This whole of the above equation refers to AR(p), the Autoregressive model of order p.

Integrated (I): This denotes the raw data differencing of raw observation to make the time series stationary to eliminate trends of seasonality [32].

Moving Average (MA): This model uses the past error rather than the past value of the forecast values in the autoregressive model. The forecast values were calculated using the actual values at each step in the time series. It measures the dependency between an observation and the residual error from a moving average model applied to lagged observations [33]. This is referred to as MA(q), the moving average of order q, and is defined mathematically in the equation below:

$$y_t = C + \epsilon_t + \theta_1 \epsilon_{t-1} + \theta_2 \epsilon_{t-2} + \ldots \theta_q \epsilon_{t-q} \tag{15.2}$$

ARIMA models are often denoted as ARIMA (p, d, q), where:

- **p**: Order of the autoregressive part.
- **d**: Degree of differencing required to make the data stationary.
- **q**: Order of the moving average part.

ARIMA can identify and forecast linear trends which is an essential approach for understanding the dynamics of various factors when applied to environmental data [34]. This includes the long-term trends in temperature increase due to climate change [34], air pollution levels in urban areas [26], or the consumption of natural resources can be effectively modeled using ARIMA [35] among others. By serving as a baseline model, ARIMA provides a foundation for more complex forecasting methods, such as machine learning models [27]. Jatinder et al. [36] highlighted that ARIMA processes are stochastic approaches used to investigate behavioral patterns of time series, their flexibility property made them handle pure autoregressive (AR), pure moving average (MA), and their combination ARMA series. Despite the numerous advantages of the ARIMA model, part of its limitation is that they are not able to capture the nonlinear trend of the series, which makes it not suitable for modeling real-world complex problems. Another shortcoming of the ARIMA basic model is they cannot represent seasonal behavior in time series, they can only capture the trends. Another derivative of the ARIMA (p,d,q) like ARIMA (p,d,q) (P, D, Q) can be used to model the seasonal pattern. To model complex nonlinear models, ARIMA serves as the baseline or standard model for most of the complex forecasting methods like hybrid models, mixture models, and machine learning Models, to improve forecasting performance especially when dealing with nonlinear trends by better capturing the complexities of environmental systems. ARIMA models were applied to analyze linear trends and seasonal variations in RH. As noted in the Literature Review, ARIMA is widely used in time series analysis for its robustness in capturing long-term patterns. Key steps include.

- Stationarity Check: The augmented Dickey-Fuller (ADF) test was conducted to determine data stationarity.
- Model Selection: Based on the autocorrelation function (ACF) and partial autocorrelation function (PACF) plots, the optimal p, d, q parameters were identified.
- Forecasting: The fitted ARIMA model provided baseline predictions for RH trends.

15.3.2.2 Advanced AI Techniques for Nonlinear Dynamics

To complement ARIMA, AI-based methods were employed to capture nonlinear dynamics and short-term oscillations:

1. **Convolutional Neural Network (CNNs)**: Extracted localized patterns and anomalies in RH using convolutional layers, focusing on short-term oscillations.

Methodologically, the key aspects of CNN architecture used convolution layers for recognition of sparse interaction and considerably fewer parameters. As such, the three major mechanisms of CNN are convolution, pooling, and fully connected layers [37]. Starting with the convolution layers for one-dimensional, two-dimensional, or n-dimensional cases, the convolution operations detect the local features in a vector [38] using filters or kernels across the input data. In this study, the one-dimensional convolution operation to identify short-term variation is given as follows:

$$h_t = \sum_{i=1}^{k} w_i . x_{t+i-1} + b \tag{15.3}$$

where

h_t is the output of the convolution at time step t.

w_i are the filter weights.

x_{t+i-1} is the input series, RH, at a position $t + i - 1$.

b is the bias term.

k is the kernel size or window size.

On the pooling layer, the dimension of convolutional output neurons is reduced with pooling operation, thus retaining the most significant features while reducing the computational time [37, 38]. The various types of pooling operations are max-pooling, average-pooling, min-pooling, and sum-pooling [39]. In the study with one-dimensional input as in [40], the max-pooling operation was used, and it is specified below:

$$p_t = max\{h_{\{i,t\}} : i = 1, \cdots, k\} \tag{15.4}$$

where:

p_t is the pooled output at time t

$h_{i,t}$ are the convolved features in the pooling window.

Lastly, the fully connected layer banked on the outcome of the pooling operation, $p_t p_t$, to learn higher level features. In this case, CNNs isolate short-term variations by focusing on localized patterns in the series to capture RH fluctuations that occur over shorter periods. Following the research of [41], the features relevant to the short-time spikes in humidity levels employ RELU activation function which is based on the follow below:

$$F_j = f(X * W_j + b_j) \tag{15.5}$$

where F_j is the jth feature map, f is the activation function (e.g., ReLU), X is the input RH vector, W_j is the filter weights of the hidden layers, b_j is the bias of the hidden layers and $*$ denotes the convolution operation. This layer reconstructs the RH series (the input vector) to have it stretched into one-dimensional form. Then, there is operation of matrix multiplication, with a combination of input activation function of the bias vector, between the RH series and the weight matrix. In the end, the prediction result for X is obtained from the fully connected layers.

Prior to deploying the CNN model on the RH series, it should be noted that CNN required data normalization to scale the attribute into the certain range required of the activation function. Following the previous studies [37, 41, 42], the RH attribute is normalized within [0, 1] using min–max scaler operation.

$$X_{tnorm} = \frac{X_t - X_{min}}{X_{max} - X_{min}} \tag{15.6}$$

where:

X_{tnorm} is the outcome of normalization to be used in CNN model operation.

X_t is the data to be normalized.

X_{min} is the minimum value.

X_{max} is maximum value.

2. **LSTM**: Captured Long-Term Dependencies and Temporal Sequences in RH Data, Addressing Limitations in ARIMA's Linear Assumptions.

LSTM is a recurrent neural network (RNN) that was established in 1980 [43]. A potent kind of artificial neural network (ANN), RNNs are mostly employed in solving time series forecasting problems. The ability of RNNs to internally preserve memory allows them to recall details from previous occurrences that can be used for prediction. However, vanishing and exploding gradients are usual problems with RNNs, which causes the model learning to stall or become overly slow. In addressing these issues, LSTMs were developed in 1997 [44], it can learn from inputs separated by length of time delays and longer memory (lags).

Concerning the research of [45], an LSTM has three gates which are input, forget, and output. An input gate determines whether to let the new input in, a forget gate that eliminates irrelevant information, and an output gate that chooses what information to output. The sigmoid function, which operates on the interval between 0 and 1, is fundamental for these three analogical gates. Figure 15.1 shows these sigmoid gates, while the cell state is represented by a visible horizontal line that passes through the cell. LSTM formulas are listed below [46].

$$\text{Input gate} : i_t = \sigma(Z^{(if)} x_t + Z^{(if)} h_{t-1}) \tag{15.7}$$

$$\text{Forget gate} : f_t = \sigma(Z^{(if)} x_t + Z^{(hf)} h_{t-1}) \tag{15.8}$$

$$\text{Output gate} : i_t = \sigma(Z^{(io)} x_t + Z^{(ho)} h_{t-1}) \tag{15.9}$$

$$\text{Process Input} : \underset{-t}{\underline{C}} = tanh(Z^{(ic)} x_t + Z^{(hc)} h_{t-1}) \tag{15.10}$$

$$\text{Cell update} : C_t = f_i * C_{i-1} + i_i * C_t \tag{15.11}$$

$$\text{Output} : y_t = h_i = O_i + i_i * tanh(C_t) \tag{15.12}$$

3. **Bi-LSTM**: In time-processing, standard RNN and LSTM often overlook information regarding the future, while the Bi-LSTM can take benefit of the future information. According to [45], the Bi-LSTM's fundamental structural concept is that each training sequence's front and back layers are two LSTM networks respectively. Moreover, one input layer and one output layer are coupled to both LSTM networks. Through the illustrated structure in Fig. 15.2, the output layer can acquire past information about each point from the input sequence and future details on each end.

Hybrid Model: Combined CNN's feature extraction capabilities with LSTM's sequence learning to analyze RH trends more comprehensively.

4. **CNN-LSTM**: This technique consists of two distinct networks of DL, viz., the CNN-LSTM Onwards and the CNN-LSTM Backwards, which are coupled in

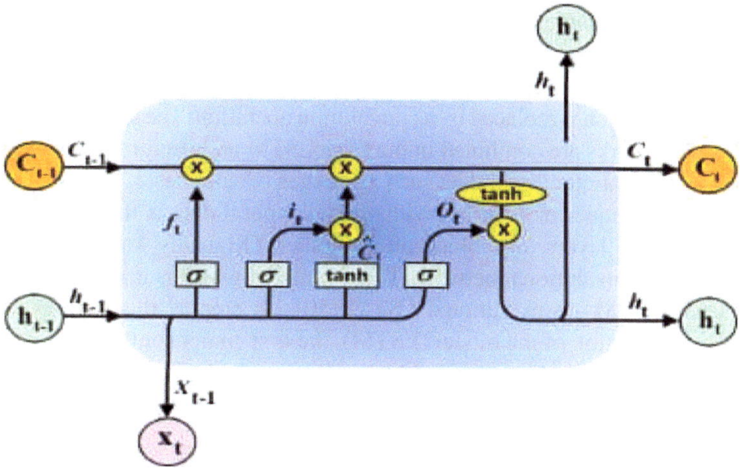

Fig. 15.1 Schematic representation of a long short-term memory (LSTM) cell, highlighting the key components: input gate, forget gate, output gate, and cell state

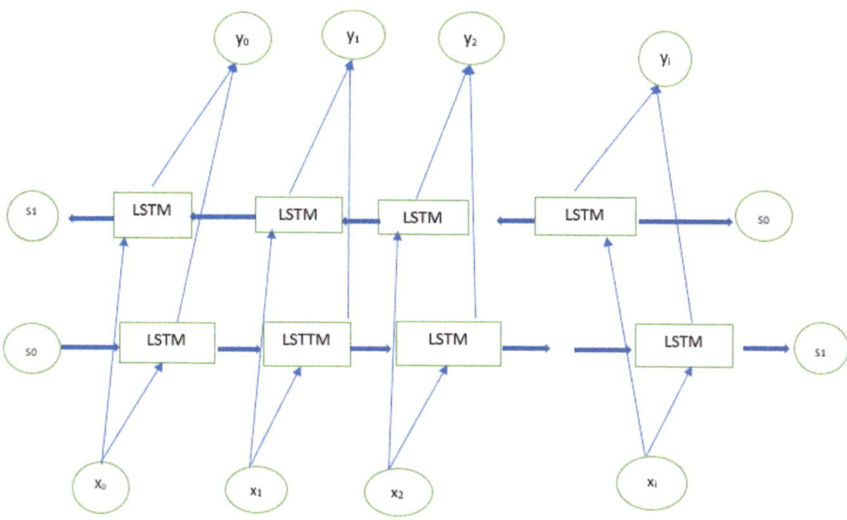

Fig. 15.2 Schematic representation of a bidirectional long short-term memory (BiLSTM) network, showing the forward and backward LSTM layers and their connections

producing the final predictions for the filling of data. Hence, both networks possess the same structure of neuron [47].

The application of the classical architecture of CNN is the best option when networks of input are 2-D or 3-D tensors such as images or videos [48]. Regarding the research of [49], the Convolutional LSTM (ConvLSTM) is a novel LSTM version created because LSTM architectures are better suited for 1-D data. The architecture of ConvLSTM retains the LSTM cell in the output layer, which includes a convolution operation and the input dimension of data. At each gate of the classical LSTM, matrix multiplication is replaced by a convolution operation. The capabilities of CNN and LSTM networks are combined in the ConvLSTM architecture. Typically, it was created for 2D spatiotemporal data such as images of satellites.

Another technique for working with spatiotemporal data is the combination of CNN and LSTM layers, one block after another. Originally known as the long-term recurrent convolutional network (LRCN) model which is now the architecture of the CNN-LSTM—convolutional LSTM [50]. To employ the findings as input for the second portion of the model (LSTM), the first component of this model uses convolutional layers to extract major features from the input data. Lastly, information must be reshaped in the original input data form before being passed into the final hidden layer [45].

15.3.3 Hybrid Model Development

The hybrid model integrated ARIMA for trend analysis with CNN-LSTM for capturing nonlinear variations. The hybridization process followed these steps:

1. **Data Preprocessing and Preparation**: The dataset used in this study consists of daily observations of average relative humidity (RH) and minimum RH spanning from January 1936 to November 2024, with a total of 32,462 records. Given its completeness (no missing values), the preprocessing steps focused on transforming the data for effective analysis and modeling. The data were prepared as described below.
2. **Data Aggregation**: The daily RH data was aggregated into monthly averages, ensuring the temporal form aligns with long-term trend analysis and model efficiency. This transformation resulted in a more manageable time series dataset suitable for detecting monthly variations in RH over nearly nine decades.
3. **Date Indexing**: A proper time-based index was assigned to the dataset, ensuring chronological order and compatibility with time series modeling techniques.
4. **Feature Engineering**: New features, such as year and month, were extracted to allow seasonal and temporal trend analysis.
5. **Data Normalization**: To enhance model performance and reduce bias, the RH values were normalized or standardized if required by the algorithms.

15.3.4 Model Performance Metrics

Model performance was evaluated using a set of statistical metrics to ensure accuracy and robustness:

1. **Mean Absolute Error (MAE)**: Measures the average absolute difference between predicted and observed values, emphasizing overall accuracy.
2. **Root Mean Squared Error (RMSE)**: Evaluates the magnitude of prediction errors, penalizing larger deviations more heavily.
3. **Mean Absolute Percentage Error (MAPE)**: Assesses model accuracy as a percentage of observed values, allowing for easier interpretation across scales.
4. **R-squared (R^2)**: Quantifies the proportion of variance in the observed data explained by the model, highlighting goodness-of-fit.

These metrics were applied to ARIMA, CNN, LSTM, and the hybrid CNN-LSTM models, with comparisons highlighting the advantages of hybridization in terms of precision and adaptability. Consistent with findings in the Literature Review, hybrid models demonstrated superior performance, particularly in handling nonlinear variability.

15.3.5 Validation and Cross-Validation

To ensure the reliability of the models:

i. **Train-Test Split**: The data were divided into training (80%), validation (10%), and testing (10%) sets.
ii. **Cross-Validation**: k-fold cross-validation was implemented for robust model assessment.
iii. **Residual Analysis**: Residuals were analyzed to confirm the absence of systematic errors.

15.4 Implementation and Results

This section provides a technical exploration of our research into the analyses of the hybrid models.

15.4.1 Integration with Climatic Insights

The hybrid methodology bridges theoretical and practical gaps, as discussed in the Introduction and Literature Review. By combining ARIMA's statistical rigor with AI models' capacity for handling complexity, the approach addresses the challenges of climate data variability. This framework not only advances the understanding of RH dynamics but also demonstrates the potential of hybrid models in environmental research. The results from this methodological framework are presented and discussed in the subsequent subsection, offering actionable insights into climate variability and sustainability.

15.4.2 Exploratory Data Analysis

Figure 15.3 above depicts the relative humidity spanning from 1936 (January) to 2024 (November). The graph shows a high degree of variability in relative humidity throughout the years. There is no clear upward or downward trend, suggesting that the relative humidity fluctuates without a consistent pattern. The fluctuations appear to be relatively small, indicating that the changes in relative humidity were not drastic.

The histogram in Fig. 15.4 shows the distribution of relative humidity over the years. The distribution shape appears to be bell-shaped, with its peak around the time 30. This suggests that humidity levels have oscillated around this mean over the period. There is a wider range of humidity values in the earlier years compared to the later years, indicating more variability in the early period. The distribution gradually declines as it moves to the right and left of the middle, which indicates that events of extremely high and low humidity are not as frequent as those of moderate humidity. Overall, the graph suggests a trend toward more stable humidity levels over time, with less extreme fluctuations in recent years.

Figure 15.5 shows the density plot of RH data. The horizontal axis corresponds to RH values, and the vertical one corresponds to the probability density. The plot of this distribution shows signs of a unimodal distribution with its peak at around 25–30. This in fact indicated the fact that most RH values fall within this range. The distribution seems to be positively skewed somewhat higher RH to the right, and the numbers to the lower right corner of the figure as $N = 889$ and $\text{Bandwidth} = 1.99$ convey information about the number of cases and the chosen density estimation bandwidth.

The graph in Fig. 15.6 provides a decomposition of an additive time series, breaking it down into its constituent components: trend, seasonal, and random. The observed series as seen above has a general tendency of an upward movement with some oscillations. The trend component gives this overall sense, smoothing out the

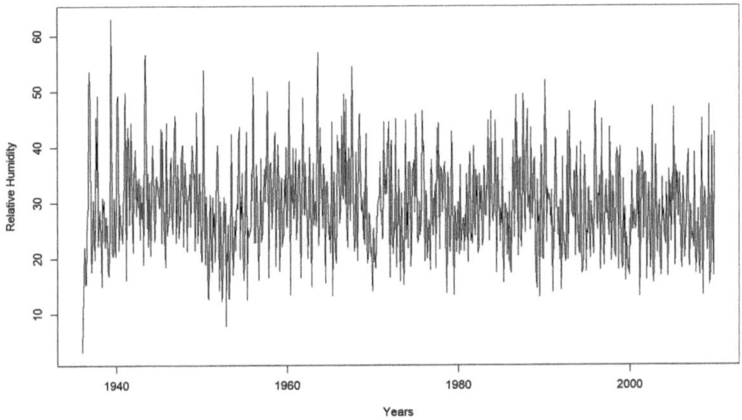

Fig. 15.3 Time plot of relative humidity

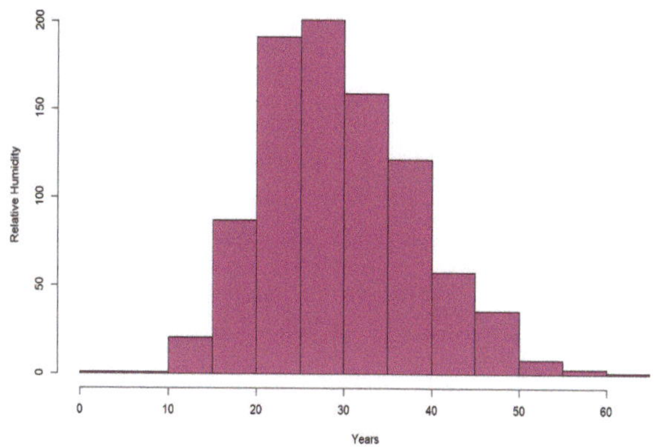

Fig. 15.4 A histogram showing the distribution of relative humidity

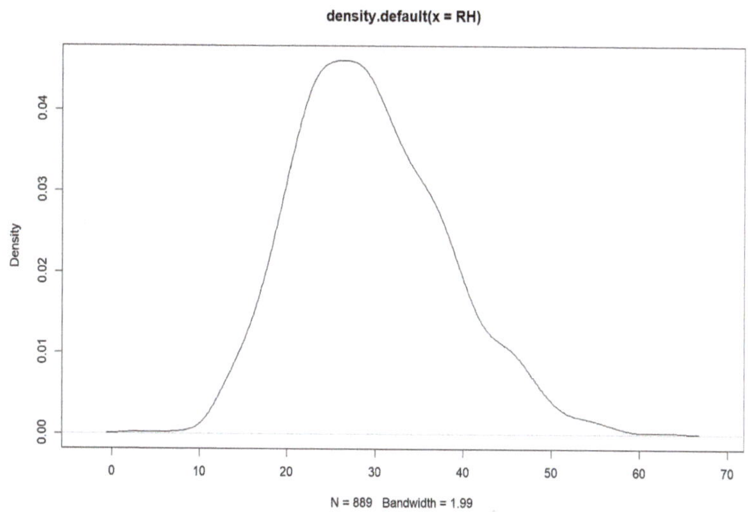

Fig. 15.5 Density plot of relative humidity

numbers to show the general movement of the data over time. Seasonal components give a cyclical feature, which may correspond to annual fluctuations. The random component gets the rest of the volatile fluctuations that cannot be associated with trends and seasons, which, therefore, comprise noise.

15.4.3 Analysis and Discussion

This section gives a discussion in relation to the findings derived from the techniques employed in establishing the sustainability of the environmental dataset.

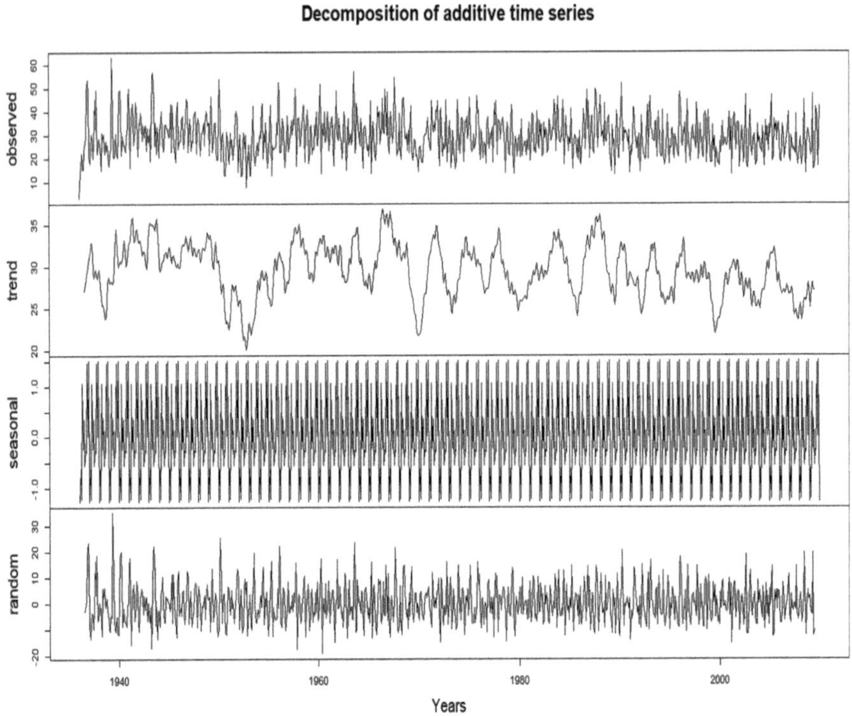

Fig. 15.6 ARIMA seasonality of relative humidity

15.4.3.1 LSTM

The performance of the long short-term memory (LSTM) model for relative humidity prediction is measured based on important evaluation metrics. Table 15.1 affirms the MAE of 5.70 and therefore means the model's prediction is on average 5.7 units away from the actual values. On the same note, an RMSE of 7.07 shows slightly larger deviations and points to some outliers or relatively large errors. A mean absolute percentage error of 22.15% indicates that the percentage of forecasts made by the model is, on average, within 22.15% of the actual value, which may be expected for time series information containing oscillations.

It can be established from Table 15.1 that the value of R-squared is 0.949 proving high reliability of the chosen LSTM model and meaning that 94.9% of fluctuations of the relative humidity depend on it. This is a very high R-squared value, indicating that much of the variability in the data is being explained by the model, nevertheless; the

Table 15.1 LSTM model metrics of relative humidity

MAE	RMSE	MAPE	R-squared
5.7024	7.0695	22.1497	0.9490

Table 15.2 Bi-LSTM model metrics of relative humidity

MAE	RMSE	MAPE	R-squared
5.6117	6.9727	21.7928	0.9520

error metrics are moderate which imply that the model is unable to capture patterns and trends in the data set. In summary, the LSTM model can be considered generally accurate at forecasting and has potential for performance refinement specifically with quantitative error reduction to increase accuracy.

15.4.3.2 Bi-LSTM

Bi-LSTM models show high accuracy in relative humidity forecast compared to LSTM and there are slight enhancements of the values for Bi-LSTM. From Table 15.2, the MAE of 5.61 shows that, in average, predictions are 5.6 units off from the true values. The RMSE of 6.97 indicates slightly greater deviations but still acceptable for the given dataset. The accuracy of the model can be described as good as the mean absolute percentage error (MAPE) of 21.79% means that the average prediction from the model is within 21.79% from the actual values.

According to evaluation in Table 15.2, the Bi-LSTM model yields an R-squared value of 0.9 which means that 95.2% variation of the data is explained by this model. All these metrics suggest that Bi-LSTM can learn both temporal features and bidirectional features in the data and gives slightly better, yet more accurate and dependable forecasts than the traditional LSTM model.

15.4.3.3 CNN

The convolutional neural network (CNN) model for capturing the short-term vertical fluctuation and local variation in the relative humidity time series is quite efficient in terms of predictive accuracy. In Table 15.3, the predicting model also has an MAE of 5.52, which means that the deviation of the estimated values from the actual ones is on average 5.5 units. Nevertheless, an RMSE of 7.12 suggests a slightly bigger deviation is present, which corresponds to the possible large prediction errors now and then. A MAPE of 22.57% indicates that predictions are, on average, at a 22.57% difference from the observed values, which can be considered good for time series data.

The extracted R-squared of 0.947 indicates that the CNN model explains 94.7% of the variances in the dataset from which it trains in Table 15.3; it shows that the

Table 15.3 CNN model metrics of relative humidity

MAE	RMSE	MAPE	R-squared
5.5185	7.1229	22.5674	0.9470

Table 15.4 Hybrid CNN-LSTM model metrics of RHs

MAE	RMSE	MAPE	R-squared
5.4181	6.9234	19.7625	0.9540

CNN model can capture local trends, and/or dependencies elegantly. As expected, while the CNN model fits the data quite well, its recurrent approach will be less suitable for modeling for longer patterns than the data than recurrent architectures like LSTM or Bi-LSTM. Nevertheless, it is a very suitable model for forecasting activities.

15.4.3.4 CNN-LSTM

The CNN-LSTM model gives a high performance for the relative humidity prediction by utilizing the CNN and LSTM properties. The model leverages the strength of CNN to capture local patterns and short temporal dependencies, and the strength of LSTM in modeling sequential and long-term dependencies, which results in high accuracy. Table 15.4 affirms an MAE of 5.42 suggesting that the predictions are on average 5.4 units off from actual values and RMSE of 6.92 indicates slightly larger deviations that are, however, still reasonable. For further details of the model's accuracy, a mean absolute percentage error of 19.76% was obtained meaning that the predictions are, on average, within 19.76% of the observed values. The model achieves a value of R-squared of 0.954 which indicates how close the actual values fitted by the model are to the real values, is 95.4% an indication that the model effectively captures the underlying patterns and trends in the data.

This proves the advantages of the hybrid approach as compared to the separate CNN and LSTM models and indicates that it may be used in effective time series forecasting. The model's integration of complementary techniques leads to successful prediction, and especially when the data we are working with involves intricate interactions, uncertainty, and dependencies over different time periods.

15.4.3.5 CNN-Bi-LSTM

Based on the results, the hybrid CNN-Bi-LSTM model shows a high level of accuracy in the relative humidity forecasting taking the advantages of convolutional neural networks (CNN) and bidirectional long short-term memory (Bi-LSTM) networks. CNN excels at modeling short-term dependencies and local patterns, while Bi-LSTM captures long-term sequential dependencies well—thus the synergy of the two approaches is best suited for time series.

In Table 15.5, when evaluated using mean absolute error (MAE), the model has a score of 5.42 which means that the disparity between estimated and true values is negligible in this case. The value of root mean square error (RMSE) is 6.77 which is a small but significant discrepancy with the data due to its complexity. Having an

Table 15.5 Hybrid CNN-Bi-LSTM model metrics of RH

MAE	RMSE	MAPE	R-squared
5.4162	6.7726	20.6357	0.96

average of 20.64%, the mean absolute percentage error (MAPE) certainty provides confidence that the model is, generally, within 20.64% of the actual values observed. Reliability of models is established by that R-squared value of 0.96 indicating that the model fits 96 percent of the variability in the dependent variable. This hybrid approach surpasses the standalone CNN and Bi-LSTM models since it takes advantage of both the local oscillating behaviors and the progressive trends, which fits the time series predictive assignment context well due to the presence of intricate patterns.

15.4.4 Model Comparisons

The evaluation of the different models for the estimation of relative humidity shows a significant divergence in the prediction ability of the models. Prediction performance of the models is quantified in terms of MAE, RMSE, and MAPE and as depicted in Table 15.6 and Fig. 15.7, the LSTM model with MAE of 5.70, RMSE of 7.07, and MAPE of 22.15% shows acceptable performance but with slightly higher error levels than other models. However, the Bi-LSTMs have a slightly better performance with an MAE of 5.61 and RMSE of 6.97, while giving a better MAPE of 21.79%, which indicates a better error ratio in comparison to the earlier models in capturing sequential dependencies. The CNN model, as a good estimator of short-term patterns, yields the highest MAE of 5.52, RMSE of 7.12, and MAPE of 22.57%.

The CNN-LSTM hybrid model is found to be better in accuracy than individual LSTM and CNN with an MAE of 5.42, RMSE of 6.92, and MAPE of 19.76%. However, the CNN-Bi-LSTM hybrid outperforms all of the other models, providing the minimum MAE of 5.42, the smallest value of RMSE 6.77, and a higher value of MAPE of 20.64% highest R-squared value of 0.96. This model performs fine in capturing short-term and long-term dependencies and is therefore the most accurate and reliable for the task, as it was seen when compared to other models.

Table 15.6 Model performance metrics

Models	Performance metrics				Remarks
	MAE	RMSE	MAPE	R^2	
LSTM	5.702	7.069	22.15	0.949	Good
Bi-LSTM	5.612	6.973	21.793	0.952	Better
CNN	5.518	7.123	22.567	0.947	Good
CNN-LSTM	5.418	6.923	19.762	0.954	Better
CNN-Bi-LSTM	**5.416**	**6.772**	**20.635**	**0.96**	**Best**

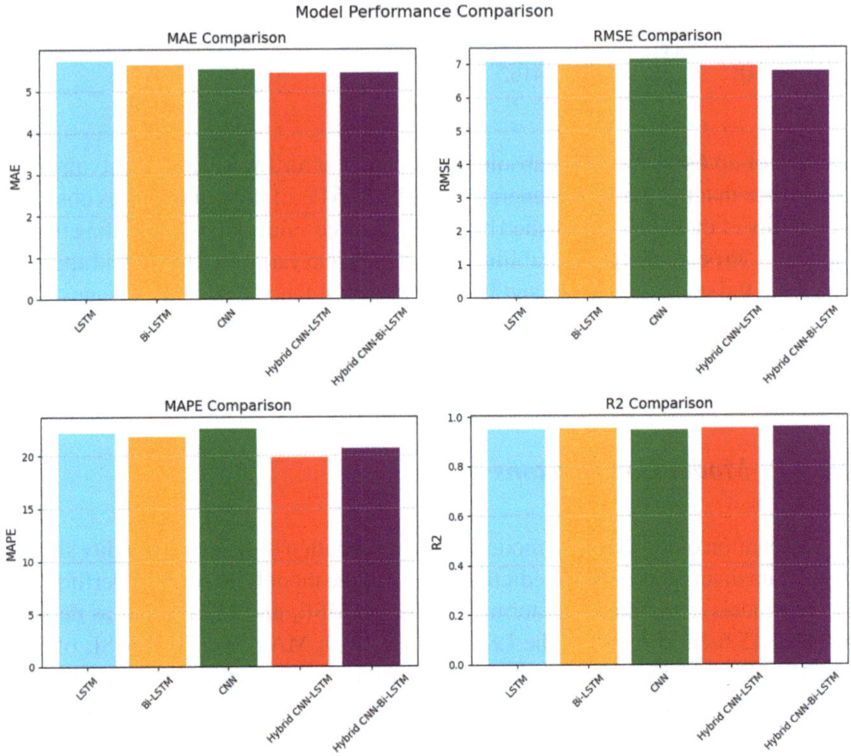

Fig. 15.7 Bar chart illustrating the performance of various machine learning models (LSTM, Bi-LSTM, CNN, hybrid architectures), as assessed by common metrics: MAE, RMSE, MAPE, and R2

15.5 Study Limitations and Recommendations

15.5.1 Limitations

The proposed AI systems to address relative humidity processes are susceptible to biases that can arise from various sources, such as biased datasets and algorithmic assumptions. These biases can compromise the accuracy of predictions and decisions, potentially leading to unfavorable outcomes in applications like environmental monitoring and climate control. Addressing these challenges requires the implementation of targeted strategies. This bias often originates from the use of historical data that may embed societal biases, resulting in skewed outputs [51, 52]. Additionally, the "black box" nature of many AI algorithms can obscure the decision-making process, making it challenging to identify and rectify biases effectively [53]. To mitigate these issues, several approaches have proven effective. For instance, employing diverse and representative datasets allows models to learn from a broader range of scenarios, reducing bias (Choudhari and Choudhari, 2024). Hyperparameter optimization is another critical strategy, as fine-tuning model parameters can improve accuracy and decrease bias by refining the model's configuration (Choudhari and Choudhari, 2024). Furthermore, random sampling techniques can counteract adversarial data tampering, enhancing the reliability of AI predictions [54]. Although these strategies can significantly reduce bias, eliminating it remains a challenge due to the inherent complexities of AI systems and their dependence on historical data [51].

15.5.2 Recommendations

- **For Researchers**: Further exploration of hybrid models in other environmental variables can enhance predictive accuracy and model robustness.
- **For Policymakers**: Implementing AI-driven models in environmental monitoring can improve decision-making and resource management.
- **For Practitioners**: Adoption of advanced AI techniques in climate prediction can lead to better preparedness and mitigation strategies for climate-related impacts.

15.6 Conclusion

Five models were developed and evaluated for relative humidity (RH) forecasting: LSTM, Bi-LSTM, CNN, CNN-LSTM, and CNN-Bi-LSTM. Performance was assessed using mean absolute error (MAE), root mean square error (RMSE), mean absolute percentage error (MAPE), and R-squared coefficients. Among these, the CNN-Bi-LSTM hybrid model emerged as the most effective, achieving the lowest MAE (5.42), RMSE (6.77), and MAPE (20.64%) while attaining the highest R-squared value of 0.96. This model leveraged CNN's strength in detecting local patterns and Bi-LSTM's ability to capture bidirectional dependencies, outperforming the other approaches.

The study further highlights that combining CNN-LSTM and CNN-Bi-LSTM models enhances RH forecasting accuracy by effectively capturing time series patterns. These hybrid models deliver superior performance compared to standalone models, thanks to the synergy between convolutional and recurrent structures. Theoretical and practical implications of these findings span various sectors.

In agriculture, improved humidity forecasts support better irrigation planning and mitigate yield losses. In energy management, accurate predictions enable precise heating and cooling estimates, enhancing energy supply efficiency. In meteorology, these models facilitate water resource optimization and drought management, contributing to climate change adaptation strategies. Additionally, accurate RH forecasts aid in outlining regional climate variability and bolstering efforts to mitigate climate change impacts.

The research demonstrates that hybrid models like CNN-LSTM and CNN-Bi-LSTM outperform traditional methods, such as ARIMA, and standalone deep learning models by capturing both short- and long-term dependencies. This capability makes them highly suitable for environmental monitoring, forecasting, and decision-making across weather-dependent industries.

15.7 Future Directions

This study demonstrates the potential of hybrid models for relative humidity (RH) forecasting, offering a strong foundation for future advancements. However, certain limitations should be noted. The models relied on historical data, and their efficacy depends significantly on data quality and representativeness.

Future work could address these limitations by incorporating additional variables, such as temperature, wind speed, or rainfall, into the models to enhance their accuracy. Furthermore, integrating advanced features like attention mechanisms or transformers could improve the models' capability to capture complex temporal patterns and dependencies.

In addition, industries with high economic feasibility like healthcare, finance, technology, engineering, and many others can employ the concept of this hybrid of AI.

The findings of this study underscore the effectiveness of combining CNN and Bi-LSTM structures, as demonstrated by the CNN-Bi-LSTM model. This approach provides a robust framework for accurately modeling RH, with broad applications in industries dependent on climate data. The results offer a promising pathway for developing less biased decision-making tools and optimizing resource allocation, further reinforcing the relevance of hybrid models in environmental monitoring and forecasting.

References

1. Zhong S, Zhang K, Bagheri M, Burken JG, Gu A, Li B, Ma X, Marrone BL, Ren ZJ, Schrier J, Shi W (2021) Machine learning: new ideas and tools in environmental science and engineering. Environ Sci Technol 55(19):12741–12754
2. Henderson K, Loreau M (2023) A model of sustainable development goals: challenges and opportunities in promoting human well-being and environmental sustainability. Ecol Model 475:110164
3. Raihan A (2023) Artificial intelligence and machine learning applications in forest management and biodiversity conservation. Nat Resour Conserv Res 6(2):3825
4. Yuan Q, Shen H, Li T, Li Z, Li S, Jiang Y, Xu H, Tan W, Yang Q, Wang J, Gao J (2020) Deep learning in environmental remote sensing: achievements and challenges. Remote Sens Environ 241:111716
5. Lamba A, Cassey P, Segaran RR, Koh LP (2019) Deep learning for environmental conservation. Curr Biol 29(19):R977–R982

6. Heydari A, Majidi Nezhad M, Astiaso Garcia D, Keynia F, De Santoli L (2022) Air pollution forecasting application based on deep learning model and optimization algorithm. Clean Technol Environ Policy 1–15
7. Sharma A, Jain A, Gupta P, Chowdary V (2020) Machine learning applications for precision agriculture: a comprehensive review. IEEE Access 9:4843–4873
8. Folorunso S, Alsmadi H, Al Kafari AS, Kandasamy G (2024) Autistic spectrum disorder screening classification with machine learning approaches. In: 2024 intelligent methods, systems, and applications (IMSA) (pp 372–377). IEEE
9. Adiaha MS, Adiaha CS (2020) Impacts of climate change on human and plant nutrition and health. Plants Environ 2:126–137
10. Folorunso MA, Folorunso SA (2022) Environmental degradation in Nigeria: the challenges of peaceful co-existence. In: Spiegel E, Mutalemwa G, Liu C, Kurtz LR (eds) Peace studies for sustainable development in Africa. Advances in African economic, social and political development. Springer, Cham. https://doi.org/10.1007/978-3-030-92474-4_18
11. Davis SM et al (2016) The stratospheric water and ozone satellite homogenized (SWOOSH) database: a long-term database for climate studies. Earth Syst Sci Data 8:461–490
12. Lundstad E, Brugnara Y, Pappert D, Kopp J, Samakinwa E, Hürzeler A, Andersson A, Chimani B, Cornes R, Demarée G, Filipiak J, Gates L, Ives GL, Jones JM, Jourdain S, Kiss A, Nicholson SE, Przybylak R, Jones P, Rousseau D (2023) The global historical climate database HCLIM. Scientific Data 10(1):44. https://doi.org/10.1038/s41597-022-01919-w
13. Hurst DF et al (2016) Recent divergences in stratospheric water vapor measurements by frost point hygrometers and the Aura Microwave Limb Sounder. Atmos Meas Tech 9:4447–4457
14. Smith JW, Haynes PH, Maycock AC, Butchart N, Bushell AC (2021) Sensitivity of stratospheric water vapour to variability in tropical tropopause temperatures and large-scale transport. Atmos Chem Phys 21:2469–2489
15. Hemmer M, Klausen A, Van Khang H, Robbersmyr KG, Waag TI (2020) Health indicator for low-speed axial bearings using variational autoencoders. IEEE Access 8:35842–35852
16. Sirota I, Dor YB, Gvirtzman Z (2024) Short-term climatic oscillations versus long-term delta propagation: Controls on sand transport into the deep Levant Basin since the Pliocene. Basin Res 36(4):e12892
17. Wang K, Xu Q, Li T (2020) Does recent climate warming drive spatio temporal shifts in functioning of high-elevation hydrological systems? Sci Total Environ 719:137507
18. Dhrubo AFH, Tamim S, Siddique S (2024) IoT towers of resilience: a revolutionary approach to weathering climate extremes and enhancing sustainability in vulnerable ecosystems. In: 2024 28th international conference on methods and models in automation and robotics (MMAR) (pp 23–28). IEEE
19. Kumar MA, Ashokkumar C, Niranjana R, Saravanan K, Sundararajan S, Narayanan KL (2023) EcoGuard: uniting IoT and AI to secure forests and combat climate change in real-time
20. Schwartz R, Schwartz R, Vassilev A, Greene K, Perine L, Burt A, Hall P (2022) Towards a standard for identifying and managing bias in artificial intelligence (Vol 3, p 00). US Department of Commerce, National Institute of Standards and Technology
21. Fernández-Duque B, Vicente-Serrano SM, Maillard O, Domínguez-Castro F, Peña-Angulo D, Noguera I, Azorín-Molina C, El Kenawy A (2023) Long-term observed changes of air temperature, relative humidity and vapour pressure deficit in Bolivia, 1950–2019. Int J Climatol. https://doi.org/10.1002/joc.8226
22. Kuradusenge M, Hitimana E, Minani JB, Ukundimana Z, Mukamanzi F, Sinayobye OJ, Nizeyimana AT, Gaju N (2024) A framework for monitoring the effectiveness of ecosystem-based adaptation strategies using internet of things and machine learning techniques. J Environ & Earth Sci
23. Moiroux-Arvis L, Royer L, Sarramia D, de Sousa G, Claude A, Latour D, Roussel E, Voldoire O, Chardon P, Vandaele R, Ameglio T, Chanet J-P (2023) ConnecSenS, a Versatile IoT platform for environment monitoring: bring water to cloud. Sensors. https://www.mdpi.com/1424-8220/23/6/2896/pdf?version=1678190005

24. Pole A, West M, Harrison J (2018) Applied Bayesian forecasting and time series analysis. Chapman and Hall/CRC
25. Singh S, Parmar KS, Kumar J (2021) Soft computing model coupled with statistical models to estimate the future of the stock market. Neural Comput Appl 33(13):7629–7764
26. Jaiswal A, Samuel C, Kadabgaon VM (2018) Statistical trend analysis and forecast modeling. J Air Waste Manag Assoc 54(7):809–814
27. Siami-Namini S, Tavakoli N, Namin AS (2018) A comparison of ARIMA and LSTM in forecasting time series. In: 2018 17th IEEE international conference on machine learning and applications (ICMLA) (pp 1394–1401). IEEE
28. Wu M, Liu X, Gui N, Yang X, Tu J, Jiang S, Zhao Q (2023) Prediction of the remaining time and time interval of pebbles in pebble bed HTGRs aided by CNN via DEM datasets. Nucl Eng Technol 55(1):339–352. https://doi.org/10.1016/j.net.2022.09.019
29. Frausto-Solís J, Hernández-González LJ, González-Barbosa JJ, Sánchez-Hernández JP, Román-Rangel E (2021) Convolutional neural network-component transformation (CNN–CT) for confirmed COVID-19 cases. Math & Comput Appl 26(2):29. https://doi.org/10.3390/mca26020029
30. Taherkhani A, Cosma G, McGinnity TM (2023) A deep convolutional neural network for Time Series classification with intermediate targets. SN Comput Sci 4(6). https://doi.org/10.1007/s42979-023-02159-4
31. Prema V, Bhaskar MS, Almakhles D, Gowtham N, Rao KU (2022) Critical review of data, models and performance metrics for wind and solar power forecast. IEEE Access 10:667–688. https://doi.org/10.1109/ACCESS.2021.3137419
32. Günel E (2020) A comparative study of classical and machine learning approaches for time series forecasting: an empirical analysis on exports in Turkey (Master's thesis, Middle East Technical University (Turkey))
33. Montgomery DC, Jennings CL, Kulahci M (2015) Introduction to time series analysis and forecasting. John Wiley & Sons
34. Dimri T, Ahmad S, Sharif M (2020) Time series analysis of climate variables using seasonal ARIMA approach. J Earth Syst Sci 129:1–16
35. Dritsaki C, Niklis D, Stamatiou P (2021) Oil consumption forecasting using ARIMA models: an empirical study for Greece. Int J Energy Econ Policy 11(4):214–224
36. Jatinder K, Parmar KS, Singh S (2023) Autoregressive models in environmental forecasting time series: a theoretical and application review. Environ Sci Pollut Res 30:19617–19641
37. Lee S, Lee Y-S, Son Y (2020) Forecasting daily temperatures with different time interval data using deep neural networks. Appl Sci (Basel, Switz) 10(5):1609. https://doi.org/10.3390/app10051609
38. Saini M, Satija U, Upadhayay MD (2020) Light-weight 1-D convolutional neural network architecture for mental task identification and classification based on single-channel EEG. arXiv [eess.SP]. http://arxiv.org/abs/2012.06782
39. Rala Cordeiro J, Raimundo A, Postolache O, Sebastião P (2021) Neural architecture searches for 1D CNNs-different approaches tests and measurements. Sensors (Basel, Switzerland) 21(23):7990. https://doi.org/10.3390/s21237990
40. Guessoum S, Belda S, Ferrandiz JM, Modiri S, Raut S, Dhar S, Heinkelmann R, Schuh H (2022) The short-term prediction of length of day using 1D convolutional neural networks (1D CNN). SensS (Basel, Switz) 22(23):9517. https://doi.org/10.3390/s22239517
41. Cao C, Huang J, Wu M, Lin Z, Sun Y (2024) A multivariate time series prediction method based on convolution-residual gated recurrent neural network and double-layer attention. Electronics 13(14):2834. https://doi.org/10.3390/electronics13142834
42. Wibawa AP, Utama ABP, Elmunsyah H, Pujianto U, Dwiyanto FA, Hernandez L (2022) Time-series analysis with smoothed convolutional neural network. J Big Data 9(1):44. https://doi.org/10.1186/s40537-022-00599-y
43. Nath S, Chetia B, Kalita S (2023) Ionospheric TEC prediction using hybrid method based on ensemble empirical mode decomposition (EEMD) and long short-term memory (LSTM) deep learning model over India. Adv Space Res 71(5):2307–2317

44. Smagulova K, James AP (2020) Overview of long short-term memory neural networks. Deep Learn Cllfiers Memristive Netw: Theory Appl 139–153
45. Bekkar A, Hssina B, Douzi S, Douzi K (2021) Air-pollution prediction in smart city, deep learning approach. J Big Data 8(1):1–21. https://doi.org/10.1186/s40537-021-00548-1
46. Bhoj N, Singh Bhadoria R (2022) Time-series based prediction for energy consumption of smart home data using hybrid convolutional-recurrent neural network. Telemat Inf 75(October):101907. https://doi.org/10.1016/j.tele.2022.101907
47. Tzoumpas K, Estrada A, Miraglio P, Zambelli P (2024) A data filling methodology for time series based on CNN and (Bi)LSTM neural networks. IEEE Access 12(February):31443–31460. https://doi.org/10.1109/ACCESS.2024.3369891
48. Le VT, Tran-Trung K, Hoang VT (2022) A comprehensive review of recent deep learning techniques for human activity recognition. Comput Intell Neurosci 2022(1):8323962
49. Xu D, Peng P, He D (2021) Short-term traffic flow prediction with missing data. Gaojishu Tongxin/Chinese High Technology Letters 31(9):934–941. https://doi.org/10.3772/j.issn.1002-0470.2021.09.004
50. Ullah K, Ahsan M, Hasanat SM, Haris M, Yousaf H, Raza SF, Tandon R, Abid S, Ullah Z (2024) Short-term load forecasting: a comprehensive review and simulation study with CNN-LSTM hybrids approach. IEEE Access 12(August):111858–111881. https://doi.org/10.1109/ACCESS.2024.3440631
51. Roselli D, Matthews J, Talagala N (2019) Managing bias in AI. In: Companion proceedings of the 2019 world wide web conference (pp 539–544)
52. Gregory S (2019) Bias in artificial intelligence. N C Med J 80(4):220–222. https://doi.org/10.18043/NCM.80.4.220
53. Sinha A, Sapra D, Sinwar D, Singh V, Raghuwanshi G (2024) Assessing and mitigating bias in artificial intelligence: a review. Recent Adv Comput Sci Commun (Former: Recent PatS Comput Sci) 17(1):1–10
54. Muat O, Nila S, George V (2021) Artificial intelligence bias minimization via random sampling technique of adversary data 1226–1230. https://doi.org/10.1109/CCWC51732.2021.9375929

Part V
AI in Social and Environmental Sustainability

Chapter 16
Artificial Intelligence in Social Work to Ensure Environmental Sustainability

Mina Margaret Ogbanga, Shashikant Nishant Sharma, Ashutosh Kumar Pandey, and Pavitar Singh

Abstract Artificial intelligence has emerged as a transformative force across various domains, with social work and environmental sustainability being no exception. The contemporary mainstream integration of AI technologies into social work practice may also lead to innovative solutions to complex challenges involving the environment. This chapter explores the potential of AI-driven approaches enhancing environmental sustainability through socially responsible interventions. It examines the intersection of AI, social work, and sustainable development, highlighting how predictive analytics, machine learning, and data-driven insights can empower social workers in environmental advocacy and resource management. The relation between the social work and the AI is increasing day by due to a wide range of data being generated from a lot of development works. The chapter also delves into ethical and social considerations, emphasizing the need for fairness, transparency, and accountability in AI systems to prevent bias and ensure equitable outcomes. Challenges such as technical limitations, resource constraints, and regulatory barriers are discussed, alongside recommendations to foster collaboration between AI researchers, social work practitioners, and policymakers. The chapter concludes by outlining future directions and opportunities to harness AI innovations for achieving the United Nations' Sustainable Development Goals (SDGs). By integrating AI into social work frameworks, this interdisciplinary approach can create a more sustainable and socially just world, where technology serves as a catalyst for environmental stewardship and community well-being.

M. M. Ogbanga (✉)
Head of Department, Track2Training, New Delhi, India
e-mail: Mina.Ogbanga@ust.edu.ng

S. N. Sharma
Maulana Azad National Institute of Technology, Bhopal, India

A. K. Pandey
JECRC University, Jaipur, Rajasthan, India

P. Singh
JECRC University, Jaipur, India

M. Nasr et al. (eds.), *Artificial Intelligence Applications for a Sustainable Environment*,
Green Chemistry and Sustainable Technology,
https://doi.org/10.1007/978-3-031-91199-6_16

Keywords Artificial intelligence · Social work · Environmental sustainability

16.1 Introduction

Social work is a profession dedicated to addressing social inequities, fostering well-being, and building resilience in individuals and communities [1]. Traditionally, social work has focused on issues like poverty, health, and inequality. However, environmental sustainability has emerged as an essential dimension of social work. Environmental sustainability is important for the long-term well-being of individuals and communities, as it directly impacts social justice and quality of life. Incorporating environmental concerns into social work practice can help address systemic issues and promote a more holistic approach to fostering resilience in society.

Environmental factors, such as resource depletion, pollution, and climate change, often exacerbate social challenges, disproportionately affecting vulnerable populations [1, 2]. Social work's role in environmental sustainability includes advocacy for policies that mitigate ecological harm, direct intervention in communities facing environmental crises, and promoting sustainable practices at the grassroots level [3]. For instance, social workers engage in programs that teach communities about waste reduction, resource conservation, and environmental justice. Additionally, social workers may collaborate with local governments and organizations to develop policies that protect the environment and promote sustainability.

Social work is a practice aimed at improving societal well-being, focusing on marginalized communities and addressing systemic inequalities [4]. However, in the face of global challenges such as climate change and environmental degradation, social work must evolve to incorporate tools and strategies that address both human and ecological crises [2, 4]. This transformation is possible through artificial intelligence (AI), which offers powerful capabilities to enhance traditional social work methods. AI provides the means to analyze complex data, predict outcomes, and optimize interventions, making it a vital tool for fostering environmental sustainability and advancing social justice.

By harnessing AI technology, social workers can more effectively identify patterns and trends in environmental issues, leading to targeted interventions and policies that promote sustainable practices. Additionally, AI can help bridge the gap between social work and environmental science by facilitating interdisciplinary collaboration and innovative solutions to complex challenges.

Artificial intelligence (AI) is increasingly becoming a vital tool for addressing complex challenges across various fields, including social work and environmental sustainability. Its ability to process large datasets, analyze patterns, and generate actionable insights makes it an invaluable resource for tackling environmental issues. From monitoring climate change to optimizing resource utilization, AI offers transformative solutions that can enhance the efficiency of sustainability efforts. Social work, with its emphasis on community engagement and systemic change, can leverage AI to create innovative strategies for fostering environmentally sustainable practices.

The objectives of this chapter are to explore the ways AI can enhance the efficiency and sustainability of environmental initiatives, with a focus on waste management and resource optimization. By analyzing case studies and practical applications, the chapter aims to highlight how AI-driven solutions, such as predictive analytics, intelligent waste sorting, and data-driven policy planning, can revolutionize waste management systems. The chapter will also discuss how social work professionals can utilize AI to engage communities, promote sustainable behaviors, and ensure equitable access to environmental benefits. Through these insights, this chapter seeks to bridge the gap between technology and social work in the pursuit of a sustainable future.

16.1.1 The Role of Artificial Intelligence (AI) in Sustainable Development

Artificial intelligence (AI) is a transformative technology that has the capability to analyze vast amounts of data, identify intricate patterns, and make predictive analyses far beyond human capacity. In the context of sustainable development, AI plays an important role in addressing pressing global challenges [5]. By integrating Internet of Things (IoT sensors and satellite data, AI systems monitor and analyze environmental conditions in real-time, providing insights into climate patterns, pollution levels, and natural resource consumption [6, 7]. This technology also powers smart systems for energy management, enabling efficient use of renewable energy sources, optimizing power grids, and reducing waste [8, 9]. Additionally, AI enhances waste recycling processes and helps in the sustainable utilization of resources by predicting demand and improving supply chain efficiency [9, 10]. These advancements support a more sustainable future by fostering proactive and efficient approaches to environmental stewardship.

In the realm of social work, AI's powerful capabilities offer opportunities to enhance advocacy and intervention efforts that align with sustainable development goals [11]. AI-driven data analysis provides insights into socioeconomic trends, enabling policymakers and social organizations to make data-informed decisions that address poverty, inequality, and social exclusion [12]. AI-powered tools can assist in identifying vulnerable populations, predicting risks, and delivering timely interventions for healthcare, education, and social welfare services [13]. Furthermore, AI facilitates the creation of platforms that connect individuals with resources, services, and support systems, enhancing community engagement and impact [14]. By harnessing AI for social good, communities can work more effectively toward achieving sustainable development objectives that balance economic growth, social well-being, and environmental protection [15].

AI empowers social workers to tackle environmental challenges through innovative and data-driven approaches, enhancing their ability to contribute meaningfully to sustainable development [16]. Predictive analytics, one of AI's powerful tools, can identify areas at risk for natural disasters by analyzing historical and real-time data [17]. This capability enables the implementation of preventive measures and disaster management strategies, reducing harm and saving lives. Machine learning models further aid social workers by processing complex socio-environmental data to reveal patterns and correlations that guide targeted interventions [18]. For example, these models can identify communities most affected by pollution or resource scarcity, helping direct resources and support where they are needed most [19].

In addition to predictive and analytical capabilities, AI-powered community platforms revolutionize outreach, education, and mobilization for environmental advocacy. These platforms facilitate engagement by connecting social workers with diverse groups, enabling widespread dissemination of information about sustainability practices and environmental policies. Interactive tools driven by AI help personalize educational content, making it more accessible and impactful for different audiences. Moreover, social workers can harness AI to coordinate grassroots campaigns, track progress toward sustainability goals, and promote collective action on a larger scale [4, 12]. By integrating AI into their efforts, social workers can drive impactful solutions that advance both environmental conservation and social equity.

AI can also assist social workers in analyzing data to identify trends and patterns that can inform more effective strategies for promoting sustainability and addressing environmental challenges. Additionally, AI can help social workers streamline administrative tasks, allowing them to focus more on direct community engagement and advocacy efforts. Furthermore, AI can enhance the efficiency of resource allocation and decision-making processes within social work organizations, ultimately maximizing the impact of their interventions. Overall, the integration of AI technology holds great potential for enhancing the effectiveness and reach of social work initiatives aimed at addressing environmental issues and promoting social justice.

16.2 The Intersection of AI, Social Work, and Environmental Sustainability

16.2.1 Defining Social Work and Its Connection to Environmental Issues

Social work is a multidisciplinary profession committed to promoting social justice, enhancing well-being, and addressing systemic issues that affect individuals and communities. It goes beyond addressing traditional social problems by engaging with broader societal challenges, including environmental degradation, which increasingly impacts vulnerable populations [1]. Environmental justice, a core tenet of modern social work, emphasizes the equitable distribution of environmental and natural benefits and burdens. Social workers recognize that disadvantaged communities often bear the brunt of environmental crises, facing greater exposure to hazards such as air pollution, water contamination, and climate-induced disasters [20]. By integrating environmental advocacy into their practice, social workers strive to create sustainable solutions that advance both human rights and ecological health.

A clear connection between social work and environmental issues can be seen in cases of urban flooding driven by climate change. Low-income communities, often situated in poorly planned or flood-prone areas, disproportionately suffer from the devastating effects of such events, including property damage, displacement, and health risks. Social work intervenes by advocating for sustainable urban planning practices that prioritize resilient infrastructure and equitable resource allocation [21]. Additionally, social workers provide direct support to affected populations through crisis management, mental health counseling, and access to essential services. This holistic approach reflects the growing recognition that environmental sustainability and social equity are intertwined, requiring collaborative efforts to build more resilient and inclusive societies.

16.2.2 Social and Environmental Justice Frameworks

Social and environmental justice frameworks provide a comprehensive approach to addressing the interconnected challenges of inequality and environmental degradation [22]. Environmental justice emphasizes the fair treatment and meaningful involvement of all individuals, regardless of race, income, or geography, in the development and enforcement of environmental laws, regulations, and policies [23]. This framework recognizes that marginalized and low-income communities are disproportionately burdened by environmental hazards, such as pollution and resource scarcity, while often having limited access to environmental benefits [24]. Social workers play a pivotal role in advancing environmental justice by advocating for equitable policies, supporting community mobilization, and addressing systemic disparities that are intensified by environmental challenges.

Eco-social work is an emerging framework that integrates environmental sustainability into the core principles of social justice and community development [25]. It acknowledges the interdependence of social and ecological systems, promoting practices that foster both human well-being and environmental health. Social workers using an eco-social approach prioritize sustainable resource management, climate resilience, and environmental education while empowering communities to participate in sustainable development initiatives [26]. This holistic perspective not only addresses immediate social needs but also ensures long-term ecological sustainability, creating a foundation for more resilient and equitable societies. By incorporating environmental justice and eco-social frameworks, social work broadens its impact, advancing comprehensive strategies to build sustainable, just, and inclusive communities.

16.2.3 Sustainability Goals and the Role of Technology

Technology, particularly artificial intelligence (AI), plays a transformative role in advancing the United Nations Sustainable Development Goals (SDGs), providing innovative solutions to global challenges. AI-driven technologies contribute significantly to clean water management (SDG 6) by optimizing water distribution systems, detecting leaks, and predicting water quality issues [9, 27, 28]. In sustainable cities and communities (SDG 11), AI enables smart urban planning, efficient public transportation, and waste management systems, making cities more livable and resource-efficient. Furthermore, AI empowers climate action (SDG 13) by enhancing climate modeling, monitoring carbon emissions, and developing adaptive strategies to mitigate the effects of climate change [29]. These technological improvements drive progress toward sustainability by being more efficient, reducing waste, and enhancing environmental monitoring and response. They also ensure a more resilient and sustainable future for generations ahead. The introduction of AI into urban planning and environmental management can help achieve the long-term goal of sustainability in cities. It makes cities adapt to change more effectively.

Social workers have a crucial role in ensuring that AI-driven sustainability solutions are accessible, equitable, and inclusive. Collaborating with technologists and policymakers, social workers can advocate for ethical AI practices that prioritize marginalized populations, who are often most vulnerable to environmental and social challenges. Their expertise in community engagement allows them to reduce the gap between technology developers and end-users, ensuring that solutions are designed with cultural relevance and inclusivity. By integrating social justice principles into technology deployment, social workers contribute to the creation of tools that not only meet technical objectives but also promote fairness, equity, and holistic community development, aligning with the broader vision of the SDGs. This approach helps to address systemic inequalities and empower communities to actively participate in decision-making processes that affect their well-being. By advocating for policies that center on human rights and social justice, social workers play a crucial role in advancing sustainable development goals and creating a more just and equitable society.

16.3 AI Technologies Applied to Social Work for Environmental Sustainability

16.3.1 AI Techniques in Data Collection and Environmental Monitoring

AI techniques play a pivotal role in modern data collection and environmental monitoring, enabling more efficient, accurate, and comprehensive assessments of ecological conditions [30]. Tools like remote sensing, drones, and Internet of Things (IoT) devices gather vast amounts of real-time data on air and water quality, deforestation rates, and pollution levels. AI-powered drones, for instance, can survey large forested areas to detect illegal logging or measure the extent of pollution in water bodies [31]. Machine learning algorithms assess these data streams to identify patterns and predict future environmental impacts, allowing for timely and proactive interventions. This technological capability improves precision, reduces manual labor, and provides insights that were difficult to obtain. By leveraging the power of AI and machine learning, environmental monitoring has become more efficient and effective in addressing environmental challenges. These technologies enable faster response times to environmental threats and help in developing sustainable solutions for a healthier planet.

Social workers can harness AI-generated environmental data to design community-specific solutions that address localized challenges. By integrating data on pollution exposure, climate risks, and resource scarcity, social workers can develop targeted programs that mitigate environmental harm and promote health and resilience within vulnerable populations. Utilizing AI technology can also enhance collaboration between social workers and other stakeholders, such as scientists, policymakers, and community members, to create comprehensive strategies for environmental protection. By leveraging the power of data-driven insights, social workers can advocate for policies that prioritize environmental justice and support communities in building a more sustainable future.

Furthermore, this data can inform policy recommendations that advocate for sustainable urban planning, equitable resource distribution, and stronger environmental protections. Collaboration between social workers, technologists, and policymakers ensures that AI-driven monitoring systems are used ethically and inclusively, enhancing the ability to respond effectively to environmental crises while fostering sustainable community development. By incorporating AI-driven monitoring systems, we can more effectively track environmental impacts on vulnerable populations and tailor interventions accordingly. This multidisciplinary approach maximizes the potential for positive outcomes in both environmental and public health realms.

16.3.2 Case Study: IBM's Green Horizon Project in China

The IBM Green Horizon Project in China is a pioneering example of how AI techniques enhance environmental monitoring and data-driven decision-making [14]. Launched to tackle the country's severe air pollution crisis, the project integrates Internet of Things (IoT) sensors, satellite imagery, and big data analytics to track pollution sources and monitor air quality in real-time. Using AI, the system analyzes vast datasets to model pollution patterns, providing actionable insights into how different factors contribute to air quality degradation. This comprehensive approach enables authorities to understand pollution dynamics better and implement evidence-based solutions tailored to local needs.

One of the project's key innovations is the use of predictive models to simulate the impact of various policy interventions. These models help policymakers assess the effectiveness of potential measures before implementation, optimizing resource allocation for maximum environmental benefit. Additionally, social workers play a vital role in the project by engaging with communities to raise awareness about air pollution and promote behavior changes that reduce environmental harm. Collaborative efforts between technologists and social workers ensure that the project's technological advancements translate into meaningful, on-the-ground improvements for public health and community well-being.

The outcomes of IBM's Green Horizon Project demonstrate the tangible benefits of AI-driven environmental monitoring. In cities like Beijing, the project contributed to a 20% reduction in the number of severe pollution days, significantly improving air quality and public health. This case study underscores the relevance of predictive analytics in addressing sustainability challenges, as historical data and machine learning models can forecast future environmental risks. Social workers, equipped with these insights, can design proactive interventions to prepare communities for potential disasters, implement effective risk mitigation strategies, and advocate for policies that foster long-term environmental resilience. By integrating AI technologies into social and environmental initiatives, stakeholders can collaboratively drive sustainable development and promote healthier, more equitable communities. These innovative approaches can help build more resilient communities that are better equipped to adjust to changing natural conditions and mitigate potential risks. By utilizing the power of AI and data-driven insights, social workers can play a crucial role in promoting sustainability and creating a more sustainable future for all. In order to prevent marginalized communities from suffering disproportionately from environmental challenges, AI technologies can also help social workers identify and address environmental justice issues. By leveraging AI, social workers can advocate for policies and practices that promote environmental equity and justice for all individuals.

16.3.3 Case Study: Flood Forecasting System in Bangladesh

The Flood Forecasting System in Bangladesh, developed in partnership with Google, showcases how AI-powered predictive analytics can mitigate the devastating effects of natural disasters. Bangladesh, prone to frequent and severe flooding, benefits from this system's ability to analyze vast amounts of hydrological, meteorological, and historical data to predict flooding patterns with remarkable accuracy. By leveraging AI technology, the system identifies potential flood events and issues detailed forecasts, enabling authorities and communities to prepare for impending dangers. This initiative exemplifies how cutting-edge technology can transform disaster management and improve resilience in vulnerable regions.

A key feature of the program is its real-time alerting system, which delivers timely warnings to at-risk populations through mobile applications and text messages. These alerts inform people about when and where flooding is expected, allowing them to take preventive measures such as evacuating to safer locations or securing valuable assets. Social workers play an instrumental role in translating this technological capability into community action. They organize preparedness training, guide evacuation efforts, and ensure vulnerable groups, including the elderly and disabled, receive necessary support. The combination of AI insights and social worker intervention creates a comprehensive, human-centered approach to disaster response that prioritizes safety and well-being.

The outcomes of this flood prediction system have been significant, with a reported 75% increase in the accuracy of early warnings. This improvement has saved lives, minimized injuries, and reduced property damage across flood-prone areas. The case study illustrates the power of predictive analytics in empowering social workers to address environmental risks proactively. Armed with reliable forecasts, social workers can implement targeted risk reduction strategies, enhance disaster preparedness, and advocate for policies that strengthen flood resilience. This collaborative model, integrating AI-driven technology with social work expertise, offers a scalable and impactful solution to environmental challenges, reinforcing the importance of interdisciplinary approaches to sustainable development.

By leveraging data-driven insights, social workers can prioritize resources effectively and allocate support to vulnerable populations most at risk of environmental hazards. This proactive approach not only minimizes the impact of disasters but also fosters community resilience and empowerment in the case of climate change. Additionally, this innovative approach allows for early intervention and preventative measures to be implemented, reducing the long-term negative effects of environmental disasters. By combining the strengths of both AI technology and social work, communities can better prepare for and respond to environmental challenges in a more holistic and efficient manner.

16.3.4 Case Study: AI-Enabled Waste Management in Seoul, South Korea

AI technologies, particularly image recognition, are transforming the way waste is sorted and recycled, offering a more efficient and automated approach to managing environmental waste. AI-driven systems can identify recyclable materials in waste streams with high accuracy, reducing human error and enhancing sorting efficiency. Social workers leverage these innovations by integrating AI tools into community recycling initiatives, helping households understand proper waste separation, and fostering habits that reduce waste sent to landfills [32]. This intersection of technology and social work not only optimizes recycling efforts but also promotes sustainable living practices in communities.

A notable case study that illustrates this transformation is Seoul's Smart Waste Management System, which utilized AI and IoT technologies to tackle the city's growing urban waste challenges [33]. Seoul faced increasing volumes of waste while striving to promote recycling and environmental sustainability. To address this, the city deployed AI-powered sensors in waste bins, which collected real-time data to monitor fill levels and automate waste collection schedules. The AI system analyzed this data to optimize collection routes and frequencies, significantly improving efficiency. This system allowed for better resource management, reducing the carbon footprint of waste collection processes.

The outcomes of Seoul's initiative were highly successful. Optimizing collection routes and schedules resulted in a reduction of carbon emissions from waste collection vehicles. Additionally, within the first two years, the city saw a 34% increase in recycling rates, demonstrating the effectiveness of the system in fostering sustainable behaviors. Moreover, social workers played an essential role in this project by using the data provided by the AI system to conduct educational campaigns on waste sorting and recycling. These campaigns raised citizen awareness and encouraged active participation in waste reduction efforts, proving that the integration of AI technology and social work can enhance both environmental sustainability and community engagement. This case highlights the powerful synergy between AI and social work, where technology aids in efficient resource management and social workers drive education and action at the community level.

By leveraging the insights generated by the AI system, social workers were able to tailor their educational campaigns to target specific areas of improvement, resulting in a more impactful and targeted approach to waste reduction. The collaboration between AI technology and social work not only improved environmental sustainability but also fostered a sense of community ownership and responsibility toward waste management practices. This partnership showcases the potential for technology to enhance the effectiveness of social work initiatives, ultimately leading to more sustainable and impactful outcomes. By combining the strengths of AI and social work, communities can address complex issues like waste reduction in a more comprehensive and strategic manner. This integrated approach allows for data-driven approach of decision-making and tailored interventions to address specific challenges within the community. Overall, the synergy between AI technology and social work demonstrates the power of innovation in creating positive change for both people and the planet.

16.3.5 Case Study: Smart Water Management in Cape Town

In the aftermath of a severe drought in 2017, Cape Town faced an urgent need to address its water scarcity and ensure sustainable water use for its residents. To tackle this challenge, the city implemented AI-driven water conservation systems that leverage data analytics to optimize water distribution and usage [34]. AI algorithms analyze real-time usage patterns, historical data, and weather forecasts to make informed decisions about water allocation. These predictive tools help ensure that water is distributed efficiently, avoiding wastage and prioritizing critical needs during periods of low supply. This innovative approach was a crucial part of the city's strategy to conserve water and manage resources in the face of environmental stress.

One of the key features of the system was its integration with social workers, who collaborated with city authorities to promote water-saving behaviors among residents [35]. Social workers played an essential role in engaging communities, raising awareness about the importance of water conservation, and educating households on how to reduce water consumption [3]. In addition, the city installed sensors throughout public infrastructure to monitor and reduce water leakage, ensuring that precious water resources were not lost due to inefficient infrastructure. These efforts, supported by AI-driven data, enabled the city to respond quickly to emerging issues and reduce overall water waste.

The outcomes of the Smart Water Management system in Cape Town were remarkable. The city successfully reduced water consumption by 50%, a significant achievement that helped avert a major water crisis. This case highlights how AI technologies can support environmental sustainability efforts, particularly in areas facing critical resource shortages. Moreover, it underscores the importance of collaboration between social workers and technological systems in fostering community engagement and sustainable behavior. Social workers, empowered with AI-driven insights, were able to guide communities toward collective action, ensuring that water-saving practices became embedded in everyday life. This case study illustrates the powerful role that AI can play in supporting both environmental goals and social work initiatives aimed at community resilience.

16.4 Findings and Discussion

Findings:

The case studies of AI-driven initiatives in environmental monitoring and resource management—such as IBM's Green Horizon Project, [36, 37] the Flood Forecasting System in Bangladesh, Seoul's Smart Waste Management System, and Cape Town's Smart Water Management—demonstrate the transformative impact of AI technologies in countering critical environmental challenges. One of the key findings across these initiatives is the significant role AI plays in improving the accuracy and efficiency of environmental monitoring systems. For example, in both the Green Horizon Project and Cape Town's water management system, AI algorithms analyze real-time data to predict pollution levels and optimize water distribution, respectively, helping authorities respond more proactively to environmental crises. Similarly, in Bangladesh, AI-based flood forecasting systems increase early warning accuracy, reducing the risk of flood-related casualties and property damage [38].

Another critical finding is the successful integration of AI with community engagement efforts through social work. In each case, social workers were instrumental in translating technological advancements into actionable, community-level interventions. Whether it was promoting water-saving behaviors in Cape Town or educating citizens about waste sorting in Seoul, social workers utilized insights from AI systems to guide public education and encourage participation in sustainability efforts. This integration of social work and AI highlights the importance of human-centered approaches in technology implementation, ensuring that solutions are inclusive, accessible, and culturally relevant.

By bridging the gap between technology and community needs, social workers were able to effectively address environmental challenges and empower individuals to make a positive impact. Their ability to leverage AI for social good demonstrates the potential for innovative collaborations that prioritize both technological innovation and human well-being. This collaborative approach not only fosters a more sustainable future, but also strengthens the connection between technology and social impact. It showcases the power of interdisciplinary partnerships in creating meaningful change for both people and the planet.

Discussion:

To address the types of IoT technologies used for assessing different steps/phases of implementing waste management technologies contributing to social justice goals, the following classifications can be considered:

1. **IoT for Waste Monitoring and Data Collection**
   - **Smart Sensors**: Deployed in bins or waste collection points, these sensors measure fill levels, type of waste (e.g., organic, recyclable, hazardous), and temperature to monitor conditions in real-time.

- **GPS-Enabled Devices**: Track waste collection vehicles for efficient routing and identifying underserved areas to ensure equitable service distribution.
- **Environmental Sensors**: Monitor air quality, methane, or other emissions at waste sites to ensure environmental justice in marginalized communities.

2. **IoT for Waste Segregation and Sorting**
   - **Vision-Based IoT Systems**: Cameras and AI-based image recognition systems integrated with IoT to classify and segregate waste at sorting facilities.
   - **RFID (Radio Frequency Identification) Tags**: Used on waste items to automate sorting processes and provide detailed data on material recovery.

3. **IoT for Waste Collection and Transportation**
   - **Fleet Management Systems**: IoT devices in vehicles to optimize collection routes, reduce fuel consumption, and minimize carbon footprints.
   - **Smart Bins**: Integrated with IoT sensors that notify collection authorities when full, preventing overflow and ensuring timely service.

4. **IoT for Waste Processing and Recycling**
   - **IoT-Driven Recycling Plants**: Machinery equipped with IoT technology to automate recycling processes and monitor efficiency in real-time.
   - **Energy Recovery Monitoring**: Sensors in waste-to-energy plants to track conversion efficiency and emission levels, ensuring compliance with regulations.

5. **IoT for Community Engagement and Social Justice**
   - **Mobile Applications**: IoT-integrated apps allow communities to report waste issues, track recycling activities, and access information on sustainable waste practices.
   - **Citizen-Centric Sensors**: Devices used by residents to monitor environmental conditions (e.g., air quality near waste sites) and hold authorities accountable.

6. **IoT for Policy and Decision-Making Support**
   - **Big Data Analytics Platforms**: Collect and analyze data from IoT devices to inform policies that address disparities in waste management access and environmental health impacts.

- **Digital Twins**: Virtual IoT-based simulations of waste systems to evaluate potential impacts of new policies or technologies before implementation.

By deploying these IoT technologies, waste management systems can enhance efficiency, minimize environmental harm, and ensure equitable access to clean and sustainable living conditions, aligning with social justice goals. Let me know if you'd like to expand or dive deeper into specific IoT applications!

The case studies also reveal the importance of predictive analytics in fostering environmental resilience. AI's ability to forecast potential environmental risks, such as air pollution, flooding, or waste overflow, enables governments and communities to prepare and respond before a crisis escalates. Predictive models used in the Green Horizon Project, for instance, simulate the effects of various policy interventions, allowing for targeted and data-driven strategies. This forward-thinking approach not only mitigates risks but also helps in the efficient allocation of resources, ensuring that responses are both timely and effective.

Finally, the successful outcomes observed in these case studies—such as reduced water consumption in Cape Town, improved air quality in Beijing, and increased recycling rates in Seoul—underscore the potential of AI to drive meaningful change in environmental sustainability. However, these results also highlight the need for ongoing collaboration between technology developers, policymakers, and social workers to ensure that AI tools are used ethically and equitably. While AI provides valuable insights and efficiencies, its true potential is realized when combined with community engagement, education, and inclusive policy-making, demonstrating that AI can be a powerful tool for advancing both environmental and social justice goals.

By leveraging AI technology, cities like Seoul can continue to improve their recycling programs and overall environmental impact. This collaborative approach ensures that AI is not only effective but also benefits all members of society, creating a more sustainable and equitable future for all. Additionally, incorporating AI into urban planning can help optimize resource allocation and infrastructure development, leading to more sustainable and resilient cities. This holistic approach to AI integration fosters a culture of innovation and collaboration, ultimately driving positive change for both people and the planet.

16.5 Conclusions and Recommendations

Social work often involves sensitive data. The use of AI necessitates stringent data privacy protocols to prevent misuse. Ethical AI frameworks must prioritize transparency and consent, ensuring community trust. Bias in AI algorithms can exacerbate inequalities. For instance, if training data excludes marginalized groups, AI tools may fail to address their needs. Social workers must advocate for inclusive datasets and fairness audits. While AI offers efficiency, it lacks the empathy central to social work. The integration of AI should complement human-led efforts, ensuring technology supports rather than replaces social workers. Deploying AI solutions in social work is hindered by a lack of technical infrastructure and expertise. Addressing this challenge requires partnerships with tech companies and capacity-building initiatives. Many social work organizations operate with limited budgets, restricting their ability to adopt advanced AI technologies. Public–private partnerships and grants can bridge this funding gap. The absence of clear regulations for AI use in social work can lead to misuse or underutilization. Policymakers must create adaptive frameworks that enable innovation while protecting vulnerable populations.

Study Limitation and Recommendation

While AI has significant potential to advance environmental sustainability, its implementation is not without challenges. One major limitation is the high cost associated with developing and deploying AI systems, which can be a barrier for resource-constrained regions. Additionally, AI applications require robust data infrastructure, which may not be readily available in all areas. Ethical considerations, such as data privacy and algorithmic bias, also pose challenges that need to be addressed to ensure equitable and inclusive outcomes.

A key recommendation is to focus on the economic feasibility of AI applications by drawing insights from successful initiatives like IBM's Green Horizon Project. This project leverages AI to optimize energy usage, reduce emissions, and forecast pollution levels, demonstrating how AI can generate cost-effective and scalable solutions. By integrating AI-driven tools with community-based social work initiatives, it is possible to achieve both environmental and economic goals. Investments in education and training programs for social work and environmental professionals to use AI technologies can further ensure the accessibility and success of these innovations.

Future research should investigate the socio-economic feasibility of leveraging machine learning and ANN-based algorithms for raising awareness about environmental risks. For instance, the practical application of ANN-based models in flood prediction should be examined not only from a technical standpoint but also in terms of affordability, accessibility, and social impact. Additionally, studies should explore how these models can effectively disseminate information to at-risk communities, minimizing flood-related casualties and property damage.

Further exploration is also necessary to assess the integration of AI tools in public policy frameworks. This includes designing policies that support the equitable adoption of AI-driven technologies while addressing potential ethical and legal challenges. By fostering interdisciplinary collaborations, the deployment of AI can ensure a more sustainable and inclusive future for environmental management and social work.

Acknowledgement Statement Authors acknowledge the constructive feedback and comments to improve the manuscript and enhance the quality of the content.

Funding Statement There is no funding for this research.

Consent to Publish Authors give consent to publish this in form of a book chapter in the Springer Nature.

Declaration of Interest There is no conflict of interests.

Contributions Statement All authors contributed equally in the research.

References

1. Flexner A (2001) Is social work a profession? Res Soc Work Pract 11(2):152–165. https://doi.org/10.1177/104973150101100202
2. Healy K (2017) Becoming a trustworthy profession: doing better than doing good. Aust Soc Work 70(sup1):7–16. https://doi.org/10.1080/0312407X.2014.973550
3. Eriksson M, Bjerge B (2019) Negotiating the social work profession. Nord Soc Work Res 9(1):1–4. https://doi.org/10.1080/2156857X.2019.1568370
4. Ricciardelli LA, Nackerud L, Quinn AE, Sewell M, Casiano B (2020) Social media use, attitudes, and knowledge among social work students: ethical implications for the social work profession. Soc Sci Human Open 2(1):100008. https://doi.org/10.1016/j.ssaho.2019.100008
5. Dwivedi YK, Hughes L, Ismagilova E, Aarts G, Coombs C, Crick T, Duan Y, Dwivedi R, Edwards J, Eirug A, Galanos V, Ilavarasan PV, Janssen M, Jones P, Kar AK, Kizgin H, Kronemann B, Lal B, Lucini B, Williams MD (2021) Artificial Intelligence (AI): multidisciplinary perspectives on emerging challenges, opportunities, and agenda for research, practice and policy. Int J Inf Manage 57:101994. https://doi.org/10.1016/j.ijinfomgt.2019.08.002
6. Gruetzemacher R, Whittlestone J (2022) The transformative potential of artificial intelligence. Futures 135:102884. https://doi.org/10.1016/j.futures.2021.102884
7. Gulzar B, Sofi SA, Sholla S (2024) Exploring personalized internet of things (PIoT), social connectivity, and artificial social intelligence (ASI): a survey. High-Confid Comput 4(3):100242. https://doi.org/10.1016/j.hcc.2024.100242
8. Dwivedi YK, Hughes L, Bhadeshia HKDH, Ananiadou S, Cohn AG, Cole JM, Conduit GJ, Desarkar MS, Wang X (2024) Artificial intelligence (AI) futures: India-UK collaborations emerging from the 4th Royal Society Yusuf Hamied workshop. Int J Inf Manage 76:102725. https://doi.org/10.1016/j.ijinfomgt.2023.102725
9. Sharma SN, Dehalwar K, Singh J (2024) Emerging techniques of solid waste management for sustainable and safe living environment. In: Nasr M, Negm A (eds) Solid waste management. Springer Nature Switzerland, pp 29–51. https://doi.org/10.1007/978-3-031-60684-7_3
10. Kumar G, Vyas S, Sharma SN, Dehalwar K (2024) Challenges of environmental health in waste management for peri-urban areas. In: Nasr M, Negm A (eds), Solid waste management. Springer Nature Switzerland, pp 149–168. https://doi.org/10.1007/978-3-031-60684-7_9
11. Segal M (2024) Confronting and managing ethical dilemmas in social work using ChatGPT. Eur J Soc Work 1–13. https://doi.org/10.1080/13691457.2024.2377786
12. James P, Lal J, Liao A, Magee L, Soldatic K (2024) Algorithmic decision-making in social work practice and pedagogy: confronting the competency/critique dilemma. Soc Work Educ 43(6):1552–1569. https://doi.org/10.1080/02615479.2023.2195425
13. Størksen HT, Nielsen B, Haga SM, Drozd F (2023) Internet interventions for parents of young children: a cross-sectional study of health, social, and welfare professionals' attitudes to their use. Child Youth Serv Rev 155. Scopus. https://doi.org/10.1016/j.childyouth.2023.107253
14. Wang K, Ke Y (2024) Social sustainability of communities: a systematic literature review. Sustain Prod Consumpt 47:585–597. https://doi.org/10.1016/j.spc.2024.04.031
15. Moallemi EA, Malekpour S, Hadjikakou M, Raven R, Szetey K, Ningrum D, Dhiaulhaq A, Bryan BA (2020) Achieving the sustainable development goals requires transdisciplinary

innovation at the local scale. One Earth 3(3):300–313. https://doi.org/10.1016/j.oneear.2020.08.006
16. Manny L (2023) Socio-technical challenges towards data-driven and integrated urban water management: a socio-technical network approach. Sustain Cities Soc 90:104360. https://doi.org/10.1016/j.scs.2022.104360
17. AlKhereibi AH, Wakjira TG, Kucukvar M, Onat NC (2023) Predictive machine learning algorithms for metro ridership based on urban land use policies in support of transit-oriented development. Sustainability (Switzerland) 15(2). Scopus. https://doi.org/10.3390/su15021718
18. Zhang X (2022) Machine learning-based psychology evaluation of college students for building innovative health service system. J Environ Publ Health Scopus. https://doi.org/10.1155/2022/6302756
19. Patel L, Friedman E, Johannes SA, Lee SS, O'Brien HG, Schear SE (2021) Air pollution as a social and structural determinant of health. J Clim Change Health 3:100035. https://doi.org/10.1016/j.joclim.2021.100035
20. Adedibu AA (1988) Measuring waste generation in third world cities: A Case study of Ilorin, Nigeria. Environ Monitor Assess 10(2):89–103. https://doi.org/10.1007/BF00401775
21. Banwell N, Montoya J, Opeña M, Ijsselmuiden C, Law R, Balboa GJ, Rutherford S, Chu C, Murray V (2016) Developing the Philippines as a global hub for disaster risk reduction—a health research initiative as presented at the 10th Philippine national health research system week celebration. PLoS Curr 8(Disasters). Scopus. https://doi.org/10.1371/currents.dis.5cf90566bb7791456dcf6b9baf6d4873
22. Walker SE, Smith EA, Bennett N, Bannister E, Narayana A, Nuckols T, Pineda Velez K, Wrigley J, Bailey KM (2024) Defining and conceptualizing equity and justice in climate adaptation. Glob Environ Chang 87:102885. https://doi.org/10.1016/j.gloenvcha.2024.102885
23. Kato-Huerta J, Geneletti D (2022) Environmental justice implications of nature-based solutions in urban areas: a systematic review of approaches, indicators, and outcomes. Environ Sci Policy 138:122–133. https://doi.org/10.1016/j.envsci.2022.07.034
24. Pineda-Pinto M, Herreros-Cantis P, McPhearson T, Frantzeskaki N, Wang J, Zhou W (2021) Examining ecological justice within the social-ecological-technological system of New York City, USA. Landsc Urban Plan 215:104228. https://doi.org/10.1016/j.landurbplan.2021.104228
25. Wang P, Altanbulag A (2022) A concern for eco-social sustainability: background, concept, values, and perspectives of eco-social work. Cogent Soc Sci 8(1):2093035. https://doi.org/10.1080/23311886.2022.2093035
26. Banks S, Bertotti T, Cairns L, Shears J, Shum M, Sobočan AM, Strom K, Úriz MJ (2024) Social work beyond the pandemic: exploring social work values for a new eco-social world. Int Soc Work 67(4):890–904. https://doi.org/10.1177/00208728241227062
27. Kemechian T, Sigahi TFAC, Martins VWB, Rampasso IS, de Moraes GHSM, Serafim MP, Leal Filho W, Anholon R (2023) Towards the SDGs for gender equality and decent work: Investigating major challenges faced by Brazilian women in STEM careers with international experience. Discov Sustain 4(1). Scopus. https://doi.org/10.1007/s43621-023-00125-x
28. Mori Junior R, Fien J, Horne R (2019) Implementing the UN SDGs in universities: challenges, opportunities, and lessons learned. Sustainability (United States) 12(2):129–133. Scopus. https://doi.org/10.1089/sus.2019.0004
29. Iacobuţă GI, Brandi C, Dzebo A, Elizalde Duron SD (2022) Aligning climate and sustainable development finance through an SDG lens. The role of development assistance in implementing the Paris Agreement. Glob Environ Change 74:102509. https://doi.org/10.1016/j.gloenvcha.2022.102509
30. Jelks NO, Hawthorne TL, Dai D, Fuller CH, Stauber C (2018) Mapping the hidden hazards: community-led spatial data collection of street-level environmental stressors in a degraded, urban watershed. Int J Environ Res Publ Health 15(4). Scopus. https://doi.org/10.3390/ijerph15040825
31. Dehalwar K, Sharma SN (2023) Fundamentals of research writing and uses of research methodologies. Edupedia Publications Pvt Ltd.

32. Olawade DB, Wada OZ, Ige AO, Egbewole BI, Olojo A, Oladapo BI (2024) Artificial intelligence in environmental monitoring: advancements, challenges, and future directions. Hyg Environ Health Adv 12:100114. https://doi.org/10.1016/j.heha.2024.100114
33. (Thanos) Bourtsalas AC, Seo Y, Tanvir Alam M, Seo Y-C (2019) The status of waste management and waste to energy for district heating in South Korea. Waste Manage 85:304–316. https://doi.org/10.1016/j.wasman.2019.01.001
34. Geetha Varma V (2022) Water-efficient technologies for sustainable development. In: Current directions in water scarcity research, vol 6. Elsevier, pp 101–128. https://doi.org/10.1016/B978-0-323-91838-1.00009-9
35. Brühl J, Visser M (2021) The cape town drought: a study of the combined effectiveness of measures implemented to prevent "Day Zero." Water Resourc Econ 34:100177. https://doi.org/10.1016/j.wre.2021.100177
36. Bakker K, Ritts M (2018) Smart earth: a meta-review and implications for environmental governance. Glob Environ Chang 52:201–211. https://doi.org/10.1016/j.gloenvcha.2018.07.011
37. Nwulu N, Damisa U (2023) Blockchain's integration in the energy industry: practical implications. In: Blockchain-based peer-to-peer transactions in energy systems. IOP Publishing, pp 1–6. https://doi.org/10.1088/978-0-7503-6295-5ch6
38. Chowdhury ME, Islam AKMS, Zzaman RU, Khadem S (2024) A machine learning-based approach for flash flood susceptibility mapping considering rainfall extremes in the northeast region of Bangladesh. Adv Space Res S0273117724010792. https://doi.org/10.1016/j.asr.2024.10.047

Chapter 17
Recommendations and Concluding Remarks for Marinating Circular and Green Economy by Artificial Intelligence

Mahmoud Nasr

Abstract More studies should focus on the contribution of artificial intelligence applications toward the achievement of smart urbanization, carbon-free industrialization, and green economic growth. This book succeeds in demonstrating various effective techniques of artificial intelligence that can sustain a low-carbon, resource-efficient, and socially inclusive finance for present and future generations. These artificial intelligence methods include artificial neural network (ANN), fuzzy logic, and adaptive neuro-fuzzy inference system (ANFIS), showing positive participation in marinating circular and green economy. These techniques can maintain pollution management, water and wastewater treatment, sustainable energy solutions, environmental monitoring and prediction, and social and environmental sustainability. The book outputs assist citizens, decision-makers, businesses, policymakers, and government institutions in using artificial intelligence techniques to overcome the challenges associated with carbon emissions, resource depletion, job dissatisfaction, the need for skilled personnel, and fund lack.

1. **Recommendations and concluding remarks for maintaining pollution management**

The findings of this book demonstrate that several artificial intelligence techniques can:

M. Nasr (✉)
Environmental Engineering Department, Egypt-Japan University of Science and Technology (E-JUST), Alexandria 21934, Egypt
e-mail: mahmmoudsaid@gmail.com; mahmoud-nasr@alexu.edu.eg; mahmoud.nasr@ejust.edu.eg

Sanitary Engineering Department, Faculty of Engineering, Alexandria University, Alexandria 21544, Egypt

M. Nasr et al. (eds.), *Artificial Intelligence Applications for a Sustainable Environment*, Green Chemistry and Sustainable Technology,
https://doi.org/10.1007/978-3-031-91199-6_17

- enhance data analysis and decision-making processes, regarding land use, forest management, agriculture, urban environments, and climate change to fulfill SDG 2 "zero hunger" [1]
- be used to collect information on land use changes, soil quality indices, and ecosystem services
- develop sensor and data transmission environments, regarding air quality, temperature, humidity, atmospheric pressure, and incoming solar radiation
- present transformative solutions to improve pollution monitoring systems hampered by inadequate infrastructure and delayed responses
- be applied in air quality control, microplastic detection, and water resource preservation
- be used to improve marine pollution detection and management, mitigating the issues related to marine pollution, coastal degradation, and climate change
- monitor sources, such as industrial discharges, plastics, and oil spills, enabling real-time data analysis and timely responses to mitigate environmental risks
- facilitate early warning of pollution events, reduce environmental harm, and prioritize resources for pollution hotspots and vulnerable ecosystems
- optimize various solid waste management processes [2], including waste collection, sorting, recycling, and disposal (SDG 12: responsible consumption and production)
- predict waste generation patterns, and improve resource recovery rates
- enable efficient waste recycling technologies, enhance waste-to-energy conversion, and promote environmental awareness, fulfilling SDG 7 "affordable and clean energy"
- minimize plastic waste dumping in landfills, rivers, and oceans, ultimately reducing plastic pollution and improving public health (SDG 3: good health and well-being).

2. **Recommendations and concluding remarks for maintaining water and wastewater treatment**

The outputs of this book indicate that several artificial intelligence techniques can:

- assist in the removal of emerging inorganic pollutants, maintain heavy metal adsorption, and design water desalination facilities
- optimize operating conditions of wastewater treatment (e.g., initial pH, catalyst dosage, pollutant concentration, and reaction time) [3], further maximizing the degradation of organic contaminants (SDG 6: clean water and sanitation)
- predict the wastewater treatment system performance under various environmental conditions [4], fulfilling SDG 14 "life below water"
- analyze data collected from surveys and interviews to ensure a safe and affordable water supply, thereby enhancing public health and reducing the burden of waterborne diseases

- assess household water treatment efficiency (e.g., sedimentation, sand filtration, and solar water disinfection), and understand public perceptions to comply with the national regulations
- unravel complex correlations within water management systems, and model the impact of climate change on water resources
- allow for remote measurement of water quality indicators, such as pH, total dissolved solids, temperature, dissolved oxygen, turbidity, and conductivity [5]
- identify key parameter interactions and optimal conditions for the sustainable treatment of landfill leachate
- provide real-time monitoring and data acquisition for dynamic process optimization; however, insufficient datasets or noisy data can negatively influence the accuracy and reliability of artificial intelligence predictions [6].

3. **Recommendations and concluding remarks for maintaining sustainable energy solutions**

The outputs of this book suggest that various artificial intelligence techniques can:

- optimize energy usage by managing inputs and balancing renewable and non-renewable energy sources [7]
- address climate change and achieve sustainable energy transitions [8], supporting carbon capture and utilization (CCU) technologies and meeting SDG 13 "climate action" [9]
- assist in discovering economically feasible pathways to convert CO_2 into high-value products, including industrial chemicals, fuels, and polymers (SDG 9: industry, innovation and infrastructure) [10]
- address data integration challenges from combining renewable feedstocks and clean energy sources [11]
- improve process automation to optimize process parameters, minimize energy consumption, and maximize product yield [12]
- enhance decision-making frameworks in balancing energy efficiency, cost, and scalability, minimizing resource depletion and waste generation
- forecast the photovoltaic (PV) power performance to optimize renewable energy systems toward a greener environment
- fulfill global clean energy goals by matching supply with demand, further reducing dependence on fossil fuels, and minimizing energy wastage [13]
- analyze and optimize biofuel production systems to mitigate reliance on conventional energy facilities [14]
- enhance the feasibility and scalability of biofuel technologies by improving the extraction of fermentable sugars from lignocellulosic biomass [15]
- use data analytics, and predictive ANN models to provide innovative solutions in optimizing feedstock selection, fermentation, hydrolysis, and transesterification
- optimize geothermal facilities, regarding resource exploration, reservoir characterization, drilling, and heat pump operation and maintenance

- handle complex datasets, recognize real-time patterns, and predict subsurface reservoir conditions and the associated fluid dynamics, achieving sustainable geothermal resource management.

4. **Recommendations and concluding remarks for maintaining environmental monitoring and prediction**

The outputs of this book demonstrate that different artificial intelligence techniques can:

- generate flood susceptibility maps, having diverse climatic and hydrological characteristics
- identify and classify flood-prone areas subjected to heavy rainfall, river overflows, and sediment accumulation, minimizing nitrate leaching into groundwater
- expect and mitigate flood impacts in vulnerable areas, with different environmental and hydrological conditions
- predict hydraulic conductivity (saturated and unsaturated) using simply measurable soil parameters, playing a crucial role in climate change control and GHG emission reduction [16]
- forecast long-term relative humidity changes, improving decision-making and resource management related to crop health, green CO_2 tracking, soil conditions, and water requirements [17]
- enable water resource optimization and drought management, promoting climate change adaptation strategies (infrastructural, institutional, behavioral, and nature-based scenarios) [18].

5. **Recommendations and concluding remarks for maintaining social sustainability**

The findings of this book demonstrate that different artificial intelligence techniques can be useful for present and future generations to:

- assess the complex relationships between water, energy, and food systems, identifying potential trade-offs and synergies to promote sustainable resource management
- empower social workers to protect environmental health and natural resources, focusing on knowledge, attitudes, and perceptions
- integrate ethical and social considerations with gender equities to foster well-being, and build resilience in individuals and communities (SDG 11: sustainable cities and communities) [19]
- foster collaboration between researchers, social work practitioners, and policymakers, which is a key aspect of circular economy promotion [20]
- fulfill the targets of the 17 United Nations' SDGs due to pollution reduction, energy generation, and job creation (SDG 8: decent work and economic growth).

Conclusions

The current book reveals that several artificial intelligence techniques can be used to minimize pollution (e.g., air, soil, and surface water), treat wastewater, provide sustainable energy solutions, monitor and predict several environmental conditions, and maintain socio-environmental sustainability. These objectives are highly linked to the fulfillment of several SDGs related to job creation and employment (SDG 1), food production (SDG2), human health protection (SDG 3), wastewater remediation (SDG 6), cleaner energy recovery (SDG 7), project profit enhancement (SDG 8), industrial facility and infrastructure increase (SDG 9), sustainable city support (SDG 11), solid waste recycling (SDG 12), CO_2 emission mitigation (SDG 13), aquatic life preservation (SDG 14), and forest and land use protection (SDG 15). The book's findings are crucial to low- and middle-income countries that suffer from deficient funding, natural resources, incomes, or infrastructure, fulfilling the three dimensions of sustainable development.

Acknowledgements The editor who wrote this chapter would like to thank the authors of the chapters for their efforts during the different phases of the book including their input in this chapter. The author acknowledges Nasr Academy for Sustainable Environment (NASE).

Ethics Declarations

Conflict of Interest The author has no relevant financial or non-financial interests to disclose.

Ethical Approval Not applicable.

Consent to Participate Not applicable.

Consent for Publication All authors consent for publication.

References

1. Akhtar S, Hollaender H, Yuan Q (2023) Impact of heat and contaminants transfer from landfills to permafrost subgrade in arctic climate: a review. Cold Reg Sci Technol 206:103737. https://doi.org/10.1016/j.coldregions.2022.103737
2. Ibrahiem H, Ismail G, Migahid M, Ghazy M, Nasr M (2024) Dual phytoremediation and biochar production by Eichhornia crassipes in hydroponic system receiving different 1,4-dioxane dosages. Int J Phytorem 26(4):546–556. https://doi.org/10.1080/15226514.2023.2253915
3. Augustine D, Abdelhaleem A, Ookawara S, Nasr M (2024) A novel adsorption/co-digestion/pyrolysis scheme for potato peel waste management to fulfill the sustainable development goals (SDGs). Waste Biomass Valorization. https://doi.org/10.1007/s12649-024-02594-1
4. Sapkota B, Pariatamby A (2023) Pharmaceutical waste management system—are the current techniques sustainable, eco-friendly and circular? A review. Waste Manag 168:83–97. https://doi.org/10.1016/j.wasman.2023.05.052
5. Dadebo D, Atukunda A, Ibrahim M, Nasr M (2023) Integrating chemical coagulation with fixed-bed column adsorption using rice husk-derived biochar for shipboard bilgewater treatment:

scale-up design and cost estimation. Chem Eng J Adv 16:100520. https://doi.org/10.1016/j.ceja.2023.100520
6. Dadebo D, Ibrahim M, Fujii M, Nasr M (2023) Sequential treatment of surfactant-laden wastewater using low-cost rice husk ash coagulant and activated carbon: modeling, optimization, characterization, and techno-economic analysis. Bioresour Technol Rep 22:101464. https://doi.org/10.1016/j.biteb.2023.101464
7. Ezz H, Ibrahim M, Fujii M, Nasr M (2023) Dual biogas and biochar production from rice straw biomass: a techno-economic and sustainable development approach. Biomass Convers Biorefinery 13(12):10807–10821. https://doi.org/10.1007/s13399-021-01879-y
8. Podder J, Patra B, Pattnaik F, Nanda S, Dalai A (2023) A review of carbon capture and valorization technologies. Energies 16(6):2589. https://doi.org/10.3390/en16062589
9. Hassan R, Kriaa K, Wahaballa A, Elsayed M, Mahmoud M, Nasr M, Tawfik A (2024) Performance assessment of up-flow anaerobic multi-staged reactor followed by auto-aerated immobilized biomass unit for treating polyester wastewater, with biogas production. Appl Water Sci 74:14. https://doi.org/10.1007/s13201-024-02129-y
10. Xu M, Sun H, Chen E, Yang M, Wu C, Sun X, Wang Q (2023) From waste to wealth: innovations in organic solid waste composting. Environ Res 229:115977. https://doi.org/10.1016/j.envres.2023.115977
11. Msemwa G, Ibrahim M, Fujii M, Nasr M (2022) Phytomanagement of textile wastewater for dual biogas and biochar production: a techno-economic and sustainable approach. J Environ Manag 322:116097. https://doi.org/10.1016/j.jenvman.2022.116097
12. Mensah-Sackey G, Shokry H, Fujii M, Nasr M (2025) Sustainable utilization of plastic-derived graphene for tetracycline wastewater treatment and its recycling for biogas and biochar production. J Water Process Eng 69:106554. https://doi.org/10.1016/j.jwpe.2024.106554
13. Eraky M, Nasr M, Elsayed M, Ai P, Tawfik A (2023) Synergistic interaction of tween 20 and magnesium@ functionalized graphene oxide nano-composite for dual productivity of biohydrogen and biochar from onion peel waste. Renew Energy 216:119082. https://doi.org/10.1016/j.renene.2023.119082
14. Alhajeri N, Tawfik A, Nasr M, Osman A (2024) Artificial intelligence-enabled optimization of Fe/Zn@biochar photocatalyst for 2,6-dichlorophenol removal from petrochemical wastewater: a techno-economic perspective. Chemosphere 352:141476. https://doi.org/10.1016/j.chemosphere.2024.141476
15. Ansari F, Nasr M, Rawat I, Bux F (2023) Meeting sustainable development goals (SDGs) through progression of pilot-scale algal system to commercial raceway pond (300,000 L). Biomass Convers Biorefinery 13(9):7663–7676. https://doi.org/10.1007/s13399-021-01661-0
16. Aboughaly M, Fattah I (2023) Environmental analysis, monitoring, and process control strategy for reduction of greenhouse gaseous emissions in thermochemical reactions. Atmosphere 14(4):655. https://doi.org/10.3390/atmos14040655
17. Heiba Y, Nasr M, Fujii M, Mohamed A, Ibrahim M (2023) Improving irrigation schemes using sustainable development goals (SDGs)-related indicators: a case study of tomato production in pot-scale experimentation. Environ Dev Sustain. https://doi.org/10.1007/s10668-023-03357-z
18. Heiba Y, Ibrahim M, Mohamed A, Fujii M, Nasr M (2023) Developing smart sustainable irrigation matrix (SIM)-based model for selection of best irrigation techniques: a framework to achieve SDGs. J Clean Prod 420:138404. https://doi.org/10.1016/j.jclepro.2023.138404
19. Zidan A, Nasr M, Fujii M, Ibrahim M (2023) Environmental and economic evaluation of downflow hanging sponge reactors for treating high-strength organic wastewater. Sustainability (Switzerland) 15(7):6038. https://doi.org/10.3390/su15076038
20. Subbarao P, D' Silva T, Adlak K, Kumar S, Chandra R, Vijay V (2023) Anaerobic digestion as a sustainable technology for efficiently utilizing biomass in the context of carbon neutrality and circular economy. Environ Res 234:116286. https://doi.org/10.1016/j.envres.2023.116286

MIX
Papier aus verantwortungsvollen Quellen
Paper from responsible sources
FSC® C105338
FSC
www.fsc.org

If you have any concerns about our products,
you can contact us on
ProductSafety@springernature.com

In case Publisher is established outside the EU,
the EU authorized representative is:
Springer Nature Customer Service Center GmbH
Europaplatz 3, 69115 Heidelberg, Germany

Printed by Libri Plureos GmbH
in Hamburg, Germany